ADVANCES IN CHEMISTRY SERIES **248**

Hydrophilic Polymers

Performance with Environmental Acceptance

J. Edward Glass, EDITOR
North Dakota State University

Developed from a symposium sponsored
by the Division of Polymeric Materials:
Science and Engineering, Inc.,
at the 206th National Meeting
of the American Chemical Society,
Chicago, Illinois,
August 22–27, 1993

American Chemical Society, Washington, DC 1996

Library of Congress Cataloging-in-Publication Data

Hydrophilic polymers: performance with environmental acceptance / J. Edward Glass, editor.

p. cm.—(Advances in chemistry series, ISSN 0065-2393; 248)

"Developed from a symposium sponsored by the Division of Polymeric Materials: Science and Engineering, Inc., at the 206th National Meeting of the American Chemical Society, Chicago, Illinois, August 22-27, 1993."

Includes bibliographical references and indexes.

ISBN 0-8412-3133-8 (alk. paper)

1. Water-soluble polymers—Congresses.
I. Glass, J. E. (J. Edward), 1937– . II. American Chemical Society. Division of Polymeric Materials: Science and Engineering. III. American Chemical Society. Meeting (206th: 1993: Chicago, Ill.) IV. Series.

QD1.A355 no. 248
[QD382.W3] 95-46764
540 s—dc20 CIP
[547.7'045422]

This book is printed on acid-free paper.

PRINTED IN THE UNITED STATES OF AMERICA

1995 Advisory Board

Advances in Chemistry Series

FOREWORD

The ADVANCES IN CHEMISTRY SERIES was founded in 1949 by the American Chemical Society as an outlet for symposia and collections of data in special areas of topical interest that could not be accommodated in the Society's journals. It provides a medium for symposia that would otherwise be fragmented because their papers would be distributed among several journals or not published at all.

Papers are reviewed critically according to ACS editorial standards and receive the careful attention and processing characteristic of ACS publications. Volumes in the ADVANCES IN CHEMISTRY SERIES maintain the integrity of the symposia on which they are based; however, verbatim reproductions of previously published papers are not accepted. Papers may include reports of research as well as reviews, because symposia may embrace both types of presentation.

ABOUT THE EDITOR

J. EDWARD GLASS is a professor of polymers and coatings at North Dakota State University. He received a bachelor of science degree in chemistry from Louisiana State University in 1959 and a Ph.D. degree from Purdue University in 1964. From 1963 to 1980, he worked for the Union Carbide Corporation at its South Charleston Research and Development Center.

His research interests have included almost all aspects of water-soluble polymers: kinetics and synthesis by free-radical processes, controlled substituent placement in the derivatization of carbohydrate polymers, solution and interfacial adsorption and viscosity behavior of both polymer types, and meaningful extrapolation of such fundamental data to the performance of water-soluble polymers in applications formulations. He has published more than 150 technical papers and received several patents in these areas of study.

CONTENTS

PREFACE

THE BEHAVIOR of surfactant-modified, water-soluble polymers in aqueous solutions has been the focus of intensive research efforts. This book reviews recent developments.

The first section covers hydrogels, which are limited to two unique polymer classes. The first class consists of totally new hydrogels based on polyphosphazene inorganic compositions; in the past, these hydrogels have been hydrolytically unstable. The second class consists of those hydrogels formed by hydrophobe-modified, ethoxylated urethane (HEUR) polymers, a family that forms hydrogels when the terminal units are fully substituted and the polymer is narrow in molecular weight or when the central unit is not hydrophobic.

The second section of this book relates to the biocompatibility and biodegradation of polymers. The subtitle of this book could have been "The Importance of Oxyethylene Segments in Water-Soluble Polymers," for there are only a few chapters in this text that do not discuss polymers containing oxyethylene segments. The ability of such polymers to inhibit blood clotting has been an area of past studies, and the role of such polymers in the behavior of proteins is discussed in Chapters 3 and 4. The concern for more biodegradable plastics through the use of natural polymers is considered in Chapters 5–8 (a theme initially addressed in *Agricultural and Synthetic Polymers: Biodegradability and Utilization*; J. Edward Glass and Graham Swift, Eds., ACS Symposium Series 433, 1990).

The third section addresses the characterization and aqueous solution properties of hydrophobe-modified, water-soluble polymers. Associating block polymers have been studied for a number of years, and in the initial stages of these water-soluble polymers, detailed analyses were conducted on poorly defined materials. That has been the case in most of the studies in hydrophobe-modified, water-soluble polymers during the past decade. In the subsection on general concepts, one of the best examples of combining detailed physical analyses on well-characterized water-soluble block polymers is described in Chapter 9. Chapter 10 defines the type of water-soluble polymers obtained in the different synthetic procedures used in the production of hydrophobe-modified, water-soluble polymers The topics discussed include hydrophobe modification of hydroxyethylcellulose (HMHEC), the chain-growth syntheses of hydrophobe-modified, alkali-swellable emulsions (HASE) and hydrophobe-modified acrylamide (RAM), and the step-growth synthesis of hydrophobe-modified, ethoxylated urethanes (HEURs). This chapter is followed by simulation studies (Chapter 10) of the influence of hydrophobe placement on the solution properties of HEURs. These polymers have found commercial acceptance in architectural coatings, cosmetics, and as airplane de-icers.

As the applications are broadened to improve the environmental acceptance in applications such as original equipment manufacturing water-borne coatings where different associative thickeners may be required, the extent of hydration in balancing the hydrophilic–lyophilic balance of hydrophilic polymers will become important. The extent of hydration of different polyether diols is discussed in Chapter 12. Surfactants such as sodium dodecyl sulfate interact with poly(oxyethylene) (POE), but the contribution of the "string of pearls" to the viscosity of aqueous solutions has not been quantified. This quantification is done in Chapter 13. One of the most important contributions of associative thickeners in water-borne coatings is their contribution to higher viscosities at high shear rates. A fundamental study of the influence of shear deformations on the viscosity of surfactants, a prerequisite for understanding the behavior of associative thickeners, is reviewed in Chapter 14.

The solution properties of associative thickeners are discussed in the second subsection of the associating polymers section. These discussions include RAM (Chapter 15), the influence of fluoro- vs. hydrocarbon hydrophobic units (Chapter 16), and HEUR polymers. HEUR thickeners can be synthesized by a direct addition of monoisocyanates to produce "uni-HEURs," which provide associative thickeners with known placement and molecular-weight control. The rheology and neutron scattering of uni-HEURs are discussed in Chapters 17 and 18. The aggregation numbers and diffusion coefficients in HEUR aqueous solutions are determined from NMR measurements in Chapters 19 and 20, and their influence on the bridging flocculation of colloidal clusters are presented in Chapter 21.

In the chapters that follow the HEUR studies, data on the solution behavior of HMHEC (Chapter 22) and HASE (Chapter 23) thickeners are described. In the last chapter of this section, the influence of HASE, HMHEC, and commercial and model HEURs in architectural coating formulations are presented and compared in Chapter 24 with their behavior in neat and surfactant aqueous solutions given in earlier chapters.

In the final chapter, a new commercial, water-soluble polymer that is not hydrophobe-modified is discussed. This water-soluble polymer offers a flexible all-carbon backbone with a combination of pendant primary amine and hydroxyl groups. This combination has promise of developing new application areas and meeting the needs of older applications where some existing conventional water-soluble polymers have fallen short.

Acknowledgments

I especially appreciate the work of the reviewers of the individual chapters of this text. I gratefully acknowledge financial support from the Division of Polymeric Materials: Science and Engineering, Inc., of the American Chemical Society, Texaco, Rohm and Haas, Hercules, S.C. Johnson, and Air Products for the speakers in the symposium from which the nucleus of this text was drawn.

The next symposium in this series will be held in Boston, MA, at the ACS national meeting the week of August 16, 1998. Those wishing to participate should contact me in the early spring of that year.

J. EDWARD GLASS
Department of Polymers and Coatings
North Dakota State University
Fargo, ND 58105

September 18, 1995

HYDROGELS

1

Water-Soluble Polyphosphazenes and Their Hydrogels

Harry R. Allcock

Department of Chemistry, The Pennsylvania State University, University Park, PA 16802

Six different classes of water-soluble polyphosphazenes have been synthesized and characterized. Several of the new polymers have been cross-linked to yield materials that absorb water to form hydrogels. This chapter examines the underlying concepts used in this work and describes the use of this chemistry to prepare new materials with controlled surface structures and with special properties such as lower critical solution temperatures, bioerodibility, or biocompatibility. The incorporation of these polymers into membranes or microencapsulles is also described.

WATER-SOLUBLE SYNTHETIC POLYMERS AND HYDROGELS are important in areas as varied as biomedicine, adhesion, membranes, and viscosity enhancement. They are possible replacements in technology and medicine for many naturally occurring polymers. Unfortunately, relatively few of the hundreds of known synthetic polymers are soluble in water. Thus the design and synthesis of new water-soluble polymers or hydrogels are subjects of considerable interest. This chapter is a review of an approach to this problem that makes use of the concepts discussed in the next section.

Concepts Used in This Work

Polymer solubility in water results from two factors. First, solubility in water may result from the presence of certain hydrophilic units in

0065–2393/96/0248–0003$13.75/0

the polymer backbone, especially units such as –O–, –NH–, or –N= that possess lone pair electrons for hydrogen bonding to water. Second, solubility in water often results from the presence of hydrophilic side groups such as –OH, –COONa, $-NH_2$, $-NHCH_3$, $-SO_3^-$, or –C=O or of amphiphilic units such as $-OCH_2CH_2O-$, etc. High concentrations of hydrophilic side groups may overcome a lack of hydrophilic units in the backbone, but a hydrophilic backbone is the best starting point for designing a water-soluble polymer.

The second concept used in this work is related to the method of polymer synthesis. Two general methods exist for bringing about variations in polymer structure: (1) the polymerization or copolymerization of different monomers and (2) macromolecular substitution reactions in which side groups already attached to a polymer chain are replaced by other groups (Scheme I). The first method is more widely used than the second, mainly because of the availability of a wide range of petrochemical monomers but also because the side group

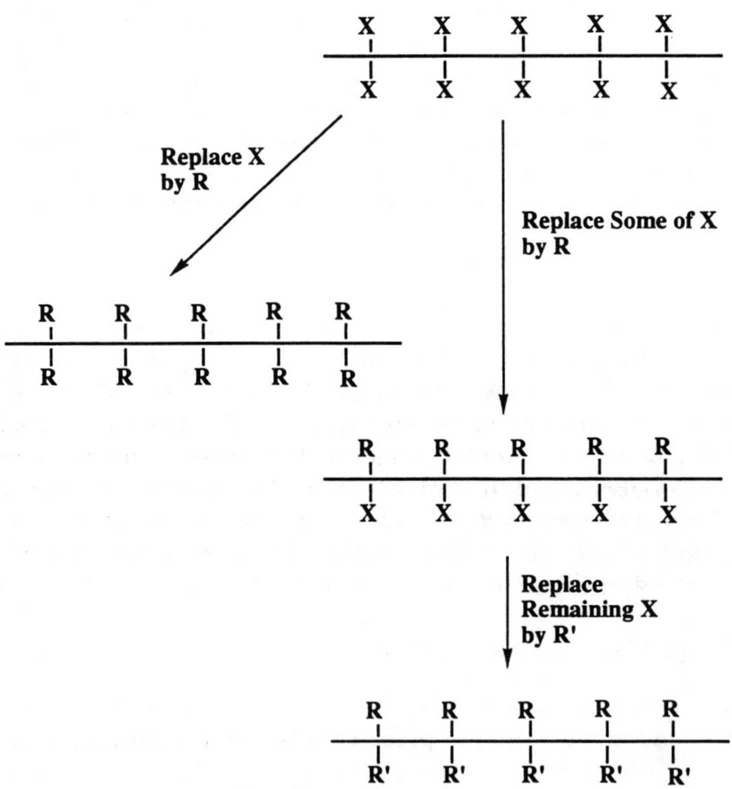

Scheme I. Macromolecular substitution.

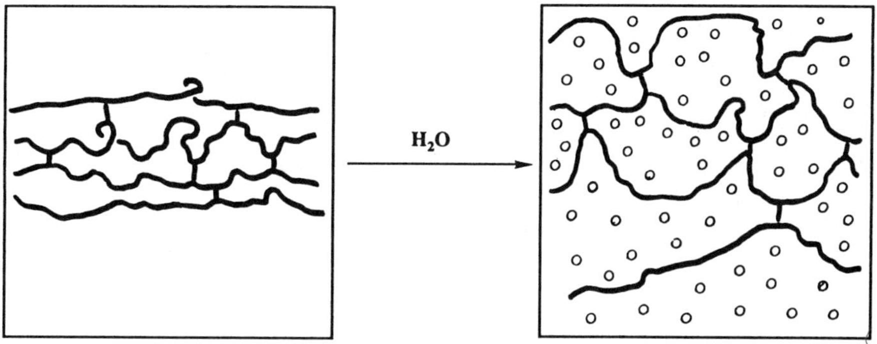

Figure 1. Cross-linking of water-soluble polymers generates hydrogels.

replacement reactions of organic polymers are often relatively inefficient. Nevertheless, as will be demonstrated, the macromolecular substitution approach is an excellent method for the synthesis of water-soluble polymers because it allows a high degree of utilization of molecular design and either extensive or subtle structural manipulation.

The third principle is this: One of the most effective routes to hydrogel formation is via the cross-linking of water-soluble polymers (Figure 1). Cross-links between hitherto water-soluble polymer molecules prevent dissolution of the polymer in water. However, the cross-linked material absorbs water and swells to an extent that is defined by the number of cross-links per chain. Thus the design of hydrogels (which are of critical importance in the field of biomedicine) depends on the development of cross-linking methods that are appropriate for side groups that impart water-solubility.

The fourth concept, which will be referred to later, concerns the stability of a water-soluble polymer or hydrogel to hydrolysis in aqueous media. In most technological applications, hydrolytic instability is considered a detrimental property. However, in biomedicine, hydrolytic breakdown of the polymer or a hydrogel may be an essential requirement if the polymer must eventually "erode" as it is replaced by living cells or after it has been used as a drug delivery platform (Figure 2). Hydrolytic instability can often be designed into a polymer by the selection of the main chain units, the side groups, or both.

The last concept to be illustrated in this chapter is that a group of polymers known as polyphosphazenes (structure **1**) have many advantages for development as water-soluble polymers or hydrogels. The backbone is hydrophilic, the chain structure has a high degree of flexibility, and (depending on the side groups) the backbone may be induced to undergo hydrolysis. However, the main advantage of these polymers is the ease with which water-solubilizing side groups can

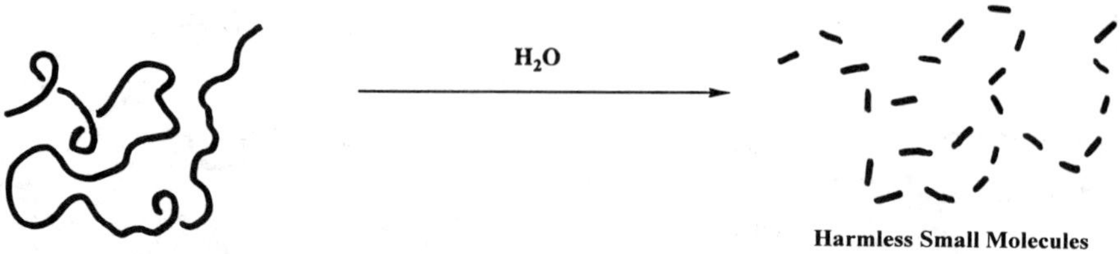

Figure 2. Hydrolytic instability may be utilized as bioerodibility.

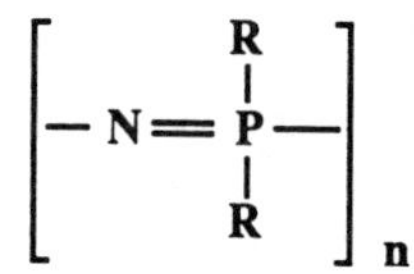

1

be linked to the chain via macromolecular substitution reactions. As will be demonstrated, the backbone is sufficiently stable to high-energy radiation that X rays, gamma rays, electron beams, or ultraviolet irradiation can be used as a clean and effective way to cross-link the polymers through the side groups in order to generate hydrogels.

Methods of Phosphazene Polymer Synthesis

The main method for the synthesis of polyphosphazenes is illustrated in Scheme II (*1–6*). The method consists of a ring-opening polymerization of a heterocyclic "monomer," shown as **2**, followed by replacement of the chlorine atoms in the resultant polymer (**3**) by organic groups through macromolecular nucleophilic substitution reactions. The chlorine replacement step can be carried out either to introduce only one type of side group or, by simultaneous or sequential substitution, to introduce two or more different types of side groups. The most important feature of this reaction is that the high reactivity of the P–Cl bonds allows all the halogen atoms to be replaced. Because the average chain length of polymer **3** is 15,000 repeating units, 30,000 chlorine atoms are replaced per polymer chain. Bulky nucleophiles (such as aryloxide) may slow this reaction to the point that elevated temperatures may be required to allow the reaction to proceed to completion. A variation on this synthesis method, in which some of the organic groups are introduced before ring-opening polymerization, is shown in Scheme III (*7–9*). Both of the routes shown in Schemes II and III were discovered and developed in our research program; they have so far led to the synthesis of more than 300 different polyphosphazenes.

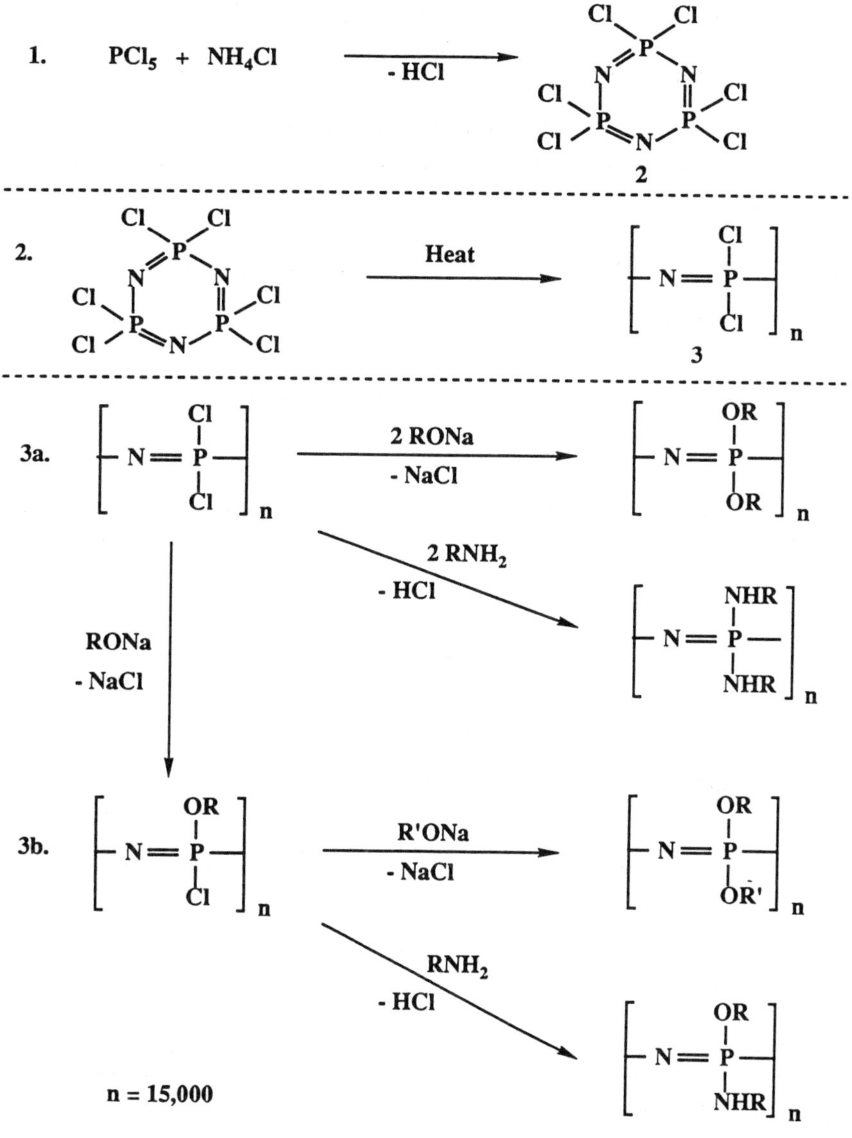
1.
PCl5 + NH4Cl
- HCl
2
2.
Heat
3
3a.
2 RONa
- NaCl
2 RNH2
- HCl
RONa
- NaCl
3b.
R'ONa
- NaCl
RNH2
- HCl
n = 15,000

Scheme II.

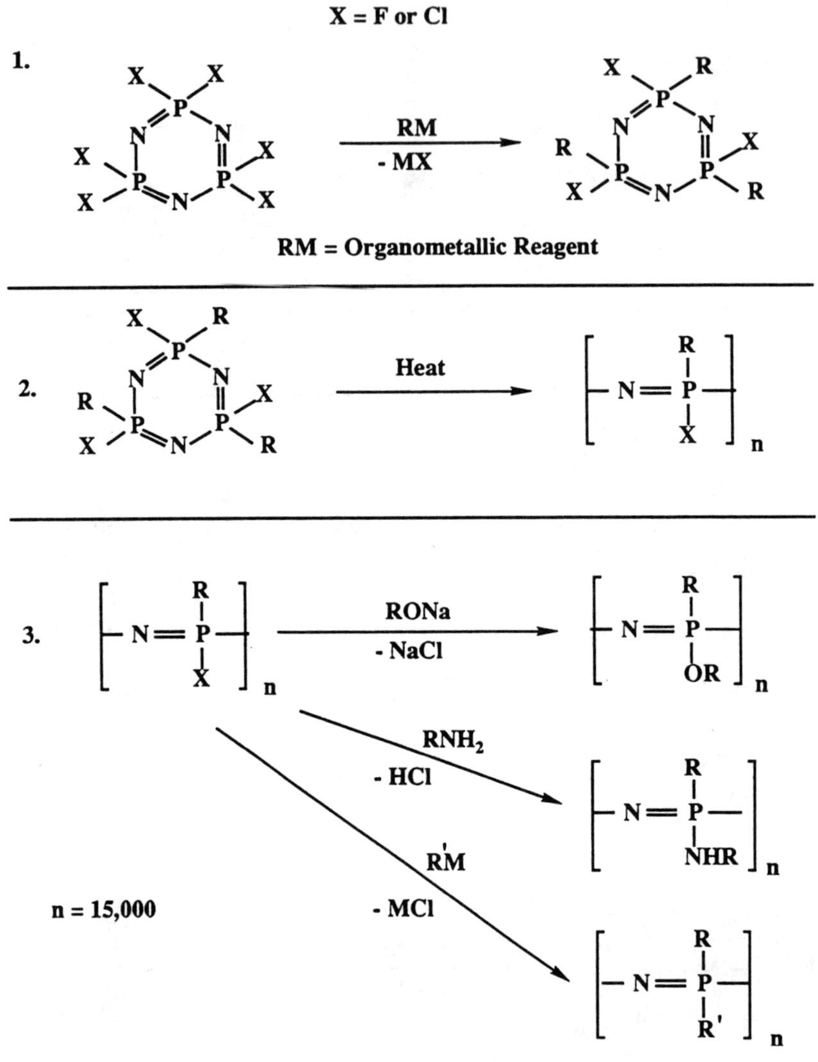

Scheme III.

Alternative synthesis routes, which involve condensation-type processes, have been developed in other laboratories. These routes are shown in Scheme IV (*10–15*).

Examples of Water-Soluble Polyphosphazenes

Chart I shows six different polyphosphazenes that are soluble in water. All of them were synthesized in our laboratory via variations of the

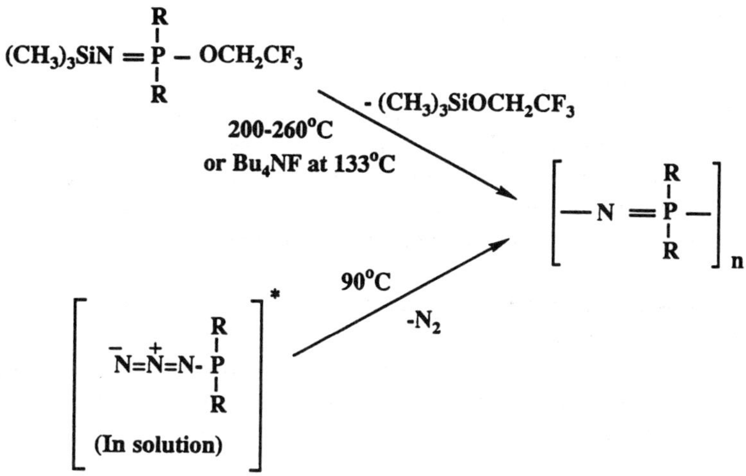

Scheme IV. The asterisk indicates a potentially explosive intermediate that is not isolated but is both prepared and decomposed in solution.

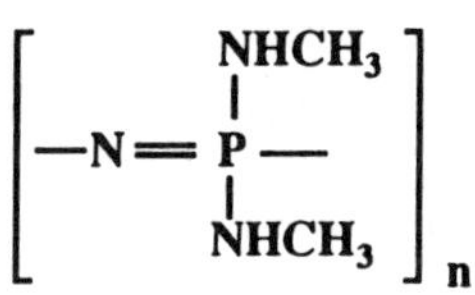
4

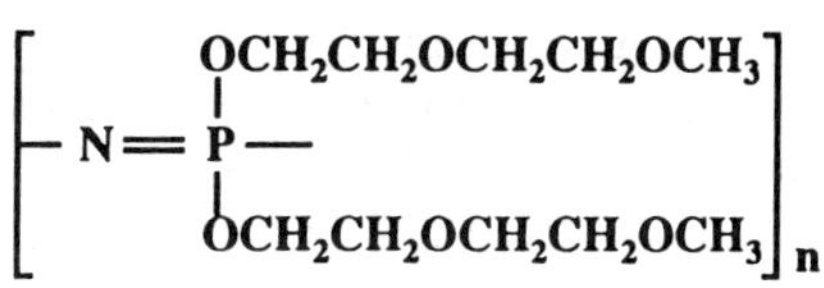
5

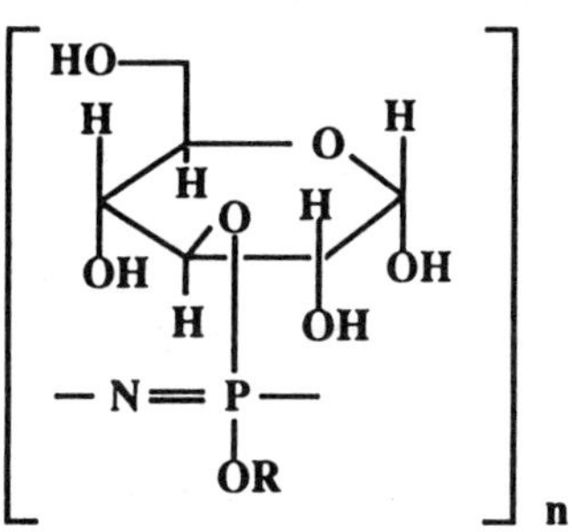
6

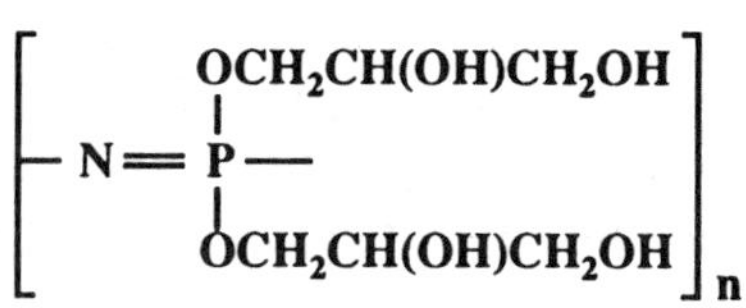
7

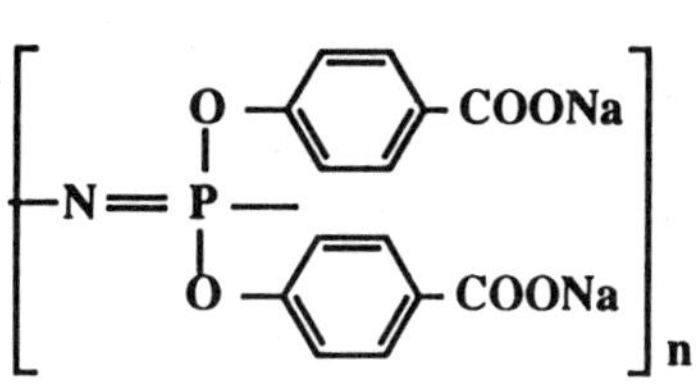
8

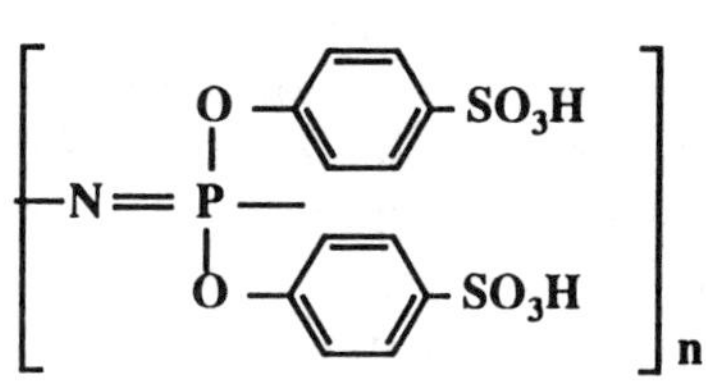
9

Chart 1. Water-soluble polyphosphazenes.

chemistry shown in Scheme II. Each has the same hydrophilic backbone structure, and all bear hydrophilic side groups. Three of them, possibly four, are stable to hydrolysis at room temperature or body temperature. Two of them (**6** and **7**) hydrolyze at detectable rates in neutral aqueous media at 100 °C and are presumed to hydrolyze slowly at lower temperatures. The discussion that follows will consider each of these examples in turn, illustrating the differences and opportunities for molecular design and property optimization.

Poly[bis(methylamino)phosphazene] (4). The polymer poly[bis(methylamino)phosphazene] was the first water-soluble polyphosphazene to be synthesized (*16*). It is prepared by the addition of a tetrahydrofuran solution of poly(dichlorophosphazene) (**3**) to a large excess of methylamine in the same solvent at 0 °C. These reaction conditions were chosen to minimize the possibility that an $-N(H)-CH_3$ side unit could cross-link the chains during synthesis through reaction with a P–Cl unit or another chain. This polymer has a glass transition temperature (T_g) of 14 °C.

The water-solubility of **4** is believed to be due to (1) the small size of the side groups, which exposes the skeletal nitrogen atoms to hydrogen bonding to water, and (2) strong hydrogen bonding between water and the NH units of the side groups. This polymer appears to be stable to neutral and basic aqueous media but hydrolyzes to phosphate and ammonium salts in strong acids.

Polymer **4** is sensitive to cross-linking when it is exposed to gamma rays (*17*). The mechanism of this process is illustrated in Scheme V. Cross-linking is believed to result from radiation-induced, carbon–hydrogen bond cleavage followed by cross-combination of the $NHCH_2\cdot$ radicals produced. This cross-linking process has been used to stabilize amphiphilic membranes prepared from polyphosphazenes that contain both methylamino and fluoroalkoxy or aryloxy cosubstituent groups (*17*).

Poly[bis(methoxyethoxyethoxy)phosphazene] (5) (MEEP). One of the most interesting and potentially most useful polyphosphazenes yet synthesized is polymer **5.** This polymer is prepared by the reaction of poly(dichlorophosphazene) (**3**) with the sodium salt of methoxyethoxyethanol in tetrahydrofuran solution (Scheme VI) (*18*). Because the polymer is infinitely water-soluble at 25 °C, it can be purified by dialysis.

MEEP has unusual solution properties in water, exhibiting the phenomenon known as a lower critical solution temperature (LCST). Polymers that possess this characteristic are soluble below a specific temperature but become insoluble at temperatures above this point

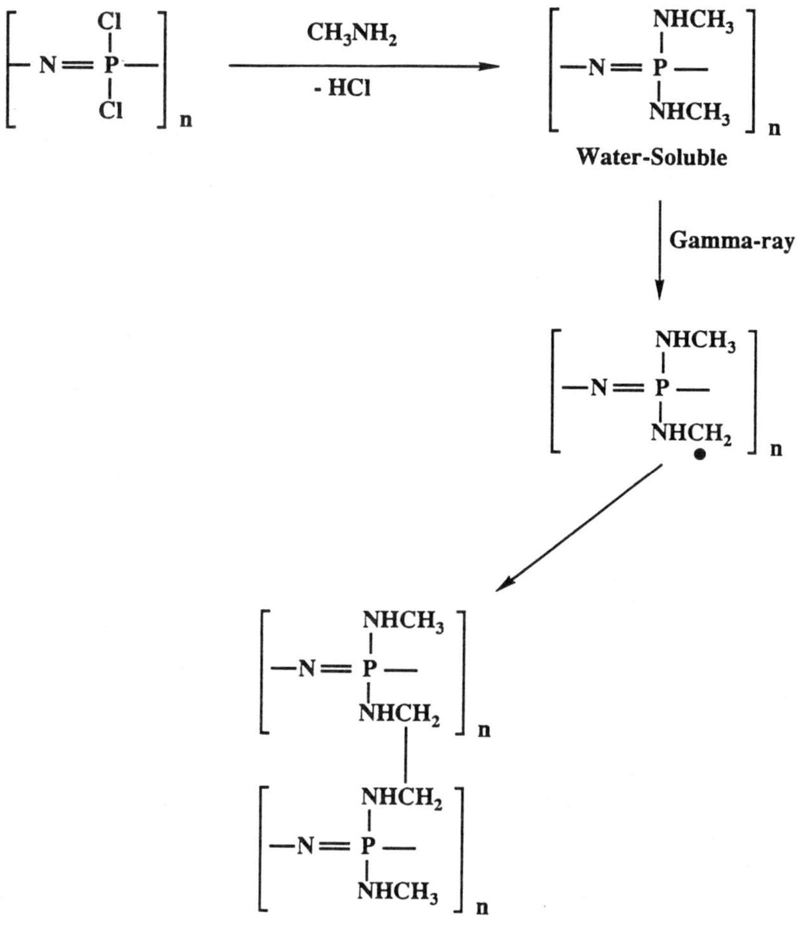

Scheme V.

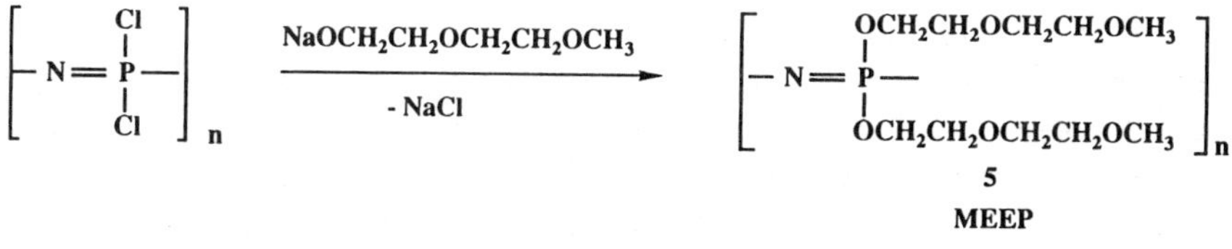

Scheme VI.

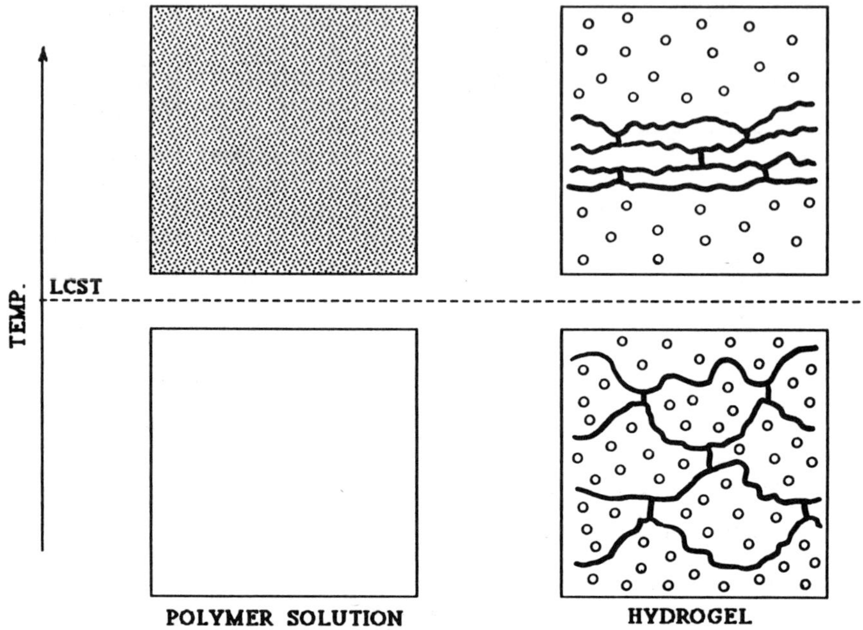

Figure 3. LCST for a polymer solution and hydrogel.

(Figure 3). MEEP has an LCST of 65 °C. A number of polymers related to MEEP but with different etheric side groups, different alkyl ether chain lengths, and different terminal alkoxy groups have also been synthesized (**10–13**). Several of these also exhibit LCSTs, as shown in Table I (*19*). Presumably the LCST behavior of these polymers reflects the dominance of the "hydrophobic" character of the CH_2CH_2 units and the alkyl end units over the hydrophilic effect of the etheric oxygen atoms above the LCST. Replacement of a hydrophobic alkyl terminal group by a hydrophilic amino unit eliminates the LCST effect.

A characteristic of MEEP-type polymers is their low T_gs, which can be attributed to the combination of a highly flexible backbone and flexible side groups. MEEP (**5**) itself has a T_g of −84 °C, polymer **10** has a T_g of −75 °C, and the T_gs for **11** and **12** are −76 and −84 °C, respectively. For **13**, with a terminal NH_2 unit at each side group, the T_g rises to −18 °C, presumably because side group hydrogen bonding restricts the thermal motions of the macromolecules.

MEEP is of interest from several points of view. First, it is an excellent solid solvent for salts such as lithium triflate. The solid solutions function as solid polymeric ionic conductors, and as such they

Table I. Lower Critical Solution Temperature (LCST) for MEEP and Related Polymers

Polymer	*LCST (°C)*
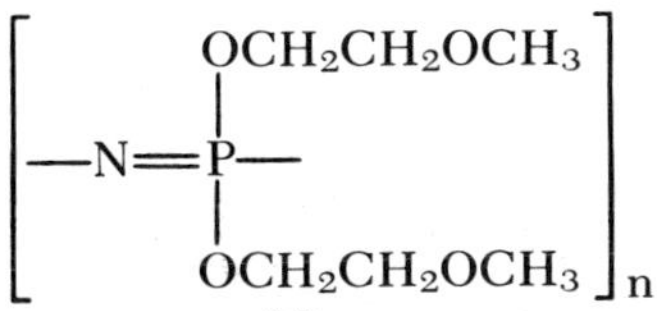 10	30
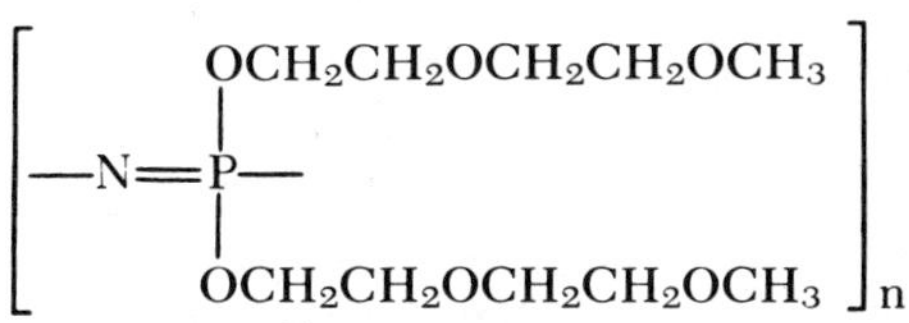 5	65
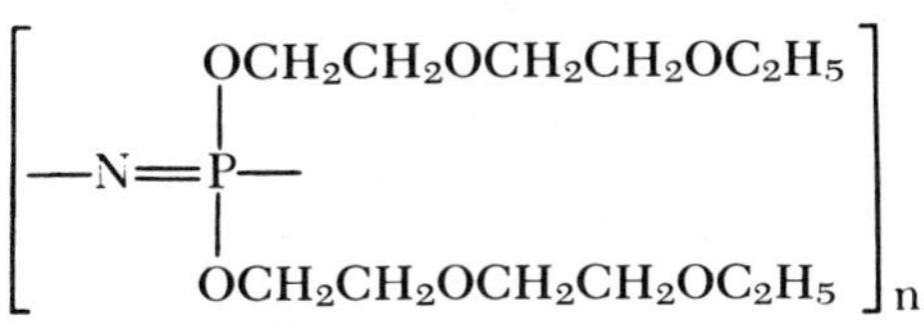 11	38
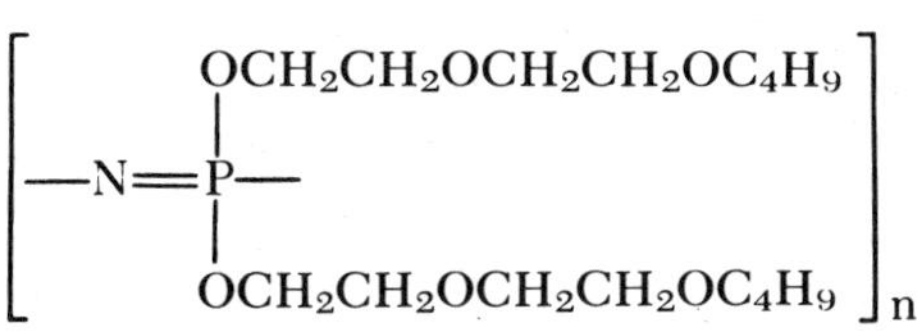 12	51
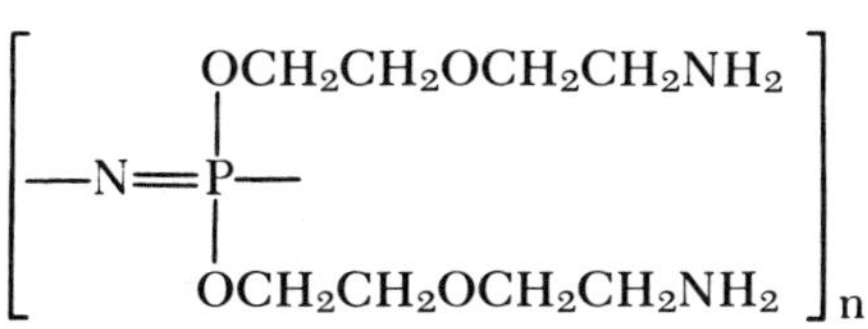 13	None

NOTE: Hydrogels extrude water at the LCST. All of the polymers shown here are water-soluble at 25 °C.

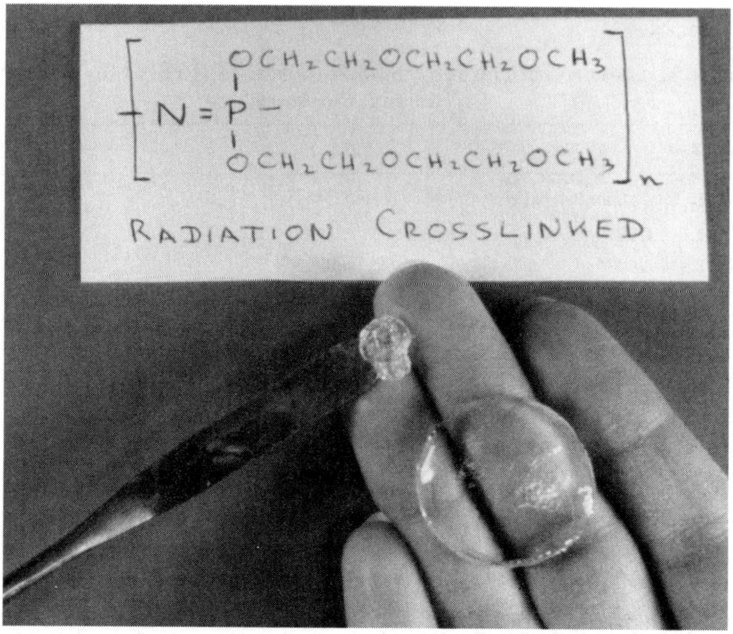

Figure 4. MEEP, cross-linked by exposure to approximately 2 Mrad of gamma rays, before and after immersion in water. (Reproduced with permission from reference 5. Copyright 1992 Prentice Hall.)

have generated widespread interest as potential electrolytes in large-area, light-weight, rechargeable lithium batteries (Figure 4) (*20–23*).

Second, MEEP can be cross-linked readily by exposure to gamma rays or ultraviolet light (*24–26*). The mechanism of this reaction, as illustrated in Scheme VII, is believed to involve C–H bond homolytic cleavage followed by cross-combination of the resultant carbon radicals. The sensitivity of MEEP to radiation cross-linking is attributed to the presence of 22 C–H bonds on every repeating unit. Cross-linked MEEP swells in water to form stable hydrogels (Figure 4), the water contents of which are a function of the degree of cross-linking. The cross-linking process has been used to entrap and immobilize enzymes with retention of enzymic activity (*27*). Diffusion–release of small-molecule solutes from the hydrogels has also been studied (*24*).

Finally, radiation-cross-linked MEEP has been converted to swollen organogels by absorption of organic vinyl monomers (Figure 5). Polymerization and cross-linking of the organic monomers has yielded a range of interpenetrating polymer network materials (*28*). Interpenetrating polymer networks prepared with acrylonitrile and acrylic acid polymers show good component compatibility, and this compatibility

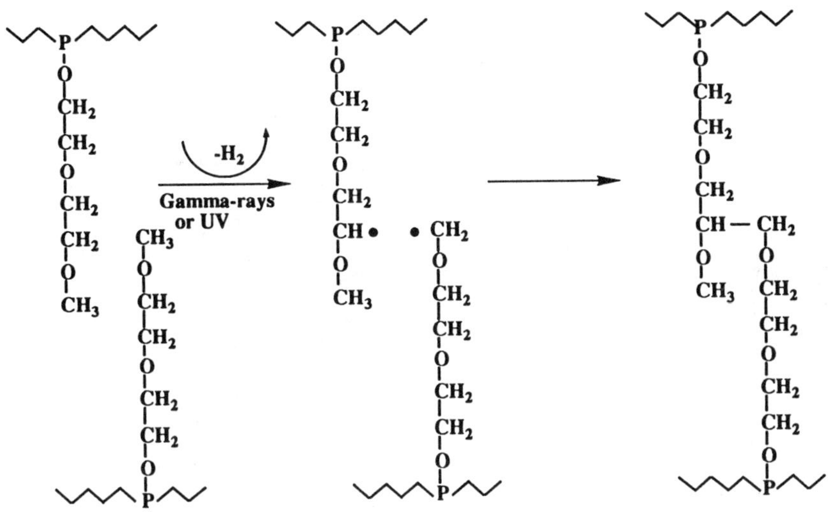

Scheme VII.

is consistent with the expected hydrophilic interactions. MEEP has also been used as the linear polymeric component in ceramic composites prepared via the sol–gel process, and these composites show a range of interesting and potentially useful properties (*29*).

Polymers with Glucosyl Side Groups. Perhaps the ultimate in hydrophilic side group interactions might be expected from glucosyl units of the type shown in polymer **6.** With four free hydroxyl groups on every side group, the opportunities for H bonding to water molecules appear to be almost unprecedented in a synthetic polymer.

The synthesis of glucosyl-substituted polyphosphazenes presents a special challenge. Glucose itself has five functional sites per molecule, and any attempt to treat poly(dichlorophosphazene) with glucose would result in extensive cross-linking and polymer precipitation long before halogen replacement was complete. Thus four of the five hydroxy units must be protected during coupling of the side group to the backbone and must be deprotected during a final step. This process is shown in Scheme VIII (*30*).

The second challenge is this: The protected diacetone glucose used in the initial macromolecular substitution is an exceedingly bulky nucleophile. Replacement of half the available chlorine atoms proceeds in a conventional manner. However, replacement of the remainder is slowed considerably by steric hindrance effects. Indeed, judging from molecular graphics simulations (Figure 6), replacement of the last chlorine in a three-repeat-unit sequence appears to be ex-

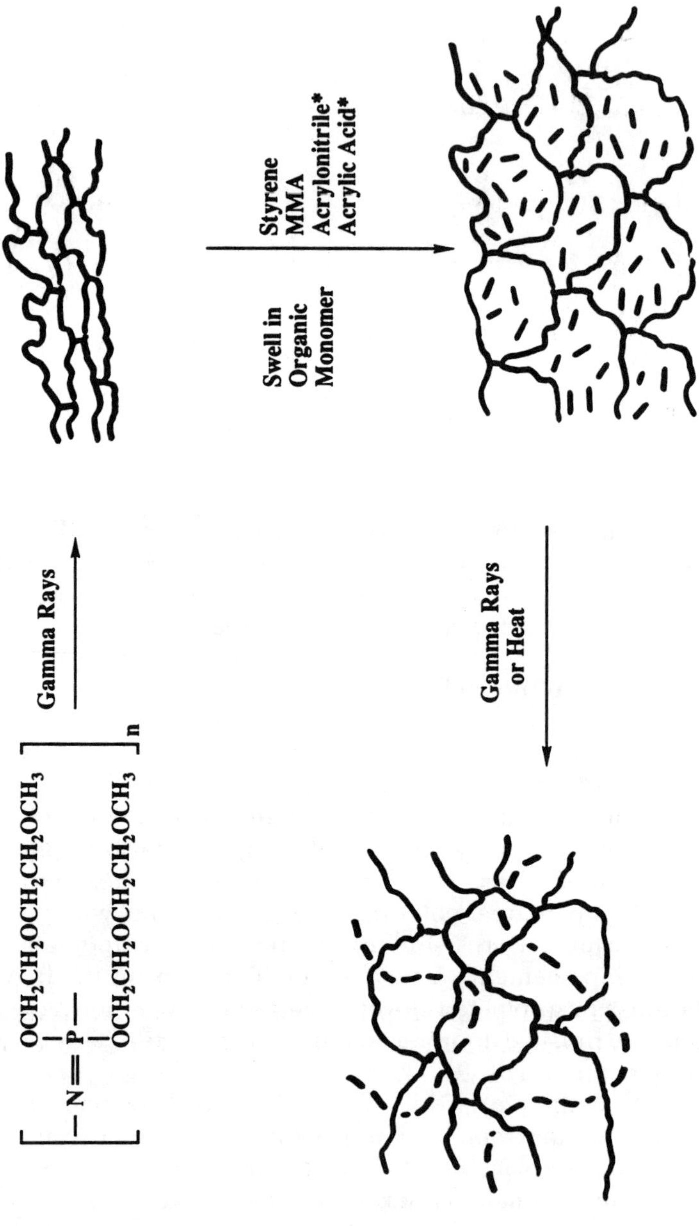

Figure 5. Interpenetrating polymer networks based on MEEP with various organic polymers. MMA is methyl methacrylate.

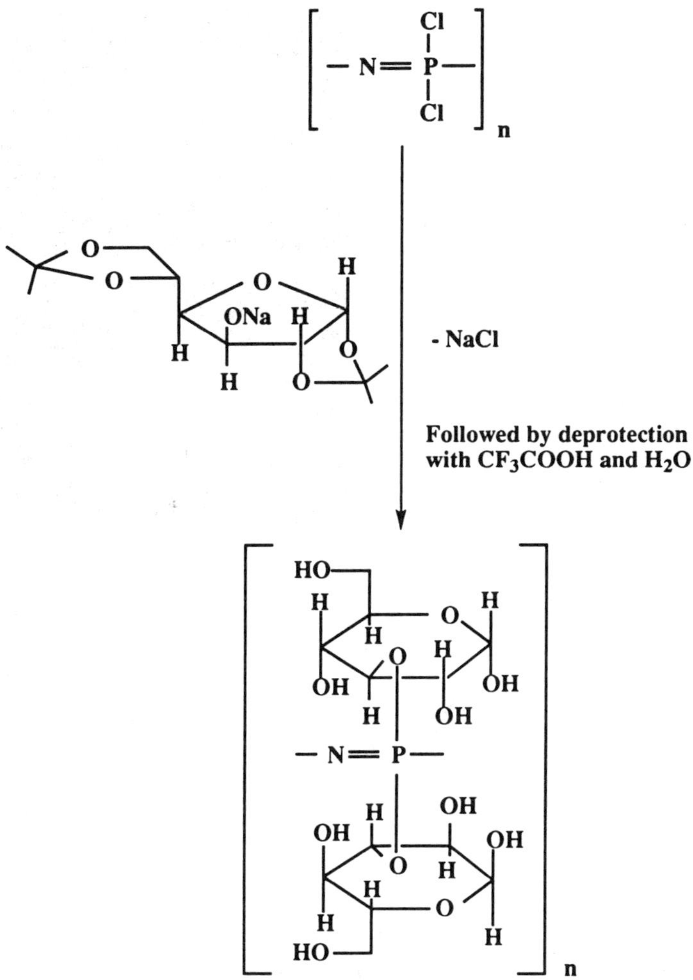

Scheme VIII.

ceedingly difficult, and it is not surprising that long reaction times and elevated reaction temperatures are needed for this final replacement.

Thus from a practical point of view, it is easier to incorporate the glucosyl units as part of a mixed-substituent polymer, with the second nucleophile being less bulky than diacetoneglucoxide (*31*). This procedure is illustrated in Scheme IX.

A great deal of additional work needs to be done with sugar derivatives of polyphosphazenes. This work will undoubtedly be carried out as the biomedical properties of the polyphosphazenes are studied in more detail.

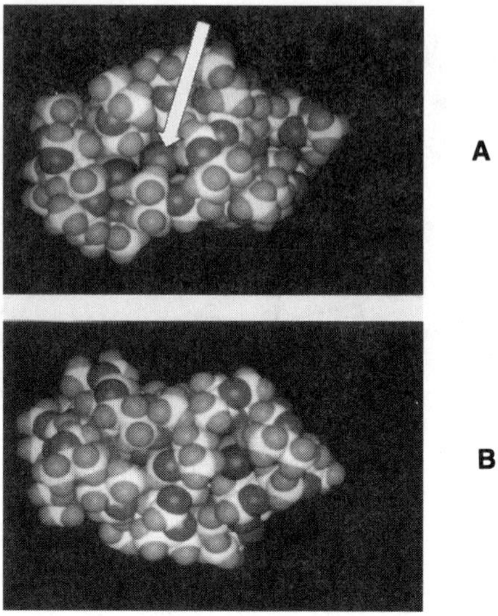

Figure 6. Molecular graphics simulation of the steric hindrance involved in the replacement of the last chlorine atom (arrow) in a three-repeat unit segment of a polyphosphazene by a diacetone glucose anion (A) and the close crowding that exists in a completely substituted segment of the chain (B).

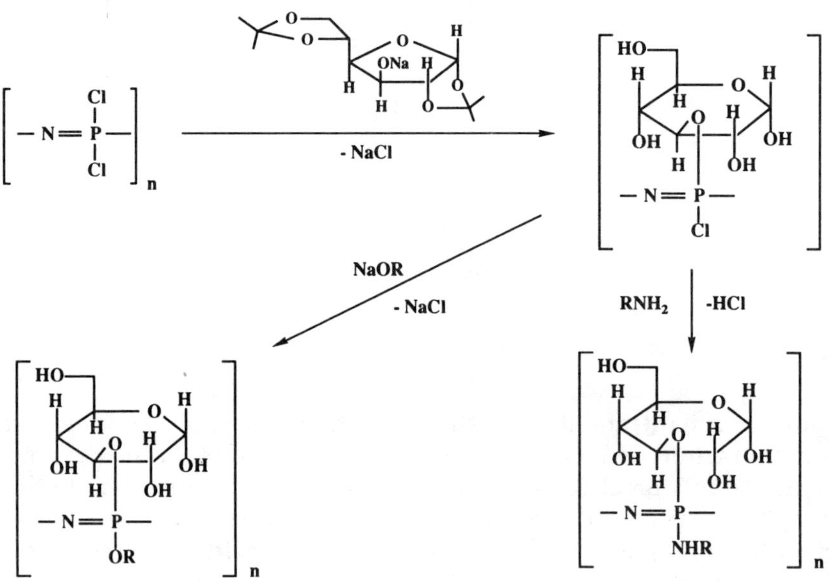

Scheme IX.

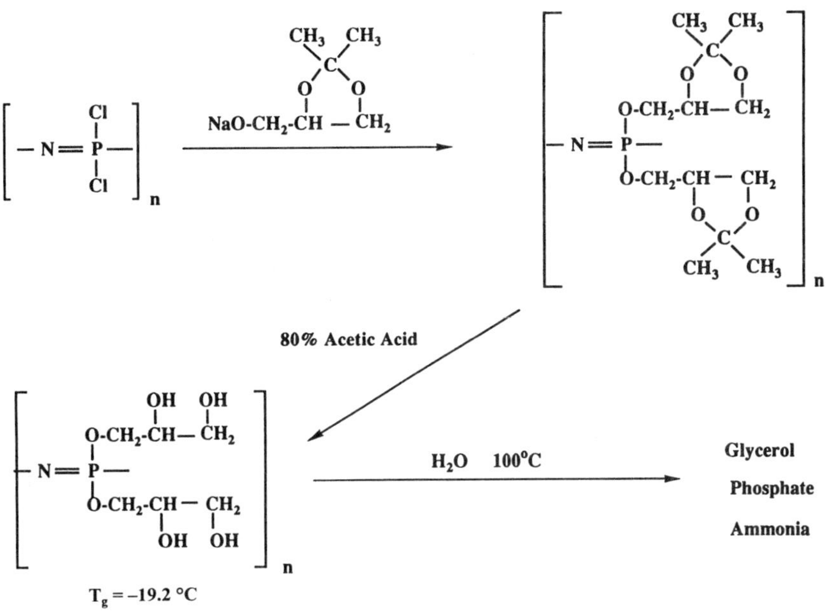

Scheme X.

Glyceryl Derivatives. The opportunities for generating polymer water solubility with glyceryl side groups are similar to those just discussed for glucosyl units. The synthetic challenges encountered are also similar, especially with respect to the multifunctionality of glycerol and the need for protection–deprotection reactions. Steric hindrance is less of a problem. The synthesis sequence is shown in Scheme X (*32*).

The glyceryl polymer, **14**, is completely soluble in water. It has a T_g of 19.2 °C.

Aryloxycarboxylic Acid Derivatives. Polymer **8** is an analog of poly(acrylic acid) or of alginate macromolecules, from which it differs by the high concentration of carboxylate groups per repeating unit. It is of interest as a water-soluble polymer, as a polyelectrolyte, and as a biomedical encapsulant.

The synthesis of **8** requires the protection and eventual deprotection of the carboxylic acid function for the reasons just discussed (*33*). The overall synthesis procedure is shown in Scheme XI. Although the parent polymer with carboxylic acid units is not soluble in water, the sodium and potassium salts have a high solubility in aqueous media.

Perhaps the most important property of the polymer is its ability to form ionic cross-links when the sodium salt is exposed to solutions

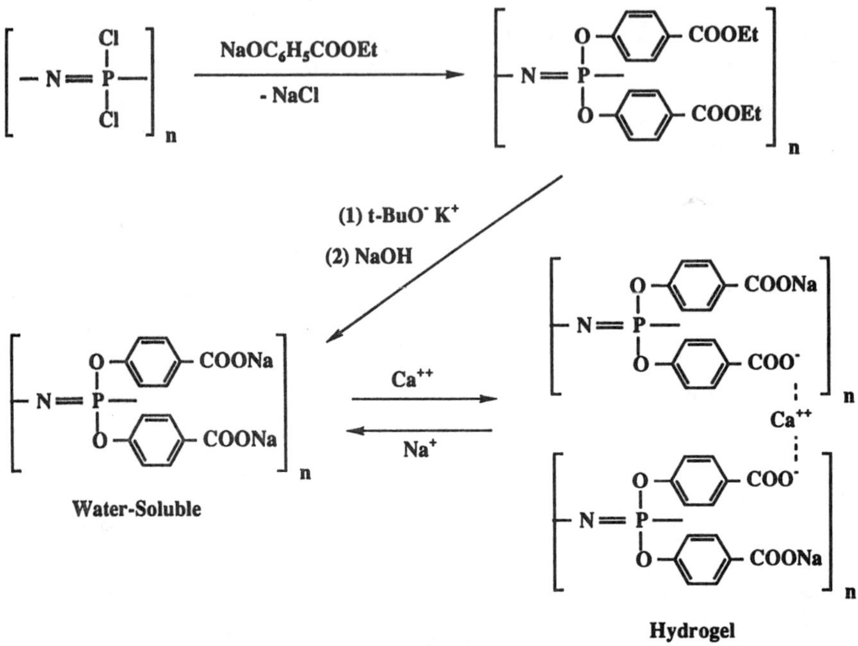

Scheme XI.

of divalent or trivalent cations, such as Ca^{2+} or Al^{3+}. The ionically cross-linked materials are hydrogels, and their physical characteristics are determined by the divalent cation concentration. Infusion with solutions of monovalent cations reverses the process and leads to dissolution of the polymers. Preliminary tests have indicated that this polymer has a low oral toxicity.

The polymer has been studied as a species for the microencapsulation of biologically active entities such as mammalian cells, microorganisms, and proteins (*34–36*). The microencapsulation procedure involves dispersion of the biological entity in an aqueous solution of polymer **8** and expulsion of the bioactive species complete with a surrounding coating of polymer solution into an aqueous solution of calcium chloride.

Liver hybridoma cells in culture, proteins, and immunostimulant species have been microencapsulated in this way. The bioerosion of the polymer has been induced by the incorporation of amino acid ester cosubstituent groups into the polymer (*37*).

Polymers with Sulfonic Acid Solubilizing Groups. Polymer **9** is representative of a range of structures with different ratios of aryl-

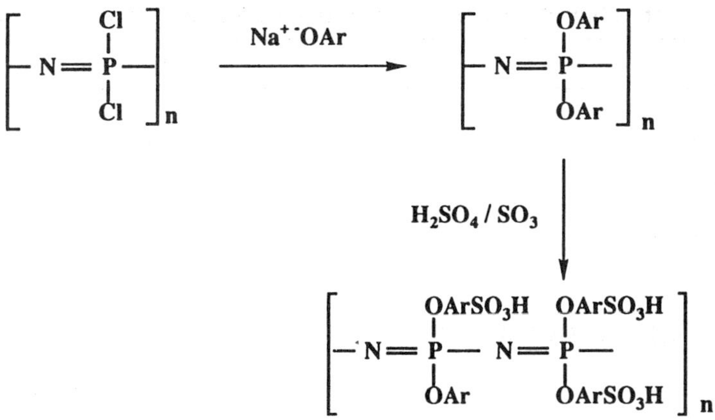

Scheme XII. n = *2000–7000 or the number of units at the surface of a polymer film.*

oxy and sulfonated aryloxy side groups. These are prepared by the sulfonation of poly[bis(phenoxy)phosphazene] (**15**), as shown in Scheme XII (*38*). Relatively small amounts (25%) of sulfonated aryloxy groups can induce water solubility. As discussed in the following section, sulfonation of the surface of an aryloxyphosphazene polymer can dramatically increase the hydrophilicity of the material. Sulfonated polyphosphazenes can also be prepared via two additional methods. First, the use of a sulfonated nucleophile in the primary macromolecular substitution process (Scheme II) yields polymers with aliphatic sulfonated units (*39*). Second, the reactions of aliphatic amino side groups with sulfones generate sulfonated polyphosphazenes (*40*).

Hydrolytic Stability

As mentioned earlier, the stability or instability of a polymer to water determines the applications, and especially the biomedical uses, for which the polymer is suited. Two of the polymer classes discussed so far are sensitive to hydrolysis in neutral pH water at 100 °C (Table II), and this sensitivity reflects a much slower hydrolysis rate at body temperature. Thus the glucosyl and glyceryl species hydrolyze slowly to phosphate, small amounts of ammonia, and glucose or glycerol. The biocompatibility of these products is obvious, and the utility of these systems in drug delivery and other biomedical applications is a subject of considerable interest. On the other hand, the methylamino-substituted polymer (**4**) and MEEP are stable to water at neutral and basic pHs but are sensitive to strong acids (*41*). The arylsulfonic acid-substi-

Table II. Hydrolytic Sensitivity at pH 7 (Possible Bioerodibility)

Polymer	*Hydrolytic Sensitivity*
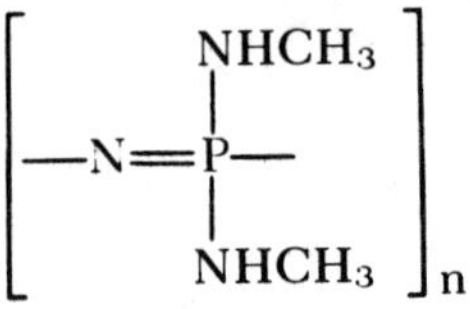	stable
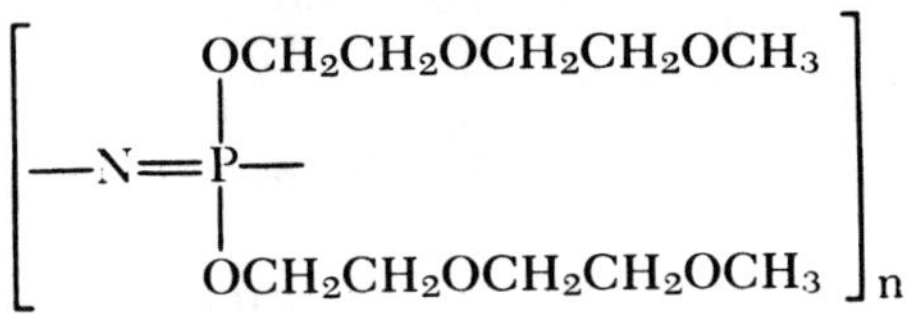	stable
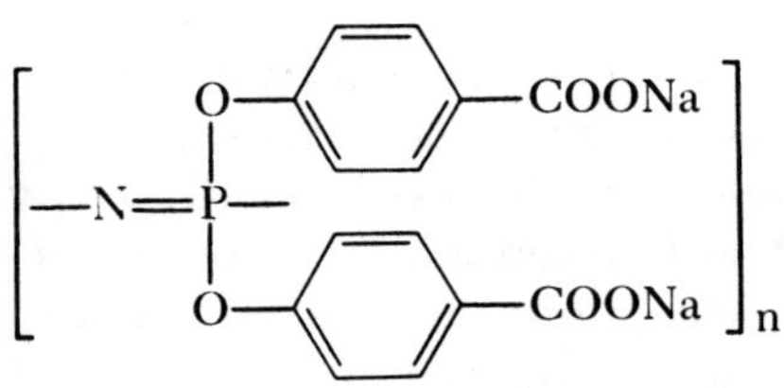	?
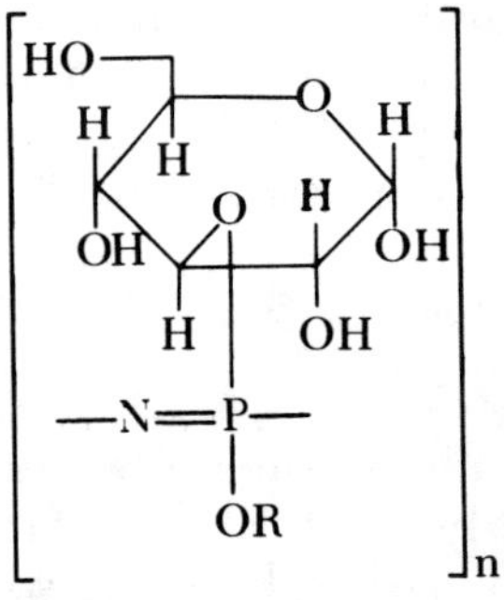	hydrolyzes at 100 °C to glucose, phosphate, ammonia, and ROH
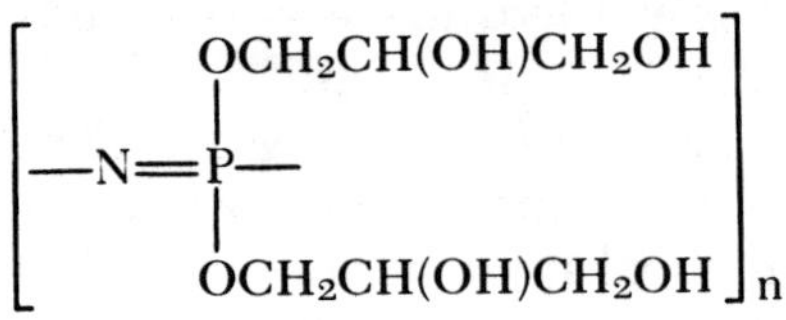	hydrolyzes at 100 °C to glycerol, phosphate, and ammonia

tuted polymers of type **9** appear to be stable to hydrolysis over a wide pH range. The hydrolysis behavior of the arylcarboxylic acid derivative (**8**) is still under investigation, although, as mentioned in the preceding section, its hydrolysis can be induced by the presence of a hydrolytically sensitizing cosubstituent group (*37*).

Hydrophilic Surfaces and Surface Hydrogels

Polymers that have a hydrophobic interior (to prevent water absorption and colonization by microorganisms) and a hydrophilic or hydrogel surface are needed in numerous biomedical applications (Figure 7). Polyphosphazenes are particularly useful starting polymers for such materials because of (1) the wide range of side groups that can be employed, (2) the chemical stability of the skeleton, and (3) the possibilities that exist for the exchange of surface side groups by a variety of hydrophilic units.

A simple solution to this problem is shown in Figure 8. Here, films of poly[bis(trifluoroethoxy)phosphazene] (**16**) undergo replacement of surface fluoroalkoxy units by hydroxyl or $-O^{-}\ {}^{+}NBu_4$ units during treatment with aqueous solutions of sodium hydroxide containing tetrabutylammonium bromide, a phase transfer agent (*42*). Hydrolysis proceeds rapidly from the surface, generating an adhesive hydrogel as it penetrates toward the interior. The contact angle to water falls from 108° to 90° as the process takes place. Similar surface reactions have been used to link functional organic units to the surface of polyphosphazenes (*43*).

Surface oxidation, as shown in Figure 9, converts *p*-methylaryloxy surface groups to units that bear carboxylic acid functions (*44*). Again, the reaction is accompanied by a sharp decrease in contact angle from 92° to 70° or to 25° for contact with basic media. Sulfonation of surface aryloxy groups, as in Figure 10, in the presence of sulfuric acid or SO_3 has a similar effect (*38*).

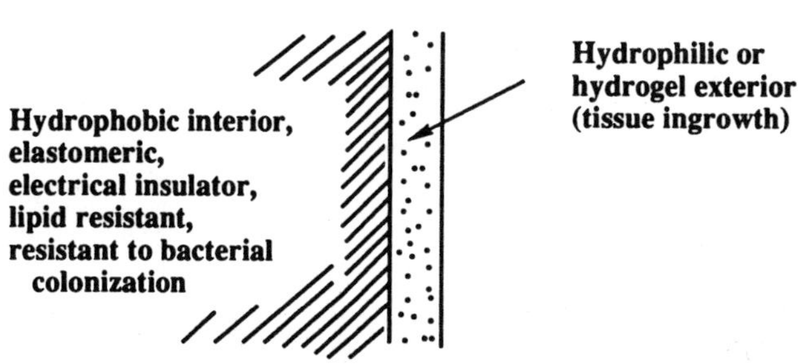

Figure 7. Cross section of a target biomedical material.

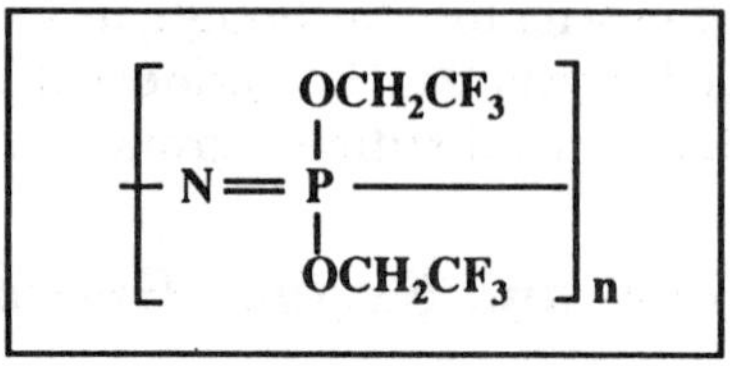

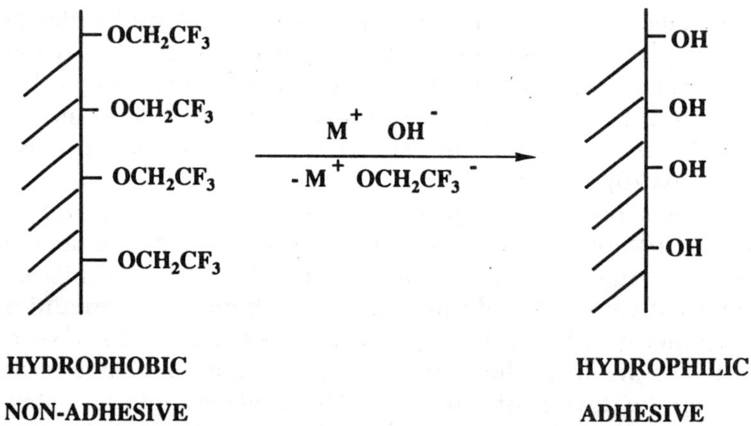

Figure 8. Treatment of the surface of poly[bis(trifluoroethoxy)phosphazene] with a strong aqueous base in the presence of a phase-transfer reagent results in a change in surface character from hydrophobic to hydrophilic.

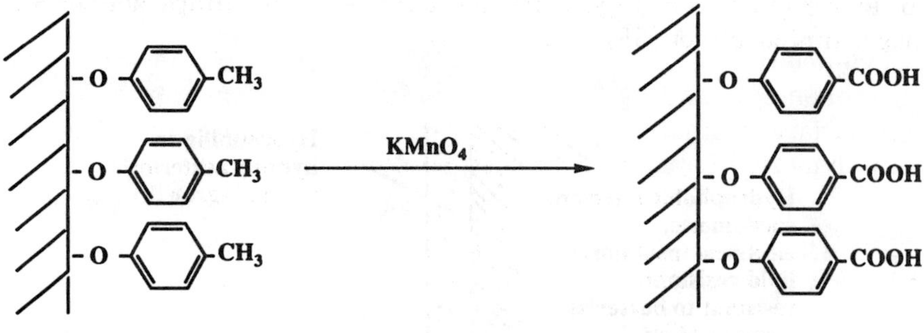

Figure 9. Oxidation of the surfaces of p-*methylphenoxy groups linked to a polyphosphazene chain yields carboxylic acid surface units.*

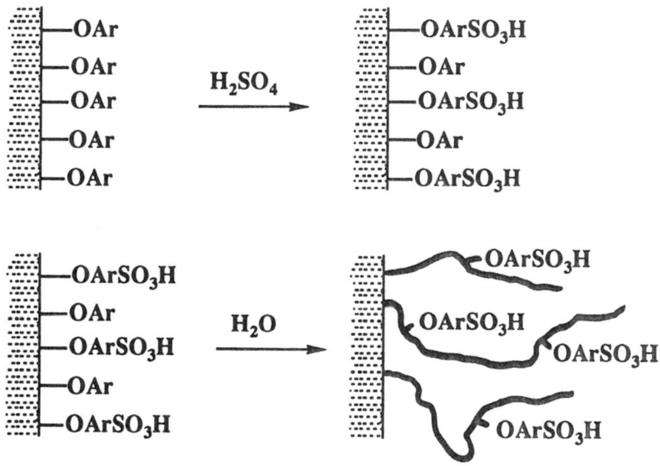

Figure 10. Sulfonation of the surface of poly[bis(aryloxy)phosphazenes] and expansion of the surface layer to form a hydrogel.

These surface reactions are important because biomedical compatibility, particularly blood compatibility, depends on the hydrophilic character of the surface and on the absence or presence of ionic species. Considerable effort is being expended in other laboratories to link poly- or oligoethylene oxide species to polymer surfaces in order to improve blood compatibility (*45*). The thicker the hydrophilic layer (within limits), the more effective is the biological effect.

Recently, we approached the same problem from a different point of view. As discussed, MEEP can be readily cross-linked by exposure to gamma rays or ultraviolet light. This same radical-induced reaction also provides a mechanism for the covalent binding of MEEP to the surfaces of polymers such as polyethylene, polypropylene, or poly(vinyl chloride). The process is illustrated in Figures 11 and 12 (*46*). Evidence that MEEP hydrogels have some antibacterial activity has also been obtained (*47*), and this evidence raises the possibility that the surface lamination process can be used to improve the resistance of polymeric biomaterials to colonization by microorganisms.

Concluding Comments

Within the field of macromolecules, relatively few polymers are soluble in water or are appropriate for conversion to hydrogels (*48*). Proteins, nucleic acids, and some polysaccharides are obvious examples of species that are water-soluble, as are linear polyphosphates and polysilicates. However, water-soluble petrochemical polymers are rel-

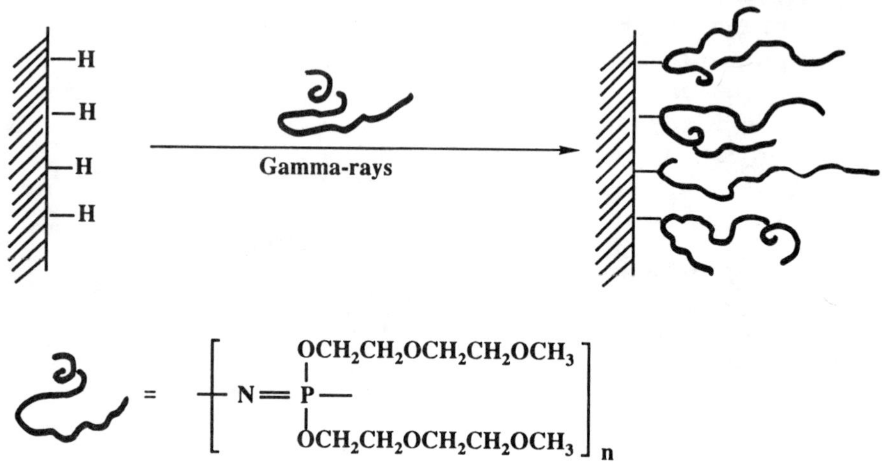

SUBSTRATE POLYMER
(CONTACT ANGLE)

POLYPROPYLENE	94°	→	27°
PVC	78°		34°
POLYCARBONATE	65°		43°
PMMA	65°		35°
MYLAR	63°		31°

Figure 11. Radiation grafting of molecules of MEEP to the surface of a solid polymer and changes in contact angles to water that occur following grafting (46). PVC is poly(vinyl chloride); PMMA is poly(methyl methacrylate).

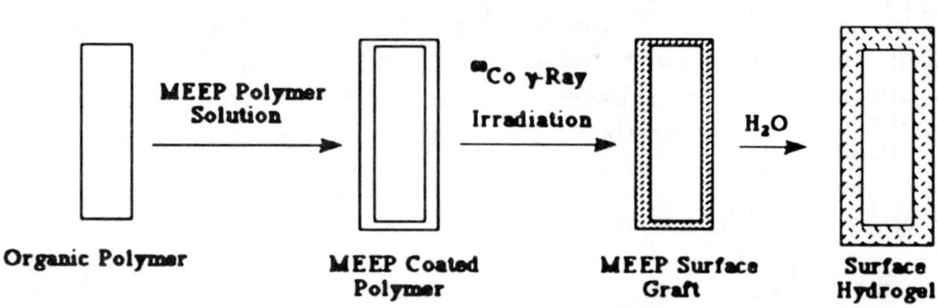

Figure 12. Generation of a hydrogel on the surface of an organic polymer following radiation-induced surface grafting of MEEP. (Reproduced from reference 46. Copyright 1992 American Chemical Society.)

atively few; examples include poly(ethylene oxide), poly(vinyl alcohol), poly(acrylic acid), poly(vinylpyrridine), and a variety of newer polymers that contain ether units in the backbone and hydroxy or carboxylate side units (*49*).

The development of polyphosphazenes as water-soluble polymers and hydrogels offers many new possibilities in this field. These polymers have high molecular weights, hydrophilic backbones, and structures that can be varied over a wide range by macromolecular substitution and cosubstitution and that, when necessary, can be designed to be bioerodible. Polymers **4–9** represent the starting point for work in this field. A much wider range of water-soluble structures can be anticipated as structure–property relationships and uses become more extensively developed.

Acknowledgments

It is a pleasure to acknowledge the contributions of a number of co-workers at The Pennsylvania State University to the development of the field of water-soluble polyphosphazenes. These co-workers include Drs. Daniel P. Mack, Angelo G. Scopelianos, Richard Fitzpatrick, Sukky Kwon, Paul Austin, Geoffrey Riding, Thomas Neenan, Marie Gebura, Karyn Visscher, Mark Welker, Eric Klingenberg, Shawn Pucher, and Michael Turner. The development of the biological microencapsulation process was part of a collaborative effort with Drs. Robert Langer, Smadar Cohen, and A. Andrianov and their co-workers at Massachusetts Institute of Technology and with colleagues at the Virus Research Institute (V.R.I.) in Cambridge, Massachusetts. Various parts of this work were funded by the National Institutes of Health, the U.S. Army Research Office, the Office of Naval Research, Corning Inc., Johnson & Johnson, and V.R.I.

References

1. Allcock, H. R.; Kugel, R. L. *J. Am. Chem. Soc.* **1965,** *87,* 4216.
2. Allcock, H. R.; Kugel, R. L.; Valan, K. J. *Inorg. Chem.* **1966,** *5,* 1709.
3. Allcock, H. R.; Kugel, R. L. *Inorg. Chem.* **1966,** *5,* 1716.
4. Allcock, H. R. *Phosphorus–Nitrogen Compounds;* Academic: New York, 1972.
5. Mark, J. E.; Allcock, H. R.; West, R. *Inorganic Polymers;* Prentice Hall: Englewood Cliffs, NJ, 1992; Chapter 3.
6. Singler, R. E.; Schneider, N. S.; Hagnauer, G. L. *Polym. Eng. Sci.* **1975,** *15,* 321.
7. Allcock, H. R. In *Ring-Opening Polymerization;* Brunelle, D. J., Ed.; Hanser Publishers: Munich, Germany, 1993; Chapter 7.
8. Allcock, H. R. In *The Chemistry of Inorganic Ring Systems;* Steudel, R., Ed.; Elsevier: Amsterdam, Netherlands, 1992; Chapter 9.

9. Allcock, H. R. In *Catalysis in Polymer Synthesis;* Vandenberg, E. J.; Salamone, J. C., Eds.; ACS Symposium Series 496; American Chemical Society: Washington, DC, 1992; p 236.
10. Flindt, E. P.; Rose, H. Z. Z. *Anorg. Allg. Chem.* **1977,** *428,* 204.
11. Wisian-Neilson, P.; Neilson, R. H. *J. Am. Chem. Soc.* **1980,** *102,* 2848.
12. Montague, R. A.; Matyjaszewski, K. J. *J. Am. Chem. Soc.* **1990,** *112,* 1980.
13. Herring, D. L. *Chem. Ind. (London)* **1960,** 717.
14. Tesi, G.; Haber, C. P.; Douglas, C. M. *Proc. Chem. Soc. London* **1960,** 219.
15. Franz, U.; Nuyken, O.; Matyjaszewski, K. *Macromolecules* **1993,** *26,* 3723.
16. Allcock, H. R.; Kugel, R. L. *Inorg. Chem.* **1966,** *5,* 1716.
17. Allcock, H. R.; Gebura, M.; Kwon, S.; Neenan, T. X. *Biomaterials* **1988,** *19,* 500.
18. Allcock, H. R.; Austin, P. E.; Neenan, T. X.; Sisko, J. T.; Blonsky, P. M.; Shriver, D. F. *Macromolecules* **1986,** *19,* 1508.
19. Allcock, H. R.; Pucher, S. R.; Turner, M. L.; Fitzpatrick, R. J. *Macromolecules* **1992,** *25,* 5573.
20. Blonsky, P. M.; Shriver, D. F.; Austin, P. E.; Allcock, H. R. *J. Am. Chem. Soc.* **1984,** *106,* 6854.
21. Blonsky, P. M.; Shriver, D. F.; Austin, P. E.; Allcock, H. R. *Polym. Mater. Sci. Eng.* **1985,** *53,* 118.
22. Blonsky, P. M.; Shriver, D. F.; Austin, P. E.; Allcock, H. R. *Solid State Ionics* **1986,** *18 & 19,* 258.
23. Bennett, J. L.; Dembek, A. A.; Allcock, H. R.; Heyen, B. J.; Shriver, D. F. *Chem. Mater.* **1989,** *1,* 14.
24. Allcock, H. R.; Kwon, S.; Riding, G. H.; Fitzpatrick, R. J.; Bennett, J. L. *Biomaterials* **1988,** *19,* 509.
25. Bennett, J. L.; Dembek, A. A.; Allcock, H. R.; Heyen, B. J.; Shriver, D. F. *Polym. Prepr. (Am. Chem. Soc., Div. Polym. Chem.)* **1989,** *30,* 437.
26. Nelson, C. J.; Coggio, W. D.; Allcock, H. R. *Chem. Mater.* **1991,** *3,* 786.
27. Allcock, H. R.; Pucher, S. R.; Visscher, K. B. *Biomaterials* **1994,** *15,* 502.
28. Allcock, H. R.; Visscher, K. B. *Chem. Mater.* **1992,** *4,* 1182.
29. Coltrain, B. K.; Ferrar, W. T.; Landry, C. J. T.; Molaire, T. R.; Zumbulyadis, N. *Chem. Mater.* **1992,** *4,* 358.
30. Allcock, H. R.; Scopelianos, A. G. *Macromolecules* **1983,** *16,* 715.
31. Allcock, H. R.; Pucher, S. R. *Macromolecules* **1991,** *24,* 23.
32. Allcock, H. R.; Kwon, S. *Macromolecules* **1988,** *21,* 1980.
33. Allcock, H. R.; Kwon, S. *Macromolecules* **1989,** *22,* 75.
34. Cohen, S.; Bano, M. C.; Visscher, K. B.; Chow, M.; Allcock, H. R.; Langer, S. *J. Am. Chem. Soc.* **1990,** *112,* 7832.
35. Bano, M. C.; Cohen, S.; Visscher, K. B.; Allcock, H. R.; Langer, R. *Biotechnology* **1991,** *9,* 468.
36. Cohen, S.; Bano, M. C.; Cima, L. G.; Allcock, H. R.; Vacanti, J. P.; Vacanti, C. A.; Langer, R. *Clin. Mater.* **1993,** *13,* 3.
37. Andrianov, A. K.; Payne, L. G.; Visscher, K. B.; Allcock, H. R.; Langer, R. *Polym. Prepr. (Am. Chem. Soc., Div. Polym. Chem.)* **1993,** *34,* 233.
38. Allcock, H. R.; Fitzpatrick, R. J. *Chem. Mater.* **1991,** *3,* 1120.
39. Ganapethiappan, S.; Chen, K.; Shriver, D. F. *Macromolecules* **1988,** *21,* 2299.
40. Allcock, H. R.; Klingenberg, E. H.; Welker, M. F. *Macromolecules* **1993,** *26,* 5512.
41. Allcock, H. R.; Fuller, T. J.; Matsumura, K. *Inorg. Chem.* **1982,** 515.

42. Allcock, H. R.; Rutt, J. S.; Fitzpatrick, R. J. *Chem. Mater.* **1991,** *3,* 442.
43. Allcock, H. R.; Fitzpatrick, R. J. *Chem. Mater.* **1991,** *3,* 450.
44. Allcock, H. R.; Fitzpatrick, R. J.; Salvati, L. *Chem. Mater.* **1992,** *4,* 769.
45. Andrade, J. D.; Hlady, V.; Jeon, S.-I. *Polym. Mater. Sci. Eng.* **1993,** 69, 60–61.
46. Allcock, H. R.; Fitzpatrick, R. J.; Visscher, K. B. *Chem. Mater.* **1992,** *4,* 775.
47. Allcock, H. R.; Pucher, S. R.; Fitzpatrick, R. J. *Biomaterials* **1992,** *13,* 857.
48. McCormick, C. L.; Bock, J.; Schulz, D. N. In *Concise Encyclopedia of Polymer Science and Engineering,* Wiley Interscience: New York, 1990; pp 1299–1306.
49. See, for example, Vandenberg, E. J. In *Catalysis in Polymer Synthesis;* Vandenberg, E. J.; Salamone, J. C., Eds.; ACS Symposium Series 496; American Chemical Society: Washington, DC, 1992; Chapter 1.

RECEIVED for review November 18, 1993. ACCEPTED revised manuscript July 1, 1994.

2

Association and Phase Separation of Amphiphilic Poly(ethylene oxide) Star Polymers

Guangbin Zhou, Xin Chen, and Johannes Smid*

Polymer Research Institute, Department of Chemistry, College of Environmental Science and Forestry, State University of New York, Syracuse, NY 13210–2786

Tri- and tetrafunctional isocyanates were reacted with methoxypoly(ethylene glycol)s (MPEG) and with nonylphenoxypoly(ethylene glycol)s to form three- and four-armed star polymers with large hydrophobic cores. Those with MPEG arms associated into micelles, and the logarithms of their critical micelle concentrations (cmc) were proportional to the hydrophile–lipophile balance of the stars. Cloud points were measured as a function of arm length and the concentration of added NaF or NaH_2PO_4. Hydrophobic species such as Coomassie brilliant blue and 1-pyrenebutyrate strongly bind to the stars even below the cmc of the stars. The stars with nonylphenoxy end-capped poly(ethylene oxide) arms (associative star polymers) phase separated in water to form a polymer-rich and a polymer-poor phase. As long as the two phases coexist, the polymer concentrations in the phases are constant and their volume fractions change proportionally with the total polymer concentration. The phase separation was studied as a function of temperature and salt content. Preliminary data on the partitioning of hydrophobic compounds between the two phases are reported.

AMPHIPHILIC POLYMERS have been extensively examined in recent years as a source of polymeric surfactants, drug carriers, and rheology

* Corresponding author

0065–2393/96/0248–0031$12.00/0

modifiers in latex paint formulations or tertiary oil recovery systems (*1–6*). Typical examples are hydrophobically modified poly(acryl amide)s, poly(acrylate)s, and cellulosics, and block or graft copolymers containing hydrophilic and hydrophobic chain segments. Amphiphilic star polymers, particularly those in which the hydrophilic arms are capped with hydrophobic end groups (*7*), have also drawn attention recently (*2, 5, 7–9*).

Our work focused on star polymers with poly(ethylene oxide) (PEO) arms and a core consisting of a well-defined hydrophobic tri- or tetraisocyanate (*10–14*). The isocyanate was made by the hydrosilylation of *m*-isopropenyl-α,α-dimethylbenzylisocyanate (*m*-TMI) (*15*). For associative star polymers, the PEO arms contained nonylphenoxy end groups (*12, 13*). We also used the more common tris(isocyanatophenyl)methane (T3PM) as the core molecule for three-armed PEO stars (*14*). The homopolymer stars aggregate in aqueous solutions, but no phase separation occurs. The associative stars, on the other hand, interact through their nonylphenoxy end groups and separate into a dilute and a concentrated polymer phase. As long as the two phases coexist, their polymer concentrations remain constant (*12, 13*).

Following a brief discussion of the properties of the homopolymers, specifically their aggregation into micelles and the effect of salts on cloud points, we report in more detail the phase separation behavior of associative stars as a function of polymer concentration, temperature, and the presence of salt. Some work on the distribution of hydrophobic compounds between the two phases is also discussed.

Experimental Details

Synthesis. The structures of the three core isocyanates, abbreviated T3TMI, D4TMI, and T3PM, are given in Chart I. The synthesis of T3TMI and D4TMI (a quantitative bulk hydrosilylation of *m*-TMI by two commercially available siloxanes) was described elsewhere (*15*); T3PM (or Desmodur R) is a Mobay product used without further purification. Homopolymer stars were obtained (*10, 11*) by reacting the respective isocyanates in toluene with different methoxypoly(ethylene glycol)s (MPEG). The associative stars (*12, 13*) were synthesized with nonylphenoxypoly(ethylene glycol)s (Igepals). The MPEGs and Igepals were acquired from Aldrich. Their respective molecular weights (MW) as determined by ^{1}H NMR are given in Table I. The catalyst used was either *N,N*-tetramethylethylenediamine or dibutyltin dilaurate. Higher-molecular-weight glycols were added in excess in order to minimize formation of allophanates, a common side reaction. The disappearance of isocyanate was monitored by IR (2260 cm^{-1}). The star and excess glycol were precipitated in cold hexane, the product was dissolved in water or a mixture of water and ethanol, and unreacted MPEG or Igepal was removed by ultrafiltration through an Amicon YM2 membrane. Water was then removed by rotaevaporation; the last traces were removed azeotropically with benzene or dioxane. Details of the syntheses were published elsewhere (*11, 13*).

Measurements. Silicon was analyzed with an ICP FMA-03 emission spectrometer. T3PM star concentrations were measured with a DMS-100 spectrophotometer. ^{1}H and ^{13}C NMR spectra were recorded on a Bruker AMX-300 spectrometer, and melting points and glass transition temperatures were measured on a Perkin Elmer DSC4 (heating rate, 20 °C/min). Gel permeation chromatograms were obtained with a Waters GPCIA instrument. Viscosities were measured with an Ubbelohde viscometer (30 °C), and surface tensions were measured with a DuNouy tension meter. Cloud points (T_p) were determined with a calibrated thermometer suspended in a 2 wt% aqueous solution of the homopolymer star. Phase separation studies were carried out by shaking an aqueous mixture of an Igepal star (with or without added salt) and then letting it equilibrate for several hours at the required temperature. After the volumes of the upper and lower phases had been measured, the concentration of star molecules with T3TMI or D4TMI core was measured by silicon ICP, and those with a T3PM core were measured by optical spectroscopy at λ_m = 246 nm. The absorptivity for T3PM homopolymer stars (T3PM-MPEG) at this wavelength is ϵ = 48,000, and that for the associative Igepal stars (T3PM-IG) is ϵ = 61,000. The higher value is due to the contribution of the nonylphenoxy end groups.

Results and Discussion

ICP silicon analysis and the ^{1}H and ^{13}C NMR spectra confirm the structures of the T3TMI and D4TMI star polymers. Gel permeation chromatography tracings show only one sharp peak. The reaction with the respective glycols is essentially quantitative, and the ultrafiltration step effectively removes excess MPEG or Igepal.

Table I lists the various star polymers and some of their properties. Their MWs were calculated from the ^{1}H NMR spectra; the core molecules were assumed to be pure starting materials. Although the core molecules T3TMI and D4TMI were indeed pure, the Desmodur compound was a mixture of multifunctional isocyanates, and the star polymers derived from this compound were also mixtures, with the main product being the tristar polymer.

Homopolymer Stars. ***Micellization.*** The amphiphilic homopolymer stars T3TMI-MPEG and D4TMI-MPEG are expected to aggregate in aqueous solution because of their large hydrophobic cores. Surface tension (γ) measurements show that they do indeed aggregate. Figure 1 depicts plots of γ (in dynes per centimeter) versus the log of the star concentration for D4TMI-MPEG stars. Micelle formation is implied by the distinct breaks in the respective plots. Critical micelle concentrations (cmc) are listed in Table II. The relationship between

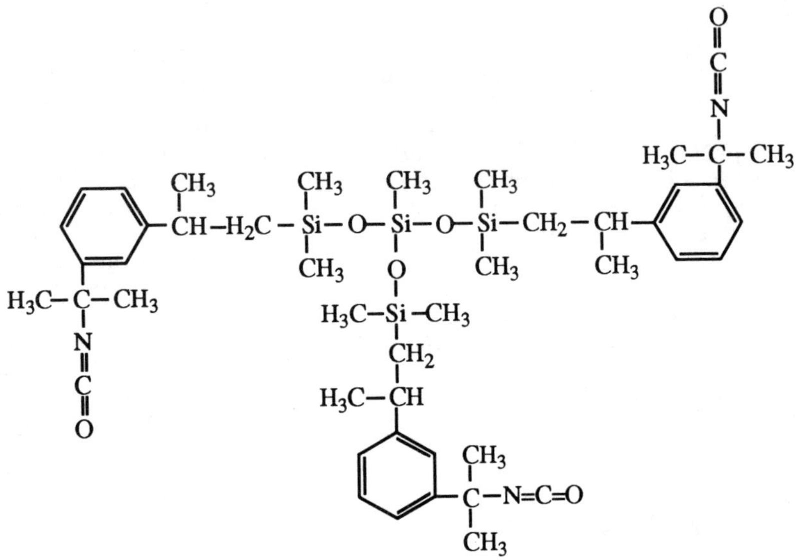

T3TMI

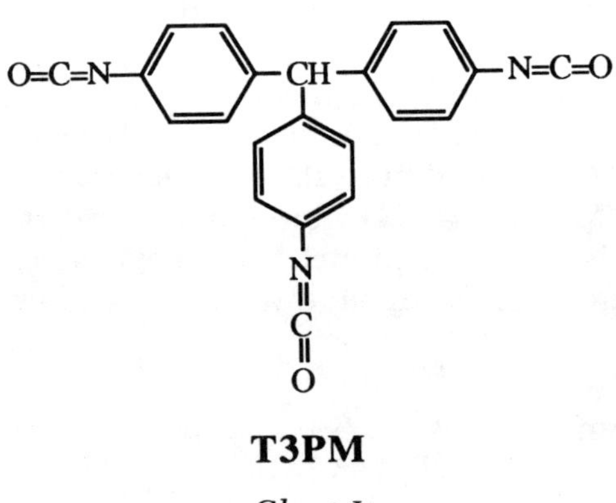

T3PM

Chart I.

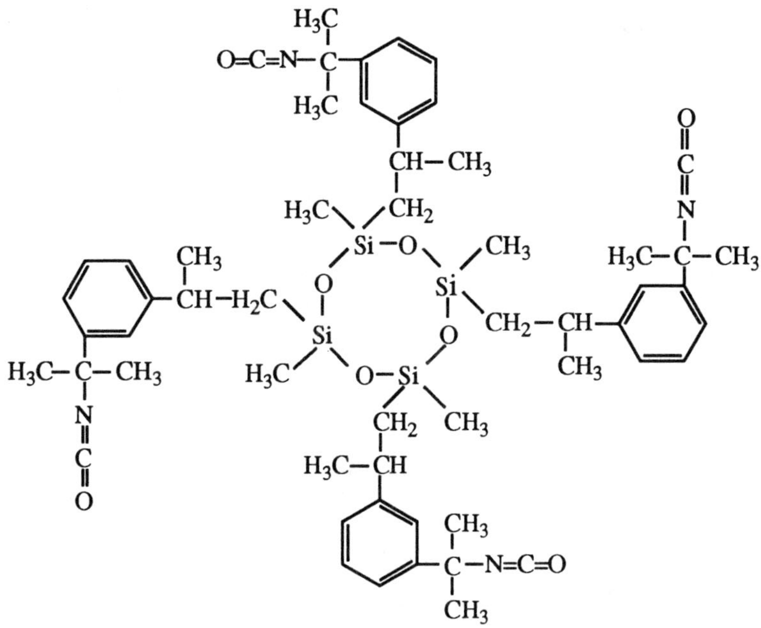

D4TMI

Chart I.

Table I. Properties of PEO Star Polymers with a D4TMI or T3TMI Core

Star Polymer	*MW*	f	T_m *(°C)*	T_g *(°C)*
T3TMI-MPEG				
350	1,900	3.0		
550	2,480	3.0		
750	2,980	3.0	34	−51
1900	6,460	2.9	57	
5000	15,300	3.0	63	
D4TMI-MPEG				
350	2,360	3.8		−41
550	3,150	4.0	13	−49
750	3,750	3.9	36	−50
1900	8,320	3.8	59	
T3TMI-IG990	12,300	2.9	49	
D4TMI-IG990	16,300	3.9	50	
T3PM-IG990	10,700	2.7	49	
D4TMI-IG890	8,900	4.1	36	

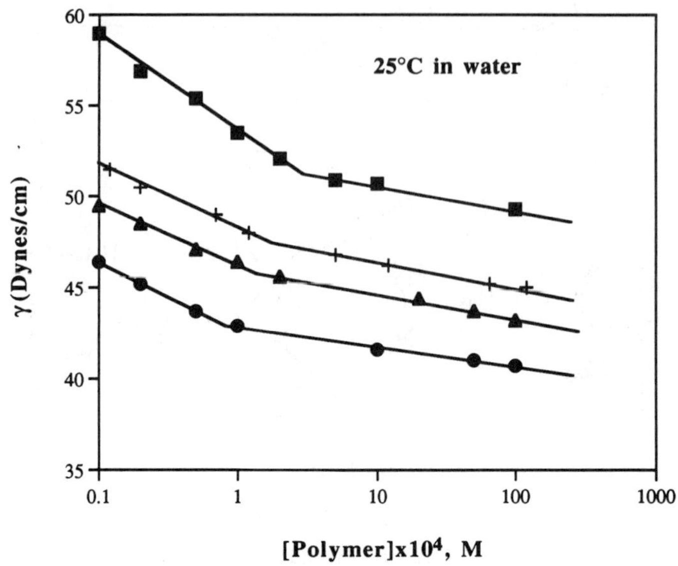

Figure 1. Micellization of D4TMI-MPEG stars in water at 25 °C. MWs of the MPEG arms are 350 (●), 550 (▲), 750 (+), and 1990 (■) (11).

ln cmc and the hydrophile–lipophile balance (HLB) number of the stars is linear:

$$\ln \text{cmc} = -11.6 + 0.20\ \text{HLB} \quad (1)$$

where HLB = 20 $M_H/(M_H + M_L)$, with M_H and M_L being the formula weights of the hydrophilic and hydrophobic parts of the molecules, respectively.

Table II. Critical Micelle Concentrations in Water at 25 °C

Star Polymer	*EO units per arm*	*cmc* $(\times 10^4\ M)$	*HLB*	γ_{cmc} *(dyn cm*$^{-1}$*)*
D4TMI-MPEG				
350	7.5	0.9	11.5	42.5
550	11.7	1.3	13.4	45.6
750	15.5	1.6	14.5	47.0
1900	43.0	3.0	17.6	51.0
T3PM-MPEG				
350	7.5	30		48.2
750	15.5	59		46.4
1900	43.0	>200		

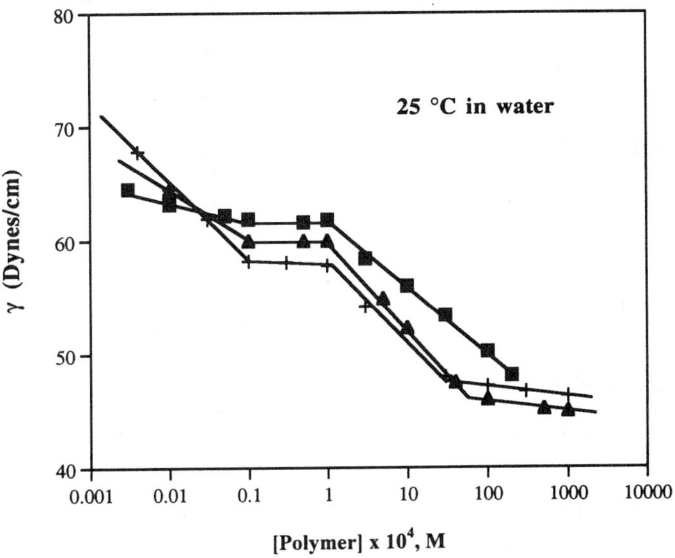

Figure 2. Dependence of surface tension on the log of the star concentration for T3PM-MPEG polymers. MWs of the MPEG arms are 350 (+), 750 (▲), and 1900 (■).

Recent surface tension studies of T3PM-MPEG polymers reveal a more complex behavior, that is, additional discontinuities in the low-concentration portions of γ versus log C plots (Figure 2). The behavior of comblike amphiphilic polymers with short PEO branches is similar (*16*). The cmcs of the T3PM-MPEG stars are nearly an order of magnitude higher than those of D4TMI stars because of the smaller hydrophobic T3PM core. A second distinct break is found around 10^{-4} M, and possibly a third occurs around 10^{-5} M. A repeat of the γ measurements for D4TMI-MPEG1900 down to lower star concentrations also yielded a second transition, which was close to 10^{-5} M. The cause of these discontinuities in the region of lower star polymer concentrations is not yet clear. The discontinuities may be related to conformational changes in polymer aggregates at the air–water interface, as was recently reported for amphiphilic polymers, including starlike octopus molecules (*17–19*).

Micelle formation is confirmed from viscosity data for the aqueous star solutions. For example, the Huggins viscosity plot of D4TMI-MPEG1900 shows a distinct break at 0.25 g/dl or 2.9×10^{-4} M. This break is close to the cmc of this star (Table II). A break in the plot for T3PM-MPEG750 is found at 1.4 g/dl or 5.7×10^{-3} M, which is close to the cmc of 5.9×10^{-3} M for this star (Table II).

Table III. Cloud Points (T_p) and K_s Values for PEO Star Polymers

Star Polymer	T_p *(°C)*	*Salt*	K_s *(deg/m)*
D4TMI-MPEG350	35.9	NaH_2PO_4	−41.4
		NaF	−29.6
D4TMI-MPEG550	69.4	NaH_2PO_4	−51.9
		NaF	−38.4
D4TMI-MPEG750	85.5	NaH_2PO_4	−55.7
		NaF	−43.1
D4TMI-MPEG1900	101	NaH_2PO_4	−62.0
		NaF	−55.1
T3TMI-MPEG350	21.5		
T3TMI-MPEG550	64.9	NaH_2PO_4	−49.1
T3TMI-MPEG750	77.1	NaH_2PO_4	−59.7
T3TMI-MPEG1900	99.4	NaH_2PO_4	−75.2

SOURCE: Data are from reference 11.

Cloud Points. Our star polymers exhibit lower critical solution temperature behavior in aqueous media that is consistent with the polymers or nonionic surfactants containing segments of ethylene oxide (EO) units (*20–22*). Some cloud points (T_p) measured for 2 wt% star solutions are listed in Table III. A plot of T_p versus HLB number shows that three- and four-star polymers with the same HLB numbers have nearly identical T_p (*11*).

Electrolytes have a profound effect on the cloud points of polymers with EO units (*20–23*). For our stars we only used two "salting out" electrolytes, NaF and NaH_2PO_4. The relationship between the observed T_p and the molal salt concentration C_s is linear (*11*):

$$T_p = T_p{}^0 + K_s C_s \tag{2}$$

which is consistent with our recent findings for comblike polymers with EO side chains (*23*) and for PEO hydrogels (*24*). $T_p{}^0$ is the T_p in the absence of salt, and the constant K_s may be interpreted as a measure of the effectiveness of the salt in modifying the stability of the star polymer solution.

The salt effect is dependent on the length of the PEO arm (*11*). As shown in Figure 3, K_s is inversely proportional to the degree of polymerization (dp) of the MPEG arm.

$$K_s = a + b/\mathrm{dp}_{\mathrm{arm}} \tag{3}$$

or

$$T_p - T_p{}^0 = (a + b/\mathrm{dp}_{\mathrm{arm}})C_s \tag{4}$$

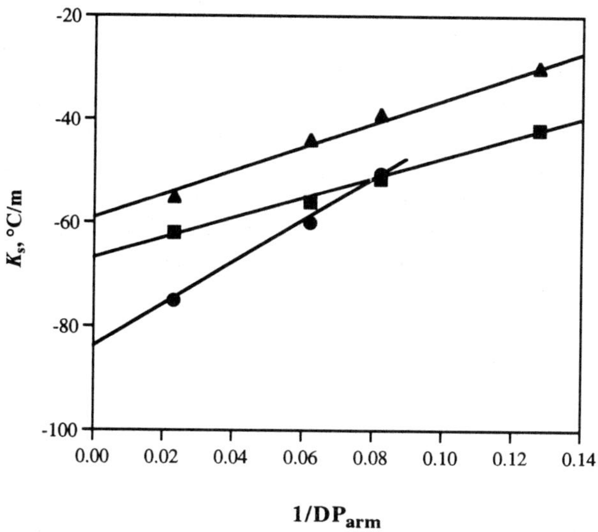

Figure 3. Dependence of K_s *on the number of EO units in the MPEG star arm for the systems T3TMI-MPEG/NaF (●), D4TMI-MPEG/ NaH_2PO_2 (■), and D4TMI-MPEG/NaF (▲) (11).*

The constant a may be defined as the K_s of a star with high-molecular-weight MPEG arms. It is 18 °C higher for the four-star polymer than for the tristar polymer. The constant b measures the change of K_s with arm length. Constant b increases more rapidly when the MPEG arm of the tristar polymer is shortened than when the MPEG arm of the four-star polymer is shortened. The $T_p{}^0$ of the two stars for dp $\rightarrow \infty$ are nearly the same as those for high-molecular-weight linear PEO, which is reported to be 99 °C (*25*). All T_p were obtained under conditions in which the star molecules exist as micelles.

Binding of Hydrophobic Molecules. The relatively large hydrophobic cores of the D4TMI-MPEG and T3TMI-MPEG stars are likely to attract hydrophobic molecules. Optical and fluorescent probes are available to explore such interactions. One such probe is Coomassie brilliant blue (CBB), which is used to assay proteins (*26, 27*). The reddish species (λ_m = 470 nm) found at low pH changes to blue (λ_m = 595 nm) or green (λ_m = 650 nm) when it binds to proteins (*27*). Adding D4TMI-MPEG550 to a 10^{-5} M CBB solution replaces the 470-nm peak by a peak at λ_m 621 nm. The absorptivity of the new peak is three times that of the peak at 470 nm, and a clear isosbestic point is present at 535 nm. The color change occurs in the star concentration range of 2×10^{-5} to 6.4×10^{-3} M.

Quantitative studies with CBB are complicated by the presence of more than one species (anionic, cationic, or neutral) in CBB. A more convenient probe for our purpose is 1-pyrenebutyrate (PB^-), which we used earlier in binding studies with polysoap-type macromolecules such as poly(vinylbenzo-18-crown-6) (*28*). When the probe binds to the hydrophobic domain of this polymer, the sharp pyrene maxima at 326 and 341 nm are replaced by peaks at 332 and 348 nm, respectively. Figure 4 shows that adding D4TMI-MPEG550 to PB shifts the two peaks barely 3 nm, to 328.5 and 344 nm, respectively, but the maxima of the free and bound species are clearly distinguishable, and so is

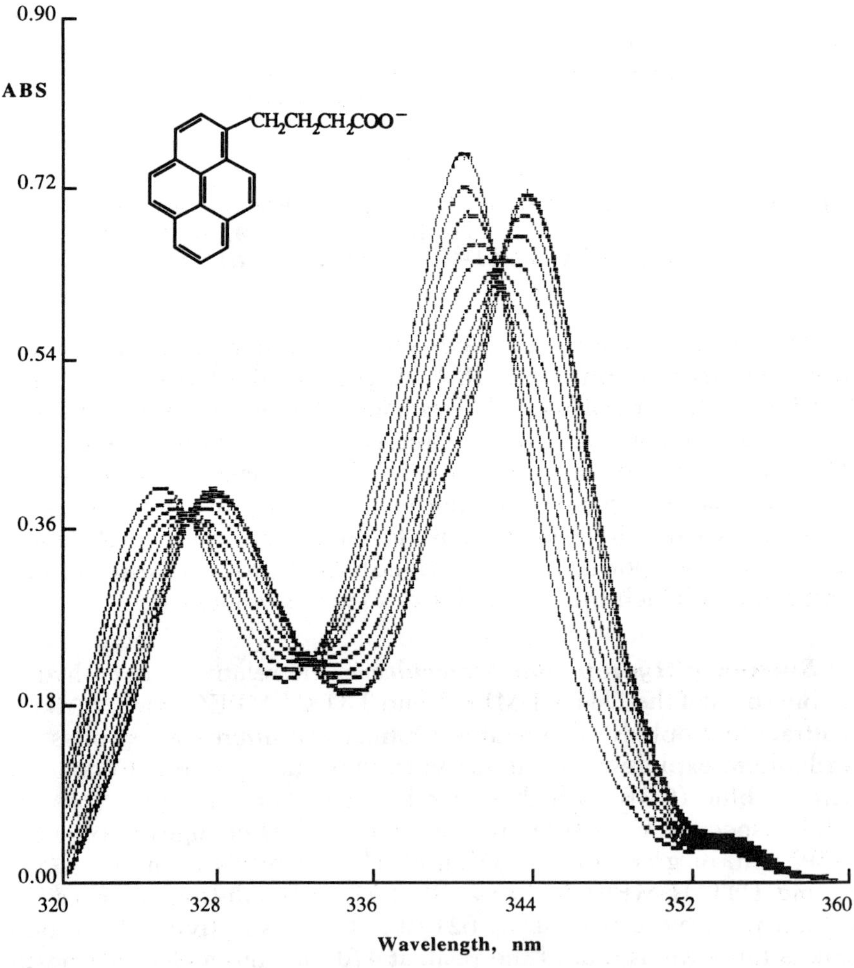

Figure 4. Changes in the optical spectrum of PB^- (2×10^{-5} M) when it binds to D4TMI-MPEG550. ABS is absorption.

the isosbestic point. Recent measurements over a wider star polymer concentration range show that for both CBB and PB^-, a typical Klotz plot shows a distinct break close to the cmc of the star polymer (*29*). For example, for D4TMI-MPEG550 and PB^-, the break is found at 1.8×10^{-4} M star, while the surface tension cmc for this star is 1.3×10^{-4} M. The first binding constants for CBB and PB^- below the cmc are of the order of 10^4 M.

Associative PEO Star Polymers. ***Phase Separation.*** When the star polymers with nonylphenoxy end-capped PEO arms are mixed with water, two sharply separated phases are formed above a critical star concentration (C^*). Most of the polymer is concentrated in the viscous and slightly hazy lower phase, whose molar concentration is denoted by C_1. The dilute, clear upper phase contains polymer at a molar concentration of C^*. When the total concentration C_t is varied, the volume fractions of the upper and lower phases (v_u and v_l, respectively) change linearly with C_t, since C^* and C_1 remain constant as long as the two phases coexist:

$$v_u = \frac{C_1}{C^1 - C^*} - \frac{C_t}{C_1 - C^*} \tag{5}$$

For $C_t > C_1$ or $C_t < C^*$, only one phase is present. Plots of v_u versus C_t for four of the star polymers are shown in Figure 5. C^* was determined by ICP for the Si-containing stars and by optical spectroscopy for the stars with a T3PM core. C_1 can be calculated from equation 5 or obtained from the intercept of the plots at $v_u = 0$.

Relevant data for five star polymers are collected in Table IV. Included are the partition coefficients $k_1 = C_1/C^*$, a measure of the distribution of the star between the two phases. Also calculated were the ratio H_2O/EO units and the water content (wt%) for the concentrated phase.

The data for the D4TMI-IG990 star (Table IV) indicate that phase separation ensues at a C^* that exceeds the cmc of a D4TMI-MPEG star with the same number of EO units in its PEO arms (in this case an average of 86 EO units). For such a homopolymer star, the cmc can be calculated from equation 1. The cmc is 3.65×10^{-4} M, which is lower than the 5.5×10^{-4} M C^* of D4TMI-IG990. At C^*, the nonylphenoxy end group concentration is four times higher, that is, 2.2×10^{-3} M. This value is seven times the cmc of 3×10^{-4} M found for IG990 itself (*13*). On the other hand, the C^* for T3PM-IG990 equals 3.7×10^{-4} M (Table IV), which is much lower than the cmcs of T3PM-MPEG stars (Table II). Phase separation for the T3PM-IG990 star

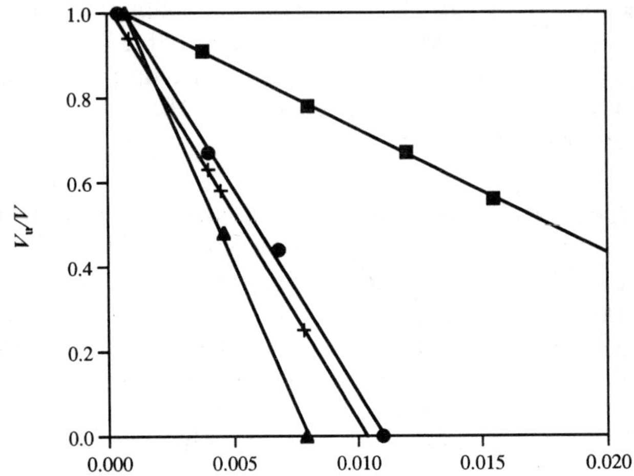

Figure 5. Phase separation of nonylphenoxy end-capped PEO star polymers in water at 25 °C. Plots are of volume fraction V_u/V of the dilute upper phase as a function of total polymer concentration C_t for D4TMI-IG990 (■), T3PM-IG990) (+), T3TMI-IG990 (●), T3PM-IG890 (■). (IG890 and IG990 contain averages of 38 and 84 EO units, respectively.)

Table IV. Phase Separation Constants for Associative PEO Star Polymers in Aqueous Media at 25 °C

Star Polymer	C^* ($\times 10^3$ *M*)	C_l ($\times 10^3$ *M*)	k (C_l/C^*)	H_2O/EO	H_2O (*wt%*)
D4TMI-IG890	1.07	22.1	21	13.2	80.3
D4TMI-IG990	0.55	7.8	14	18.5	87.3
T3TMI-IG990	0.32	11.0	34	17.4	86.5
T3PM-IG890	0.38	41.0	108	8.8	74.9
T3PM-IG990	0.37	9.0	24	22.0	88.9
T3PM-IG990 (0.13 M Na_2SO_4)	0.11	12.0	109	16.1	85.6
T3PM-IG990 (0.45 M Na_2SO_4)	0.028	19.0	678	9.1	77.1
T3PM-IG990 (1 M Na_2SO_4)	0.0045	33.0	7330	3.7	60.3

occurs at a nonylphenoxy end group concentration of 1.1×10^{-3} M, which is about 3.7 times the cmc of IG990. Formation of the polymer-rich phase chiefly results from intermolecular association of the hydrophobic end groups. Below C^*, intramolecular association may prevail, but at C^*, calculations (*13*) imply that the star molecules begin to overlap. Core–core interactions leading to micelle formation in the homopolymer stars may not significantly contribute to phase separation for Igepal stars, probably because the cores are partially shielded by the hydrophobic arm ends. In addition, the water contents of the condensed phases of the three IG990 stars are nearly identical, being approximately 88 wt%, or 18 water molecules per EO unit. Both numbers decrease when the PEO arm is shortened (compare in Table IV the IG990 star with the IG890 star, which has an average of 38 EO units per arm). To a degree, the properties of the polymer-rich phase resemble those of a hydrogel. The hydrophobic domains act as physical cross-links, and the amount of water content in this phase is controlled by its EO content.

Salt Effect. As pointed out earlier for the homopolymer stars, many electrolytes alter the solubility of polymers with EO units; the effect is largely controlled by the nature of the anion. Sulfates substantially lower the cloud points of EO-containing macromolecules (*23*). We used these salts to check the role of electrolyte in the phase separation process of T3PM-IG990.

The effect of Na_2SO_4 is schematically shown in Figure 6, and the data are recorded in Table IV. When the electrolyte is added, C^* rapidly decreases from 3.7×10^{-4} M in the absence of salt to 4.5×10^{-6} M in 1 M Na_2SO_4, a result that is consistent with the effect of the salt

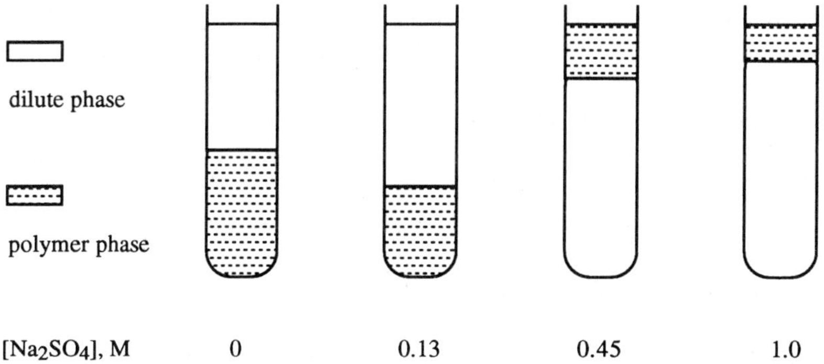

Figure 6. Effect of Na_2SO_4 on phase separation of the tristar polymer T3PM-IG990. Striped area is the polymer-rich phase. See *Table IV for polymer concentrations.*

on the solubility of polymers with EO units. At the same time, C_1 increases from 9.0×10^{-3} to 33×10^{-3} M, and the partition coefficient k thus increases from 24 to 7330 (Table IV). Figure 6 also shows that the volume fraction of the polymer-rich phase decreases and that somewhere between 0.13 and 0.45 M Na_2SO_4, the polymer-rich phase changes from being the lower phase to being the upper phase. This reversal is caused by differences in the salt concentrations of the two phases. By using ICP, we found that the Na_2SO_4 concentration in the dilute phase is seven times that of the polymer-rich phase of 1 M Na_2SO_4. The resulting changes in the densities of the two phases cause the polymer-rich phase to float on top. The large difference in electrolyte content also causes water to flow by osmosis from the concentrated to the dilute phase. At 1 M Na_2SO_4, the ratio H_2O/EO units in the former phase is down to 3.7, which is not far from the composition of the relatively stable trihydrates found in mixtures of poly(ethylene glycol) and water (*30*).

Effect of Temperature. The temperature dependence of the phase separation of aqueous mixtures of T3PM-IG890, T3PM-IG990, and T3TMI-IG990 was checked between 0 and 80 °C. Plots of v_u versus the total polymer concentration were linear at every temperature. For each system, C^* and C_1 were determined. Some of the results are shown in Figure 7. C^* rapidly decreases with temperature; for T3PM-IG990, C^* decreases from 9.1×10^{-4} M at 0 °C to 8.0×10^{-6} M at 80 °C. This decrease is consistent with the reported lower critical solution temperature behavior of our homopolymer stars. At the same time, the polymer concentration C_1 of the polymer-rich phase increases. This phase loses water at elevated temperatures, largely because of the exothermicity of the hydrogen-bonding interactions between water and the EO units. The C_1 versus T plot exhibits a peculiar break around 50 °C for both T3PM-IG990 and T3TMI-IG990. The mp of a poly(ethylene glycol) of 4000 MW (90 EO units) is 58 °C. The break may conceivably result from the formation of crystalline domains below the mps of the two star polymers. No break in the C_1 versus T plot is found for T3PM-IG890, which has only 38 EO units per arm. Formation of crystalline domains would be more difficult for this polymer in the water-swollen state because of the shorter PEO segments.

Partitioning of Hydrophobic Compounds. Recent work on PEO networks with D4TMI cross-links demonstrated a high affinity for hydrophobic compounds such as bromophenol blue, 8-anilinonaphthalene-1-sulfonate (ANS), picrate, and tetraphenylboron (*24*). The binding sites in these hydrogels are the large hydrophobic cross-links. The amphiphilic character of the polymer-rich phase is expected to favor

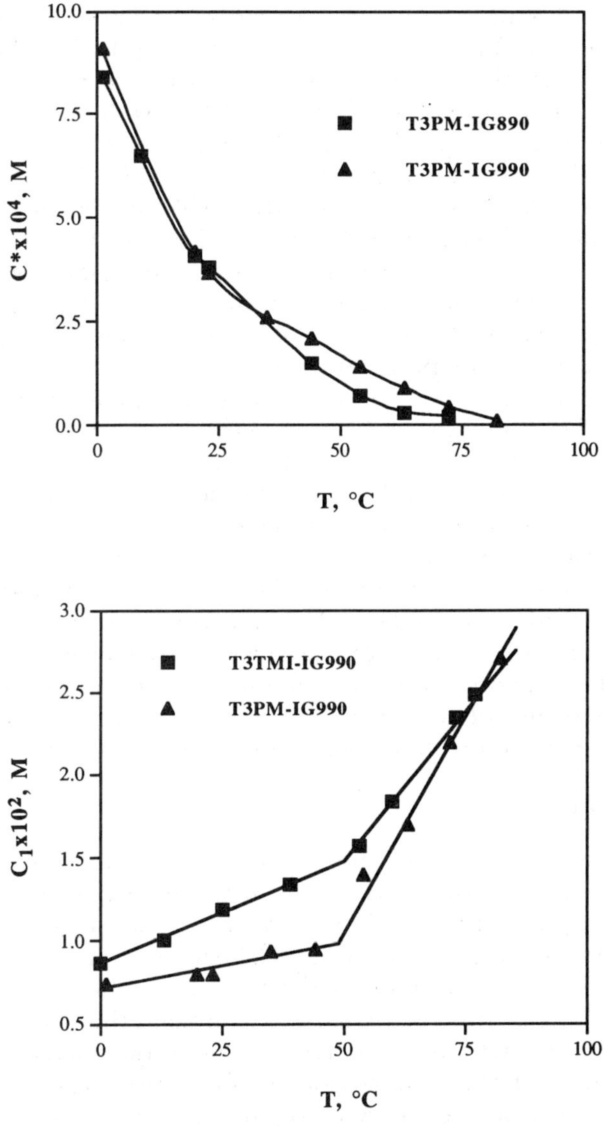

Figure 7. Temperature dependence of the polymer concentrations C* *(dilute phase) and* C_1 *(polymer-rich phase) in the phase separation of T3PM-IG890, T3PM-IG990, and T3TMI-IG990.*

the absorption of hydrophobic compounds. Indeed, when picrate or ANS is added to an aqueous mixture of an associative star, most of the dye can be found in the concentrated polymer layer after phase separation (29). With UV light, the ANS-enriched phase is a deep blue, implying that the fluorescent ANS is present in the hydrophobic domain. The ratios of the dye concentrations in the polymer-rich and dilute phases with T3TMI-IG990 were 10 for sodium picrate and 16 for the more hydrophobic NaANS (29). Under the same conditions the partition coefficient for sodium picrate with T3PM-IG990 as the star was 2.6. The hydrophobic content of the T3PM-IG990 star in the polymer-rich phase is less than that of the T3TMI-IG990 star. The more hydrophobic BPh_4^- easily displaces the picrate anion from the polymer-rich phase; thus the associative star system may be useful in separating compounds with different hydrophobicities. Dodecyl sulfate also accumulates in the polymer phase, but the two-phase system becomes homogeneous when too much surfactant is added, because water from the dilute phase diffuses to the electrolyte-rich phase. As mentioned earlier, inorganic electrolytes such as Na_2SO_4 and NaH_2PO_4 are largely excluded from the polymer-rich phase because of their incompatibility with the polymer.

The partition coefficients are quite sensitive to the concentration of the hydrophobic compound, and the coefficients increase with decreasing chromophore content. The partition coefficients are much higher when salts such as Na_2SO_4 are present or when the temperature is raised. Both factors lower the polymer content in the dilute phase while raising the polymer concentration of the polymer-rich phase. Details of these studies have been published elsewhere (29).

Conclusion

Amphiphilic PEO star polymers with hydrophobic cores such as D4TMI, T3TMI, or T3PM associate in water to form micelles. End capping the arms with nonylphenoxy groups causes a polymer-rich and a polymer-poor phase to form. The volume fractions of the two phases are proportional to the total star concentration, but their polymer concentrations remain unchanged as long as the two phases coexist. The volume fractions and polymer concentrations are sensitive to changes in temperature and to added electrolyte. The associative stars may be useful in the recovery or separation of hydrophobic species from aqueous solutions.

Acknowledgments

We gratefully acknowledge the financial support of the National Science Foundation's Polymers Program (Grant No. DMR 8722245) and of the Petroleum Research Fund, administered by the American Chemical Society (Grant No. 25836AC7P).

References

1. Landoll, L. M. *J. Polym. Sci., Polym. Chem.* **1982,** *20,* 443.
2. *Polymers in Aqueous Media;* Glass, J. E., Ed.; Advances in Chemistry Series 223; American Chemical Society: Washington, DC, 1989.
3. McCormick, C. L.; Block, J.; Schulz, D. N. *Encyclopedia of Polymer Science and Engineering;* John Wiley: New York, 1989; Vol. 17, pp 730–784.
4. Xu, R.; Winnik, M. A.; Hallet, F. R.; Ries, G.; Croucher, M. D. *Macromolecules* **1991,** *24,* 87.
5. Kanaoka, S.; Sawamota, M.; Higashimura, T. *Macromolecules* **1991,** *24,* 5741.
6. Tanaka, R.; Meadows, J.; Williams, P. A.; Phillips, G. O. *Macromolecules* **1992,** *25,* 1304.
7. Brown, R. G.; Glass, J. E. *Proc. Polym. Mater. Sci. Eng.* **1987,** *57,* 709.
8. Gnanou, Y.; Lutz, P.; Rempp, P. *Makromol. Chem.* **1988,** *189,* 2885.
9. Saunders, R. S.; Cohen, R. E.; Wong, S. J.; Schrock, R. R. *Macromolecules* **1992,** *25,* 2055.
10. Zhou, G. B.; Smid, J. *Polym. Prepr. (Am. Chem. Soc., Div. Polym. Chem.)* **1991,** *32*(3), 613.
11. Zhou, G. B.; Smid, J. *Langmuir,* **1993,** *9,* 2907.
12. Zhou, G. B.; Smid, J. *Polym. Prepr. (Am. Chem. Soc., Div. Polym. Chem.)* **1993,** *34*(1), 822.
13. Zhou, G. B.; Smid, J. *Polymer* **1993,** *35,* 5128.
14. Zhou, G. B.; Chen, X.; Smid, J. *Proc. Polym. Mater. Sci. Eng.* **1993,** *69,* 102.
15. Zhou, G. B.; Smid, J. *J. Polym. Sci., Polym. Chem.* **1991,** *29,* 1097.
16. Maltesh, C.; Xu, Q.; Somasundaran, P.; Benton, W. J.; Nguyen, H. *Langmuir* **1992,** *8,* 1511.
17. Conner, M.; Kudelka, I.; Regen, S. L. *Langmuir* **1991,** *7,* 982.
18. Zhu, J.; Eisenberg, A.; Lennox, R. B. *Macromolecules* **1992,** *25,* 6556.
19. Cha, X.; Yin, R.; Zhang, X.; Shen, J. *Macromolecules* **1991,** *24,* 4985.
20. Schick, M. J. *J. Colloid Sci.* **1962,** *17,* 801.
21. Bailey, F. E., Jr.; Callard, R. W. *J. Appl. Polym. Sci.* **1959,** *1,* 56.
22. Florin, E.; Kjellander, R.; Erikson, J. C. *J. Chem. Soc., Faraday Trans.* **1985,** *80,* 2889.
23. Nwankwo, I.; Xia, D. W.; Smid, J. *J. Polym. Sci. Polym. Phys.* **1988,** *26,* 581.
24. Xia, D. W.; Smid, J. *Polym. Prepr. (Am. Chem. Soc., Div. Polym. Chem.)* **1991,** *31*(1), 168.
25. Sacki, S.; Kuwahara, N.; Nakata, M.; Kaneko, M. *Polymer* **1976,** *17,* 685.
26. Read, S. M.; Northcote, D. H. *Anal. Biochem.* **1981,** *116,* 53.
27. Compton, S. J.; Jones, C. G. *Anal. Biochem.* **1985,** *151,* 369.
28. Roland, B.; Smid, J. *Polymer* **1984,** *25,* 1166.
29. Chen, X. M.Sc. Thesis, State University of New York at Syracuse, April 1995.
30. Graham, N. B.; Zulfigar, M.; Nwachuku, N. E.; Rashid, A. *Polymer* **1989,** *30,* 528.

RECEIVED for review April 16, 1994. ACCEPTED revised manuscript April 7, 1995.

Biocompatible and Biodegradable Polymers

3

Poly(ethylene oxide) and Protein Resistance

Principles, Problems, and Possibilities

J. D. Andrade, V. Hlady, and S.-I. Jeon

Department of Bioengineering, University of Utah, Salt Lake City, UT 84112

Poly(ethylene oxide) (PEO)-based protein-resistant surfaces function principally by a steric exclusion mechanism involving very high surface mobility and surface dynamics of the PEO chains. For such a surface to be effective, the dynamics and mobility of the chain must be maximized and, contradictorily, the underlying surface must be entirely covered by the PEO chains. Because of geometric constraints, these criteria are optimally met on highly curved surfaces; PEO probably cannot be used to make ideally flat surfaces as optimally protein-resistant as surfaces with low radii of curvature. A curved surface simply has more room for end-attached polymer chains than a flat surface.

SURFACES RESISTANT TO PROTEIN ADSORPTION AND CELL ADHESION are needed, particularly in the health care product and biotechnology industries. Although much has been done in the preparation, characterization, and even application of poly(ethylene oxide) (PEO) surfaces over the last 15 years, controversy in the field is considerable, and most of the key scientific questions are still open. A volume edited by J. M. Harris provides a concise, up-to-date, authoritative presentation of the field (*1*).

0065–2393/96/0248–0051$12.00/0

Background

PEO and poly(ethylene glycol) (PEG) are used for a wide variety of interface engineering applications. Higher-molecular-weight PEO is widely used to stabilize aqueous colloids and dispersions, generally by means of physical adsorption followed by steric repulsion of the modified particles. Lower-molecular-weight PEG, roughly in the 1000 to 4000-Da range, is commonly used as a prepolymer in the synthesis of polyurethanes, epoxies, silicones, and other polymers. Low-molecular-weight PEG can also be readily coupled to hydrophobic chains to make a wide variety of nonionic surfactants that are widely used in the chemical industry, biochemistry, and the biotechnology industry (*1*).

Polymerized ethylene oxide is a somewhat anomalous molecule. It is both hydrophilic and hydrophobic, because it is soluble in both aqueous and nonpolar solvents. In solution it tends to be highly dynamic, and yet it can readily pack and form crystalline solids. Despite its dynamics and mobility, it can complex and aggregate, develop specific helical and near-helical conformations, and interact and complex with a variety of ionic and hydrogen-bonding structures. PEO, as a molecule alone and as part of other molecules, is generally nontoxic and is considered safe for a wide variety of cosmetic, food, and biomedical applications. PEO and its derivatives are readily available in a range of purities and molecular weights and are relatively inexpensive and easy to obtain. Here we focus on interface modification by PEO and PEO-based polymers and the optimization of the protein resistance of such surfaces.

The heterogeneity and dynamics characteristic of proteins are also characteristic of many solid surfaces, particularly those of synthetic polymers (*2*). Hydrophobic solid surfaces may be relatively homogeneous, as in poly(dimethylsiloxane), or very heterogeneous, as in semicrystalline polyethylene or block polyurethanes. All polymer surfaces are highly heterogeneous because of the sizes of polymer molecules, which are of the same order as the sizes of the individual protein molecules. Because of steric exclusion and a tendency to satisfy entropic concerns, polymer chains tend not to interpenetrate very effectively, and this failure further enhances the macromolecular granularity of polymeric surfaces. Polymers have a range of molecular dynamics and molecular relaxation processes, including the glass transition and side chain relaxations, that further contribute to the complexity of such interfaces. The time scales of relaxation processes for polymers are of the same order as those for proteins, which is not surprising, because both polymers and proteins are macromolecules. Although the terminology is different, the mechanisms and processes are basically the same (*3*).

The dynamics of both the protein and the polymer and the wide repertoire of intermolecular interactions possible between proteins and polymeric surfaces lead to the conclusion that most proteins adsorb on most interfaces (*4*). Proteins are polymeric surfactants and are not only adsorbed but are also generally conformationally altered as a result of interfacial activity (*2*). The major interest in PEO surfaces is therefore in finding a way to minimize or eliminate the tendency for protein adsorption. This problem can be almost completely avoided by developing protein-resistant surfaces or interface-resistant proteins.

Protein-Resistant Surfaces

Minimizing protein adsorption requires some knowledge and understanding of the structures of proteins (*5*) and their interfacial behavior (*2, 4, 6*).

Regardless of whether the underlying substrate is highly hydrophobic, highly ionic, or highly hydrogen bonding, the protein has regions on its surface that can indeed interact with the substrate (*5, 6*). The protein itself has loops, tails, helices, and sheets that can make their way through the PEO layer and interact with the substrate below. A variety of bridging, pinning, and related processes can then further complicate the problem (*7*).

The major interactions that drive the interfacial activity and adsorption of proteins are the water structure-driven hydrophobic effect, electrostatic interactions, and strong hydrogen-bonding interactions characterized by cooperative, multiple hydrogen bonds. A typical isolated hydrogen bond does not play much of a role in aqueous solutions, because that bond is largely satisfied by interactions with the 55 M water that is present. Isolated, random hydrogen bonds are generally unimportant. They become important when a multiplicity of such bonds, either acceptors or donors, occurs and the complementary component is on the other surface. Multiple hydrogen bonds that match up in space to form a cooperatively interacting structure consisting of 3 to 5 or more hydrogen-bonding units can be a strong, effective means of interfacial adhesion, as they are in biorecognition. Such matching requires multiple bonds *and* stereo complementarity.

That neutral, highly hydrophilic polymers tend to have minimal or very weak interactions with most aqueous proteins has been well known for over half a century. The development of the dextrans (Sephadex) and the agaroses (Sepharose) for protein chromatography and electrophoresis demonstrates that such matrices have relatively weak protein interactions. These interactions are weak because the matrices are generally nonionic, thereby minimizing electrostatic interactions, and highly water soluble and hydrophilic, thereby minimizing hydrophobic interactions. Although the gels and surfaces produced by such polymers are extensively hydrogen bonded, they tend to be highly dynamic and random; therefore, cooperative hydrogen-bonding processes are not generally a problem. However, some proteins are indeed retained on such gels, often because of hydrogen-bonding interactions and in some cases because of residual charge or a hydrophobic character.

A neural, highly hydrophilic polymer that is also very dynamic at the surface has another mechanism by which to minimize protein interactions. By being neutral and hydrophilic, the polymer has already minimized enthalpic interactions, but by also being highly dynamic, the polymer has provided the interface with a high entropy. Any process that tends to decrease or minimize this interface entropy, such as by decreasing the dynamics or mobility of the polymer chains at the interface, will be unfavorable from a free-energy perspective. Adsorption on such a surface will therefore pay a high free-energy penalty, which must generally be paid in enthalpic currency. If no enthalpic interactions are available, such surfaces are said to be repulsive by an entropic, surface dynamics mechanism. This mechanism can also be related to steric exclusion and osmotic pressure. These processes have been extensively modeled and discussed by de Gennes (*8*) and are now being widely applied in the biomaterials/biotechnology community (*1*, *6*).

Figure 1 presents a schematic but comprehensive summary of PEO surfaces.

Figure 1a represents low-molecular-weight PEO (1000 to 4000 Da)

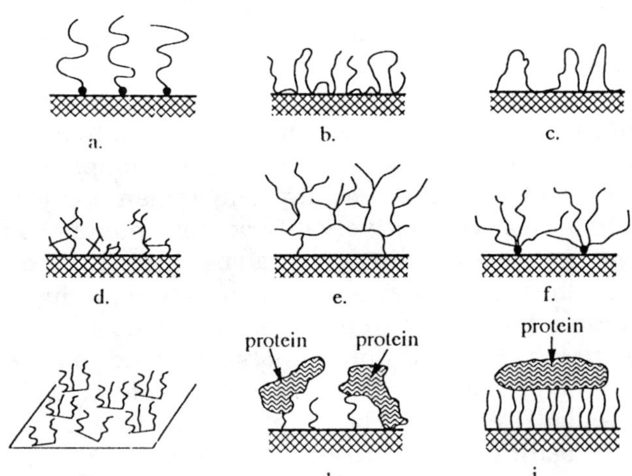

Figure 1. Some of the many structures and configurations that have been suggested for PEO and PEO-derived polymers attached to surfaces. See *text for details.*

tethered at one end to a particular surface. A wide variety of surface modification technologies and PEO derivatives are available for such surface modification. That the chains extend into solution as indicated is highly unlikely. In most studies, achieving a very high density of chains on the surface is difficult.

Figure 1b is the common illustration for high-molecular-weight PEO adsorbed onto particles or other surfaces. Here the very high molecular weight and the highly cooperative nature of polymer segmental adsorption lead to loops, tails, and trains that have been extensively characterized and modeled. The loops and tails provide a means of steric repulsion between two particles containing adsorbed PEO, although the dynamics of adsorption can clearly also lead to bridging and thus to colloidal aggregation rather than stabilization. Figure 1b also illustrates the adsorption of PEO block copolymer surfactants; an adsorbable block pins the molecule to the surface and the PEO block (loops or chains) extends into solution (*9*). Another variation is a graft copolymer, for example, with PEG chains on a hydrophobic backbone, which results in adsorption at a hydrophobic surface and PEG chains extending into solution (*10*).

Figure 1c represents a PEO chain bound to the surface, by both ends; that is, it forms a loop. This structure may occur in many types of block copolymers containing PEO block segments. It may also be part of many PEO surface modification reactions in which the PEO reagent is homobifunctional rather than heterobifunctional, as is required for the ideal situation in Figure 1a.

Figure 1d represents ethylene oxide attached to an activated surface and a PEO-like network growing out from the surface (*11*). It could also represent the plasma polymerization of ethylene oxide films (*12*). Such a film would be expected to be highly cross-linked and much less dynamic than the others indicated.

Figure 1e is an example of so-called surface amplification in which PEO is tethered to multifunctional entities such as carbohydrates or polysaccharides, which in turn are tethered to the surface (*13*). Although in principle this amplification leads to a much larger number of binding sites per unit area for the PEO chains, in practice the steric constraints imposed by the mobility and steric repulsion characteristics of PEO probably limit the number of binding sites to the same extent as in Figure 1a.

Figure 1f represents another version of surface amplification: the star polymer geometry. This polymer could be thought of as a sort of hybrid between those in Figure 1b and 1e in which a nucleus, often containing a multihydroxyl carbohydrate, is used to grow ethylene oxide chains from each reactive functional group, thereby producing a PEO star. The center or base of the star can then be appropriately attached to a surface, or the entire process can be initiated from the surface (*11*). This structure is also reminiscent of the Tetronic family of polymeric surfactants, in which PEG chains extend from four poly(propylene oxide) chains attached to a tetrafunctional nucleus (*9*).

Figure 1g represents a block or graft copolymer designed for optimum adsorption (*10*) that is surface cross-linked between the chains or between the polymer blocks and the surface either by specific cross-linking reactions or by plasma reactions (*14*).

As Figure 1h shows, PEG chains are often used to provide a tether between a protein or other biomolecule and the surface (*15*). This approach is being widely applied in biosensors in which an antibody must function as if it were in solution and yet be tethered within several hundred angstroms of an interface to provide a means of transducing a binding event into a signal (*16*). The covalently coupled protein sitting on the end of a dynamic and mobile chain, however, has extensive mobility and dynamics of its own and will interact with the underlying substrate unless the surface is exceptionally well covered and passivated by PEO or some other means.

If we could prepare a maximally dense PEO surface, we might have a packed "crystal" of PEO that would then adsorb proteins, as Figure 1i shows. Such a surface would of course not be mobile or dynamic and would not sterically or entropically exclude or resist protein adsorption.

Although the reactions in Figures 1e and 1f have the advantage of leading to a very well covered surface and avoiding the potential problems of a bare substrate, there may well turn out to be little difference between the reactions in Figures 1a, 1e, 1f, and possibly even 1d because the excluded volumes of the chains themselves prevent a very high local concentration of PEO. If this excluded volume is decreased by solution "tricks" (*17*), then after equilibration in water, the final surface will probably be less mobile and less dynamic than is required for optimum protein resistance.

Another concern with Figure 1 is that we have assumed a particular surface structure that is homogeneous, that is, not patchy. We have little evidence to indicate that such homogeneity is indeed the case. Thus the problem is even more complex than is sketched in Figure 1.

Clearly, *the surface must be fully covered by PEO to minimize protein interaction* with the underlying surface. However, if the surface is "overcovered," as in Figure 1i, then the surface becomes adsorptive.

It is, therefore, not surprising that even crude, simplistic models of hypothetical spherical proteins interacting with ideal PEO brush surfaces suggest that protein resistance is a function of protein radius,

PEO molecular weight, and the number of PEO chains per unit area on the surface (*18*).

With all this complexity, one might ask, Why PEO? Why not consider other approaches to the passivation of surfaces with respect to protein adsorption? Protein-resistant surfaces tend to be neutral, thereby minimizing electrostatic interactions, and highly hydrophilic, thereby minimizing hydrophobic interactions (*2, 4*). Of all the neutral, hydrophilic, water-soluble/swellable polymers readily available, PEO appears to be the most mobile, the most dynamic, and the least interactive (*6, 9, 10, 18*).

What are the disadvantages to PEO? The long-term stability of PEO on a surface is somewhat questionable; that is, it may be susceptible to local oxidation processes. The fact that PEO may weakly complex with proteins, particularly charged proteins, as it does with certain types of charged polymers, is also of some concern (*19*). PEO also has a tendency to form weak complexes with certain ions, particularly potassium. In fact, PEG has been called a "poor man's crown ether" (*20, 21*). Nevertheless, of all the polymers we know, PEO appears to have the highest potential for the development of truly protein-resistant surfaces (*1, 6, 22*).

A very major factor in this potential is the way in which the hydrophilic polymer chains interact with water. Although PEO solutions do not behave as ideal solutes and certainly do perturb the structure of water somewhat, they are apparently the least perturbing of all of the common neutral hydrophilic polymers. Although the nonbonding oxygen orbitals in PEO provide hydrogen-bonding capacity and indeed are largely responsible for the solubility of the molecule, this hydrogen bonding requirement is easily satisfied by water without significant perturbation of the structure of water (*22, 23*). A lack of significant perturbation in the structure and the fact that the ethyl moieties in the PEO chain are largely accommodated by the water structure minimize hydrophobic interactions. These two facts suggest that PEO indeed has minimal interactions with other solutes in aqueous solutions. In addition, the PEO chain is highly mobile and dynamic, thereby creating an entropic "insurance" that can more than compensate for any weak attractions that may be present (*18*). The end result is a weak or sometimes quite strong repulsive interaction between proteins and many types of PEO surfaces that results in very low protein adsorption. This interaction is what we define as protein resistance.

Direct measurements of the steric repulsion between PEO surfaces (*24, 25*) and between a PEO surface and a protein surface (*26*) are now available, thanks to the surface forces apparatus (*25, 26*). Direct measurement of steric exclusion and the imaging of surfaces via steric exclusion were accomplished in our group by atomic force microscopy

(*24*). Prime and Whitesides (*27*) recently presented a study of the adsorption of four different proteins on oligo(ethylene oxide) self-assembled monolayers with various oligo(ethylene oxide) surface concentrations. The protein resistance was roughly proportional to increasing surface coverage and increasing oligo(ethylene oxide) molecular weight.

Conclusions and Summary

The ideas presented in this discussion are not without controversy and criticism. Many studies in the literature argue that PEO surfaces are not particularly biocompatible. Other studies argue that if a PEO surface is resistant to one protein, it may not be very resistant to another protein (*6, 18*). Is there a specificity to PEO's protein resistance? Others argue that PEO surfaces may not be stable and in time may be degraded or otherwise deteriorated and thereby lose their passivity or protein resistance (*11–13*).

PEO-based protein-resistant surfaces function principally by a steric exclusion mechanism involving very high surface mobility and surface dynamics of the PEO chains. For such a surface to be effective, the dynamics and mobility of the chain must be maximized and, contradictorily, the underlying surface must be entirely covered by the PEO chains. Because of geometric constraints, these criteria are optimally met on highly curved surfaces; ideally flat surfaces probably cannot be made as optimally protein-resistant with PEO as surfaces with low radii of curvature. A curved surface simply has more room for end-attached polymer chains than a flat surface.

Acknowledgments

Many parts of this work were supported by the Center for Biopolymers at Interfaces at the University of Utah and by an NIH grant. S.-I. Jeon thanks KOSEF for a fellowship at the University of Utah.

References

1. *Poly(ethylene glycol) Chemistry;* Harris, J. M., Ed.; Plenum: New York, 1992.
2. Andrade, J. D.; Hlady, V.; Feng, L.; Tingey, K. In *Interfacial Behavior of Bioproducts;* Brash, J. L., Ed.; Dekker: New York, in press.
3. Chan, H. S.; Dill, K. A. *Phys. Today* **1993,** *Feb,* 24–32.
4. Andrade, J. D.; Hlady, V. *Adv. Polym. Sci.* **1986,** *79,* 1.
5. Branden, C.; Tooze, J. *Introduction to Protein Structure;* Garland: New York, 1991.
6. Andrade, J. D.; et al. *Clin. Mater.* **1992,** *11,* 67–84.

7. Johnson, H. E.; Granick, S. *Science* (Washington, D.C.) **1992,** *255,* 966.
8. de Gennes, P. G. *Ann. Chim.* **1987,** *77,* 389–410.
9. Lee, J. H.; Kopecek, J.; Andrade, J. D. *J. Biomed. Mater. Res.* **1989,** *23,* 351.
10. Lee, J. H.; Kopeckovaa, P.; Kopecek, J.; Andrade, J. D. *Biomaterials* **1990,** *11,* 455.
11. Merrill, E. W. in reference 1.
12. Lopez, G. P. *J. Biomed. Mater. Res.* **1992,** *26,* 415.
13. Kim, S. W. Personal communication.
14. Sheu, M.-S. *Trans. Soc. Biomater.* **1992,** *18,* 228.
15. Kopeckova, P.; Kopecek, J.; Andrade, J. D. *New Polym. Mater.* **1990,** *1,* 289–297.
16. Andrade, J. D. In *Biosensor Technology;* Buck, R. B.; Hatfield, W. E.; Umana, M.; Bowden, E. F., Eds.; Dekker: New York, 1990; p 219.
17. Antonsen, K. P.; Hoffman, A. S.; Holmberg, K.; In reference 1.
18. Jeon, S. I.; Andrade, J. D. *J. Colloid Interface Sci.* **1991,** *142,* 159.
19. Tsuchida, E.; Abe, K. *Adv. Polym. Sci.* **1982,** *45,* 1.
20. Balasubramanian, D.; Chandani, B. *J. Chem. Educ.* **1983,** *60,* 77.
21. Okada, T. *Analyst* **1993,** *118,* 959.
22. Nagaoka, S.; Mori, Y. *J. Biomater. Appl.* **1987,** *2,* 219, and references therein.
23. Golander, G. C. In reference 1.
24. Lea, A. S. Ph.D. Thesis, University of Utah, 1993.
25. Luckham, P. F.; Klein, J. *J. Chem. Soc. Faraday Trans.* **1990,** *86,* 1363.
26. Malmsten, M.; et al. *J. Colloid Interface Sci.* **1992,** *151,* 579, and work in progress in the Claesson laboratory.
27. Prime, K. L.; Chu, Y.-H.; Schmid, W.; Seto, C. T.; Chen, J. K.; Spaltenstein, A.; Zerkowski, J. A.; Whitesides, G. M. In *Macromolecular Assemblies in Polymeric Systems*; Stroeve, P.; Balazs, A. C., Eds.; ACS Symposium Series 493; American Chemical Society: Washington, DC, 1992; pp 227–239.

RECEIVED for review April 16, 1994. ACCEPTED revised manuscript April 24, 1995.

4

Adsorption of Poly(ethylene oxide)-Containing Block Copolymers

A Route to Protein Resistance

Jenq-thun Li,[1] Jan Carlsson,[1] Shao-Chie Huang,[2] and Karin D. Caldwell[2,3]

[1]Pharmacia Diagnostics, Uppsala, S–75182, Sweden
[2]Center for Biopolymers at Interfaces, University of Utah, Salt Lake City, UT 84112

In aqueous media, poly(ethylene oxide) (PEO)-containing block copolymers of the PEO-poly(propylene oxide) (PPO)-PEO type are stably adsorbed to nonpolar surfaces via their hydrophobic PPO center block. When we vary the PEO block length and keep the PPO block nearly constant in length, we find that surface concentrations on a given polystyrene latex substrate are independent of the size of the PEO blocks but that the adlayer thickness increases with increasing PEO block length. This increase is coupled with enhanced mobility of the PEO chains. The substrate curvature strongly affects surface concentration, adlayer thickness, and PEO chain mobility. Though increasing the PEO block length increases protein resistance of the coated surface, surfactants with a given PEO block size appear to be most effective as protein repellents when adsorbed to small latex particles, where they are less concentrated and form thin but highly dynamic coatings. Biospecific affinity adsorbents with low nonspecific uptake of protein are constructed by linking biorecognition moieties to surfaces via a PEO tether. For this purpose, functionalized PEO-containing surfactants were prepared and adsorbed to polymeric sub-

[3] Corresponding author

0065–2393/96/0248–0061$12.00/0

strates prior to derivatization. This procedure allowed us to select the stoichiometric ratio of functionalized to underivatized surfactant prior to adsorption.

POLY(ETHYLENE OXIDE) (PEO) has emerged as a desirable coating material in the past decade because of its unique ability to shield surfaces in aqueous environments from the deposition of macromolecules and particles present in the surroundings. In the biomaterials arena, the first evidence of this ability was given by Nagaoka et al. (*1*), who grafted PEO of systematically varied chain lengths to surfaces that were to come in contact with whole blood. A reduced deposition of platelets on these surfaces was readily observed, and this suppression was increasingly effective with increasing polymer chain length up to around 100 monomer units. A somewhat different use of PEO was pioneered by Abuchowski et al. (*2*), who found that by covalently attaching PEO chains to proteins of various kinds, these proteins could be made nonimmunogenic and therefore suitable for intravenous use even in species vastly different from the organism in which they had been synthesized. This high tolerance is probably due to suppressed binding of the immunoglobulins in plasma, and the technique of "stealth-coating" proteins has since been adapted to the surface treatment of liposomes (*3*) and other particulates (*4, 5*) intended for use in intravenous delivery of drugs.

The coating of particulates with bulky polymers, whose interaction with the liquid suspension medium is characterized by a large, negative free energy of solvation, has long been used to suppress flocculation and provide "steric stabilization" of the suspension. A thermodynamic model for this phenomenon was formulated early on by Flory and Kriegbaum (*6*), who expressed the free energy of particle–particle interaction as a function of the thickness of the stabilization layer, the polymer concentration in this layer, the polymer–solvent interaction parameter χ, and the radii and spatial separation of the particles. More recent work by Fleer and Scheutjens (*7*) resulted in a lattice model for the adsorption of diblock copolymers to colloidal particles and the resultant interparticle repulsion. Because the interaction of proteins with polymer-coated surfaces is similar in some ways to the steric stabilization of colloids, an examination of the factors responsible for the effectiveness of a surface coating as a suppressor of protein adsorption may logically begin with those factors considered in the Flory–Kriegbaum treatment. For the purpose of this discussion, it is therefore important to determine accurately the layer thickness and the surface concentration of PEO chains on surfaces and to attempt to relate these parameters to the protein repulsion characteristics of the surfaces.

Unlike the pairwise repulsion of two identical, polymer-coated particles whose coatings prefer to remain solvated rather than associate with their counterparts on the opposite particle, the protein–polymer interaction is somewhat more complex, as discussed recently by Jeon et al. (*8*, *9*). Being uncharged, a coating of PEO is unable to interact with an approaching protein by coulombic forces. Furthermore, because of its unique ability to fit inside the water lattice without imposing significant levels of new ordering of the solvent (*10*), the polymer has only weak tendencies to interact hydrophobically with proteins in its surroundings. In terms of its ability to engage in dispersive interactions, the PEO is also relatively unreactive, being of unusually weak polarizability to judge from its low refractive index.

Insofar as the polymer coating effectively covers the underlying substrate, enthalpy makes no pronounced contributions to the interaction between protein and coating, yet the clear repulsion of protein from the surface is evidence of more than an absence of attractive forces. The work of Andrade and others (*9*, *11*) suggests that a strong entropic component is responsible for the repulsion. According to this concept, the approach to the surface by a compactly folded macromolecule many times the size of a polymer chain attached to this surface severely restricts the motion of this chain and thereby reduces its configurational entropy. The more mobile the chain, the larger would be its loss in entropy and the more effectively would it repel the intruder from the surface. In the light of this notion, it is essential to attempt a characterization of polymer surface mobility in addition to the layer thickness and concentration parameters suggested by Flory and Kriegbaum in order to understand better the factors that control the effectiveness of a PEO coating in suppressing protein adsorption.

Although a covalent grafting of PEO was the approach taken by Nagaoka et al. (*1*) for the early demonstration of this coating's effectiveness in suppressing platelet adhesion, this approach suffers from difficulties in creating comparable surface concentrations for polymers of different molecular sizes. The attachment is also for the most part a multistep process requiring activation of both surface and polymer and is therefore labor intensive in addition to being difficult to control fully. In the present work, we have instead chosen to modify hydrophobic surfaces, notably those of polystyrene (PS) latex particles, by adsorbing PEO-containing block copolymers (Pluronics) of the general structure PEO-poly(propylene oxide) (PPO)-PEO, whose hydrophobic blocks are responsible for adsorption of the copolymers to the hydrophobic particle surface. The selected set of polymeric surfactants, which are commercially available under the trade name Pluronic, was chosen to contain central PPO blocks of comparable sizes, while the flanking PEO blocks varied nearly fourfold in length. This wide variability in polymer composition made it possible to form adsorption complexes between the polymeric surfactants and an array of PS nanoparticles of different sizes. These complexes gave rise to stable surface coatings of different thicknesses and concentrations and were therefore suitable model systems for evaluating the relationship between polymer surface configuration on the one hand and nonspecific adsorption of protein on the other.

Although in many situations protein uptake by polymeric surfaces can severely impair the use of the polymers as biomaterials, we will limit the present discussion to one particular area of applications in which nonspecific adsorption of protein can be highly detrimental. Our concerns focus mainly on using the surface as part of an immunodiagnostic device with an expected sensitivity in the femtomolar range. Such devices are typically built around a surface to which immunoglobulins are attached and to which the immunoglobulins are expected to bind their complementary antigen in proportion to its presence in the surrounding analyte. Once bound, the antigen can be quantified by the adsorption of a labeled second antibody. Any nonspecific attachment of that antibody affects the accuracy of the analysis. The design of diagnostically suitable surfaces will therefore require presenting the primary, surface-linked antibody, in the form of either an intact immunoglobulin G (IgG) molecule or its Fab′ fragment, in such a way that its antigen binding is maximized while the binding of all other interfering substances is suppressed. A more detailed discussion of this strategy is given later in this chapter.

The possibility of introducing a mat of potential tethers to a hydrophobic surface by means of adsorption of a PEO-containing block copolymer with a certain fraction of activated end groups opens a new approach to biosurface derivatization. In the following sections, we discuss the effects of PEO and PPO chain length on the physical arrangement of Pluronic-type block copolymers on the surfaces of particulate substrates. We then examine how this arrangement affects protein adsorption to the coated surface and, ultimately, how a surface that has a predetermined concentration of affinity ligands but remains inert toward proteins other than those with specific affinity for the ligand might be constructed.

Effect of PEO Chain Length on Pluronic Surface Density

The triblock surfactants chosen for this study are listed in Table I. They were selected because they have PEO blocks of various lengths but comparably sized hydrophobic center blocks. Thus any effect on the polymer surface density caused by adsorption from a fixed solution

Table I. Physical Properties of Selected Pluronics

Trade Name	M_w	*PEO/PPO/PEO*	*HLB*
Pluronic P105	6,500	37/56/37	12 ~ 18
Pluronic F68	8,400	76/30/76	>24
Pluronic F88	11,400	104/39/104	>24
Pluronic F108	14,600	129/56/129	>24

NOTE: Data are given by BASF Company.

concentration (4% w/v) would be the result of the differing PEO chain lengths.

The adsorption behavior of the higher-molecular-weight Pluronic surfactants on PS latex particles has been studied by others with various results. Kayes and Rawlins (*12*) found F108 to adsorb with a lower surface concentration than F68, but Baker and Berg (*13*) found the opposite. In the present work, the polymer surface densities on particles of different sizes were determined by a variety of different techniques. First, we developed a new approach to sedimentation field-flow fractionation (SdFFF) that allowed a determination of the mass increase per particle from observed differences in retention between the bare and coated colloids (*14*). The SdFFF technique is an analytical separation method that relies on the coupled influences of an applied sedimentation field with a perpendicular flow of sample through a thin channel. The technique allows an exact determination of particle mass from the observed level of retention of a particulate sample under a given field, provided the densities of particle and suspension medium are known. If the particle is allowed to adsorb a coating of known unsolvated density, the added mass translates into an increased retention under the chosen field. Through its ability to produce measurable retention shifts for coated samples, the SdFFF acted as a very sensitive microbalance. Since the surface area per particle was determined from the retention of bare particles under a given field strength, the mass per unit area was readily calculated from these measurements for all particles in the 100–300-nm size interval. The lower limit here is set by the maximum field strength the system can sustain and thus the minimum mass retainable by our SdFFF unit. The upper limit is set by the precision in the measurement; for particles above this limit, the fractional mass increase due to the adlayer becomes too small to be accurately determined.

An entirely different approach to the measurement of surface densities comes from the controlled adsorption of surfactants that have undergone some form of labeling. In the present work, three different labels were introduced to different batches of surfactants. The primary labeling reaction involved the radioiodination (^{125}I) of a tyrosine residue previously grafted, as part of a Bolton-Hunter reagent, onto a small fraction of the end hydroxyls of the PEO chains (*15*). A few of the data points derived from the two techniques described thus far were verified by adsorption of labeled surfactant containing either 1-pyreneisothiocyanate fluorescent probe (*16*), directly quantifiable from the measured fluorescence intensity (excitation at 345 nm, emission at 376 nm), or 2-pyridyl disulfide (PDS) (*15*), measured from the extinction at 343 nm following a reductive cleavage of the disulfide bond. The amounts of surfactant determined in this manner were then related to the amount of colloid present with the polymer.

The agreement between the various techniques was good (*14, 16*). As a result, we trust our analytical data, which show two important trends. First, for any one substrate diameter, there is no significant difference between the surface densities generated through adsorption of the different surfactants (*see* Figure 1a). This lack of difference is an indication that the flanking PEO blocks are reaching out from the surface rather than hovering in its immediate vicinity and that the hydrophobic PPO blocks govern the close packing. Second, for any given surfactant, with the exception of the micellizing P105, the surface density increases as the substrate flattens (*see* Figure 1b). This curvature effect on adsorption behavior is often overlooked, as surface densities are generally assumed to apply interchangeably to substrates of different kinds (*12, 13, 17, 18*).

Effect of PEO Arm Length on Adlayer Thickness

According to scaling law arguments (*19, 20*), the spatial extension of hydrophilic polymer chains tethered to a surface for which they have no affinity increases with increasing number of monomers in the chain. For PEO-containing surfactants, this postulate has been qualitatively verified by several authors (*18, 21, 22*). In work with colloidal substrates, the photon correlation spectroscopic technique (PCS) is frequently used to gain this information, and the expected trend of thicker layers with longer PEO arms is determined from the difference in the Stokes diameters of bare and coated particles. In the present study, we also relied on PCS as one technique for assessing layer thickness. However, reasons for caution were obvious, as the light-scattering-derived sizes of several adsorption complexes were larger than expected. Flow field-flow fractionation or SdFFF of the complexes clearly revealed the presence of small amounts of aggregate, which had skewed the determination in the direction of larger sizes. The prefractionated coated singlets, however, were amenable to accurate sizing by PCS, and the layer thickness was reproducibly determined

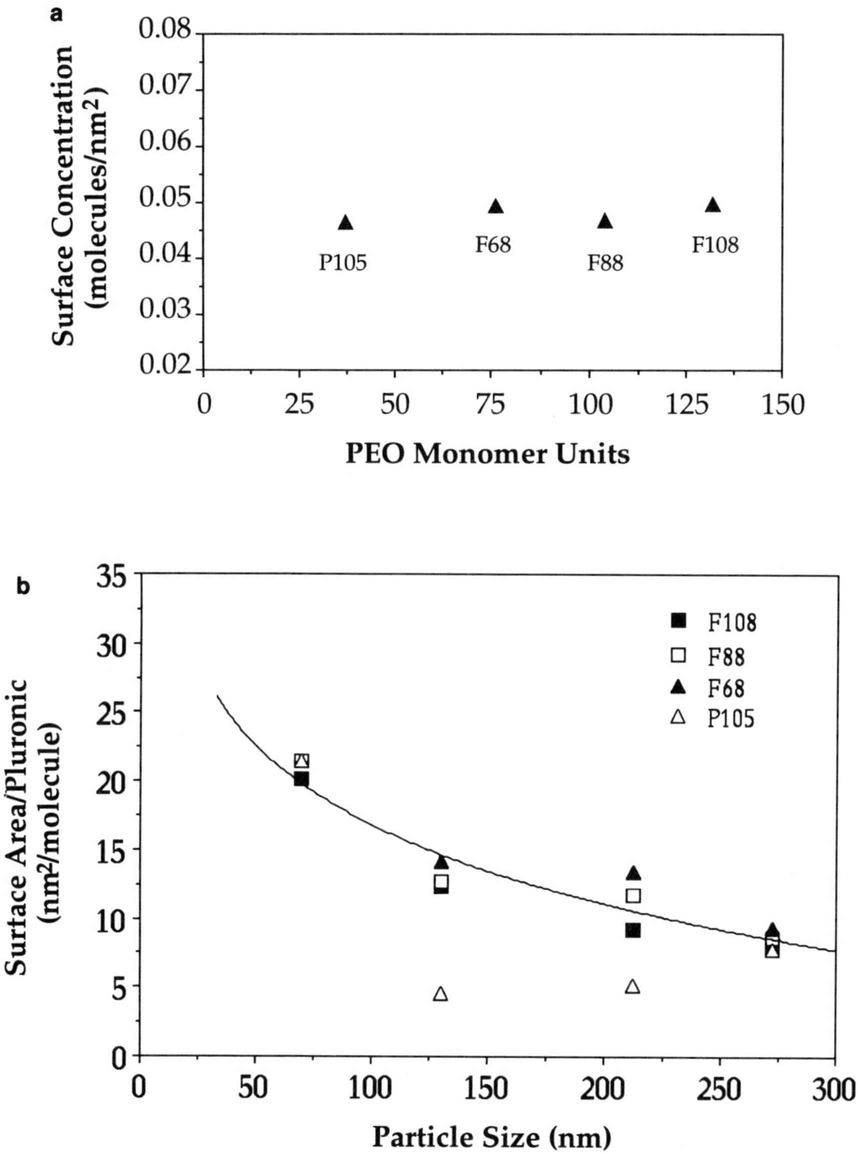

Figure 1. Top: Surface concentration (molecules per nm²) of Pluronic surfactants with various PEO chain lengths. The substrate is 69-nm PS latex particles (Seradyn) coated at 25 °C with 4% (w/v) solutions of the selected surfactants in 0.15 M phosphate buffered saline (pH 7.4). Bottom: Surface densities of Pluronic surfactants adsorbed to PS particles of different sizes.

from the difference in size between bare and coated particles. Because the retention in flow field-flow fractionation is a direct measure of a sample's diffusion coefficient and therefore its size (*23*), the layer thickness could be independently calculated from the measured retentions of bare and coated particles. The two sets of data are in good agreement (largest deviation was 3%). By analogy with the discussion of adsorption density above, the measured layer thicknesses should be examined both as a function of PEO chain length for complexes formed with particles of a given size (Figure 2a) and as a function of particle size for a given surfactant (Figure 2b).

The curvature effect on adsorption density that is demonstrated in Figure 1b has its direct counterpart in the set of layer thickness data shown in Figure 2b. Thus the lowest surface density, which is associated with the smallest particles, also gives rise to the most collapsed adlayer. Conversely, the highest packing densities, which are generated by the largest particles, lead to thicker and more brushlike coatings, a result in accord with the relationships between chain distance and chain extension postulated by de Gennes (*19*). Chains whose attachment points are spaced more than one molecular diameter (D = 2 radii of gyration [R_g]) apart on a surface with which they do not interact are unaffected by their neighbors. In the language of de Gennes, they are said to form "mushroom"-like structures with a height of R_g. Indeed, for the PS latex particle of 69-nm diameter, agreement between layer thickness on the one hand and dimensions of the different PEO side arms on the other appears to be good. This agreement is evident in Figure 2a, which shows both the observed layer thickness and the random coil diameter ($2\,R_g$) for each respective PEO side arm, calculated according to Bhat and Timasheff (*24*), for the surfactants listed in Table I.

Layer Thickness and Chain Mobility

Chains in a mushroom configuration are expected to be more mobile than those confined to a more compact, brushlike surface arrangement (*20*). In order to verify this assumption, the PEO end hydroxyls were partly labeled with a spin probe in the form of the 3-carboxy proxyl radical (*25*). This label, whose presence does not affect the adsorption process, allows the measurement of the surfactants' electron spin resonance (ESR) in an applied magnetic field. The observed correlation times at high probe dilution (*25, 26*) give one a direct insight into PEO chain mobilities whether the surfactant is in free solution or is adsorbed to a particle surface. The effect of chain length on mobility is illustrated in Figure 3, which for the sake of comparison also contains the layer thickness data from Figure 2a. The expected trend of

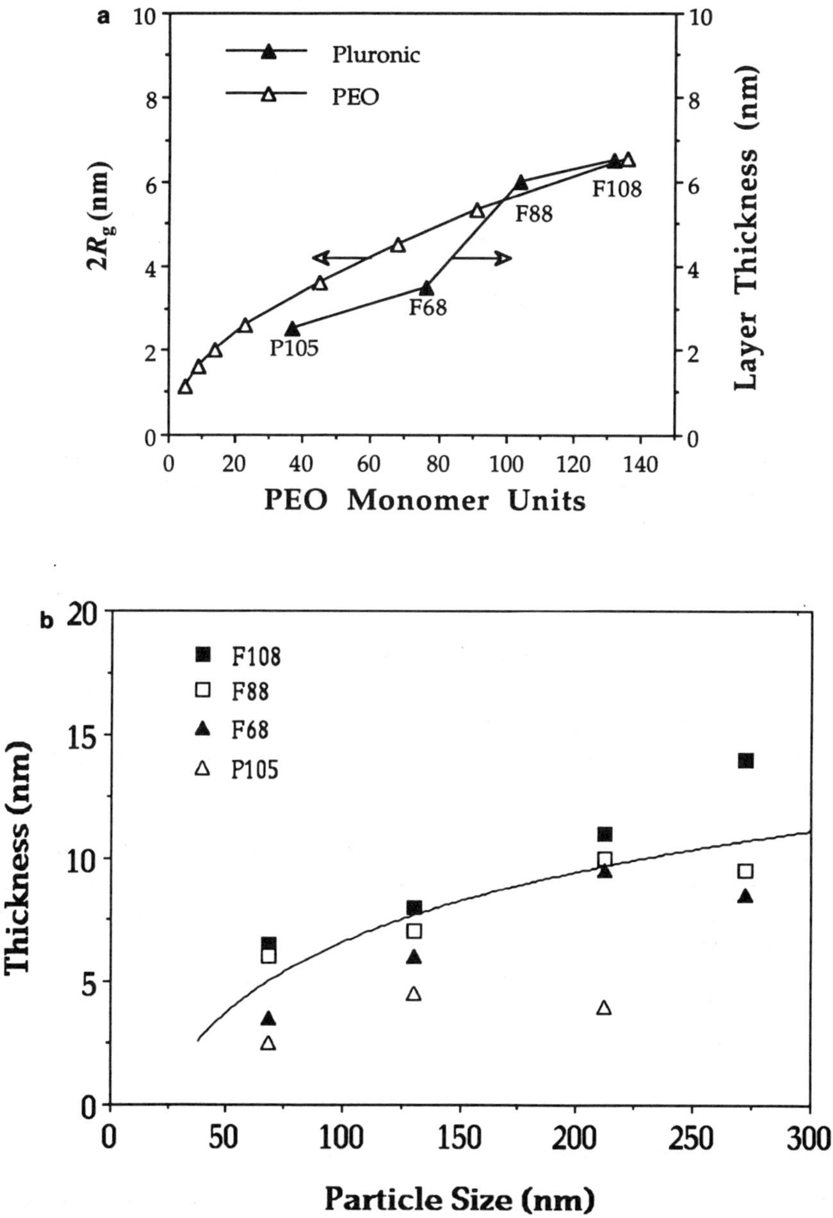

Figure 2. Top: Adlayer thicknesses of different Pluronic surfactants on a 69-nm PS latex particle. Also shown are the estimated solution sizes (23) of PEO molecules with molecular weights comparable to the PEO blocks of the different surfactants. Bottom: Layer thicknesses of Pluronic surfactants on PS particles of different sizes.

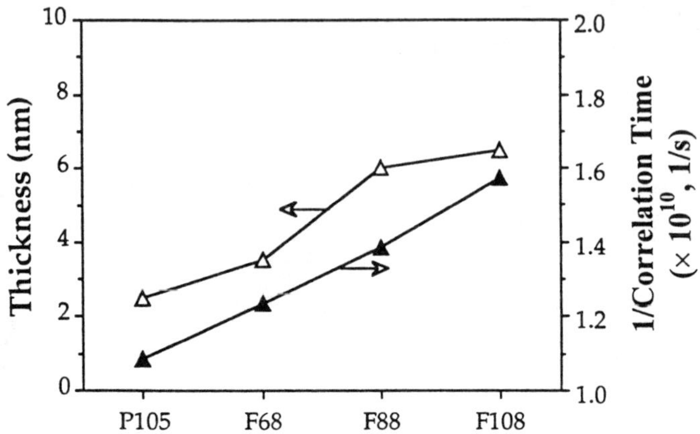

Figure 3. Effect of PEO chain length on chain mobility and adlayer thickness for different Pluronic surfactants adsorbed on 69-nm PS latex particles.

greater mobilities farther away from the surface is well documented by these data for surfactants of different compositions adsorbed to particles of a fixed size (69 nm). The curvature effects, documented for the layer thickness of F108 in Figure 2a, are also seen in the mobilities of the spin probe. Figure 4 shows clearly that the 69-nm particle, with its more collapsed and mushroomlike coating, offers significantly larger freedom of motion for the PEO end groups than does the 272-nm analog, whose polymer chains are crowded into a brushlike arrangement. It is interesting that the end group mobilities for Pluronic F108 adsorbed to 69-nm latex particles are not much less than those found in the dissolved surfactant or even for the free spin probe; the inverse rotational correlation times are 1.6×10^{10}, 1.7×10^{10}, and 2.9×10^{10} s^{-1} for the PS 69-nm complex, the F108–proxyl, and the proxyl–carboxylic acid alone, respectively. These values are in contrast to the low value of 0.9×10^{10} s^{-1} for the F108–PS 272 complex (27).

PEO Block Length and Protein Resistance

The effects of the various Pluronic coatings in terms of their abilities to suppress protein adsorption were measured in a series of incubation experiments in which bare and coated PS 272-nm particles were given a 24-h exposure to a solution of human fibrinogen. The plasma protein fibrinogen is notorious for adsorbing to virtually all surfaces, and this adsorption is frequently the trigger of thrombus formation, platelet activation, and other events associated with bioincompatibility. Figure

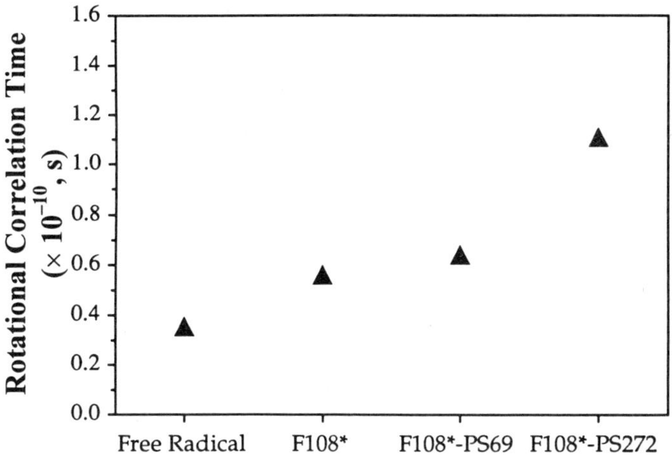

Figure 4. Rotational correlation times τ for the proxyl spin probe as determined by ESR. As indicated, the probe is either free in solution or attached to the F108 surfactant, which in turn is observed either free in a dilute solution (F108) or adsorbed to PS latex particles that are 69 nm (F108*–PS69) or 272 nm (F108*–PS272) in diameter. The mobility (τ^{-1}) is only moderately reduced by the attachment to F108 and the adsorption of the modified surfactant to PS69, but adsorption to the larger particle causes a substantial slowdown.*

5 illustrates fibrinogen uptake as a function of the number of monomer units in the PEO blocks (*15, 16*). The protein analysis in this case is based on measured levels of depletion and is therefore of limited sensitivity at low adsorption levels. The bare PS particle adsorbs as much as 3.3 mg/m^2 from the 1.0 mg/mL protein solution, but an increase in the PEO block length clearly suppresses fibrinogen uptake to the point that F108 with its 129 EO units per block shows no uptake at this level of sensitivity. However, a more accurate amino acid analysis protocol applied to the particles themselves (*28, 29*) actually reveals the presence of 0.04 mg of fibrinogen per m^2 on this substrate (Table II).

When observations made as parts of the general characterization of these complexes are combined with those made in conjunction with the evaluation of the protein uptake of the complexes, a few conclusions emerge. First, the data set in Table II clearly shows that neither a high surface concentration nor a large coating thickness are in and of themselves responsible for the observed protein repulsion. This conclusion springs from the fact that the F108 coating on the 69-nm bead, which is half as thick and 40% as concentrated as that on the 272-nm particle, is about twice as effective as a protein repellent. Second, one parameter that in this early phase of our investigation appears

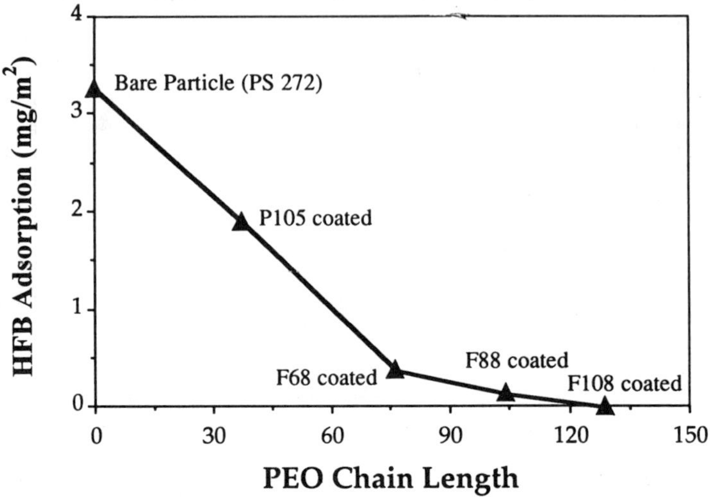

Figure 5. Human fibrinogen (HFB) uptake by 272-nm PS latex particles coated with different types of Pluronic. The data were obtained by depletion analysis using a 1 mg/mL HFB solution and a 1:5 protein:particle weight ratio.

to correlate with protein repulsion is the mobility of the PEO chains, and this correlation supports the notion of Nagaoka and Andrade mentioned in the introductory paragraphs of this chapter.

The encouraging repulsion effects discussed earlier were the results of simple coating procedures, and in order to assess the usefulness of these coatings, one must examine their stability in various environments. For this purpose, partly radioiodinated F108 surfactant was adsorbed from a 4% solution onto 272-nm latex beads overnight at room temperature, in accordance with the standard adsorption protocol used throughout this study. After careful washing through multiple suspension–centrifugation cycles, all soluble radioactivity was removed, and the coated particles could be suspended in the chosen test environments, which included phosphate-buffered saline, a solu-

Table II. Effect of Substrate Size on Properties of Adsorbed Pluronic F108

Particle Diameter (nm)	*Adlayer Thickness (nm)*	*Surface Coverage (nm²/molecule)*	*Mobility ($\times 10^{10}\ s^{-1}$)*	*Amount of Adsorbed HFB (mg/m²)[a]*
69	6.5	20.1	1.57	0.021 ± 0.003
272	14.0	7.9	0.90	0.038 ± 0.002

[a] The HFB adsorption on bare PS particles is 3.3 mg/m² (*see* Figure 5).

tion of human serum albumin (8 mg/mL), and whole human plasma. Although no isotope leakage was observed during 3 days in either of the first two solutions, the whole plasma rapidly gave rise to a significant release of ^{125}I. Amino acid analysis of the coated particulates after 3 days of plasma incubation indicated a 10% protein uptake compared to that of bare particles under the same conditions. The nature of the plasma components responsible for this behavior is not yet known. However, unlike the bare PS spheres, the coated particles are the same size before and after exposure to human plasma (*15, 16*). A related finding is that upon injection into rats, F108-coated 272-nm PS latex particles have a significantly prolonged clearance time compared to that of their bare counterparts (half-lives of 16 h versus a few minutes) (*16*). Although uncoated particles aggregate extensively in plasma, the coated spheres show no evidence of aggregation, even after 24 h in vivo. Whatever the nature of the protein deposit on these particles, it does not appear to perturb the protein repulsion of the coating, which remains largely (to 70%) intact after 3 days of exposure to plasma.

PEO-Coated Surfaces for Immunodiagnostic Use

As briefly mentioned early in this chapter, the high sensitivities required of today's immunodiagnostic devices mandate a low nonspecific background adsorption of protein. In a recent study (*30, 31*) we compared the specific antigen-binding capabilities of the monoclonal (9-40) antifluorescein IgG when it was attached nonspecifically directly to amine groups on an aminosilane derivatized silica surface and when it was attached specifically to the carbohydrate in the hinge region of the molecule. The attachment of IgG to carbohydrate was mediated by a 5000-Da dihydrazide–PEO tether, at that time covalently linked to the surface. Briefly, the outcome of this comparison showed a twofold increase in specific binding for the PEO-tethered antibody with a twofold decrease in nonspecific protein binding to the surface: all in all, a fourfold advantage for the site-specific PEO-tethered coupling.

Although the nonspecific protein binding was reduced in the PEO-mediated coupling just discussed, this binding was still significantly higher than that of the Pluronic-coated surfaces discussed previously in conjunction with Table II. An attempt to further suppress this binding was therefore made. The attempt involved dilution of the divalent PEO tethers with monovalent methoxy–PEO chains. The resulting mixed surface was then reacted with the heterobifunctional succinimidyl 4-(*N*-maleimidomethy)cyclohexane-1-carboxylate SMCC reagent (*32*), and this reaction allowed the subsequent site-directed attachment of Fab′ fragments of the monoclonal anti-fluores-

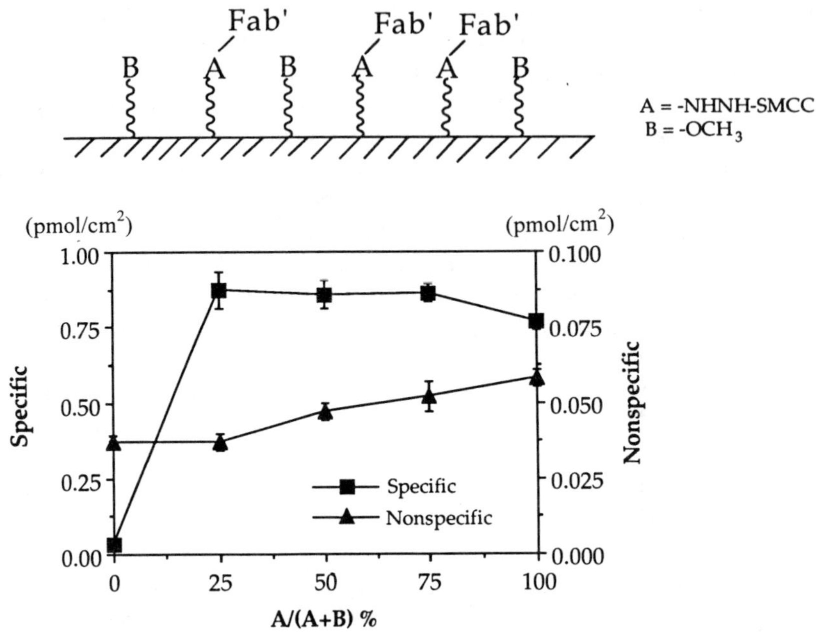

Figure 6. Specific affinity of an immunosorbent consisting of FAB′ fragments tethered to the substrate via PEO chains with molecular weights of 5 kDa. The active structures (A) are mixed in with blunt-ended PEO (B) of the same molecular weight. The Fab′ derives from a monoclonal antifluorescein antibody (9-40), and the specific binding is determined by means of a ^{125}I-labeled fluorescein-BSA conjugate. Nonspecific binding is measured as the adsorption of ^{125}I-labeled BSA.

cein (9-40) antibody via their free C-terminal thiols. The data shown in Figure 6 make it clear that nonspecific adsorption (of radioisotope-labeled bovine serum albumin [BSA]) is reduced with a reduced surface concentration of Fab′ fragments. At zero surface concentration, that is, for a surface exclusively covered with PEO, the level of adsorption is comparable to that determined for the F108-coated PS particles (*see* Table II).

Remarkably, a dilution of ligand-binding sites to 25% of maximum does not appear to reduce the amount of antigen, in the form of ^{125}I-labeled albumin–fluorescein, that binds per unit surface area. On the contrary, the dilution appears to increase this binding somewhat, presumably because of reduced steric hindrance. The nonspecific binding in this case is virtually identical to that of the surface coated with pure PEO. These findings are leading us to take a closer look at the effect of ligand density not only on specific binding of analyte but also on the kinetics of binding and displacement, all of which appear to favor the sparse surface arrangement.

Although the covalent linkage of PEO tethers produces surfaces that are stable in a variety of solvents and under different temperature conditions, a necessary requirement in some instances, these surfaces are more difficult to generate than the adsorption-coated surfaces discussed earlier. In order to allow the attachment of ligands to such surfaces, the PEO blocks of the surfactant had to be chemically modified prior to adsorption in a manner that did not interfere with the ability of the surfactant to stably coat the surface. The active structures needed to be hydrolytically stable to remain in place even after extended coating procedures, and they had to allow an efficient ligand attachment under conditions in which adsorption to the surface was of unperturbed strength and the structure of the protein ligand remained intact. These rather stringent requirements led to the choice of the PDS structure, which is known to provide a good leaving group for the coupling of thiols under physiological conditions (*33*). The active structure was introduced into the *p*-nitrophenyl-activated F108 surfactant by allowing it to react with 2-(2-pyridyl dithio)ethylamine in methanol. After purification by dialysis against water and then lyophilization, the derivatized surfactant was stored in a desiccator for months with retained activity (*15*).

Surface coatings with F108–PDS gave the same close packing as coatings with the underivatized F108, and following adsorption, the reactive PDS structures were readily replaced by thiol-containing ligands that either occurred naturally, as in β-galactosidase, or were introduced via a thiolation reagent. This introduction was used to provide anti-IgE antibodies with the handle required for immobilization to PS microtiter plates previously coated with F108–PDS. This coupling, which took place under physiological conditions, resulted in a specific binding of antigen around eight times higher than that for hydrophobically adsorbed (passively coated) antibody (*34*). In light of the favorable effects on both binding and displacement kinetics as well as on the equilibrium levels of specific and nonspecific adsorption, which result from a dilution of the protein ligands on the surface (*see* Figure 6), we were interested in producing surfaces with a wide range of ligand concentrations. Such surfaces are potentially produced by selecting the desired ratio of derivatized to underivatized F108 when the coating mixture is made. That the coating proceeds without distinction between the two types of surfactant molecules is seen from Figure 7, which relates the concentration of F108–PDS in the supernatant to the surface concentration of reactive structures on a PS latex substrate. The figure's linear relationship between solution and surface concentrations suggests that desired ligand densities can easily be produced

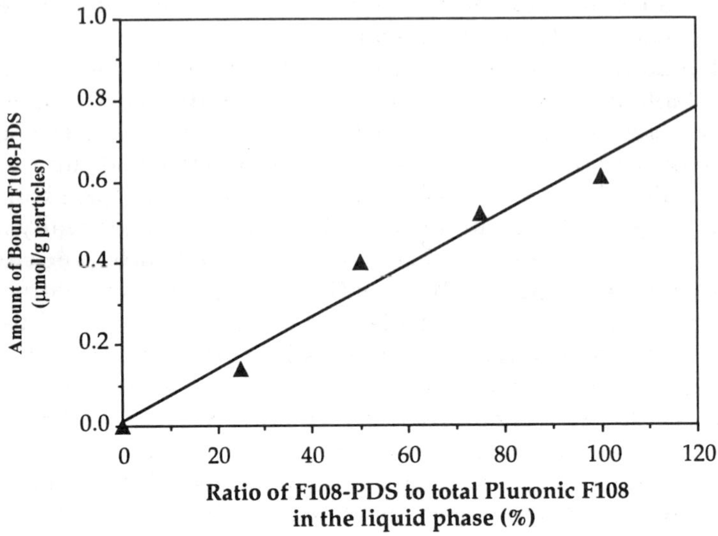

Figure 7. Adsorption of PDS-labeled Pluronic F108 (F108–PDS) on PS. The graph represents the relationship between the fraction of reactive PDS end groups in solution and the amount adsorbed on the particle surface.

by adsorption of a mixture of F108–PDS and F108 with a selected composition. The further derivatization and evaluation of these mixed surfaces are presently under investigation in our laboratory.

Conclusion

By simple adsorption of PEO-containing block copolymeric surfactants, it is possible to provide hydrophobic surfaces with stable coatings that are highly protein repellent. Specifically, the PEO-PPO-PEO triblocks form stable adsorption layers whose surface concentrations, in general, are constant for a given PPO block length, regardless of the lengths of the PEO flanking blocks. In adsorption onto particulate substrates, the radius of curvature of the substrate strongly affects the close packing on the surface, so that small particles become more sparsely covered than large ones, even when adsorption is from one and the same solution. For each specific particle size, these surfactants form adlayers that become thicker as their PEO chains become longer. As the block length increases, so does the mobility of its end group. Conversely, for any given block length, the least crowded surfaces on the smallest particles provide the best environment in terms of chain mobility.

The PEO block length has a clear influence on protein repulsion, with chains in excess of 100 monomer units reducing fibrinogen adsorption by 2 orders of magnitude compared to adsorption to the bare PS substrate. This reduction is in agreement with observations by others of surfaces with covalently attached PEO chains. Although the smallest particles in this study have the lowest surfactant coverage and the most collapsed adlayer when coated with the F108 surfactant, they show the greatest ability to prevent protein adsorption. Surface density and layer thickness may therefore be of less importance than originally thought, and chain mobility appears to play a fundamental role in the repulsion mechanism.

The protein-repelling qualities of PEO can be utilized to reduce nonspecific adsorption to surfaces designed to biospecifically recognize and bind analytes of interest. By tethering the recognizing ligand to the substrate via PEO spacer arms with lengths of around 100 monomer units, the substrate can be made less prone to nonspecific uptake of protein than it is when the ligand is attached directly and nonspecifically to the surface, and thus the accuracy of its analytical response is improved. In a further refinement of the surface's response to analyte, this ligand-binding PEO can be diluted with blunt-ended PEO molecules of the same size as the tethers. This dilution enhances the ligand's ability to specifically bind its complementary analyte and simultaneously suppresses nonspecific uptake to levels characteristic of the most effectively passivated Pluronic-coated surfaces.

Acknowledgments

This work was supported in part by grant GM 38008-05 from the National Institutes of Health. J. Carlsson, on sabbatical leave from Pharmacia of Uppsala, Sweden, gratefully recognizes support from this corporate source.

References

1. Nagaoka, S.; Mori, Y.; Takiuchi, H.; Yokota, Y.; Tanzawa, H.; Nishiumi, S. In *Polymers as Biomaterials;* Shalaby, S. W.; Hoffman, A. S.; Ratner, B. D.; Horbett, T. A., Eds.; Plenum: New York, 1984; p 361.
2. Abuchowski, A.; Van Es, T.; Placzuk, N. C.; Davis, F. F. *J. Biol. Chem.* **1977,** *252,* 3578.
3. Klibanov, A. L.; Maruyama, K.; Torchilin, V. P.; Huang, L. *FEBS Lett.* **1990,** *268,* 235.
4. Illum, L.; Davis, S. S. *FEBS Lett.* **1984,** *167,* 79.
5. Müller, R. H. In *Colloidal Carriers for Controlled Drug Delivery and Targeting;* Müller, R. H., Ed.; CRC Press: Boca Raton, FL, 1991; p 19.
6. Hiemenz, P. C. *Principles of Colloid and Surface Chemistry,* 2nd ed.; Marcel Dekker: New York, 1986; p 664.

7. Fleer, G. J.; Scheutjens, J. M. H. M. *Colloids Surf.* **1990,** *51,* 281.
8. Jeon, S. I.; Lee, J. H.; Andrade, J. D.; de Gennes, P. G. *J. Colloid Interface Sci.* **1991,** *142,* 149.
9. Jeon, S. I.; Andrade, J. D. *J. Colloid Interface Sci.* **1991,** *142,* 159.
10. Kjellander, R.; Florin, E. *J. Chem. Soc., Faraday Trans. 1* **1981,** *77,* 2053.
11. Nakao, A.; Nagaoka, S.; Mori, Y. *J. Biomater. Appl.* **1987,** *2,* 219.
12. Kayes, J. B.; Rawlins, D. A. *Colloid Polym. Sci.* **1979,** *257,* 622.
13. Baker, J. A.; Berg, J. C. *Langmuir* **1988,** *4,* 1055.
14. Li, J.-T.; Caldwell, K. D. *Langmuir* **1991,** *7,* 2034.
15. Li, J.-T. *Plasma Protein Interactions with Copolymer Stabilized Colloids;* Ph.D. Dissertation, University of Utah, Salt Lake City, UT, 1993.
16. Tan, J. S.; Butterfield, D. E.; Voycheck, C. L.; Caldwell, K. D.; Li, J.-T. *Biomaterials* **1993,** *14,* 823.
17. Garvey, M. J.; Tadros, T. F.; Vincent, B. *J. Colloid Interface Sci.* **1976,** *55,* 440.
18. Baker, J. A.; Pearson, R. A.; Berg, J. C. *Langmuir* **1989,** *5,* 339.
19. de Gennes, P. G. *Macromolecules* **1980,** *13,* 1069.
20. Milner, S. T. *Science (Washington D.C.)* **1991,** *251,* 905.
21. Killman, E.; Maier, H.; Baker, J. A. *Colloids Surf.* **1988,** *31,* 51.
22. Lee, J.; Martic, P. A.; Tan, J. S. *J. Colloid Interface Sci.* **1989,** *131,* 252.
23. Giddings, J. C.; Caldwell, K. D. In *Physical Methods of Chemistry;* Rossiter, B. W.; Hamilton, J. F., Eds.; John Wiley: New York, 1989; p 867.
24. Bhat, R.; Timasheff, S. N. *Protein Sci.* **1992,** *1,* 1133.
25. Hommel, H.; Facchini, L.; Legrand, A. P.; Lecourtier, J. *Eur. Polym. J.* **1978,** *14,* 803.
26. Ouada, H. B.; Hommel, H.; Legrand, A. P.; Ballard, H.; Papirer, E. *J. Colloid Interface Sci.* **1988,** *22,* 441.
27. Li, J.-T.; Rapoport, N.; Caldwell, K. D. *Langmuir* **1994,** *10,* 4475.
28. Caldwell, K. D.; Li, J. M.; Li, J.-T.; Dalgleish, D. G. *J. Chromatogr.* **1992,** *604,* 63.
29. Li, J.-T.; Caldwell, K. D., University of Utah, Salt Lake City, UT, unpublished results.
30. Huang, S.-C. *Site-Directed Immobilization of Immunoglobulins;* M.S. Thesis, University of Utah, Salt Lake City, UT, 1992.
31. Huang, S.-C.; Caldwell, K. D.; Lin, J. N.; Herron, J. N., University of Utah, Salt Lake City, UT, unpublished results.
32. Yoshitake, S.; Imagawa, M.; Ishikawa, E.; Niitsu, N.; Urushizaki, I.; Nishiura, M.; Kamazawa, K.; Kurosaki, H.; Tachibana, S.; Nakazawa, N.; Ogawa, H. *Biochem. J.* **1982,** *92,* 1413.
33. Carlsson, J.; Drevin, H.; Axén, R. *Biochem. J.* **1978,** *173,* 723.
34. Li, J.-T.; Carlsson, J.; Caldwell, K. D., University of Utah, Salt Lake City, UT, unpublished results.

RECEIVED for review April 16, 1994. ACCEPTED revised manuscript August 17, 1994.

5

A Review of Synthetic Approaches to Biodegradable Polymeric Carboxylic Acids for Detergent Applications

Y. H. Paik, E. S. Simon, and G. Swift

Rohm and Haas Company, 727 Norristown Road, Spring House, PA 19477

Polymeric carboxylic acids such as poly(acrylic acid) and co-poly(acrylic–maleic acid)s are widely used as cobuilders in detergent powder formulations. They were introduced in the early 1980s in combination with zeolites as partial replacements for polyphosphates, which were removed from detergent formulations because wastewater treatment plants did not remove them from wastewater. The polyphosphates promoted the eutrophication of rivers and lakes and thereby promoted environmental imbalance and the death of certain aquatic life-forms. Although the nonbiodegradability of the currently used polymeric carboxylic acids eliminates the possibility of eutrophication of the aqueous environment, it results in an increasing accumulation of these acids in the environment. No known toxicity is associated with these polymers, but because negative results are always accompanied by uncertainty, questions as to the fate of the acids and their effects in the environment remain. Therefore the prudent option is to develop biodegradable replacements for the current products, and this goal has been recognized, desirable, and difficult almost since the time polymers were introduced. In this review chapter we trace the major synthetic approaches that have been tried and have failed to meet the requirements for performance and biodegradation, reach some conclusions on the current status of re-

0065–2393/96/0248–0079$12.00/0

search, and suggest what may be the most fruitful future directions for research in the quest for biodegradable detergent polymers.

THIS CHAPTER REVIEWS SYNTHETIC APPROACHES TO DEVELOPING biodegradable polymeric carboxylic acids (PCAs) for use in detergents, assesses the current status of these approaches, and considers the implications for future research. In order to present a comprehensive review that is easily readable by those not overly familiar with the field of biodegradable PCAs and detergents, this chapter is divided into several sections. These sections cover the background on the need for biodegradable detergent polymers, a perspective on currently used commercial polymers, some important considerations in the development of biodegradable detergent polymers, approaches that have been tried, conclusions, and suggestions for future research directions.

PCAs used in detergents represent a large percentage of the volume of all such polymers. PCAs are used in water treatment and as pigment dispersants, oil field applications, thickeners, etc. Other polymers containing water-solubilizing functionalities such as hydroxyl, amino, and ether groups are also widely used in a variety of applications, and ultimately all will probably be expected either to be removed from effluent streams or to be biodegradable; otherwise their concentration in the aqueous environment will continue to increase. Hence the continuing environmental responsibility shown by the detergent industry over many years will eventually have to be accepted by all other industries using water-soluble polymers. For the record, the history of the detergent industry's effort in environmental protection is summarized in Table I.

Background

The need for biodegradable detergent polymers in particular and all water-soluble polymers in general is a result of the difficulty in re-

Table I. Summary of Environmental Responsibility of the Detergent Industry

Period	*Action by Detergent Industry*
1960s	Branched-chain surfactants replaced because of nonbiodegradability
1970s	Eutrophication problem of polyphosphates recognized
1980s	Polyphosphates replaced by PCAs
1990s	PCAs reevaluated

Surface water/estuaries/oceans → Sediments
↗
Use → Disposal → Sewage treatment
↘
Sludge → Landfill/compost/incineration

Figure 1. Environmental exposure to water-soluble polymers.

covering them after use and the fact that they are disposed of either directly into the environment or into public and/or industrial sewage treatment plants. The polymers are almost always in very low concentration aqueous solutions (in the range of a few parts per million) in wastewater streams, and unless they are biodegradable or are trapped by adsorption on sewage sludge, they pass through treatment plants, permeate the aqueous and sedimentous environments, and slowly increase in concentration. Potential routes into the environment for a nonbiodegradable water-soluble polymer are shown schematically in Figure 1.

The key objectives in the design of biodegradable detergent PCAs, therefore, are to have them biodegrade in the sewage treatment plant and thus to have their environmental impact confined. In order to have a longer exposure to the sewage treatment plant and to increase the opportunity for biodegradation, the polymers should also be adsorbable on sewage sludge. This adsorbability will increase their residence time in the plant from a few hours (5–10 h) to several days (6–12 days).

Current Detergent Polymers

The two major polymer compositions currently used in detergent formulations are polyacrylic acid and copolyacrylic–maleic acids. The polyacrylic acids have molecular masses in the 4000–5000-Da range and are used in the United States and Europe, whereas the copolymers have molecular masses in the 70,000-Da range and are used exclusively in Europe. The need for two polymers of different compositions and molecular masses is attributable to differences in United States and European laundry conditions and water hardness levels. Both polymers are classed as cobuilders that to some degree control pH of the wash bath, detergency, dispersion of soil particles, and sequestration of hardness ions.

The wash performance of the acrylics is very sensitive to molecular weight, as indicated in unpublished work by Freeman (*1*) at Rohm and Haas. Detergency performance peaks at a molecular mass of about 5000 Da, and the whiteness index peaks closer to a molecular mass

Table II. Biodegradability and Removability of PCAs

Molecular Weight	Acrylic Acid		Acrylics and Maleics	
	% CO_2	% Removal	% CO_2	% Removal
1,000	45	45		
2,000	20	21		
4,500	9	40[a]		
10,000	16	58		
12,000			31	83
60,000		93		
70,000			20	95

[a]At polymer levels typically found in sewage treatment plants, this value is >80%.

of 10,000 Da. Since detergency, the removal of soil, is a more important function of the polymer than the whiteness index (dispersion of soil), the polymers in use in the United States are closer to 5000 Da in molecular mass. The performance of the copolymers peaks at a molecular mass of around 70,000 Da.

The fate of these polymers in wastewater streams and the environment after use in the laundry was recently reviewed by Freeman (2) and is only briefly touched on here. Table II contains a summary of the biodegradability (% CO_2) and sewage sludge adsorption (% Removal) data for a molecular mass range of the two different polymers labeled with ^{14}C. The concentration of the polymers in water for this study was 20 ppm. It is clear that the biodegradability of polyacrylic acid is inversely related to the molecular mass and that the adsorption on sewage sludge, or bioremovability, is directly related to the molecular mass of the polymer. The currently used product (molecular mass of 4500 Da) is not considered biodegradable but is largely bioremovable by adsorption, especially in the realistic concentration range of 1–2 ppm, and hence does not enter the greater aqueous environment. Instead, it remains on the sewage sludge for land filling, land application, or incineration.

Copolymers are highly adsorptive in the molecular mass ranges evaluated, and the level of biodegradation appears to be higher for them than for polyacrylic acid. However, copolymers of these two monomers have some low-molecular-mass fractions that are more susceptible to biodegradation than the major high-molecular-mass polymer component. The two presently used commercial polymers are therefore similar in that they are not biodegradable to any extent and are not released into the aqueous environment in large quantities. However, these qualities do not preclude the need for the development of fully biodegradable replacements. Such replacements are needed because some sludge is contaminated with polymer and because some waste releases do not go through the sewage treatment plants. Some appreciable quantity of nonadsorptive, nonbiodegradable polymer is also still entering the environment.

Requirements for Biodegradable PCAs

Having discussed the limitations of the current products, we can now define the performance, biodegradation, and bioremovability requirements for biodegradable detergent PCAs.

To be acceptable, performance must be at least equivalent to that of current products. The new product must provide equal performance at the same cost whether it is used at a higher or a lower level in the detergent formulation. This restriction, of course, limits the complexity of the chemistry and the raw material base that can be used to manufacture PCAs. The polymers must also be stable for a time (up to 1 h) in the basic laundry medium (pH 10), frequently at temperatures as high as 60 °C, and must consistently perform the expected functions.

The biodegradation requirement for detergent polymers is total biodegradation to gaseous products (CO_2 in aerobic environments and CO_2–CH_4 in anaerobic environments), minerals, and biomass such that no organic residue remains in the environment. The rate of biodegradation is expected to be faster than the rate of disposal into an environment, so that no buildup is observed. Preferably, the biodegradation rate is faster than the residence time in a sewage treatment plant, the most common disposal environment, thus ensuring no entry into the greater environment. In a sewage treatment facility, hydraulic turnover occurs in a few hours (5–10 h) and sludge turnover occurs within days (8–14 days). Therefore if a PCA is water-soluble and nonadsorbable on sewage solids, the biodegradation rate should be fast enough that biodegradation reaches completion in a few hours, whereas an adsorbable polymer will have several days to biodegrade and may even complete its biodegradation after removal from the treatment plant in, for example, land application or composting facilities. The advantage in PCAs being both adsorbable and biodegradable is therefore obviously in the control of their release into the environment.

For PCAs to be accepted as totally biodegradable as described above, they must meet severe accountability requirements and not just pass the standard tests, such as those for the American Society of Testing and Materials, the Organization for Economic Cooperation and Development, the Ministry of International Trade and Industry, the French Standards, the German Standards, and the International

Standards Organization, which are regulatory recommendations rather than true assessments of biodegradability. The stricter requirements for total biodegradation are best exemplified by considering the carbon balance equations for any organic PCA:

Aerobic biodegradation

$$C + O_2 \rightarrow CO_2 + C_{biomass}$$

Anaerobic biodegradation

$$C \rightarrow CO_2 + CH_4 + C_{biomass}$$

Total accountability of the carbon is essential in any biodegradation evaluation, regardless of the environmental exposure. For total, acceptable biodegradation, the carbon introduced into the environment, C, must be accounted for as gaseous carbon, that is, carbon dioxide and/or methane, and as carbon incorporated into biomass. No residue should be left. All the levels are readily measurable by a variety of techniques, and radiolabels are possible if it is desirable to work at the PCA environmental exposure levels usually observed (i.e., ppm level). Correctly carried out, the experiments will indicate the rate and extent of biodegradation, and these parameters bear on the rate of removal from the disposal environment, usually the sewage treatment plant. Adsorption on sludge can be studied by exposing the PCAs to sewage sludge, directly determining the amount adsorbed at different aqueous concentrations, and then extrapolating to the environmental concentration. Such studies with PCAs that biodegrade rapidly measure adsorbability and biodegradability in combination.

Approaches to Biodegradable PCAs

Two major approaches to the synthesis of PCAs are available and have been explored: (1) synthetic polymers are prepared from appropriate monomers, and (2) polymers are formed by the modification of renewable resources. Preparing polymers from monomers has received more attention up to this point, but renewable resource chemistry is fast becoming fashionable and, as we shall see, may be the best approach to obtaining fully biodegradable PCAs that is consistent with the goals of the industry.

Synthetic Polymers. Synthetic polymers are of two types: vinyl, or addition, polymers and condensation polymers. Because of their availability, familiarity, and ready polymerization, much effort has been expended in trying to utilize acrylic and maleic acid monomers to synthesize biodegradable detergent polymers. Vinyl polymerizations are therefore discussed first.

Vinyl Polymerizations. Because vinyl-polymerized carbon chain PCAs based on acrylic and maleic acids at a molecular mass that allows performance do not biodegrade, vinyl polymerization approaches attempt to take advantage of the biodegradability of their oligomers. In order to pursue these approaches, the definition of the highest-molecular-mass oligomer that will totally biodegrade must be established.

Since oligomers do not have useful properties in detergents, they must be chain extended to a useful molecular weight range consistent with detergent performance by using bonds susceptible to degradation by some mechanism (biodegradation, hydrolysis, etc.) after disposal of the polymer in a given environment, leaving a biodegradable short-chain oligomeric acid. This molecular mass building is represented here schematically, with X as the susceptible chemical link between oligomer units of carboxyl monomers.

------X------X------X------X------X------X------

Oligomer Biodegradation. Biodegradation of acrylic and maleic acid oligomers has received considerable attention in industrial and academic laboratories. A major problem has been the lack of consistency in measuring the extent of biodegradation. All too frequently, only imprecise measurements have been made. These measurements include biochemical oxygen demand (BOD), which is not a direct measurement of biodegradation (*3*) because it does not measure carbon removal or balance, and loss of oligomer content by gel permeation chromatography without establishing the oligomer concentration sensitivity range of the method. In spite of this lack, agreement that oligomers of acrylic acid are not biodegradable above a degree of polymerization (dp) of ca. 6–8 is general. In initial work by Lever Brothers (*4*), who used alcohol chain transfer agents to control molecular mass, substantial biodegradation (30–50%) occurred in BOD testing if molecular mass was less than ca. 1000 Da (dp ~ 15). These results indicate that a substantial portion of low-molecular-mass polymer or oligomer was present. Matsumura et al. (*5*) chemically synthesized several acrylic oligomers with different molecular masses and determined by BOD that oligomers with a dp of less than 7 were biodegradable. Subsequently, fractionated oligomers were studied by Larson et al. (*3*), scientists from Nippon Shokubai (*6*), and I. Kawai (*7, 8*). Careful biodegradation studies in all cases substantiated the earlier conclusion that only oligomers with a dp of less than ca. 8 are biodegradable.

Similar results with oligomeric maleic acids have been observed in our laboratories, but the dp for complete biodegradation appears to be lower than that for acrylics (that is, ca. 4) (9). This result appears to indicate that carboxylic acid density may play a role in inhibiting the biodegradation of PCAs, since the dp 4 maleic oligomers have the same number of acid groups as the dp 8 acrylic oligomers.

The oligomer results show that for any approaches based on this chemistry, precise molecular mass control must be maintained during synthesis, and this control is not easy to achieve, because most radical-promoted oligomerization processes give a molecular mass distribution and not a unimodal oligomer. In order to comply with the rule of dp 8, the molecular masses of the oligomers must be considerably below a dp of 8 to ensure that the molecular mass envelope does not exceed that upper acceptable limit.

Oligomer Chain Extensions. A number of patents have been obtained in the area of oligomer chain extensions, because many attempts have been made to control oligomer chain length in a polymer backbone that has built-in weak links. A patent granted to American Cyanamid (*10*) utilizes the copolymerization of ketene acetals with acrylic acid, a chemistry that is similar to that pioneered by the late W. J. Bailey (*11*) as a route to introducing biodegradable ester linkages into polyethylene and other polyolefins and rendering the products biodegradable. The chemistry is outlined in Scheme I.

The concept is good, but the weaknesses of the approach are that the ketene acetal is unstable in aqueous acidic solution and the copolymerization must be conducted by using the sodium salt of acrylic acid. In addition, the copolymerization parameters of the two monomers are not known, and control of oligomer chain length and the ester weak link–spacer is therefore unlikely. These problems may be the reasons that no biodegradation data are reported in the patent.

Deutsche Golde-und-Silber (*12*) free radical copolymerized acrylic acid with acrolein in the presence of *n*-dodecylmercaptan as the chain transfer agent to control molecular weight in the oligomer range. The polymerization was claimed to proceed through both the aldehydic and the vinyl groups of acrolein. This product was then

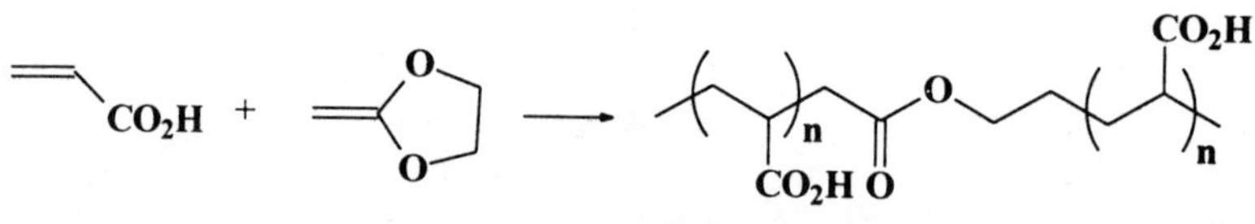

Scheme I.

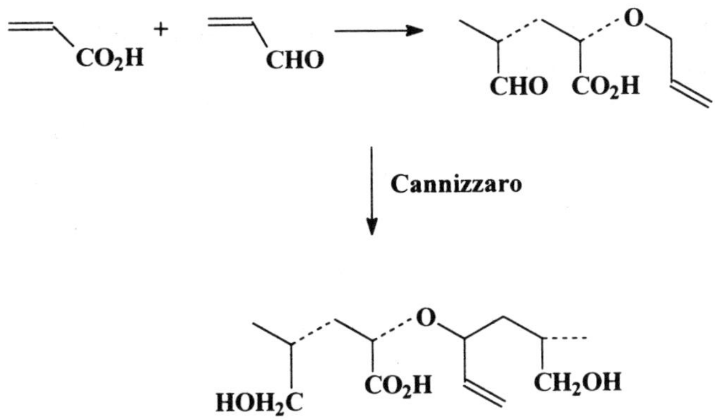

Scheme II.

further reacted in a Cannizzaro or a cross-Cannizzaro reaction with formaldehyde to produce oligomeric chains containing acrylic acid blocks with aldehydic and alcoholic functional groups to promote biodegradation, as shown in Scheme II.

The polymers are claimed on the basis of BOD determinations to be 50–60% biodegradable. However, no information on the level of low-molecular-mass materials present is given, and the true biodegradability of these oligomers cannot be assessed. Copolymerizing divinyl monomers with acrylic and maleic acids in ratios that give statistical oligomer chain lengths that are known to be biodegradable has been evaluated by Rohm and Haas (*13*) and BASF (*14*). The object is to have the divinyl monomer of sufficient chain length and lability that the branching links will be readily biodegradable and the whole network structure shown in Scheme III will collapse to oligomeric carboxylic acids that are themselves biodegradable. Unfortunately, control of the degree of branching, which is detrimental to biodegradation, and the oligomeric acid chain length is difficult, and at best only partial biodegradability is achieved. The polymers are adsorbable and bioremovable in sewage treatment facilities. They are thus equivalent to current products but offer no major advantages.

Two other attempts to utilize oligomers of acrylic acid are worth noting. A Rohm and Haas patent (*15*) reported that the free radical grafting of acrylic acid onto poly(ethylene glycol) produced partially biodegradable detergent polymers. The inability to control the acrylic oligomer side chain molecular mass was the limitation on the extent of biodegradation; complete biodegradation could not be achieved. Ecolab holds a patent (*16*) on a process in which molecular mass control by the use of a mercaptan chain transfer agent was attempted, and

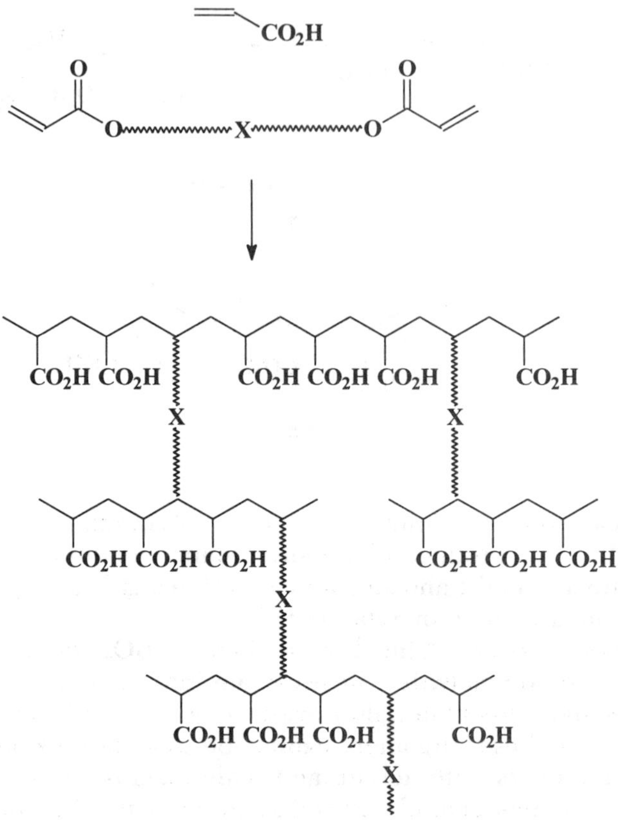

Scheme III. Branched oligomers.

the acrylic oligomer chain was attached through esterification of the poly(ethylene glycol) chain ends, in essence to produce a ladder polymer. Biodegradation was claimed, but no supporting evidence was given for biodegradation of the acrylic portion of the polymer. These two similar approaches are outlined in Schemes IV and V. These approaches, based on control of the molecular masses of free radical-initiated oligomerizations either in grafts or in block copolymerization, are technologically limited, because it is not possible to achieve the narrow molecular mass distribution of oligomers that is necessary for their total biodegradation.

Vinyl Polymerizations. Vinyl polymerization attempts to copolymerize the unsaturated acid monomer with another vinyl monomer so that the polymer chain produced will have weak links and a susceptibility to biodegradation. That poly(vinyl alcohol) is the only biode-

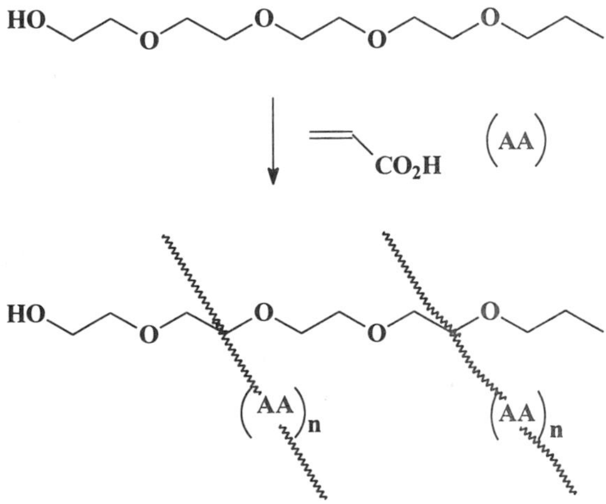

Scheme IV. Poly(ethylene glycol) graft chemistry.

gradable carbon chain polymer is consistent with the failure of this approach. Matsumura (*17*) has essentially copolymerized all available acrylic esters with acrylic acid to no avail; no composition with a molecular mass in the range useful in detergents has significant biodegradability. Even copolymers of vinyl alcohol fail to balance biodegradation and detergent properties. Polymers with high vinyl alcohol content and low acid monomer, acrylic, maleic, and fumaric contents are biodegradable but lack useful properties, and the reverse is true for polymers containing low levels of vinyl alcohol (*18, 19*).

A few years ago, work at Solvay et Cie (*20*) and Sandoz (*21*) indicated that poly(α-hydroxyacrylic acid) prepared by the hydrolysis of poly(α-chloroacrylic acid) was biodegradable and had good detergent properties. The reasoning was that the hydroxyl functionality provided a weak link at which biodegradation could be initiated. However, subsequent work indicated that the rapid carbon loss observed in testing, which was considered an indication of biodegradation, resulted in large part from adsorption onto sewage sludge and removal from the environment rather than true biodegradation.

More recently, successful conversion of poly(vinyl alcohol) into acid-containing polymers by reaction with maleic anhydride and sub-

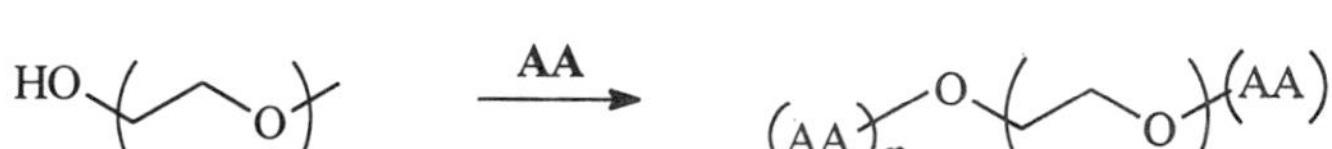

Scheme V. Poly(ethylene glycol)–oligomer esters.

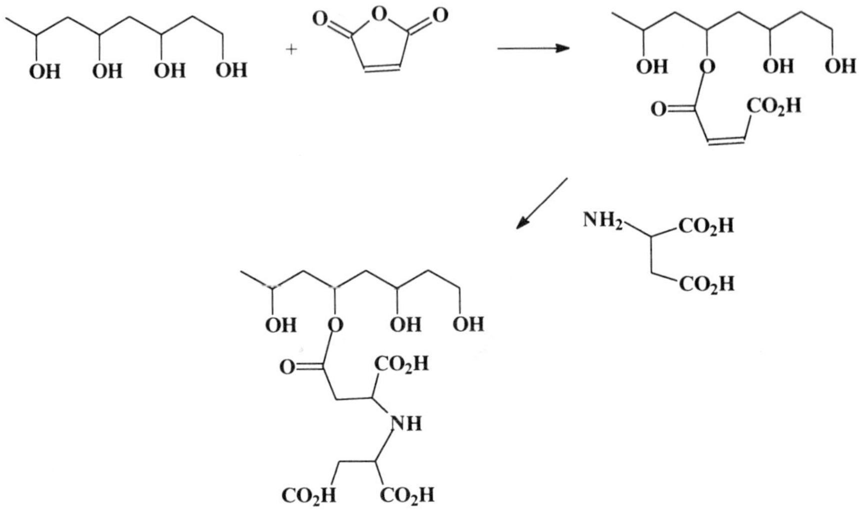

Scheme VI.

sequently aspartic acid through a Michael reaction of the amine functionality to the maleic ester double bond has been reported by Procter and Gamble (22). The products should be biodegradable, but no indication is given in the patent. The chemistry is illustrated in Scheme VI.

Matsumura et al. (23) reported homopolymers of vinyloxyacetic acid as biodegradable detergent polymers and gave evidence to substantiate their claims. The polymer is a derivative of poly(vinyl alcohol) (*see* structure **1**), which is postulated to be the intermediate in the biodegradation pathway formed through ether cleavage. The drawbacks to this elegant approach are the cost and availability of the monomer.

Condensation Polymers. Condensation polymers containing a carboxyl functionality are the second class of synthetic polymers that have been evaluated as detergent polymers. They are more difficult

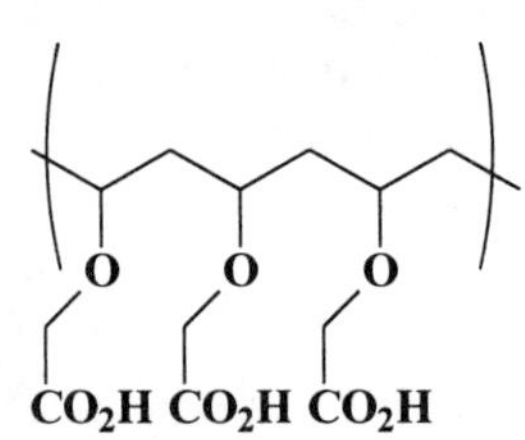

1

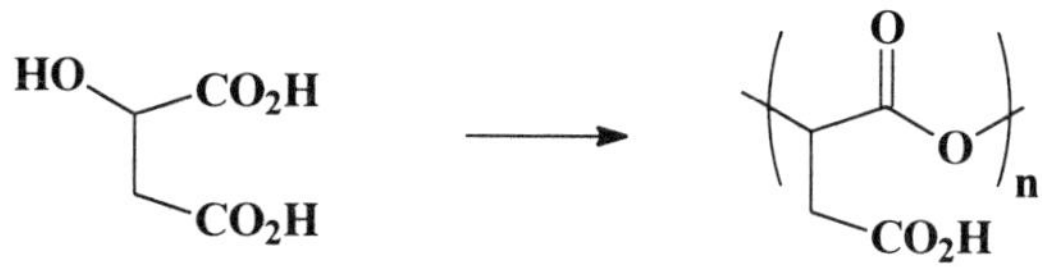

Scheme VII. Poly(α-malic) acid.

to synthesize than free-radical polymers containing a carboxyl functionality, but they are more readily controlled as far as functionality placement and molecular mass are concerned. Also, because they contain chemical bonds that are found in nature, they are expected to be more easily biodegradable. They have the general structure –R–X–R–X–R–X–R–X–R–, where X is the condensation linkage that includes ester [–C(=O)–O–], amide [–C(=O)–NH–], ether [–O–], and acetal–ketal [O–C–O], and R is an alkyl group from C_1 containing a carboxyl functionality.

Polyesters. Polyesters with a carboxyl functionality have only recently been evaluated in detergents; a recent patent granted to BASF (*24*) suggests their potential. The polyesters are based on citric acid, tartaric acid, 1,2,3,4-butanetetracarboxylic acid, and suitable polyols including ethylene glycol and higher homologs. The patent contained no biodegradation data beyond bioremovability on sewage sludge. The polyesters appear to have laundry performance, but because the butane polyacid is required to boost acidity, the cost of the polymers is probably unrealistic for commercial use.

Earlier, Lenz and Vert (*25*) prepared poly(α-malic acid), and Matsumura et al. (*26*) prepared the random poly(α/β-malic acid). Matsumura demonstrated that his polymers had poor laundry performance and rapid biodegradability. Both polymer types are very unstable and hydrolyze rapidly in aqueous basic solution (Schemes VII and VIII). They are therefore unlikely to be useful in the highly alkaline detergent bath.

Polyamides. A plethora of recent patents has been based on poly(amino acids). These acids are predominantly based on poly(aspartic acid) and poly(glutamic acid) (**2** and **3**). This activity probably stems in large part from the observation by Sikes and Wheeler (*27*)

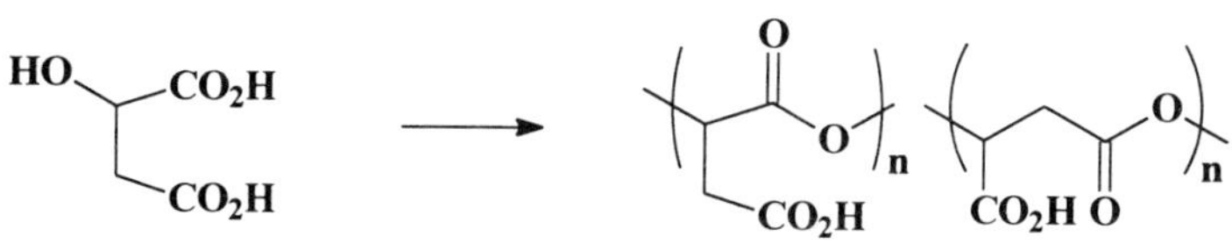

Scheme VIII. Poly(α/β-malic) acid.

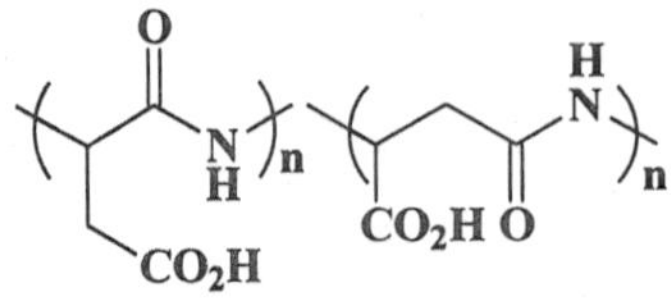

2, poly(aspartic acid)

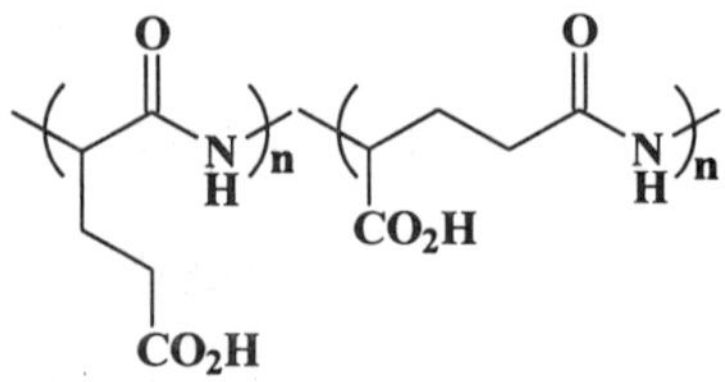

3, poly(glutamic acid)

that mixed poly(amino acids) isolated from crustaceans are good at mineral dispersal, which is a key property for detergent polymers. Companies active in this research include Donlar (*28*), Cygnus (*29*), Rhone-Poulenc (*30*), and Enichem (*31*), which focus on poly(aspartic acid), and Procter and Gamble (*32*), which has an interest in poly(glutamic acid). The poly(aspartic acids) are prepared by two different methods: the condensation polymerization of L-aspartic acid and the thermal addition–condensation polymerization of ammonia and maleic anhydride–maleic acid. The molecular masses of the products are different: masses of polymers from L-aspartic acid range from 1,000 to 70,000 Da, whereas the maleic process gives polymers with molecular masses ranging from 2000 to 4000 Da. Results to date indicate the incomplete biodegradation of polymers from either process; an unidentified residue is always present (*33, 34*) at the end of testing. Laundry performance appears to be satisfactory.

Polyacetals and Polyketals. Polyacetals (**4**) and polyketals containing acid functionality represent a very elegant approach to biodegradable detergent polymers. Monsanto scientists (*35*) developed the salts of polyacetal and polyketal carboxylates as a combination hydrolysis and biodegradation approach to biodegradable detergent polymers. Esters of glyoxylic acid and pyruvic acid are readily polymerized anionically through their carbonyl functionality, and the products are stable in basic media; after the esters have been used as detergent polymers at high pH, the pH falls rapidly as they are discarded into a wastewater stream, and hydrolysis to their monomeric salts is rapid. Biodegradation of the monomers is rapid and complete. The polymers have never been used commercially because the cost per performance is too high. However, they have been the wellspring of several similar approaches, all based on the following general structure:

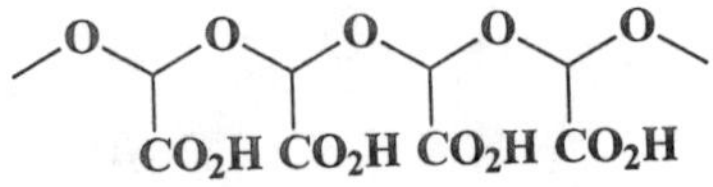

4, polyacetals

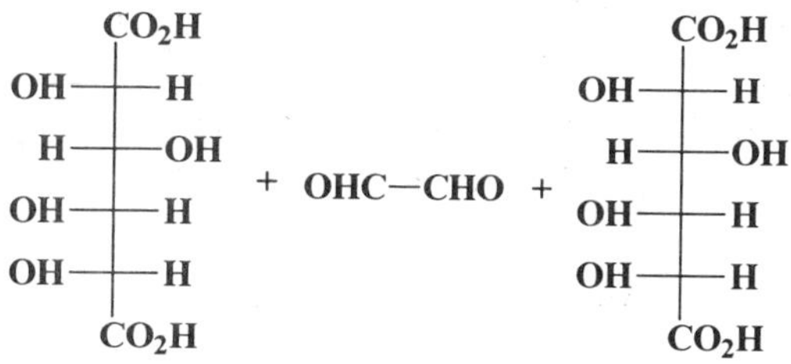

5, polyethercarboxylic acids

BASF recently extended this ketal–acetal chemistry and patented two variants on the Monsanto theme. The first variant is based on using glutaconic acid and dialdehydes such as glyoxal and glutaraldehyde to form acetals (*36*) (**5**); the other variant uses long-chain keto acids and polyfunctional alcohols to form ketals (*37*).

The biodegradability of poly(ethylene glycol) is well established (*37*), and several attempts to prepare carboxylated polyethers as biodegradable detergent polymers have been made. The initial reports on this approach were from Monsanto. Crutchfield et al. (*38*) evaluated the anionic polymerization of epoxysuccinic acid. Matsumura et al. (*39*) thoroughly researched this chemistry, establishing the molecular weight and structural relationships with biodegradation. They prepared and cationically polymerized several epoxy esters (represented by the general structures in Scheme IX) to molecular masses in the range of a few hundred to several thousand daltons. Not unexpectedly, after hydrolysis, bulky substituents were detrimental to biodegradation and molecular mass was also important.

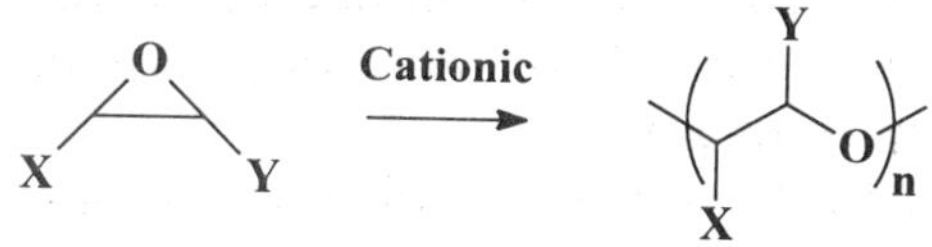

Scheme IX.

where X and Y are both carboxyl groups or one is an alkyl group.

Renewable Resource Chemistry: Natural Polymers. Renewable resource chemistry or chemistry based on the use of natural polymers will become increasingly important over the next few years, particularly in the area of biodegradable polymers. The advantages are that the polymers are generally accepted as biodegradable or at least not harmful to the environment. They are available, and they are in some cases, for example, starch, very inexpensive. Their disadvan-

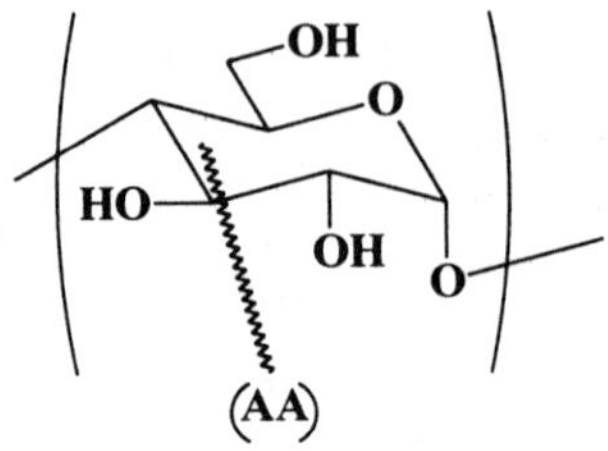

6

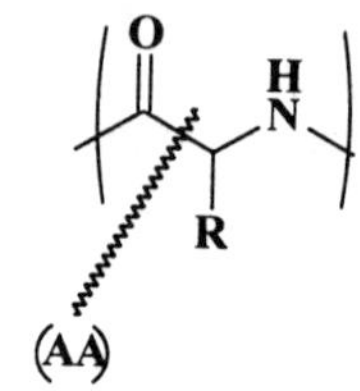

7

tages are that (1) in no case is a natural polymer usable as isolated in the detergent applications under consideration; (2) introducing the carboxyl functionality is often not easy and is a modification that can add cost; and (3) the effects on biodegradation may be negative, depending on the level of modification required. The three approaches that have received the most attention are grafting vinyl monomeric carboxylic acids onto suitable substrates; oxidation of polysaccharides, particularly starch, because of its availability and low cost; and chemical derivatization, for example, by monoacylation with polycarboxylic acids. These approaches are all briefly discussed in this section.

Graft Chemistry. Several attempts (*40–44*) at free-radical grafting of acrylic acid onto starch have been made. The major problem encountered in every case was the lack of control of the molecular mass of the acrylic polymer. As previously stated, such control is very difficult to achieve by free-radical polymerization. Attaching the "right" number of oligomer chains to the substrate is also essential to ensure that the graft will be active in detergency and will retain its biodegradability. Not surprisingly, structures **6** and **7** have not met these stringent requirements.

BASF (*45*) substituted proteins such as casein and soy proteins for starch but also were able to demonstrate biodegradability only for the natural substrate. The problem that is seen in starch grafts, that is, molecular mass control of the acrylic polymerization, also pertains here.

Starch Oxidation Chemistry. The inexpensive nature and the biodegradability of starch and other polysaccharides have attracted much activity in the area of oxidation in an effort to introduce carboxyl functionality at either the 6 position from the primary hydroxyl or at the 2 and 3 positions by ring cleavage. These oxidations have been achieved predominantly by chemical means, with periodate–chlorite

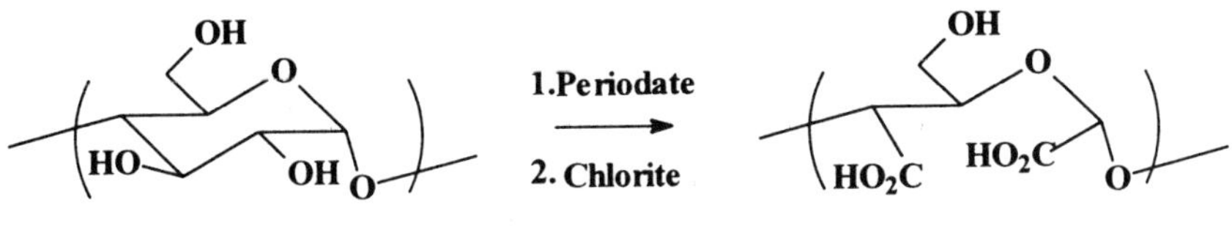

Scheme X.

and hypohalite particularly favored (Scheme X), and only recently has evaluation of catalytic oxidation begun.

Chemical oxidation has been reported by several academicians (*46–49*) and industrial laboratories (*50–52*). No PCA that combines the properties required for detergent performance and biodegradation has yet been produced. Matsumura (*48*) has the most balanced perspective on this aspect of the oxidized polysaccharides. He showed that as the carboxyl content increases in the polysaccharide, biodegradation decreases and detergent performance increases. Whether the two effects can ever be balanced to produce a biodegradable detergent polymer is not clear. From the commercial standpoint, other potential problems with the chemical oxidation of starches are that the reactions are done in very dilute solution and that large quantities of inorganic salt byproducts are generated. An obvious solution to the production of inorganic salts is to develop catalytic processes, and this effort is now receiving attention.

Catalytic processes using noble metal catalysts and air (Scheme XI) appear to be promising, but little control over the rates and degree of oxidation has been achieved. However, it is becoming clear that such processes are an excellent way to develop economically feasible commercial processes, and recent patent activity underscores this possibility (*53, 54*).

Chemical Derivatization. From the very early days of the search for biodegradable detergent polymers, the monoacylation of polysaccharides with polyfunctional carboxylic acids, especially as the anhydrides of maleic and succinic acids, has been a favored route. The acylation reaction is very well understood, and it seems to be an easy and commercially attractive process. However, the difficulties are that

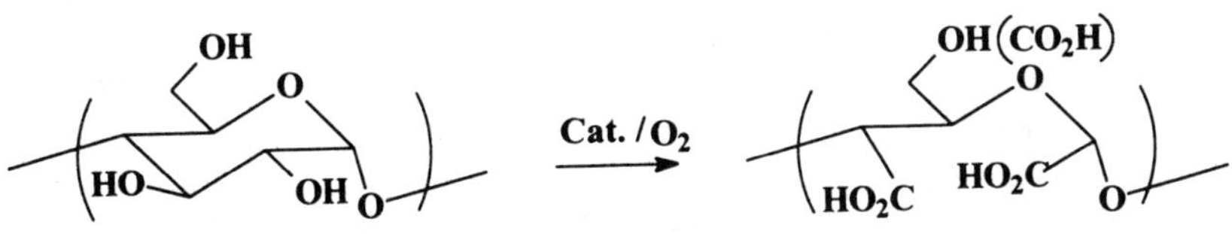

Scheme XI.

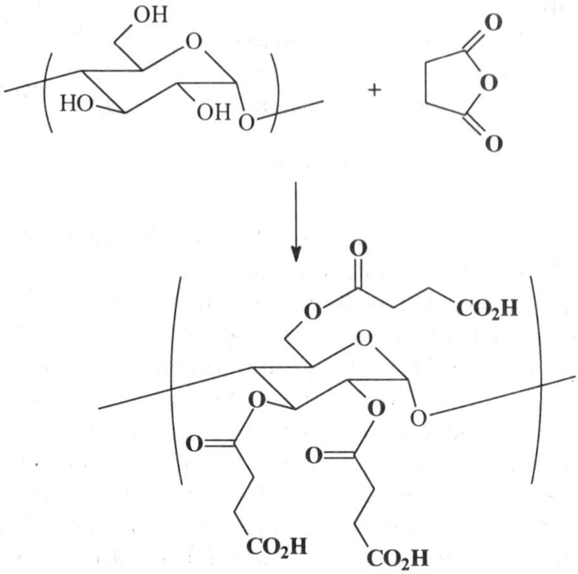

Scheme XII. Acylation of starch.

starch, for example, is not readily soluble in organic solvents and that in the neat reaction, starch has some tendency to cross-link through esterification of two carboxyl functionalities in the same monomer, especially as the degree of substitution increases to the level of carboxyl that is needed for performance. The approaches patented by FMC (55) and Procter and Gamble (56) and shown schematically in Scheme XII do not indicate that the products are fully biodegradable.

Conclusions

The need for biodegradable detergent polymers is very real, and this need is supported not only by regulatory expectations but by the detergent industry itself as a self-policing body that has always been concerned about and responsive to environmental pollution. Because of this concern for the environment, the industry expects a totally biodegradable polymer that meets exacting standards of degradation achievable in a short time while maintaining a cost-per-performance balance consistent with that of current products. The number of approaches that have been tried is very large, and no polymer to date meets the industry requirements.

Free-radical polymerized vinyl carboxylates are unlikely to meet the goal because of their molecular mass limitation for biodegradability: performance requires a higher molecular mass, and biodegradation occurs only at a low molecular mass. Combinations of renewable resources and vinyl monomers also seem to be precluded because of the same strict molecular mass requirements of the graft arms from the vinyl monomer.

Current polymers based on acrylic acid and maleic acid, being bioremovable in wastewater treatment plants, are a better interim solution to the potential environmental problems than are some of the indicted partially biodegradable polymers. Switching to the partially biodegradable polymers would introduce uncertainty about the residues that would require extensive and expensive fate and effects evaluation.

Future Directions

As the foregoing discussion makes clear, the most reasonable approaches to totally biodegradable water-soluble polymers for detergents are modification of renewable resources such as starch to balance cost per performance and development of condensation polymers with carboxyl functionality. In order to be acceptable, such polymers must be totally biodegradable, as described in this chapter, that is, from the standpoint of accountability in test protocols that are accepted widely, and not merely able to pass some of the current test protocols, such as that of the Organization for Economic Cooperation and Development, that were developed for legislative rather than scientific purposes.

References

1. Freeman, M. B. Presented at the 80th American Oil Chemists Meeting, Cincinnati, OH, May 1994.
2. Freeman, M. B. *Environ. Technol.* **1993,** *14,* 101–112.
3. Larson, R. J.; Swift, G.; Williams, R. *Polym. Mater. Sci. Eng.* Washington, Fall 1992.
4. Lever Brothers. U.S. Patents 3 922 230, 4 095 035, and 4 132 735.
5. Matsumura, S.; Maeda, S.; Takahashi, J.; Yoshikowa, S. *Kobunshi Ronbunshu* **1988,** *45*(4), 317–324.
6. Hayashi, T.; Mukou, M.; Tani, Y. *Appl. Environ. Microbiol.* **1993,** 59, 1555–1559.
7. Kawai, F., Kobe Commercial University, Kobe, Japan, personal communication.
8. Kawai, F. *Appl. Microbiol. Biotechnol.* **1993,** *39,* 382–385.
9. Swift, G.; Yocom, K., Rohm and Haas, unpublished work.
10. American Cyanamid. U.S. Patent 4 923 941, 1990.
11. Bailey, W.; et al. *Contemp. Top. Polym. Sci.* **1979,** *3,* 29.
12. Deutsche Gold-und-Silber. U.S. Patents 3 896 086 and 3 923 742.
13. Rohm and Haas. Eur. Patent 430 574.

14. BASF. DE 3 818 426.
15. Rohm and Haas. U.S. Patent pending.
16. Ecolab. U.S. Patent 4 874 540.
17. Matsumura, S.; et al. *Yukagaku* **1986,** *35,* 167–175 and 937–944.
18. Matsumura, S.; et al. *J. Am. Oil Chem. Soc.* **1993,** *70*(7), 659–665.
19. Rohm and Haas. U.S. Patent 5 191 048.
20. Solvay et Cie. U.S. Patents 4 107 411 and 4 182 806.
21. Sandoz. DE 37 111 304.
22. Procter and Gamble. U.S. Patent 5 221 711.
23. Matsumura, S.; et al. *Macromol. Chem. Rapid Commun.* **1988,** *9,* 1–5.
24. BASF. U.S. Patent 5 217 642.
25. Lenz, R. W.; Vert, M. *Polym. Prepr. (Am. Chem. Soc., Div. Polym. Chem.)* **1972,** *20,* 608.
26. Matsumura, S.; et al. *Yukagaku* **1986,** *35,* 937–944.
27. Sikes and Wheeler to U.S. Alabama. U.S. Patents 4 534 881, 4 866 161, and 50 511 401.
28. Donlar Corporation. U.S. Patents 5 057 597, 5 116 513, and 5 221 733.
29. Cygnus. U.S. Patent 5 219 986.
30. Rhone-Poulenc. Eur. Patent 511 037.
31. Enichem. Eur. Patent 454 126.
32. Procter and Gamble. WO 9 306 202.
33. Pettigrew, C.; Wheeler, A. P. *Abstracts,* Bio/Environmental Polymer Society Annual Meeting, Chicago, IL; August 1993.
34. Lowe, K. C.; Koskan, L. *Polym. Mater. Sci. Eng.* **1993.**
35. Monsanto. U.S. Patents 4 146 495, 4 114 226, 4 887 033, etc. (acetals) and 4 146 934 (ketals).
36. BASF. Eur. Patents 280 223-A1 (glutaconic acid) and DE 4 106 354-A1 (keto acids).
37. Kawai, F. *Crit. Rev. Biotechnol.* **1987,** *6*(3), 273–307.
38. Crutchfield, M. M. *J. Am. Oil Chem. Soc.* **1978,** *55,* 58.
39. Matsumura, S.; et al. *Yukagaku* **1987,** *36*(110), 874–881.
40. Sanyo, J. P. U.S. Patents 6 131 497 and 6 131 498, 1986.
41. Grillo Werke. U.S. Patent 496 329 and DE 3 834 237, 1988.
42. Mitsubishi Petrochemicals. Jpn. Patent 4 028 709.
43. Rhone-Poulenc. Eur. Patents 465 286 and 465 287.
44. Taechung Moolsan. WO 9 302 118.
45. BASF. DE 4 016 002.
46. van Bekkum, H.; et al. *Prog. Biotechnol.* **1987,** *3,* 157–161.
47. van Bekkum, H.; et al. *Starch* **1985,** *37,* 192.
48. Matsumura, S. *Angew. Macromol. Chem.* **1993,** *205,* 17–29.
49. Besemer, A. Ph.D. Thesis, University of Delft, 1993.
50. Procter and Gamble. Eur. Patent 542 496.
51. Ferruzi. Eur. Patent 472 042.
52. Zielke, R. *Tenside Deterg.* **1977,** *14*(5), 250–256.
53. Novamont. WO 9 218 542 and Eur. Patent 385 252.
54. Roquette Freres. Eur. Patents 511 081 and 455 522 and U.S. Patent 4 985 553.
55. FMC Corporation. U.S. Patents 3 941 771 and 4 029 590.
56. Procter and Gamble. U.S. Patent 3 919 107.

RECEIVED for review November 18, 1993. ACCEPTED revised manuscript September 20, 1994.

6

Commercial Poly(aspartic acid) and Its Uses

Kim C. Low,[1] A. P. Wheeler,[2] and Larry P. Koskan[1]

[1]Donlar Corporation, 6502 South Archer Avenue, Bedford Park, IL 60501
[2]Department of Biological Sciences, Clemson University, Clemson, SC 29634

The demand for biodegradable chemicals is increasing as chemical producers and consumers anticipate more stringent standards for degradability in new and existing products. Poly(aspartic acid) (PAA) is an example of a chemical that shows promise as a biodegradable replacement for polyacrylic acid, a nonbiodegradable polymer that is used in a wide variety of products and applications. In order to meet the potential demand for PAA, we have developed processes for commercial production. These processes utilize solid-phase condensation polymerization of aspartic acid to polysuccinimide in a commercial reactor. The polysuccinimide is then base hydrolyzed to PAA. The performance of PAA made by these methods is comparable and in some cases superior to that of polyacrylic acid in many mineral scale inhibition and particulate dispersion tests. Biodegradation studies of PAA show that it is significantly biodegradable. Preliminary toxicity studies of PAA indicate that it is nontoxic and environmentally compatible.

ANIONIC POLYMERS are used widely as mineral scale inhibitors in water treatment applications and as dispersing agents in detergents, paints, and papermaking processes. An example of such a polymer is polyacrylic acid (PAC), which accounts for about 20% of the 3.3 billion pounds of global acrylic acid demand per year, and the polymer demand is growing by as much as 10% per year (*1*). Although PAC is

0065–2393/96/0248–0099$12.00/0

generally considered nontoxic (2), it is also, at the molecular weights of the polymer usually employed, essentially nonbiodegradable (3). Therefore developing an anionic polymer that is both nontoxic and biodegradable in the environment into which it is released is desirable.

The efficacy of poly(aspartic acid) (PAA), an anionic polypeptide, as an antiscalant and a dispersing agent has been documented (*4, 5*). In many of the earlier studies, automated solid-phase peptide synthesis was used to make the polymer. Such a methodology would be prohibitively expensive on a large scale. Moreover, it uses relatively toxic reagents, for which disposal is problematic.

An alternative method of synthesis of PAA is described by Boehmke (*6*). This process involves mixing maleic acid or maleic anhydride with aqueous ammonia and then heating the mixture for 6 to 8 h, during which period water is removed by distillation. This method of manufacturing PAA is cumbersome and energy intensive and gives low yields (*7*).

Small-scale synthesis of PAA from L-aspartic acid by means of thermal condensation is well documented (*8, 9*). In the first step, crystalline aspartic acid is thermally polymerized (usually in laboratory glassware) to polysuccinimide (Figure 1). This process is attractive because it is inherently simple and clean: It uses only L-aspartic acid, and the only apparent byproduct is water of condensation. Subsequently, the polysuccinimide is hydrolyzed with a stoichiometric quantity of a base such as sodium hydroxide to form the active polymer (Figure 2). Since the ring of the polysuccinimide can be opened at two different sites, a copolymer with amide bonds formed from either the α or the β carboxyl groups is the result. The high efficacy of the thermally condensed PAA as an antiscalant and a dispersing agent was recently described (*10, 11*). However, the syntheses described to date were usually small scale and, if thermal exchange conditions were not properly contained, resulted in low yields, typically in the range of 50 to 68% (*12*). In order to meet the challenges of producing PAA on a larger

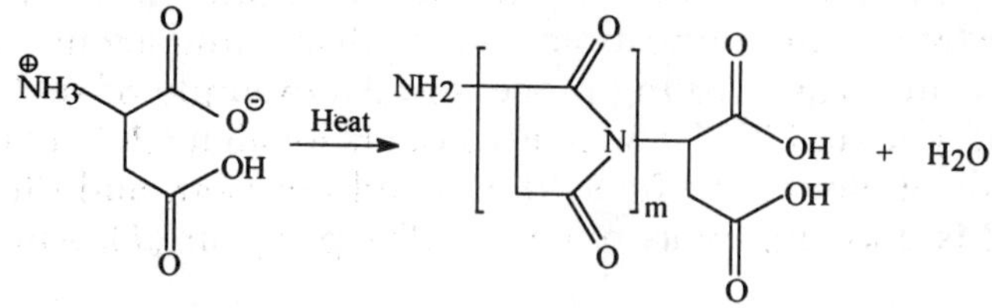

Figure 1. Thermal polycondensation of aspartic acid to polysuccinimide.

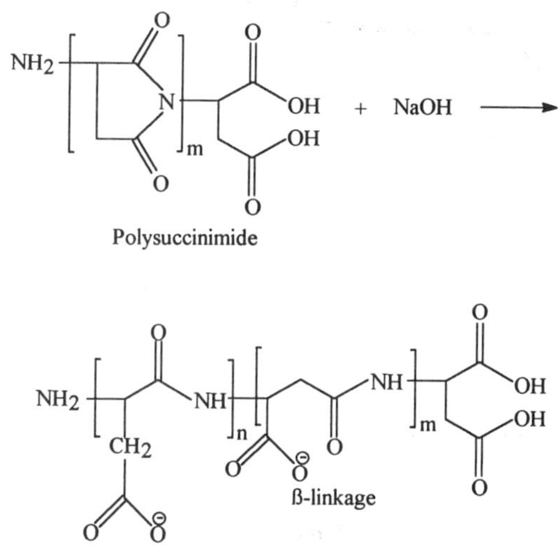

Figure 2. Base hydrolysis of polysuccinimide to PAA.

scale with high yields, we developed a method to produce polysuccinimide by using a commercial reactor (*13*).

Commercial Production of PAA

Solid-Phase Thermal Polymerization. A pilot-scale reaction of PAA was designed to incorporate a relatively high heat transfer rate and mechanical agitation to help drive off water of condensation. These conditions were adopted to achieve a more rapid and complete polymerization. In order to meet these conditions, a commercial mixer–dryer was employed (DVT-130, Littleford Brothers, Inc., Florence, KY). The jacketed dryer utilizes a heat transfer fluid (oil) and a plough blade-type impeller. It has a stack open to the atmosphere and heat transfer surface area of 0.929 m^2. Reactions were performed over a range of temperatures. In a typical run, the reactor was preheated to a jacket temperature of 260 °C (500 °F), which resulted in a product temperature of approximately 227 °C (440 °F). Product and oil temperatures were monitored by thermocouples, and the percent conversion was determined by weighing unreacted aspartic acid following extraction of the succinimide in dimethylformamide. Figure 3 depicts the relationship of the product and oil temperatures. The curves were characterized by an initial rapid rise in product temperature followed by an endotherm signaling the beginning of the reaction. During this

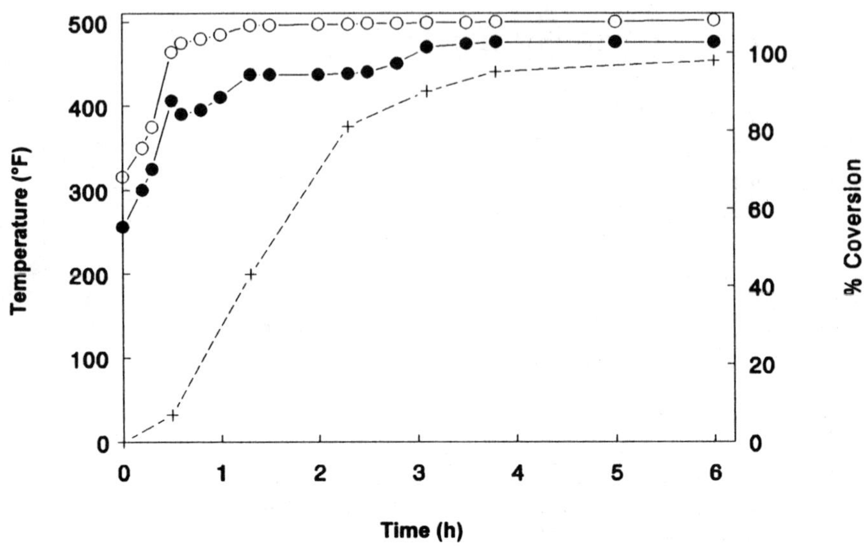

Figure 3. Temperature profile of a commercial production of polysuccinimide. The reactor oil was preheated to 260 °C. The reactor was charged with 110.4 lb (ca. 50 kg) of L-aspartic acid (Monsanto Chemical Company). The impeller speed was 155 rpm. Key: ●, *product* T; ○, *oil* T; +, *percent conversion.*

period, the product temperature dropped as steam resulting from a condensation process began to evolve. The conversion at this juncture was typically about 2–5%. Then the product temperature began to rise again steadily until it reached another plateau; conversion at this point was about 40%. During the next 1.5 h, steam evolved, and the conversion increased in a more or less linear fashion. Finally, thermal equilibrium was established. Immediately after this point, steam ceased to evolve. Typically, conversion at the end of the reaction was more than 97%.

The color of the product varied as the reaction proceeded and, compared to other analytical methods, was a good qualitative estimate of the percent conversion and thus the progress of the reaction. The color of the product changed from white to light pink, pink, tannish pink, light yellow, and yellow as the reaction progressed.

Hydrolysis. Base hydrolysis of polysuccinimide to PAA was based on the method described by Kovacs et al. (*14*). A stoichiometric quantity of base is added to hydrolyze the polysuccinimide, and the final pH is about 9.5 (*15*). Bases other than sodium hydroxide can be used depending on the final salt form desired. Suitable bases include

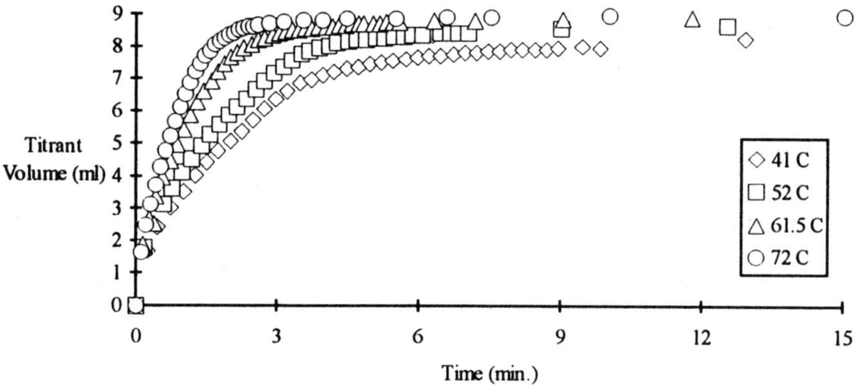

Figure 4. Base consumption in the hydrolysis of polysuccinimide as a function of hydrolysis temperatures. Polysuccinimide (2.3 g) was slurried in 5 mL of distilled water in a temperature-controlled vessel and hydrolyzed by autotitration of 2.5 M NaOH. The pH set point was 9.5. The stirring speed was 1200 rpm. (Reproduced with permission from reference 16.)

potassium hydroxide and other alkaline earth hydroxides or carbonates.

The rate of hydrolysis of polysuccinimide increases with the concentration of hydroxide ion (*16, 17*) and roughly doubles with every increase in temperature of 10 °C between approximately 40 and 70 °C, as shown in Figure 4 (*16*). In that the hydrolysis reaction is exothermic, the reaction can be run at temperatures significantly above ambient temperature with little if any energy input. Even at temperatures of approximately 70 °C and with a pH in the range of 10–11, no hydrolysis of the backbone was observed. However, at pH 12–13 and 80 °C, noticeable molecular weight changes in the final product were observed in 1–2 h.

Characterization of Synthetic PAA

Carbon-13 NMR of commercially produced PAA revealed that about 70% of the amide bonds are formed by the β-carboxyl groups. This percentage is in agreement with the findings for laboratory-synthesized PAA (*18*). Polarimetry analyses show that PAA is composed of a racemic mixture of D- and L-aspartic acid, although optically pure L-aspartic acid is used in the reaction. This result agrees with that of Kokufuta et al. (*9*). The weight average and number average molecular weights of PAA are about 5200 and 1800, respectively, by gel permeation chromatography with standards of PAC obtained from Poly-

Table I. Characteristics of Commercial PAA

Characteristic	*Result*	*Method*[a]
α/β configuration	0.3/0.7	^{13}C NMR[b]
D-/L-aspartic acid	1.0, no rotation	acid hydrolysis and polarimetry
Molecular weight	M_w, 5200; M_n, 1800	size exclusion HPLC
Polydispersity	2.9	M_w/M_n
Residue composition	>99% Asp	amino acid analysis
Primary structure	peptide backbone	FTIR
Yield (% conversion)	>97%	perchloric acid titration

[a]Abbreviations: HPLC, high-pressure liquid chromatography; FTIR, Fourier transform infrared spectroscopy.
[b]Performed as described by Pivocova et al. (*20*).

sciences Inc. Of particular interest is the fact that the polymer molecular weight did not change with either the reaction temperature or the time and was consistent throughout the reaction. The entire range of molecular weights of the polymer was 4500 to 5500 in all syntheses and did not vary in a systematic way with either temperature or time of synthesis. The composition of PAA as determined by amino acid analysis following acid hydrolysis showed no other residues and confirmed the percent conversion. Structural features of PAA were characterized by Fourier transform infrared spectroscopy. The spectra showed typical amide peaks and were nearly identical to those previously published for fully ionized PAA (*19*). The yield, defined as the percent conversion of aspartic acid to polysuccinimide, was determined by perchloric acid titration of the polysuccinimide for free amine. These data agreed with those obtained by dimethylformamide extraction and potentiometric titration for amine in PAA solution. The results of these characterizations are summarized in Table I.

Antiscalant Activity of PAA

An antiscalant is any compound that prevents the formation of insoluble minerals. Unchecked formation of insoluble minerals such as calcium carbonate, calcium sulfate, and calcium phosphate can be problematic in a variety of industrial applications (*21*). The inhibitory activity of anionic polymers is due to their high affinity for crystal surfaces, especially growth sites (*22, 23*), or for crystal nuclei. Once adsorbed, the polymers prevent growth of the mineral. In addition, they add an additional electrostatic charge that disperses the existing crystals, thus reducing the potential of the crystals for aggregating onto surfaces (*24*).

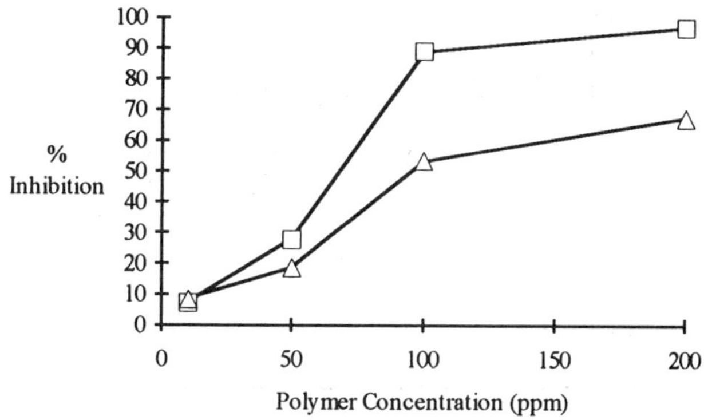

Figure 5. Calcium sulfate inhibition test. All assays were performed for 5 h at 70 °C following the mixing of 0.5 M $CaCl_2 \cdot 2H_2O$ and 0.5 M Na_2SO_4 solutions. The percent inhibition was determined from the concentration of free Ca^{2+} remaining in the supernatant or from the amount of calcium sulfate formed gravimetrically compared to that in samples without inhibitor. The concentration of polymer is that of the total solid bases. PAA was Donlar polyaspartate, produced using the pilot-scale method described in this chapter. PAC was Acrysol LMW-45 (Rohm and Haas). Key: □, PAA; △, PAC.

The activity of PAA as a mineralization inhibitor has been tested in a wide variety of assays and compared to that of PAC (Rohm and Haas 4500) (*25, 26*). For example, Figure 5 indicates that PAA is superior to PAC in inhibiting calcium sulfate precipitation. For calcium phosphate inhibition (Figure 6), the performance of PAA is comparable to that of PAC as determined by the length of the lag phase for the conversion of amorphous calcium phosphate to hydroxyapatite. The two polymers have comparable calcium carbonate inhibitory activities because they require similar volumes of 0.1 N sodium hydroxide to induce the sudden precipitation of calcium carbonate from calcium bicarbonate solution (Figure 7).

Dispersion Activity of PAA

Dispersion activity of anionic polymers is due to their adsorption onto individual particle surfaces; sufficient electrostatic charge is created to cause the particles to repel each other. This repulsion prevents particulates from forming larger aggregates, and as previously mentioned, it is also one of the mechanisms involved in their antiscaling activity (*27*). PAA can also be used for keeping particulate contaminants such as iron oxide dispersed in cooling water and boiler systems. Polyanionic dispersants are also used as detergent additives because

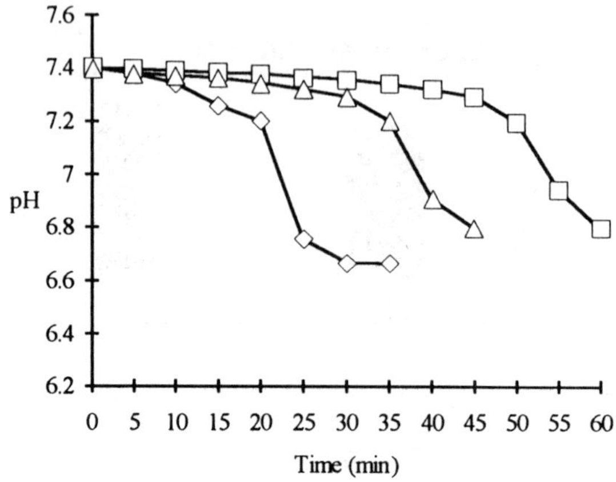

Figure 6. Calcium phosphate inhibition. All assays were performed at 20 °C at an inhibitor concentration of 30 ppm. The assays were initiated by adding 0.1 mL of 1.32 M $CaCl_2 \cdot 2H_2O$ and 0.90 M NaH_2PO_4 to 29.8 mL of water. The solution was quickly adjusted to pH 7.40 with microliter quantities of 0.1 N NaOH. The conversion of amorphous calcium phosphate to hydroxyapatite is indicated by the sudden drop in pH. The concentration of polymer is that of the total solid bases. PAA was Donlar polyaspartate, produced using the pilot-scale method described in this chapter. PAC was Acrysol LMW-45 (Rohm and Haas). Key: ◇, control; □, PAA; △, PAC.

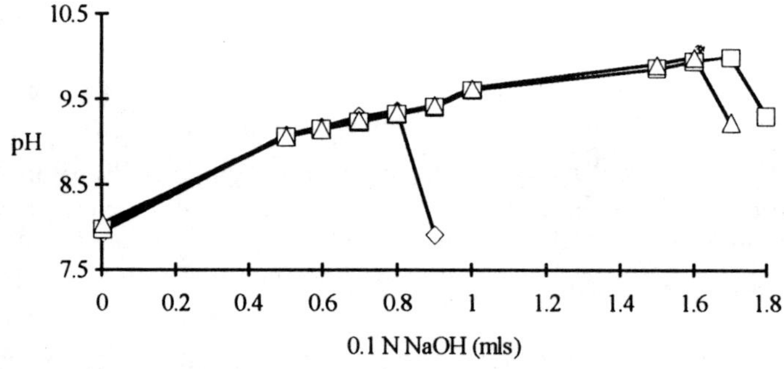

Figure 7. Calcium carbonate inhibition test. Assays were performed at 20 °C, and 48.5 mL of water, 0.3 mL of 1.0 M $CaCl_2 \cdot 2H_2O$, 2.5-ppm inhibitor, and 0.6 mL of 0.4 M $NaHCO_3$ were added in that order. The solution was titrated with 0.1 N NaOH until $CaCO_3$ precipitation occurred, as indicated by the sudden pH drop. The concentration of polymer was that of the total solid bases. PAA was Donlar polyaspartate, produced using the pilot-scale method described in this chapter. PAC was Acrysol LMW-45 (Rohm and Haas). Key: ◇, control; □, PAA; △, PAC.

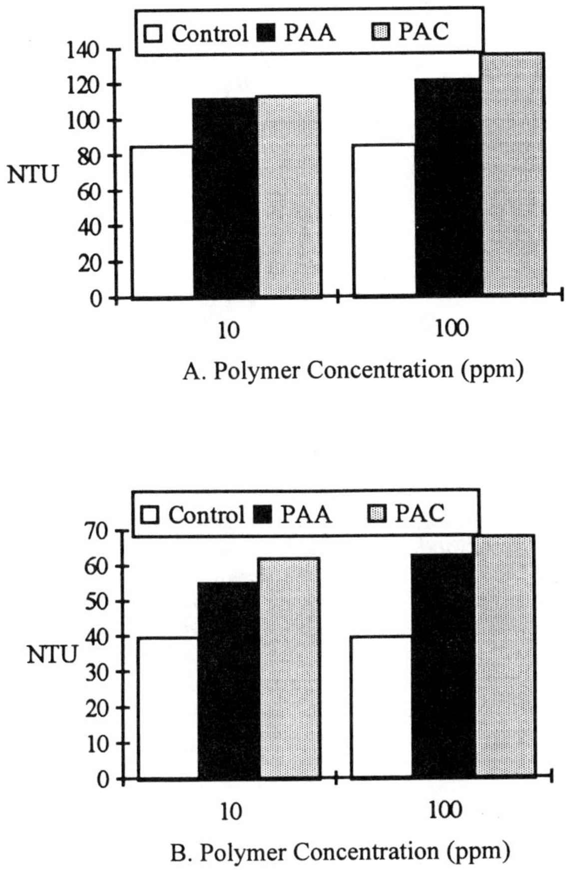

Figure 8. Soil dispersion after 2 h (A) and 24 h (B) of incubation. Soil (1 mg/mL) at pH 10.5 at ambient temperature was settled in a 100-mL measuring cylinder. The dispersing ability of the sample was determined by measuring the turbidity of 10 mL of suspension drawn from the middle of the cylinder. The concentration of polymer was that of the total solid bases. PAA was Donlar polyaspartate, produced using the pilot-scale method described in this chapter. PAC was Acrysol LMW-45 (Rohm and Haas).

they can prevent redeposition of soil on surfaces (*3*) and as dispersants of pigments in paints and in paper fillers and coatings.

The dispersing efficacy of PAA is comparable to that of PACs (Rohm and Haas Acrysol LMW-45) in a wide variety of assays. Figures 8A and B indicate that PAA and PAC have comparable garden soil dispersing efficacy at 2 and 24 h. A similar comparative efficacy was shown for Fe_2O_3 dispersion (Figure 9). For kaolin dispersion, PAA

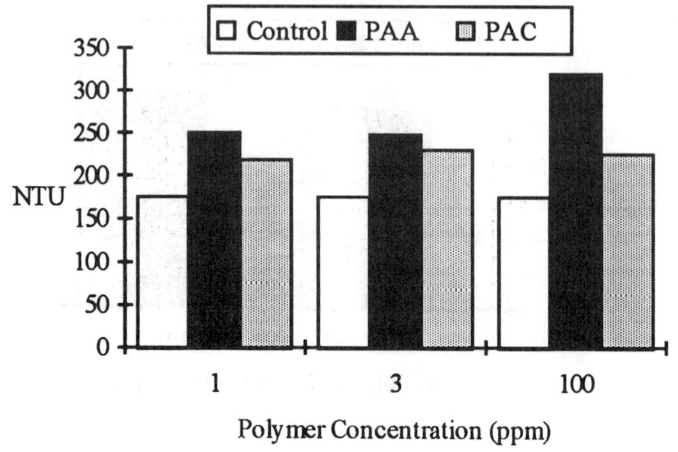

Figure 9. Ferric oxide dispersion at 24 h. Fe_2O_3 (700 ppm) with 200-ppm Ca^{2+} was stirred, adjusted to pH 10.5, and settled in a 100-mL measuring cylinder. The turbidity of the sample was measured nephometrically as described in the legend to Figure 8. The concentration of polymer was that of the total solid bases. PAA was Donlar polyaspartate, produced using the pilot-scale method described in this chapter. PAC was Acrysol LMW-45 (Rohm and Haas).

has a higher efficacy than PAC (Figure 10). The similarity of PAA to PAC in dispersing many of these same minerals also was reported by Garris and Sikes (*28*).

Biodegradability of PAA

The biodegradability of commercial PAA was determined by Wheeler et al. (*29*), who used a variety of assays, many of which were modified from Organization for Economic Cooperation and Development (OECD) guidelines (*30*). For example, the polymer has a substantial oxygen demand when it is exposed to an inoculum of the effluent of a wastewater treatment plant (biological oxygen demand assay; OECD 301D). In contrast, PAC (Rohm and Haas Acrysol LMW-45) showed no demand under the same conditions. In other low-biomass inoculum studies, acclimated sludge from the treatment plant was used in CO_2 evolution assays (SCAS-Modified Sturm assays; OECD 302A-301B). The results of these studies indicate that more than 70% of the PAA carbon can be liberated as CO_2 in 28 days. The general conclusion from the low-biomass studies is that a significant portion of PAA is labile.

When full-strength sludge was incubated with low concentrations

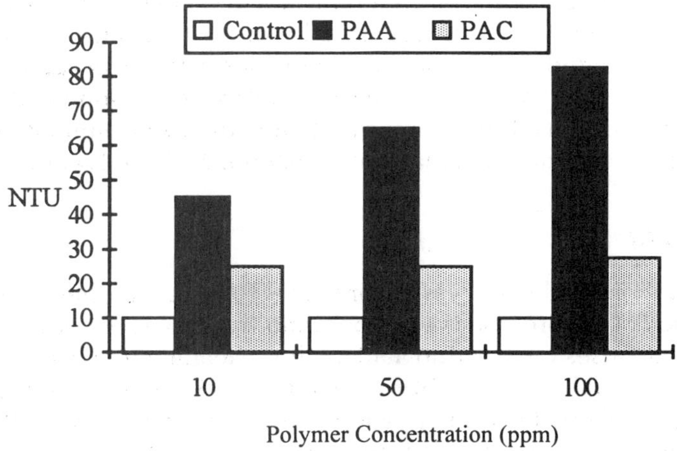

Figure 10. Kaolin dispersion at 24 h. Kaolin (1 mg/mL) with 200-ppm Ca^{2+} as $CaCO_3$ was stirred and adjusted to pH 8.0. Sample was added to a 100-mL measuring cylinder. The turbidity of the sample was measured nephometrically as described in the legend to Figure 8. The concentration of polymer was that of the total solid bases. PAA was Donlar polyaspartate produced using the pilot-scale method described in this chapter. PAC was Acrysol LMW-45 (Rohm and Haas).

(1 ppm) of [^{14}C]PAA, 30–50% of the label was liberated as CO_2 in 3–5 days. Most of the rest of the label was adsorbed to the sludge. The results of these assays reasonably accurately reflect the fate of PAA if the polymer is released into typical waste treatment systems. Such would be the case, for example, if PAA were used in detergent formulations. If the polymer is released directly into environmental waters, its fate cannot be predicted. However, given that PAA adsorbs to the surfaces of a variety of particulates, including soil (see above), much of the polymer might be removed from the soluble phase under these conditions.

Toxicity Studies

Toxicity tests of commercially produced PAA with mammalian test models such as Sprague-Dawley and Wistar Albino rats and New Zealand Albino rabbits and environmental test models such as *Daphnia pulex* (water flea), *Pimephales promelas* (fathead minnow), *Selenastrum capricornutum* (algae), and *Photobacterium* (microtox) are in progress. The 14-day acute oral toxicity in rats test was conducted according to the guidelines of the Environmental Protection Agency and the Toxic Substances Control Act (40 CFR 798.1175) (*31*). The limit test, which was performed on Sprague-Dawley rats, resulted in a 50% lethal dose of more than 5 g/kg of body weight. No lethality was demonstrated within the limits of the test and no gross lesions were seen at necropsy. Preliminary results of environmental toxicity studies indicate that the polymer will meet the Environmental Protection Agency criteria for release of a water-soluble polymer.

Summary

A commercial-scale process to produce PAA in excellent yield was developed. This process is environmentally attractive because it uses no solvents, uses only amino acid as the reactant, and produces water as the principal byproduct. Because high purity (>99%) L-aspartic acid can be purchased from many sources for approximately $1/lb (0.454 kg) and because its synthesis is extremely cost-effective and can easily be scaled up to thousand of pounds per batch, the polymer is economically viable. A variety of laboratory activity tests indicate that PAA produced in this way is at least comparable and in some instances is superior in performance to the commercially produced PACs. Biodegradability studies show that PAA undergoes significant breakdown under conditions that simulate those of wastewater treatment plants, an environment to which much of the waste polymer would be exposed. Preliminary aquatic environmental and mammalian studies on test animals suggest that PAA is nontoxic. All of these factors show that commercially prepared PAA is an attractive alternative to PAC when mineral scale inhibitor or dispersion is required.

Acknowledgments

We thank Ann Marie Atencio for performing the pilot plant synthesis. We are also indebted to Gary Lichtfield and Kim Ivey for performing spectroscopic analyses. Portions of this study were supported by a grant from the South Carolina Sea Grant Consortium to A. P. Wheeler.

References

1. Wood, A. *Chem. Week* **1993,** *December* 22/29, 22–24.
2. Chiaudani, G.; Poltronien, P. *Ing. Ambientale* **1990,** 11.
3. Hayashi, T.; Mukouyama, M.; Sakano, K.; Tani, Y. *Appl. Environ. Microbiol.* **1993,** *59,* 1555–1559.
4. Sikes, C. S.; Wheeler, A. P. U.S. Patent 4 534 881, 1985.
5. Sikes, C. S.; Yeung, M. L.; Wheeler, A. P. In *Surface Reactive Peptides and Polymers: Discovery and Commercialization;* Sikes, C. S.; Wheeler, A. P., Eds.; ACS Symposium Series 444; American Chemical Society: Washington, DC, 1991; pp 50–71.

6. Boehmke, G. U.S. Patent 4 839 461, 1989.
7. Koskan, L. P.; Meah, A. R. U.S. Patent 5 219 952, 1993.
8. Fox, S. W.; Harada, K. In *A Laboratory Manual of Analytical Methods of Protein Chemistry;* Alexander, P.; Lundgren, H. P., Eds.; Pergamon: Oxford, England, 1966; Vol. 4, pp 127–151.
9. Kokufuta, E.; Suzuki, S.; Harada, K. *Bull. Chem. Soc. Jpn.* **1978,** *51*(5), 1555–1556.
10. Sikes, C. S.; Mueller, E. M.; Madura, J. D.; Drake, B.; Little, B. J. *Corrosion* **1993,** Paper 465.
11. Wheeler, A. P.; Koskan, L. P. *Mater. Res. Soc. Symp.* **1993,** *292,* 277–283.
12. Little, B. J.; Sikes, C. S. In *Surface Reactive Peptides and Polymers: Discovery and Commercialization;* Sikes, C. S.; Wheeler, A. P., Eds.; ACS Symposium Series 444; American Chemical Society: Washington, DC, 1991; pp 263–279.
13. Koskan, L. P. U.S. Patent 5 057 597, 1991.
14. Kovacs, J.; Kovacs, N.; Konyves, I.; Csaszar, J.; Vajda, T.; Mix, H. *J. Org. Chem.* **1961,** *26,* 1084–1091.
15. Koskan, L. P.; Low, K. C.; Meah, A. R.; Atencio, A. M. U.S. Patent 5 221 733, 1993.
16. Mosig, J. Diplomarbeit, Clemson University, Clemson, SC, and RWTH Aachen, Germany, 1992.
17. Hoagland, P. D.; Fox, S. W. *Experientia* **1973,** *29,* 962–964.
18. Pivcova, H.; Saudek, V.; Drobnik, J.; Vlasik, J. *Biopolymer* **1981,** *29,* 1605–1614.
19. Saudek, V.; Strokrova, S.; Schmidt, P. *Biopolymer* **1982,** *21,* 1011–1020.
20. Pivcova, H.; Saudek, J.; Drobnik, J. *Polymers* **1982,** *23,* 1237–1241.
21. *The Nalco Water Handbook,* Kemmer, F. N.; McCallion, J., Eds.; McGraw-Hill: New York, 1979.
22. Wheeler, A. P.; Low, K. C.; Sikes, C. S. In *Surface Reactive Peptides and Polymers: Discovery and Commercialization;* Sikes, C. S.; Wheeler, A. P., Eds.; ACS Symposium Series 444; American Chemical Society: Washington, DC, 1991; pp 72–83.
23. Low, K. C. M.S. Thesis, University of South Alabama, Mobile, AL, 1990.
24. Fivizanni, K. P.; Hoots, J. E.; Cloud, R. W. In *Surface Reactive Peptides and Polymers: Discovery and Commercialization;* Sikes, C. S.; Wheeler, A. P., Eds.; ACS Symposium Series 444; American Chemical Society: Washington, DC, 1991; pp 280–315.
25. Koskan, L. P.; Low, K. C. U.S. Patent 5 116 513, 1992.
26. Koskan, L. P.; Low, K. C.; Meah, A. R.; Atencio, A. M. U.S. Patent 5 152 902, 1992.
27. Freedman, A. J.; Laronge, T. M. *Ind. Water Treatment* **1992,** *May/June,* 17–28.
28. Garris, J.; Sikes, C. S. *Colloids Surf. A: Physicochem. Eng. Aspects* **1993,** *80,* 103–112.
29. Alford, D. D.; Wheeler, A. P.; Pettigrew, C. A. *J. Environ. Polym. Degrad.* **1994,** *2*(4), 225–236.
30. *OECD Guidelines for Testing of Chemicals;* Organization for Economic Cooperation and Development, 1981.
31. *Oral Toxicity in Rats;* EPA/TSCA guidelines (40 CFR 798.1175).

RECEIVED for review February 8, 1994. ACCEPTED revised manuscript August 3, 1994.

7

Biodegradable Plastics Derived from Cellulose Fiber and Chitosan

M. Nishiyama, J. Hosokawa, K. Yoshihara, T. Kubo, H. Kabeya, T. Endo, and R. Kitagawa

Shikoku National Industrial Research Institute, 2217-14 Hayashi-cho, Takamatu, 761-03 Japan

RECENTLY, ENVIRONMENTAL POLLUTION CAUSED BY WASTE plastics has become a global problem. One countermeasure has been the development of biodegradable plastics that are decomposed within soil, and some biodegradable plastics have been proposed. Such biodegradable plastics must have nearly the same strength as ordinary plastics, must decompose in soil after use and not cause environmental pollution, must be controllable for the period of decomposition, and must be producible at a low cost.

Natural polymers such as cellulose are decomposed by microorganisms in soil, and the decomposed materials do not pollute the environment. Judging from their decomposability, natural polymers, which exist in nature abundantly, should be excellent biodegradable plastics. We developed novel biodegradable plastics by making natural polymers into composites. The materials are finely divided cellulose and chitosan, which is produced by the deacetylation of chitin. The novel biodegradable plastics that we developed and their applications are reported here.

Biodegradable Film Derived from Cellulose–Chitosan

Cellulose is the major component of the cell walls of plants and the most abundant polymer on earth. Chitin is contained in the shells of

0065–2393/96/0248–0113$12.00/0

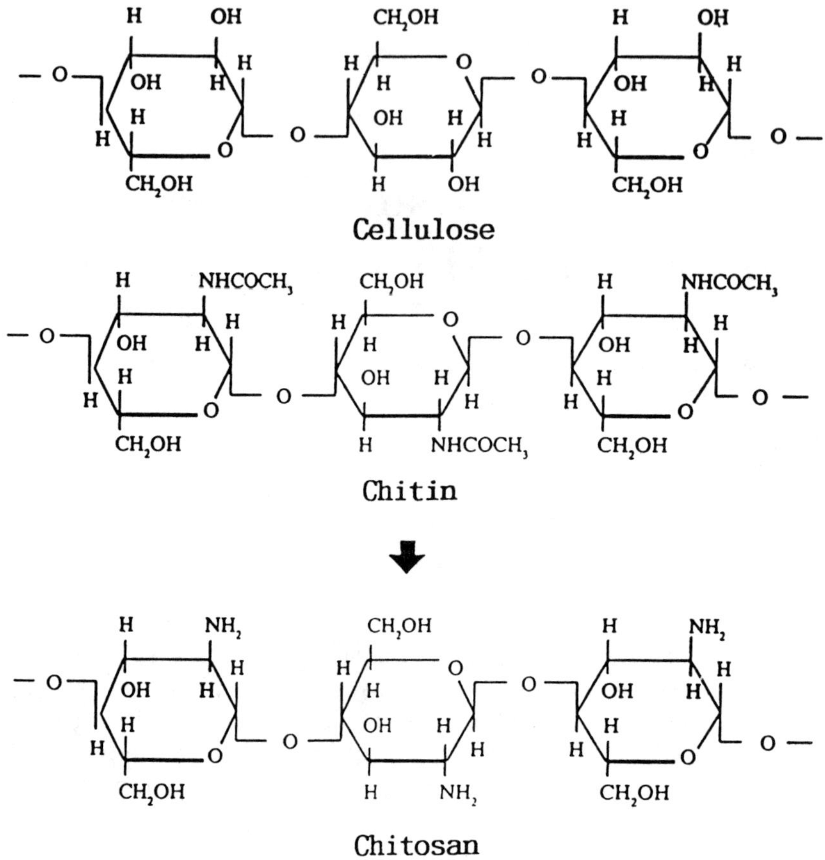

Figure 1. Chemical structures of cellulose, chitin, and chitosan.

shrimp and crabs and the cell walls of fungi and insects; the global production of chitin is believed to be second only to that of cellulose. As shown in Figure 1, the structure of chitin is similar to that of cellulose; the chitin molecule is a linear natural polymer that corresponds to cellulose in which C-2 is replaced by an acetylamino group. Chitin is converted to chitosan by deacetylation with concentrated alkali. Both chitin and chitosan have various functions such as biological compatibility, antibiotic activity, and film-forming capability that are attractive, and chitin and chitosan are being studied for use in medical and industrial materials and in foods.

Cellulose suspended in water becomes anionic at the surface from the actions of hydroxyl groups and a trace amount of carboxyl groups. Chitosan, on the other hand, is insoluble in water, and its salts, such as those with acetic acid, become water-soluble and cationic. Chitosan has good affinity for cellulose because of the similarity in their structures. Taking advantage of these characteristics, we developed biodegradable plastics derived from cellulose and chitosan.

Materials. The raw material cellulose was obtained as Micro Fibril Cellulose (MFC-100) from Daicel Chemical Industries Ltd. (8-1, Kasumigaseki 3-chome, Chiyoda-ku, Tokyo 100, Japan) and selected from samples with diameters of 0.1 μm or less, lengths of 100–500 μm, and surface areas of 200 m^2/g. The raw material chitosan had 99.8% deacetylation and was produced by Katokichi Co., Ltd. (2-35, Sanbonmatsu 4-chome, Kannonji 768, Japan), from shrimp shell.

Film Formation. The composite cellulose–chitosan film was prepared by following the process outlined in Figure 2. A solution of chitosan in acetic acid was blended with an aqueous dispersion of the Micro Fibril Cellulose; after the generated foam had been removed, the blend was spread over a flat plate. Thereafter, drying and thermal treatment were applied to produce a translucent film. Thermal drying was necessary to make both materials into a water-resistant composite film of high strength (*1, 2*).

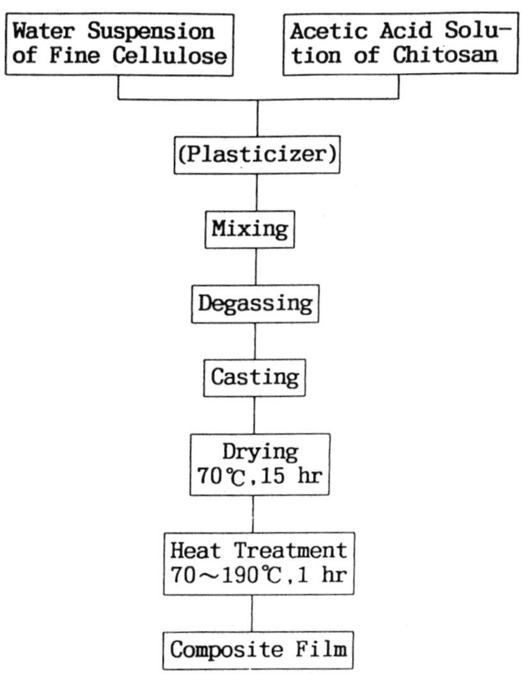

Figure 2. Method of molding biodegradable film.

Properties of the Composite Film. How the amount of chitosan influences the tensile strength of the film is shown in Figure 3. In dry conditions, the strength increases as the chitosan content increases; strength reaches about 1000 kg/cm^2 when the chitosan content is 5% or more. This level is several times as strong as commercial polyethylene films. Although finely divided cellulose alone or chitosan alone is very weak in wet conditions and is difficult to mold, the maximum wet strength of the composite film occurred when chitosan was present (at 10–20%). The 10–20% content makes the film economically reasonable despite the relatively higher price of chitosan compared to that of cellulose. The development of the higher strength in the composite may be understood from several standpoints: (1) The affinity between cellulose and chitosan is excellent, and cross-links are formed by the amino groups in chitosan and the carboxyl and carbonyl groups on the surface of cellulose; (2) chitosan salts are converted to water-insoluble amine types; and (3) self-condensation of chitosan occurs.

This composite film is not flexible. Addition of plasticizers to im-

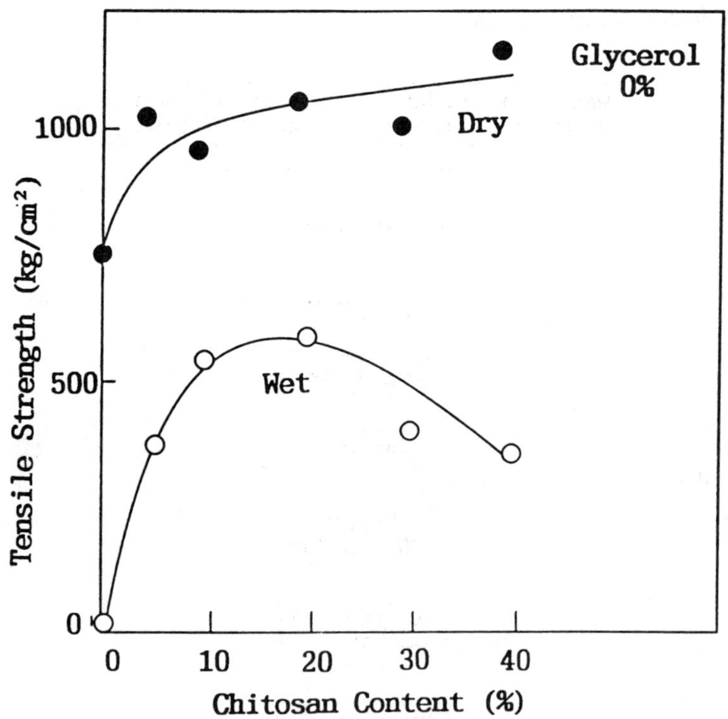

Figure 3. Effect of chitosan content on tensile strength of a composite film.

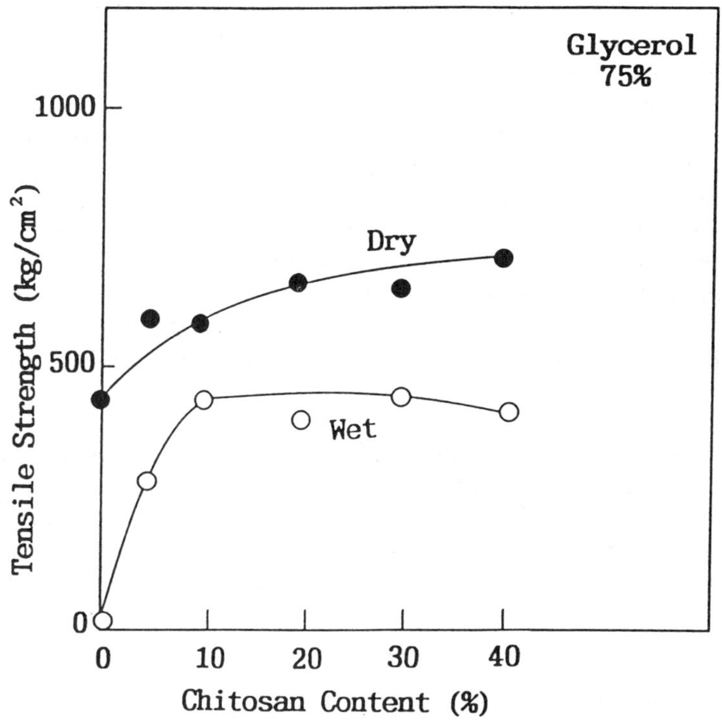

Figure 4. Tensile strength of a composite film containing 75% glycerol.

prove flexibility was studied. Figure 4 shows the results when 75 parts of glycerol is added to 100 parts of cellulose and the mixture is blended with chitosan. Although the strength decreases to some extent, a film with good flexibility is obtained.

Thermal analysis of this composite film did not show any noticeable endothermic or exothermic peak up to 280 °C; how the thermal treatment temperature influences the composite was examined. Although no substantial influence on tensile strength was revealed, the degree of swelling with water decreased as the thermal treatment temperature rose. These facts suggest an increase in cross-linkings in the composite film (*1*). Thus simple procedures such as thermal treatment cause cross-links to form. More cross-links are effective in decreasing the hydrophilic property, preventing attacks by microorganisms and enzymes, and extending the biodegradation period.

Accelerated Degradation Test of Biodegradable Film

Although no standard degradation test for biodegradable plastics has been established, the degradation period is practically important, because this composite film is expected to biodegrade completely in soil within several months. We adapted an accelerated biodegradation test using fungi that degrade the cellulose and chitosan constituents of the film (*1*). Chemical analysis of the composite film in the early stages of degradation proved that chitosan is degraded preferentially. In searching for microorganisms that degrade chitosan, we found a chitosan-degrading bacterium in soils in various places in Japan (*3*). This bacillus has a single flagellum and was identified as belonging to the genus *Pseudomonas*. It was named *Pseudomonas* sp. H-14 (called H-14 for short). H-14 grows on chitosan as the only carbon and only nitrogen source and does not use cellulose for nutrition. The early stage of degradation of the composite film is thus probably brought about by chitosanase.

Accelerated Degradation Test Method. In an accelerated degradation test, a pure culture of H-14 was transplanted under the conditions shown in Figure 5, and the culture was continued under vibrating conditions until the film was finely divided; the time from transplantation to fine division was recorded. Glass beads provided

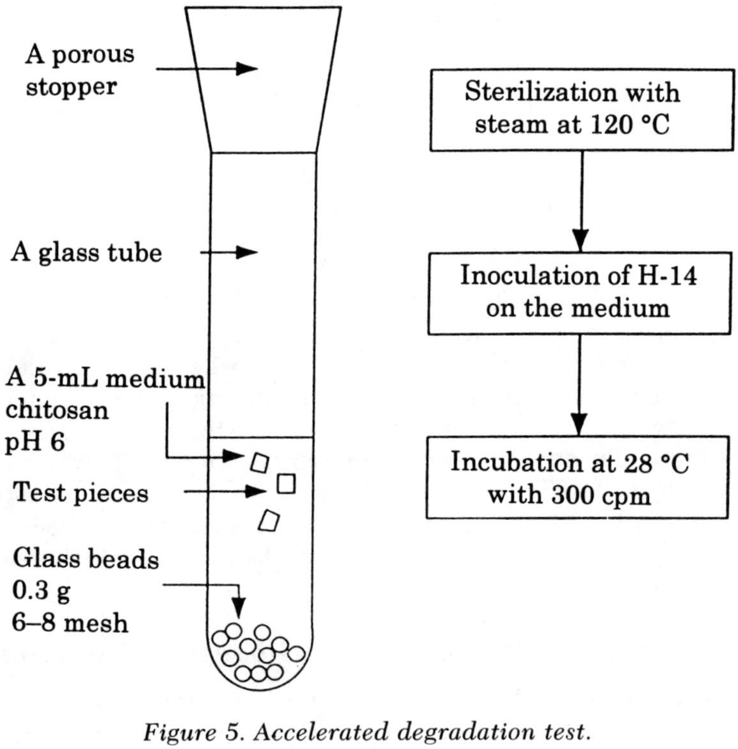

Figure 5. Accelerated degradation test.

light shocks to the film, and chitosan was added to help H-14 multiply and to promote chitosanase secretion. Three test pieces per test tube were used. Basic films of the composite film derived from cellulose and chitosan collapsed into finely divided pieces within about 3 to 4 days. This accelerated test made it possible to observe degradation occurring 10 times as fast as the actual rate in nature. Furthermore, degradation was also observed when the accelerated degradation test used a commercial cellulase. In the absence of H-14 or cellulase, no film degradation was observed.

Biodegradation of Composite Film. The decomposition period varies depending on composite production conditions and the blending ratio of the raw materials; some test results are shown in Figure 6. Higher-temperature thermal treatment and more oxidative groups in the cellulose increase the time until degradation initiation because increased cross-linking in the composite film makes it difficult for microorganisms and enzymes to attack the film. The influence of carboxyl and carbonyl groups in the raw material cellulose on the degradation period was studied in detail (Figure 7). Results indicate that the carbonyl groups have a strong correlation with the degradation period but the carboxyl groups do not. These facts suggest that the carbonyl groups in cellulose and the amino groups in chitosan form Schiff bases that cross-link (*2*). The degradation period may be controllable by manipulation of these characteristics.

Joint Research

These materials, unlike petroleum-derived plastics, are not thermoplastic and therefore cannot be processed by conventional machines for plastic molding or processing. To overcome this problem, new fabricating apparatus and commercialization are being developed jointly with private enterprises. This work is summarized below.

Biodegradable Films. Two private companies participated in joint research on forming films. Okura Industrial Co., Ltd. (1515, Nakatsu-cho, Marugame, 763 Japan), succeeded in the continuous production of films in a rolled form by using a newly developed test plant machine; the test products include molded materials such as trays. Modification of the machine to permit commercialization, improvement of film formation, and reduction of manufacturing cost is now under way. Aicello Chemical Co., Ltd. (45 Koshikawa, Ishimakihonmachi, Toyohashi, 441-11 Japan), is involved in a 3-year project, scheduled to be completed in 1995, to develop a film 500 m long and 1 m

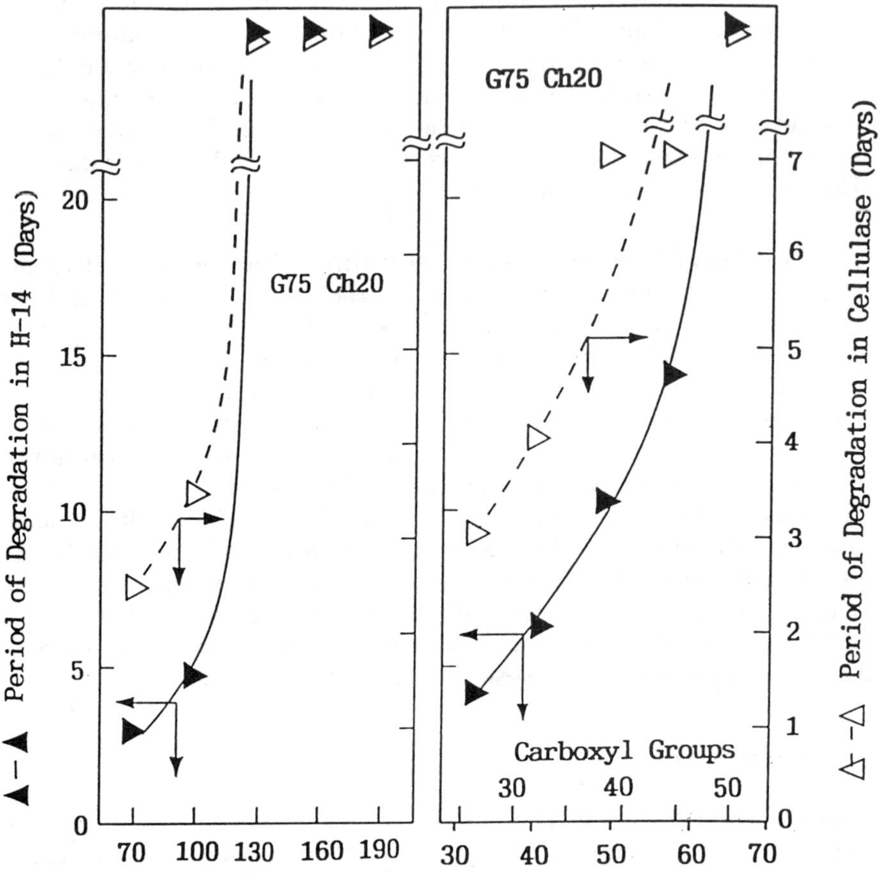

Figure 6. Period of degradation of a composite film by H-14 and cellulase.

wide. This work has been selected as a development project by the Research Development Corporation of Japan.

Biodegradable Nonwoven Fabrics. Kanai Juyo Kogyo Co., Ltd. (2–9, Dojima 1-chome Kitaku, Osaka, 530 Japan), developed a process for using these materials as a binder for dry nonwoven fabrics. As shown in Figure 8, nonwoven fabrics thus manufactured that contain a natural fiber such as cotton or rayon are biodegradable and may be produced with only a small additional cost because production with existing equipment is possible and the amount of binder required is

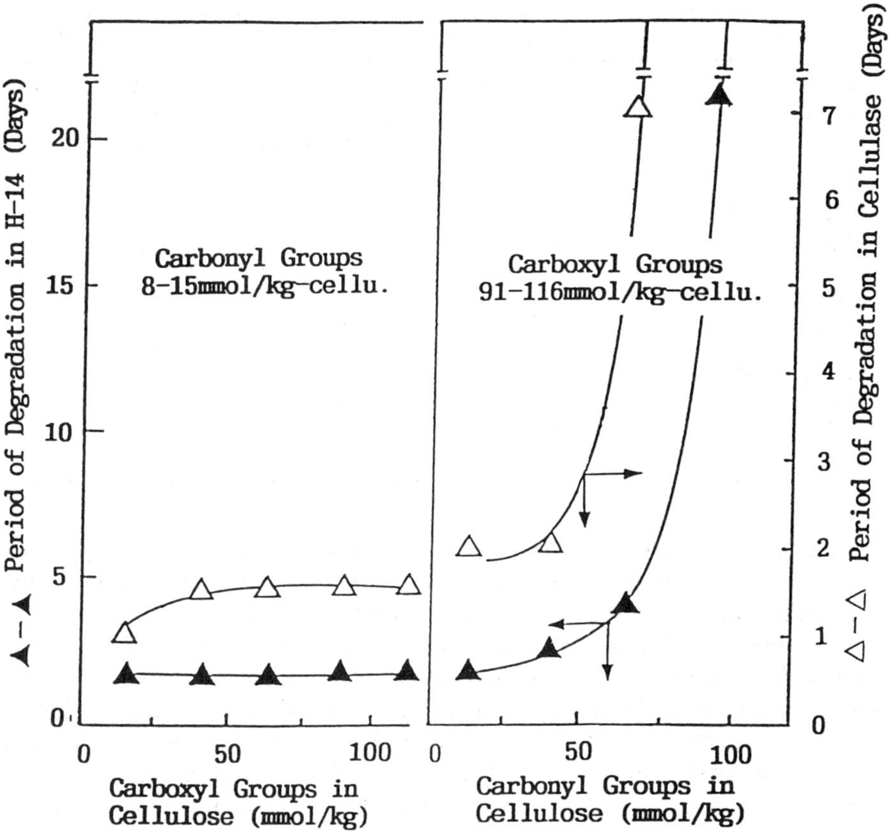

Figure 7. Period of degradation of a composite film by H-14 and cellulase.

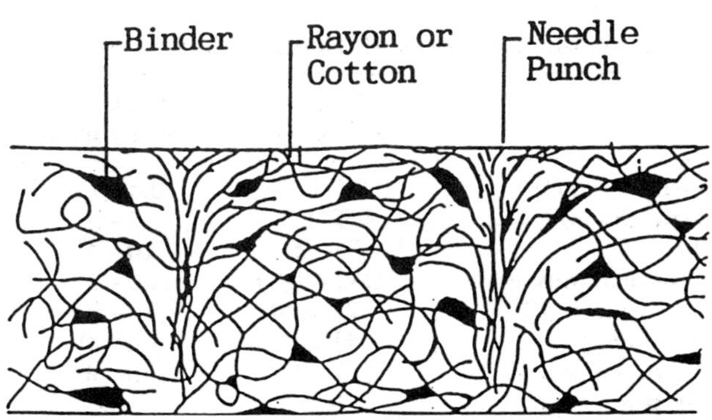

Figure 8. Cross section of biodegradable nonwoven fabric.

Table I. Properties of Biodegradable Nonwoven Fabrics

Property	*BD-30D*	*BD-100S*	*BD-30W*
Fiber	100% rayon	100% rayon	100% rayon
Structure	cross dipping	cross needle punch spray	cross water needle punch dipping
Weight (g/m^2)	30	100	32
Thickness (mm)	0.2	0.8	0.2
Tensile strength (kg/5 cm)			
MD	0.6	1.6	4.5
CD	0.3	1.8	2.2
Tear strength (kg/5 cm)			
MD	0.1	0.8	0.1
CD	0.1	0.7	0.2
Wet/dry strength (%)	54	210	53

small. Such nonwoven fabrics, when buried in soil, completely degraded in 1 month in summer and 2 months in winter. The degradation period for these nonwoven fabrics is controllable: Fabrics that degrade in 1 week or in 1 to 3 years have been developed. These fabrics have excellent porosity, permeability, and water-resisting properties and can be fabricated. Various biodegradable nonwoven fabrics with different characteristics have also been developed (Table I). Pots and sheets for gardening, sanitary materials such as paper diapers, packing materials, and wound dressings, among others, are targets of the applications.

Biodegradable Foams. Nishikawa Rubber Co., Ltd. (2-2-8, Misasa-machi, Nishi-ku, Hiroshima, 733 Japan), developed biodegrad-

Table II. Properties of Biodegradable Foams

Property	*Open-cell foams*	*Closed-cell foams*	*Film*
Apparent density (g/cm^3)	0.02–0.05	0.1–0.3	1.3
Tensile strength (kg/cm^2)	0.5–10	5–20	1000
Elongation (%)	15–40	10–20	10
Water-absorbing capacity (%)	1000–4000	50–200	
Water-holding capacity (%)	50–400	10–50	
Calorific value (cal/g)	4150	4010	
Hardness, degree (Asker F)	5–70	60–80	
Biodegradable time (days[a])	4–6	9	15–18

[a] Acceleration test, SNIRI Method.

able open-cell and closed-cell foams by using materials derived from cellulose and chitosan (Table II). The open-cell foams are flexible, like a cosmetic puff, and are highly water-absorbing, being able to absorb several orders of magnitude more water than typical foams can. The closed-cell foams are like a honeycomb and have nearly the same hardness and strength as conventional plastic foams. Their excellent characteristics of lightness, thermal insulation, permeability, and water absorption are put to use in agricultural and industrial applications. Their biological adaptability is also useful for new applications.

Conclusions

The most difficult problem to overcome for cellulose–chitosan-derived biodegradable plastics was processing, because these materials are not thermoplastic. However, this problem has been solved, and biodegradable plastics in various forms are being commercially produced. Biodegradable plastics degrade to carbon dioxide and water and may turn to compost in the process. Compost is a valuable carbon source for microorganisms in soil and seems to aid in activating soil. Cost reduction and quality improvement of the composite remain subjects for future attention.

We believe that biodegradable plastics that return to nature when discarded not only solve waste problems but also may serve as novel functional materials.

References

1. Hosokawa, J.; Nishiyama, M.; Yoshihara, K.; Kubo, T. *Ind. Eng. Chem. Res.* **1990,** *29,* 800–805.
2. Hosokawa, J.; Nishiyama, M.; Yoshihara, K.; Kubo, T.; Terabe, A. *Ind. Eng. Chem. Res.* **1991,** *30,* 788–792.
3. Yoshihara, K.; Hosokawa, J.; Kubo, T.; Nishiyama, M. *Agric. Biol. Chem.* **1990,** *54,* 3341–3343.

RECEIVED for review April 16, 1994. ACCEPTED revised manuscript April 7, 1995.

8

Reaction Parameter Effects on Substituent Distributions in the Heterogeneous Synthesis of Cellulose Ethers

Knowledge for a More Biodegradable Polymer from a Renewable Source?

Stephen D. Seneker[1] and J. Edward Glass

Department of Polymers and Coatings, North Dakota State University, Fargo, ND 58105

Polymer derivatives of cellulose that are highly substituted are utilized commercially as thermoplastics. They are hydrophobic because most of the three hydroxyl groups of the repeating glucopyranosyl units of cellulose are replaced. In Chapter 5 of this volume, water solubility was suggested as a contributor to biodegradation, yet hydrophilicity retards the rate of processing of thermoplastics. To achieve biodegradation, the C-2 hydroxyls on contiguous repeating rings of glucopyranose must remain unsubstituted, preferably in 5- or 6-segment runs of C-2 unsubstitution. Reaction parameters that control substituent placement and the costs of producing viable products are discussed.

[1] Current address: ARCO Chemical Company Technical Center, Building 740, P.O. Box 38007, South Charleston, WV 25303

0065–2393/96/0248–0125$12.00/0

CELLULOSE, A RENEWABLE RESOURCE, is the world's most abundant polymer (Figure 1A). The beta linkage (i.e., equatorial–equatorial bonding) of glucopyranosyl (GP) units projects a planar macromolecule that is ca. 70% crystalline when obtained from cotton (*1*). Inter- and intramolecular hydrogen bonding, primarily by the C-6 and C-3 hydroxyls of the repeating GP units, results in a water-insoluble polymer. Cellulose, as ester and as ether derivatives, is used in many applications. Cellulose esters, generally prepared by the esterification of the GP hydroxyl groups with anhydrides under acidic catalysis, are fully substituted (i.e., the hydroxyl groups at the C-2, C-3, and C-6 positions are replaced with acetate, propionate, or butyrate groups). As such, these esters are hydrophobic and can be processed in many thermoplastic applications.

Cellulose ethers are prepared under caustic conditions, and generally less than half of the hydroxyl groups are replaced. Cellulose treated with sodium hydroxide is reacted with methyl chloride and the sodium salt of α-chloroacetic acid to form methyl cellulose (MC)

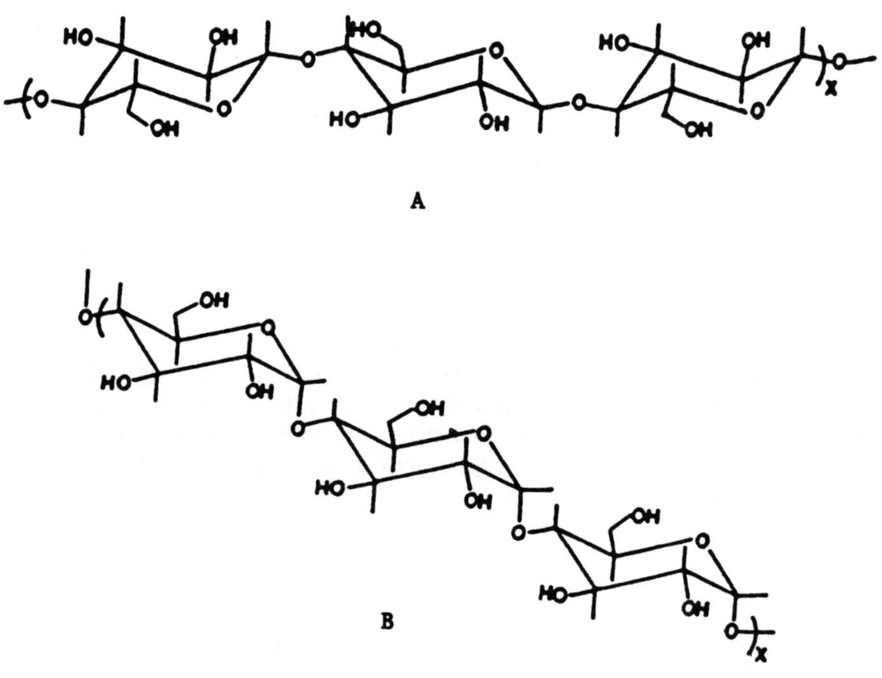

Figure 1. Structures of cellulose (β-(1,4)-D-glucopyranosyl units) (A) and amylose (α-(1,4)-D-glucopyranosyl units), the basic component of starch (B).

and carboxymethyl cellulose (CMC). CMC derivatives are more hydrophilic than cellulose and are not readily usable in fabrication processes that require thermoplastic properties. Both CMC and MC are prepared by a paste process. In MC, methyl chloride adds predominantly to the C-2 position (at a 5:2 preference over the C-6 position); addition at C-2 imparts a greater stability to enzymatic degradation. The reaction of oxiranes to alkali cellulose is exothermic and hence is carried out in a slurry process. The reaction of an oxirane, such as ethylene oxide or propylene oxide, with alkali cellulose generates oxyanions that are generally more reactive than those attached to the repeating GP rings of cellulose. This reactivity results in chaining of the oxirane units rather than uniform addition to the GP main-chain units.

In the substitution of alkali cellulose with propylene oxide, >96% of the adduct is opened through the primary carbon, and a secondary oxyanion results. This reactive species is less reactive than the primary oxyanion, which is the only oxyanion possible when ethylene oxide is the adduct. Thus less chaining and more uniform addition occur in hydroxypropyl cellulose (HPC). With a molar substitution (MS) of ~4.0, a thermoplastic HPC comparable to cellulose esters can be produced; however, none of these derivatives is biodegradable. To obtain biodegradation, the C-2 hydroxyl must remain unsubstituted (*2*), preferably in a sequence of five to six contiguous GP repeating units unsubstituted at the C-2 position (*3*). By using a slurry process and a high caustic concentration (>4 N), a water-soluble commercial hydroxyethyl cellulose (HEC) is produced within an MS range of 2.0–2.5. In a commercial HEC at this MS level, approximately 20% of the GP units are unsubstituted because of the excessive amount of chaining. This type of HEC, with poor thermoplastic properties, is not suitable for fabrication. In the two sections that follow, a brief historical review of methods used to characterize substituent distribution is given, and the basic studies of methyl-substituted D-β-glucoses that define the reactivities of the hydroxyl units in carbohydrate polymers are discussed. The concluding section examines the structures of carbohydrate polymers and the derivative approaches that could lead to thermoplastics with biodegradability that are based on a renewable resource.

Methods of Determining Substituent Distribution Patterns

Substituent distributions on carbohydrates have been the subject of many investigations (*4–9*). Cellulose has been the focus of most of these studies (*10*). Substituent distributions were first determined by measuring monomeric products with gel permeation chromatography, a method that involves hydrolysis of the substituted polysaccharide into mixtures of unsubstituted and substituted monosaccharide units. Paper (*11, 12*) and column (*13*) chromatography have also been used. With the development of volatile derivatives, that is, by trimethylsilylation, gas-liquid chromatography was also utilized (*14*). A set of characterized substituted monosaccharide samples is required. By coupling a mass spectrometer with a gas-liquid chromatograph, individual substituted monosaccharide units can be identified by their characteristic mass spectrometer spectra (*15–17*). ^{1}H NMR spectroscopy has been used with some success, but it is limited to a few substituents (e.g., carboxymethyl and acetyl) (*18, 19*). ^{13}C NMR spectroscopy has also been utilized to determine substituent distributions (*20–22*), including hydroxypropyl distribution in high-MS commercial polymers (*19, 20*). With low molecular weights and elevated temperatures (to lower the solution viscosities), direct analysis without hydrolysis is possible with these substituents. Reactivity ratios of the various hydroxyl groups of a GP unit can be calculated from the substituent distribution measurements described above when the increment of reaction is small (i.e., when polysubstitution on a GP unit is negligible).

In early studies, most investigators utilized a statistical model developed (*23*) by Spurlin in 1939. This model was designed to determine reactivity ratios from direct measurements of the individual mono-, di-, and trisubstituted monosaccharide fractions. Direct measurement of the polysubstituted monosaccharide fractions is especially difficult with substituents that can react with their adducts, for example, ethylene oxide and propylene oxide. An understanding of this complexity is important and led to the development of a stochastic process that describes the distributions for a set of reactivity ratios in terms of percent unsubstituted vicinal diol units and percent unsubstituted monosaccharide units as a function of the amount of substitution (*3*) of adduct per GP unit (MS). Determination of the two GP experimental values for a set of variable-MS products allowed determination of relative reactivities. Recently, Reuben solved Spurlin's statistical equations in terms of these same quantities (*24*). To place this effort in perspective, control of substituent placement in monomeric analogs of cellulose is reviewed in the next section.

Structural and Reaction Parameter Influences on Substituent Distributions

The reaction scheme representative of the synthesis of MC and CMC is illustrated in Scheme I. The reaction representative of HPC and

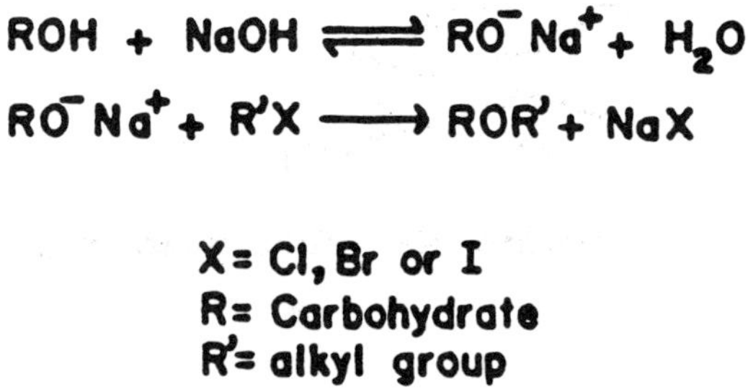

Scheme I. Reaction of an alkoxide with an alkyl halide (Williamson synthesis).

HEC synthesis is given in Scheme II. To evaluate the reactivity of the hydroxyl functions of the repeating GP units of cellulose, selective methyl-substituted monomeric GP units were studied with the non-chaining adduct *N,N*-diethylaziridinium chloride (DAC; Scheme III). The greater reactivity (4:1) of DAC with the primary hydroxyl of 1-propanol relative to the secondary hydroxyl of 2-propanol suggested that the primary hydroxyl at the aliphatic C-6 position should be more reactive than the secondary alicyclic hydroxyls in a GP ring; however,

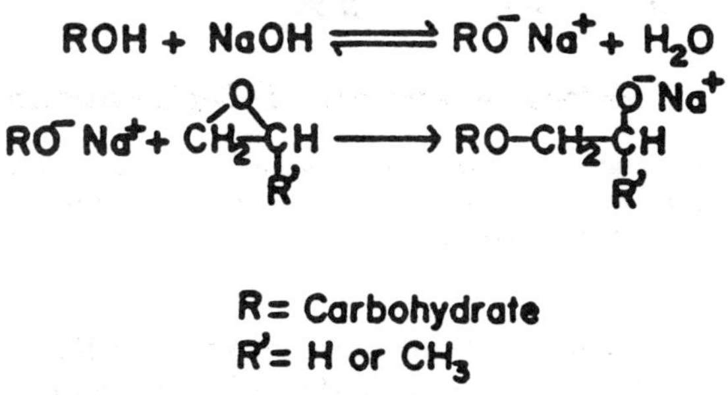

Scheme II. Reaction of an alkoxide with an oxirane when R' = CH_3, a secondary oxyanion, is formed.

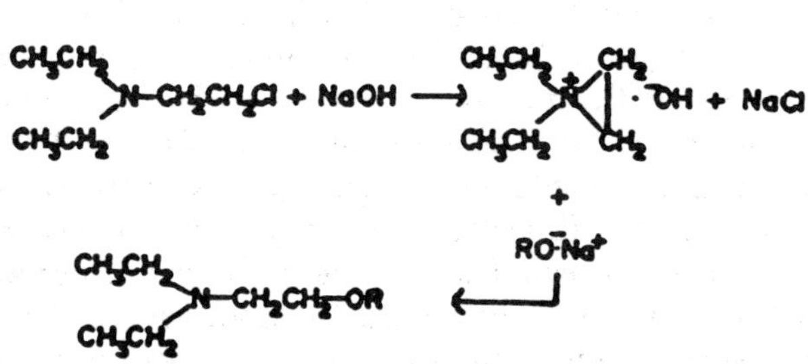

Scheme III. Reaction of an alkoxide with DAC.

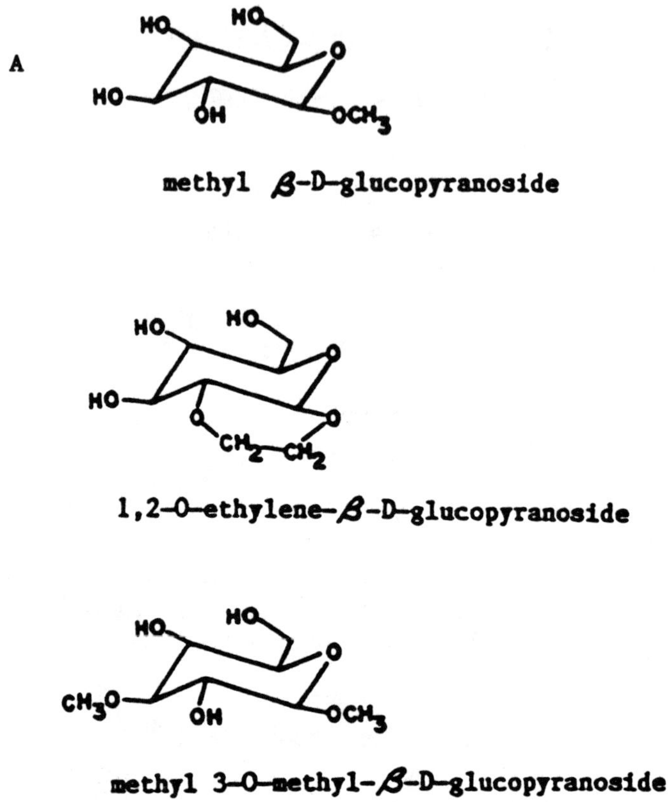

Figure 2. Structures of model β-D-glucopyranosides: (A) methyl-β-D-glucopyranoside; (B) 1,2-O-ethylene-β-D-glucopyranoside; and (C) methyl 3-O-methyl-β-D-glucopyranoside.

the relative order of substituent placements of DAC in methyl-β-D-glucopyranoside (Figure 2A) revealed the following order of hydroxyl reactivity: C-2 > C-3 > C-6 > C-4. This order was initially attributed to acidity differences related to an inductive effect influenced by distance from the C-1 acetal linkage. Distribution of DAC units at the C-3, C-4, and C-6 hydroxyls in 1,2-*O*-ethylene-β-D-glucopyranoside (Figure 2B) revealed the C-6 hydroxyl to be the least reactive, but with methyl 3-*O*-methyl-β-D-glucopyranoside (Figure 2C), in which no contiguous ring hydroxyls are available, the C-6 hydroxyl was significantly more reactive than the secondary alicyclic hydroxyls. The relative rate data are given in Table I. The inductive effect was revised to include hydration sheets around the C-6 hydroxyl and intramolecular hydrogen bond acceptor models for vicinal hydroxyl group (*25*) reactivities at low base concentrations (*26*) (Figure 3).

Table I. Effect of Substituents in β-D-Glucopyranosides on Relative Oxyanion Formation Among Hydroxyl Groups

	Oxyanion Formation Relative to C-6 Hydroxyl			
β-D-Glucopyranoside[a]	*2-OH*	*3-OH*	*4-OH*	*6-OH*
Methyl	2.08	1.11	0.75	1.00
1,2-*O*-Ethylene-	—	2.45	2.00	1.00
Methyl 3-*O*-methyl-	0.25	—	0.00	1.00

[a]Homogeneous reaction with DAC in 0.1 N aqueous sodium hydroxide at 25 °C.

In cellulose, hydrogen bonding reduces the accessibility of GP hydroxyl groups within elementary fibrils and of individual hydroxyl groups on the surfaces of fibrils (*9, 10*). Derivatization of high-molecular-weight *solubilized* cellulose at a reasonable concentration would be ineffective because of transport problems in a viscous medium. Isolation of the derivatized polymer would also be difficult. Commercial derivatization of cellulose is therefore conducted in heterogeneous slurries. In the derivatization of cellulose, the crystallinity of the polymer is also an important parameter in the substitution pattern (*27, 28*) (Table II). Aqueous sodium hydroxide (>4 N) is effective in swelling and disrupting the crystalline matrix in a heterogeneous process with an appropriate amount of "available" water. In catalyzed reactions at high caustic concentrations, the C-6 hydroxyl was observed both in model glucoside and with cellulose to be more reactive than the C-2 hydroxyl group. This difference led to an alkali–diol–adduct

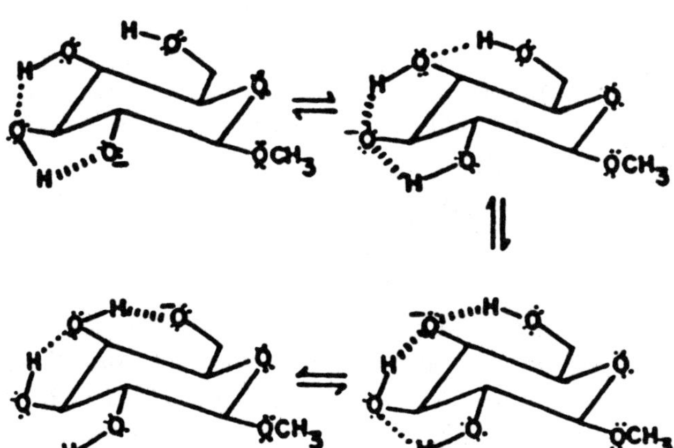

Figure 3. Idealized concept of intramolecular hydrogen bonding as proposed by Rendleman (26).

Table II. Relative Reactivity Ratios for Heterogeneous Reactions of Various Crystallinity Cellulose with DAC in 2.0 N Sodium Hydroxide

Cellulose Sample	*Relative Reactivity Ratios,* $r_2:r_3:r_6$
Noncrystalline	1.4:0.5:1.0
Fibrous	2.1:0.3:1.0
Highly crystalline	2.8:0.4:1.0

concept that is ill-defined mechanistically but does define the placements observed with additions catalyzed by high caustic concentrations.

Oxyethylene distributions in commercial HEC prepared with 4 N NaOH are approximately the same as those observed in ethoxylated low-molecular-weight (<10 repeating GP units) water-soluble "cellulose" at high caustic concentrations. Once there is sufficient disruption of the crystalline cellulose matrix, a lower caustic concentration can be used if the sample is maintained in a dispersed state, and the distribution of oxyethylene chains in a heterogeneous process (*3*) is identical to those observed in a low-caustic-concentration-catalyzed ethoxylation of low-molecular-weight "water-soluble cellulose" (*29–31*). The effect of base concentration on reactivity ratios of GP hydroxyls in both model monomeric derivatives and cellulose shows the following major trends:

1. The reactivity of secondary vicinal hydroxyl groups decreases relative to that of the C-6 primary hydroxyl group as the base concentration is increased for methyl-β-D-glucopyranoside, 1,2-*O*-ethylene-β-D-glucopyranoside (Figure 4), and cellulose (Figure 5).
2. The reactivity ratios of the hydroxyl groups in methyl 3-*O*-methyl-β-D-glucopyranoside are independent of the sodium hydroxide concentration. This compound contains no vicinal hydroxyl groups (Figure 6).
3. The degree of oxyanion formation at the C-2 and C-3 secondary vicinal hydroxyl groups of cellulose becomes independent of the sodium hydroxide concentration at the same level, while oxyanion formation at the C-6 primary hydroxyl group continues to increase. Oxyanion formation at the C-6 primary hydroxyl position appears to level off at higher base concentrations (Figure 7), and

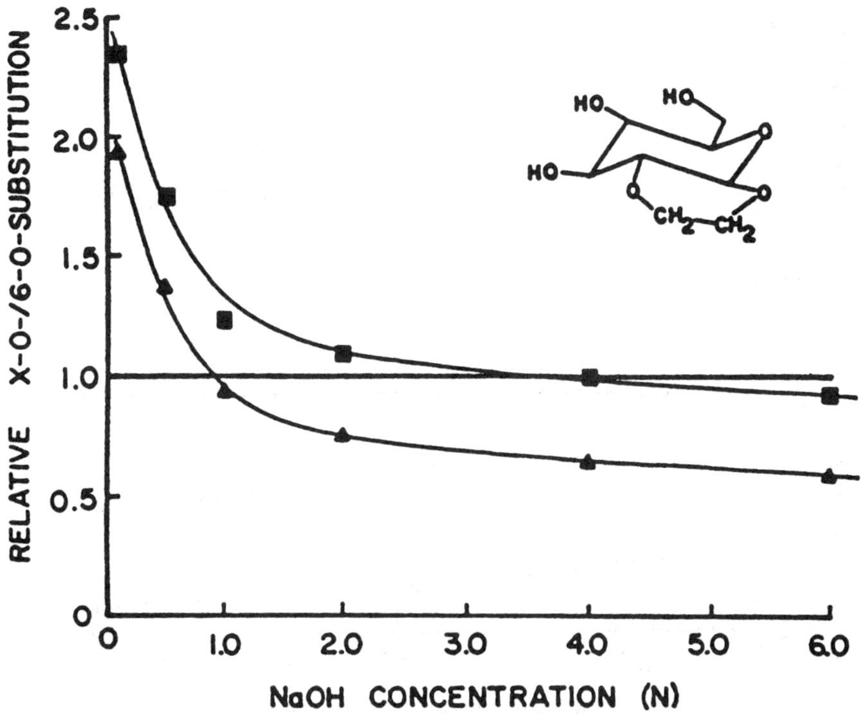

Figure 4. Relative x-O-/6-O-*substitution ratios of 2-(diethylamino)ethyl groups on 1,2-O-ethylene-β-D-glucopyranoside as a function of sodium hydroxide concentration at 25 °C. Key:* ■, 3-O-/6-O-*substitution ratio;* ▲, 4-O-/6-O-*substitution ratio.*

the rate of addition to the oxyethylene anion shows no detectable change with increasing caustic concentration.

Use of the Concepts

When polymers lack thermoplastic properties, the rate of production of fabricated materials drops catastrophically. This problem was a major deterrent in the development of biodegradable starch/polyethylene blends, although it was almost never mentioned. It was more of a factor than the fact that the starch segment made inaccessible in the polyethylene blend did not biodegrade (*32*). Derivatives of starch in which the GP units are linked through axial–equatorial (alpha) bonding (Figure 1B) were briefly touted as processable, biodegradable plastics, but specifics regarding the technical nature and placement of such derivatives were not reported (*33*).

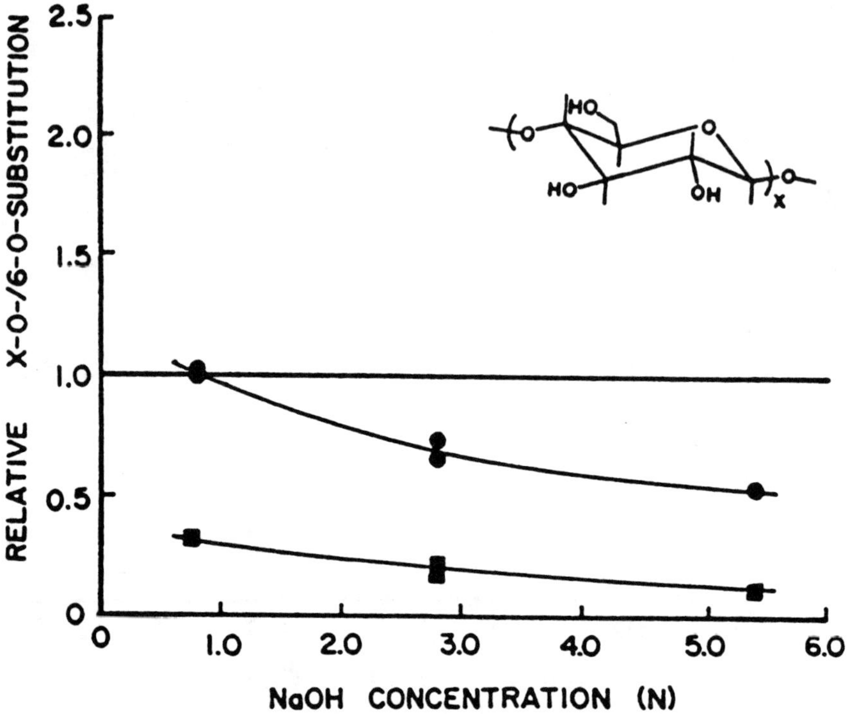

Figure 5. Relative x-O-/6-O-*substitution ratio of 2-hydroxyethyl groups on alkali-soluble, low-molecular-weight cellulose as a function of hydroxide concentration at 20 °C. Key:* ●, 2-O-/6-O-*substitution ratio;* ■, 3-O-/6-O-*substitution ratio.*

In the discussion of cellulose reactivity concepts in the previous section, the economic aspects of production were not considered. More units can be positioned preferentially at the C-6 position to achieve the goal of a *more* processable and *more* biodegradable cellulose derivative, but within current manufacturing practices, the cost is significant. The economics of production change with the substituents. For example, in the Williamson synthesis of MC by a paste process, the addition efficiency is 40%. In this process, the methyl groups prefer (by a factor of 2.5) C-2 addition because of the high crystallinity of the initial reaction matrix. This preference would decrease the biodegradability of the product. However, when methyl chloride is added to alkali cellulose in an isopropanol/water heterodispersion, its addition efficiency is only 20%, a very limiting economic factor. When propylene oxide is added to cellulose in a 4 N NaOH alkali cellulose slurry, the addition efficiency also is 20%; a 40% efficiency is achieved

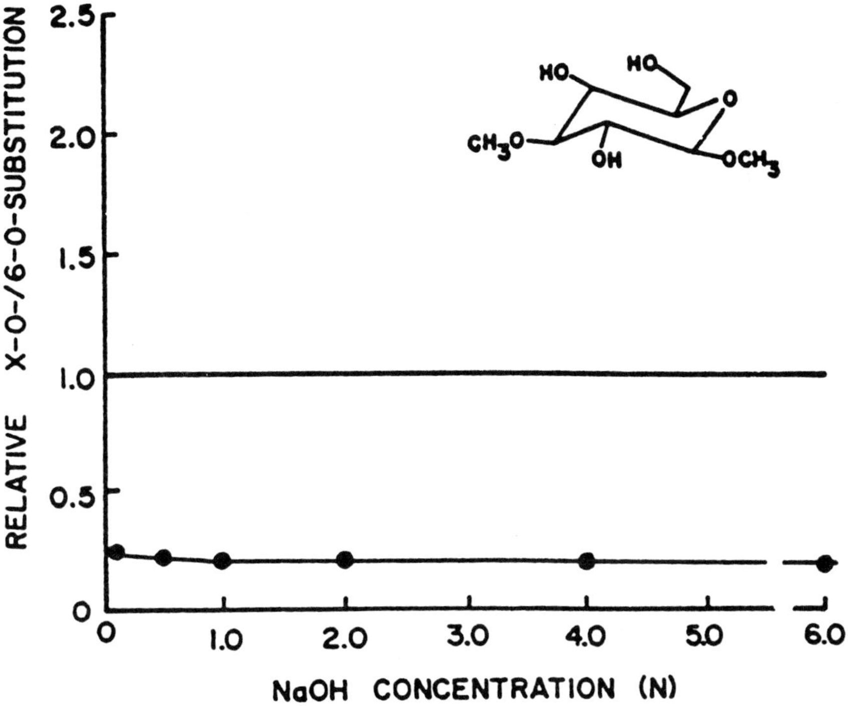

Figure 6. Relative x-O-/6-O-*substitution ratio of 2-(diethylamino)ethyl groups on methyl 3-O-methyl-β-D-glucopyranoside as a function of sodium hydroxide concentration at 25 °C. 4-O-substitution was below the level of detection. Key:* ●, 2-O-/6-O-*substitution ratio.*

only at a lower caustic concentration, and extensive substitution is required to obtain a water-soluble polymer that is thermoplastic. The low reactivity of the secondary oxypropylene anion facilitates this substitution, but the product is not readily biodegradable. To achieve biodegradability, as many of the C-2 hydroxyl positions as possible must be left unsubstituted, and the polymers should be water soluble (*see* Chapter 5).

The routes to achieving the goal could be many (from an academic viewpoint), but close scrutiny of the economics casts doubt on all scenarios. For example, at the high caustic concentration needed to achieve disruption of the crystallinity and participation of the alkali vicinal diol complex, the addition efficiency of all adducts except ethylene oxide is ca. 20%. The addition efficiency of ethylene oxide is 40%. When other adducts (e.g., propylene oxide) are added to a low-MS HEC, they add preferentially to the oxyanions generated by the ethylene oxide addition and not to the GP ring. Perhaps the best sce-

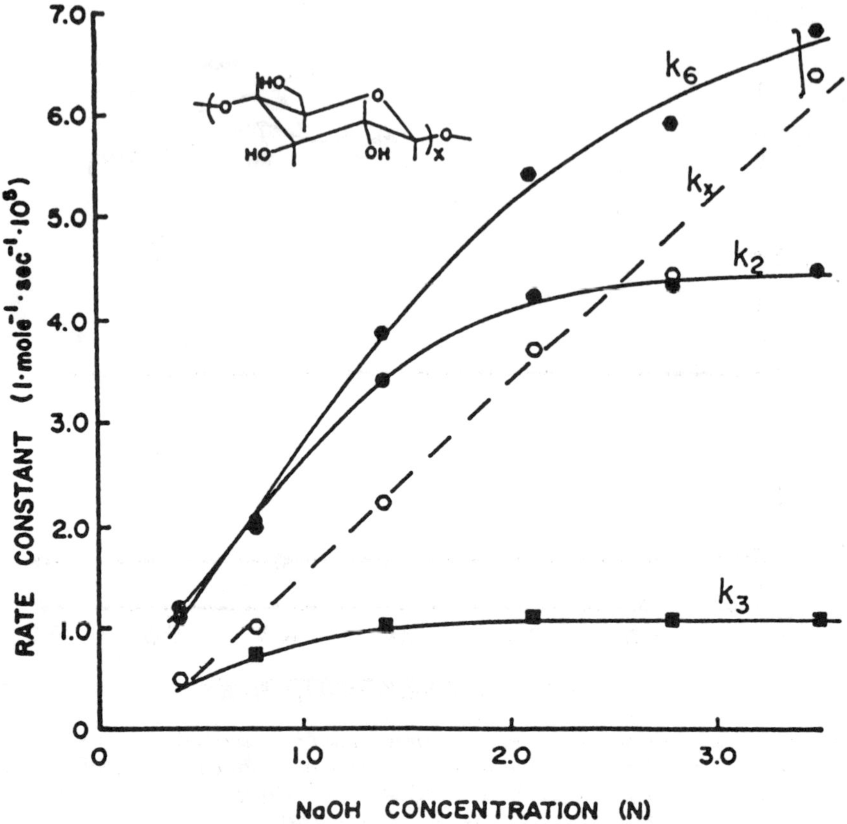

Figure 7. Reaction rates of hydroxyl groups on cellulose with ethylene oxide as a function of sodium hydroxide concentration at 20 °C. Key: ●, k_2; ■, k_3; ⬢, k_6; ○, k_x.

nario would be addition of the sodium salt of α-chloroacetic acid under high caustic concentration followed by esterification of most of the attached acid groups and acetal binding of the 2,3-hydroxyls to the extent needed to complement the carboxymethyl ester for thermoplastic properties. All such groups would be hydrolyzable to provide a water-soluble cellulose substrate for degradation.

This chapter focuses on how knowledge of reaction parameters influences substituent placement in carbohydrate polymers in current production techniques and very briefly summarizes how the use of such parameters would influence economic factors. This chapter is offered in the hope of stimulating new approaches to the production of thermoplastics from renewable resources that are readily biodegradable and thermoplastic.

References

1. Rowland, S. P. In *Modified Cellulosics;* Rowell, R. M.; Young, R. A., Eds.; Academic: New York, 1987; Chapter 5.
2. Klop, W.; Kooiman, P. *Biochim. Biophys. Acta* **1965,** *99,* 102.
3. Glass, J. E.; Buettner, A. M.; Lowther, R. G.; Young, C. S.; Cosby, L. A. *Carbohydr. Res.* **1980,** *84,* 245.
4. Timell, T. E. *Svenske Papperstidn.* **1953,** *56,* 483.
5. Sugihara, J. M. *Adv. Carbohydr. Chem.* **1953,** *8,* 1.
6. Purves, C. B.; Eddy, E. B. *Chem. Can.* **1960,** *12,* 12.
7. Croon, I. *Svenske Papperstidn.* **1960,** *63,* 246.
8. Jones, D. M. *Adv. Carbohydr. Chem.* **1964,** *19,* 219.
9. Haines, A. H. *Adv. Carbohydr. Chem.* **1976,** *33,* 11.
10. Rowland, S. P. In *Encyclopedia of Polymer Science and Technology;* Bikales, N. M., Ed.; Wiley: New York, 1976; Suppl. 1, pp 146–175.
11. Lemieux, R. U.; Bauer, H. F. *Can. J. Chem.* **1953,** *31,* 814.
12. Lenz, R. W.; Holmberg, C. V. *Anal. Chem.* **1956,** *28,* 7.
13. Ramnas, O.; Samuelson, O. *Svenske Papperstidn.* **1968,** *71,* 674.
14. Sweeley, C. C.; Bentley, R.; Makita, M.; Wells, W. W. *J. Am. Chem. Soc.* **1963,** *85,* 2497.
15. Peterson, G.; Samuelson, O. *Svenske Papperstidn.* **1968,** *71,* 731.
16. McNeil, M.; Albersheim, P. *Carbohydr. Res.* **1984,** *131,* 131.
17. McNeil, M.; Szalecki, W.; Albersheim, P. *Carbohydr. Res.* **1984,** *131,* 139.
18. Horton, D.; Lauterbach, J. H. *J. Org. Chem.* **1969,** *34,* 86.
19. Ho, F. F.-L.; Klosiewicz, D. W. *Anal. Chem.* **1980,** *52,* 913.
20. Parfondry, A.; Perlin, A. S. *Carbohydr. Res.* **1977,** *57,* 39.
21. Lee, D.-S.; Perlin, A. S. *Carbohydr. Res.* **1982,** *106,* 1.
22. Reuben, J.; Conner, H. T. *Carbohydr. Res.* **1983,** *115,* 1.
23. Spurlin, H. M. *J. Am. Chem. Soc.* **1939,** *61,* 2222.
24. Reuben, J. *Macromolecules* **1984,** *17,* 156.
25. Rowland, S. P. *Cellulose Chem. Technol.* **1980,** *14,* 423.
26. Rendleman, J. A., Jr. In *Carbohydrates in Solution;* Isbell, H. S., Ed.; Advances in Chemistry 117; American Chemical Society: Washington, DC, 1971; p 51.
27. Rowland, S. P.; Roberts, E. J.; French, A. D. *J. Polym. Sci., Polym. Chem. Ed.* **1974,** *12,* 445.
28. Rowland, S. P.; Wade, C. P. *J. Polym. Sci., Polym. Chem. Ed.* **1980,** *18,* 2577.
29. Ramnas, O.; Samuelson, O. *Svenske Papperstidn.* **1968,** *71,* 829.
30. Ramnas, O.; Samuelson, O. *Svenske Papperstidn.* **1973,** *76,* 569.
31. Ramnas, O. *Acta Chem. Scand.* **1973,** *27,* 3139.
32. Gould, J. M.; Gordon, S. H.; Dexter, L. B.; Swanson, C. L. In *Agricultural and Synthetic Polymers: Biodegradability and Utilization;* Glass, J. E.; Swift, G., Eds.; ACS Symposium Series 433; American Chemical Society: Washington, DC, 1990; Chapter 7, and Cole, M. A. *Ibid.,* Chapter 8.
33. National Starch Corporation. Presented at the Symposium on Degradable Plastics, sponsored by the Corn Products Utilization Council, Indianapolis, IN, November 1989.

RECEIVED for review March 21, 1995. ACCEPTED revised manuscript May 10, 1995.

Associating Polymers
A. General Concepts

9

Fluorescence Studies of Pyrene Capture by Naphthalene-Labeled Diblock Copolymer Micelles in Aqueous Media

Sharon L. Fox, Jack C. Chan, Douglas J. Kiserow, Chittamuru Ramireddy, Petr Munk, and Stephen E. Webber*

Department of Chemistry and Biochemistry and Center for Polymer Research, University of Texas at Austin, Austin, TX 78712

Fluorescence techniques are used to characterize the diffusion of pyrene into several naphthalene-labeled diblock copolymer micelles as a function of pH. The capture of pyrene by the micelles is composed of a fast and a slow process, as shown by intensity and spectral changes in pyrene fluorescence observed when a solution containing pyrene is mixed with a solution containing labeled micelles. The most important pH effect was at low pH: The corona is collapsed. Sensitized pyrene fluorescence results are in accord with a model in which solubilization occurs within the micellar core. Equilibration time of the sensitized fluorescence suggests a pyrene diffusion constant on the order of 10^{-16} cm²/s. Several anomalous photophysical features suggest that the micelle core may have a complex structure.

POLYMERIC MICELLES have recently received considerable attention because of their potential usefulness in many areas as a matrix for controlled release and uptake of small molecules in aqueous and organic environments (*1, 2, 4, 5;* for a review, *see* reference *3*). The fact

* Corresponding author.

0065–2393/96/0248–0141$12.50/0

that diblock copolymers can self-assemble into well-defined micelles when they are solubilized in a solvent selective for one of the blocks is a well-established phenomenon (*6–10*). Micelles formed in this manner are generally categorized as having two main regions: a compact core resulting from aggregation of the insoluble blocks and a loose outer corona composed of the solubilized block. These micelles tend to be monodisperse in size with a well-defined aggregation number. In some solvent combinations this micellization is a reversible process with an equilibrium between unimer and micelle. In a strongly precipitating solvent for the core the micelles appear to be kinetically inert. The unimers in this study are diblock polystyrene (PS)–poly(methacrylic acid) (PMA) copolymers that have been selectively tagged with naphthalene chromophores at each end of the styrene block or at the interface between the two blocks.

Each micelle is assumed to be composed of a hydrophobic core of PS surrounded by a hydrophilic corona of PMA. The thickness of this corona can be adjusted by altering the pH of the surrounding solution. When the pH is raised, the acid moieties in the corona deprotonate, causing the PMA strands that make up the shell to stretch out in order to minimize ionic interactions. Lowering the pH has the opposite effect, causing the acid groups to become more compatible with each other but less soluble in the surrounding solution and allowing the corona to collapse onto the core (*11*). Altering the pH of the solution does not alter the physical or photochemical properties of the core in any observable manner.

This chapter uses energy transfer between chromophores to examine how the structure of the corona at different pHs affects the diffusion of hydrophobic molecular probes into the micellar core. After the micelle solution is added to a saturated pyrene solution, energy transfer from the naphthalene to the pyrene groups is monitored over time by steady-state fluorescence, time-resolved steady-state (TRSS) fluorescence, and time-correlated single-photon-counting (SPC) decay. Changes in the intensity of naphthalene and pyrene emissions and in the vibronic structure of the pyrene emission that permit characterization of the effects of corona extension on small-molecule diffusion into the micelle are monitored and characterized. Pyrene is chosen as an acceptor chromophore not only because of its excellent spectral overlap with naphthalene ($R_0 \approx 30$ Å) (for 2-methylnaphthalene and pyrene, $R_0 = 30.74$ Å, and for naphthalene and pyrene, $R_0 = 28.97$ Å) (*12*) but also because of the well-known effects of environmental polarity on the vibronic structure of the pyrene emission spectra (*13–16*). This property was exploited to establish the average type of environment the pyrene chromophore was experiencing while it was diffusing through different regions of the micelle.

Experimental Details

Polymers. The naphthalene-labeled diblock copolymers used in this experiment were prepared by anionic polymerization of styrene and *tert*-butyl methacrylate at −78 °C in tetrahydrofuran with cumylpotassium as initiator. Polymerization was followed by hydrolysis in a mixture of HCl and 1,4-dioxane. Details of the synthesis and tagging procedure are presented elsewhere (*17, 18*). Two types of labeled polymers were used in this study. They are very similar, except that one polymer is tagged with a naphthalene chromophore at the end of the PS block and one polymer is tagged with a naphthalene group between the styrene and methacrylic acid blocks (denoted N–S–M and S–N–M, respectively). These polymers and the resultant micelles are described in Table I. A small change in polymer properties can strongly affect the polymer micelles' aggregation number and the resultant hydrodynamic diameter.

Block Copolymer Micelles. Polymeric micelles were prepared by dissolving the copolymer samples in an 80% 1,4-dioxane–20% water solvent mixture (v/v) and vortexing the mixture for 30 min. The concentration of copolymer in the solution was approximately 3 g/L for all micelle samples. The resulting solution was clear with a slightly bluish tint caused by light scattering. The micellar solution was then dialyzed stepwise against increasing water concentrations to remove all traces of 1,4-dioxane from the outer solution. Dialysis was performed by placing the micelle solution in a Spectra-por molecular porous membrane tubing (molecular weight cutoff = 6000–8000) and dialyzing it stepwise against outer solutions of increasing water concentration (20% by volume) for at least 3 h per step. The final dialysis into pure water from the 10% 1,4-dioxane solution was carried out over several days; the outer solution was replaced with fresh water once a day. The micelles were then dialyzed against appropriate aqueous pH buffers at constant ionic strength (I = 0.1 M).

Some problems were encountered in handling the micelle solutions at low pH (pH < 3.5). These solutions were apparently only metastable and had a tendency to precipitate from solution when attempts were made to filter them or to transfer the solutions to different containers. These micelle solutions were centrifuged only when this process was absolutely necessary to eliminate dust and particles from the solution for appropriate light-scattering analysis. At higher pHs the solutions appear to be stable under all conditions used in these experiments. The intensity of the naph-

Table I. Characteristics of Diblock Copolymers

Sample[a]	M_n (g/mol × 10^3)	PS: MW × 10^3	PS: mol%	M_n/M_w	N_{agg}	D_H (nm) (pH 11)
S–N–M (SN1A–1)	54.4	22.8	48.0	1.15	70	68.6
N–S–M (N1SA–4)	67.2	34.0	58.5	1.07	262	87.3

[a] Designations in parentheses correspond to the general notation in earlier publications (*19*).

thalene fluorescence in the absence of pyrene was very reproducible (*see* discussion of the pyrene solutions below).

Solvents. Spectral-grade 1,4-dioxane was used as purchased from Aldrich Chemical. Glacial acetic acid was used as purchased from Mallinckrodt. Deionized water was filtered through a 0.2-μm-pore-size cellulose nitrate filter and used in the preparation of all aqueous solutions. All water solutions used in this experiment were maintained at approximately constant 0.1 M ionic strength with 99.99+% pure hydrated lithium chloride ($LiCl\cdot(H_2O)_n$) as purchased from Aldrich.

Solutions. pH 7 and pH 11 Hydrion buffers were purchased dry from Aldrich and prepared as the manufacturer directed to reproducibly control the degree of ionization of the methacrylic acid groups. The ionic strengths of these buffers were then adjusted by dilution to 0.1 M. The pH 3.5 solution was prepared by dissolving 6.0 g of glacial acetic acid into 1.0 L of filtered, deionized water.

Pyrene Solutions. The pyrene used in these experiments (Kodak) was recrystallized 3 times from benzene. Pyrene solutions used in kinetic analysis of the micelle structure were prepared by saturating a 5% 1,4-dioxane–95% buffered H_2O (by volume) solution with pyrene by stirring overnight. The solution was then centrifuged for 30 min to settle all particles and undissolved pyrene crystals. The supernatant was then removed and stored as a pyrene stock solution for that pH. The addition of this amount of 1,4-dioxane improved the solubility of pyrene and helped stabilize the stock solutions. The optical density at 337 nm was ca. 0.06 for these solutions, a level that implies a pyrene concentration of 1.1×10^{-6} M (*20*). (In reference *1*, the extinction coefficient value used was 56,000 M^{-1} cm^{-1} at 342 nm. This solubility exceeds the normal solubility of pyrene in water (0.7×10^{-6} M).)

Despite the presence of 5% dioxane, in this experiment we encountered reproducibility problems that are probably a result of the instability of these pyrene solutions. This instability made it impossible to maintain a constant or reproducible pyrene concentration in these solutions. Thus we experienced difficulty in reproducing the absolute magnitude of pyrene fluorescence or naphthalene quenching. The kinetic behavior, however, appeared to be unaffected.

Light Scattering. Quasielastic light scattering (QELS) was used to measure the apparent hydrodynamic radii (R_H) of the micelles during dialysis. Measurements were performed using a Brookhaven BI 2030 apparatus with a 72-channel correlator. The temperature was maintained at 25 °C, and the scattering angle was 90°. The light source used was a He–Ne laser operating at 632.8 nm. The equipment and the method of calculation of D_H are described in detail elsewhere (*21*). QELS was used to monitor the changes in diameter of the micelles at different pHs and to confirm acceptable polydispersity during dialysis of the micelles. It also served to establish the effect of pH on the corona width.

Fluorescence Spectroscopy. The steady-state data were acquired with a Photon Technology International model LS-100 spectrophotometer. The details of this equipment have been given elsewhere (*17, 18*). The samples were excited at 293 or 337 nm for naphthalene and pyrene, respectively, and the fluorescence emission was monitored from 300 to 500 nm with all emission and excitation slits set at a 4-nm bandpass. The shorter excitation wavelength minimizes direct excitation of the pyrene and corresponds to the excitation wavelength used for the lifetime measurements.

Lifetime measurements were recorded using the method of time-correlated SPC with an Nd:YAG pumped-dye excitation source. This system is described in a previous publication (*21*). The decays were monitored at a right angle. The polarized excitation wavelength was 293 nm, and emission decays were monitored at either 340 nm (naphthalene monomer fluorescence) or 393 nm (pyrene monomer). All measurements were made with an emission polarizer set at the "magic angle" (54.7°) to avoid any fast decay due to polarization effects.

The time dependence of the steady-state fluorescence after mixing was recorded on a SPEX Fluorolog fluorimeter system, which is described elsewhere (*22*). All samples were excited at 293 nm. Emission intensity decays were monitored at both 340 and 393 nm.

All solutions used in the SPC measurements except the pH 3.5 micelle samples were degassed by bubbling nitrogen through them for 30 min. The pH 3.5 micelle samples were degassed for only 10 min because precipitation was induced by longer degassing times. Micelle solutions at this pH are not robust and must be handled carefully. All other fluorescence measurements were performed on solutions that had been degassed by 5 min of sonication under a constant flow of argon.

Experimental Procedure. For all sets of data, a reference solution was prepared by mixing 1.68 mL of buffered H_2O and 0.32 mL of buffered micelle solution in order to establish the naphthalene emission intensity and the spectra of the micelles in the absence of pyrene. Similarly, a solution containing 1.68 mL of pyrene stock solution was added to 0.32 mL of buffered water in order to establish the magnitude of the pyrene emission in the absence of naphthalene. Finally, a third solution of 1.68 mL of buffered pyrene and 0.32 mL of buffered micelle solution was prepared to examine the kinetics of solubilizing the pyrene into the micelles. All the samples were excited at 293 nm, a wavelength that provides direct excitation of the naphthalene with minimal interference from pyrene. (Placing the pyrene solution described in the excitation path of a micelle solution excited at 293 nm did not decrease the naphthalene emission intensity beyond what could be accounted for from reflection losses.)

TRSS curves were also obtained for observation wavelengths of 340 and 393 nm. After the dark count was established, the slits were opened to establish the initial intensity of the pyrene or naphthalene emission. The solutions were then mixed, and the fluorescence intensity was monitored as a function of time at 393 or 340 nm. This set of experiments was designed to characterize the behavior of the solution during the first few minutes after mixing.

Results and Discussion

In general, absolute intensity reproducibility was virtually impossible to maintain. As mentioned in **Experimental Details**, this failure is probably a result of the metastability of the pyrene solutions (*23, 24*). Because the absolute concentration of pyrene varies, the magnitudes of the pyrene emission intensity and the naphthalene quenching vary from trial to trial. The fluorescence intensity for naphthalene-tagged micelles is highly reproducible in the absence of pyrene. The steady-state data presented later in this chapter are representative of the spectra collected, but the magnitude of the pyrene and naphthalene intensity ratios may vary as much as 50%. Despite these problems, useful qualitative conclusions can be drawn by comparing rates of change of the relative intensity of naphthalene and pyrene emission within the same kinetic experiment.

Direct Excitation of Pyrene (λ_{exc} = 337 nm). Pyrene is a popular fluorescent probe because its spectrum reflects local polarity, usually referred to as the I/III ratio (Figure 1) (*16*). In general, in a polar environment (e.g., H_2O), this ratio increases. Though the precise value depends on the spectral resolution of the instrument used, the trends should be clear. The I/III ratio for pyrene was measured in a PS film cast from toluene, in a solution of PMA at pH 3.5, in a solution of PMA at pH 7.0, in buffered water (ionic strength I = 0.1 M), and in a micelle–pyrene solution 8 h or more after mixing. The I/III values for pyrene in water or in different micelle types show no pH depen-

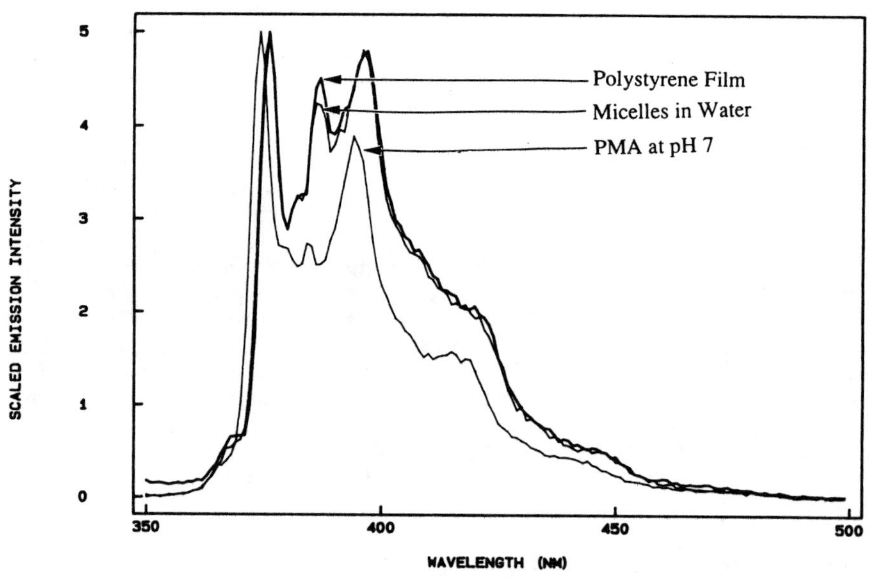

Figure 1. Pyrene fluorescence in PS film, micelle in H_2O, and PMA at pH 7 (λ_{exc} = 337 nm).

Table II. Environmental Effects on Pyrene I/III Ratio

Environment	*I/III Ratio*
Micelles[a]	1.12 ± 0.05
H_2O[b]	1.57 ± 0.05
PMA (pH 7)	1.82
PMA (pH 3.5)	0.95
PS film	1.11

[a] Average for both types of micelles.
[b] Average for different pH aqueous values.

dence (Table II). The I/III ratio for pyrene in our micelles is indistinguishable from that of PS films and is very similar to that for pyrene solubilized in low-pH PMA. A decrease in this ratio occurs within the first measured time interval (ca. 30 s) after the micelle and pyrene solutions are mixed, and the ratio does not change with time within experimental error. These time dependencies are not a function of pH, so the pyrene must be incorporated into the micelle immediately upon mixing. From the present information we cannot distinguish capture by the corona, which seems likely at lower pH, or by the core, which we believe is preferred at high pH.

The intensity of directly excited pyrene does display some surprising pH dependence. If the intensity of the pyrene is corrected for the dilution factor when pyrene is mixed with the micelle stock solution, then the pyrene intensity initially increases dramatically at pHs 7 and 11 and then more slowly over a time scale of hours. On the other hand, the addition of pyrene to a pH 3.5 micelle solution results in a decrease in pyrene fluorescence intensity (Figure 2). It is well-known that PMA at low pH can solvate pyrene (*14*), so the pyrene is probably initially captured in the corona region of the micelle. In a separate experiment, PMA and pyrene were mixed at pH 3.5 and the pyrene fluorescence went up by a factor of ca. 2.1–2.3, as implied by the experiments of Chen and Thomas (*14*). The fluorescence lifetime of pyrene was not changed significantly upon absorption into micelles at various pHs, a result that implies that this loss of fluorescence intensity is the result of a static quenching process. The following were checked as possible artifacts leading to this observation: (1) No measurable enhanced light scattering by the micelles occurs at low pH. Such scattering would decrease the amount of excitation light reaching the chromophore. This observation also discounts the possibility that pyrene induces micelle aggregation. (2) A small shift occurs in the excitation spectrum for pyrene in a hydrophobic environment, but this shift is negligible (ca. 1–2 nm) compared to the excitation bandwidth (ca. 4 nm). In any

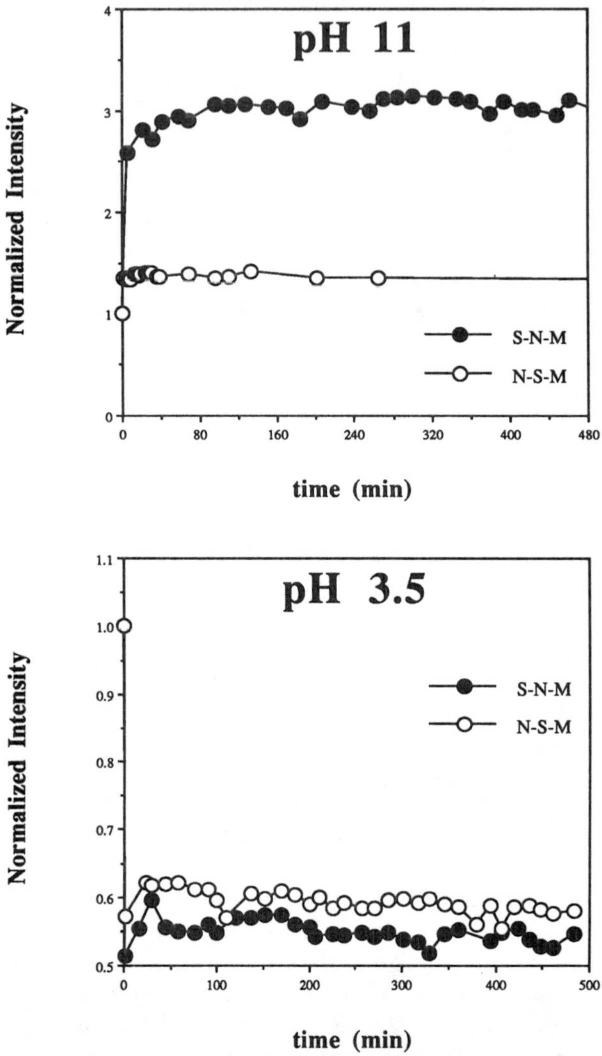

Figure 2. Pyrene fluorescence intensity after addition of micelle solution at indicated pH levels (λ_{exc} = 337 nm).

case, this effect would be present at higher pHs. Thus we must conclude that the quantum yield of pyrene decreases in this environment. This decrease in ϕ_{fl}^{Py} will be discussed later in this chapter in the context of sensitized fluorescence experiments.

From these simple experiments we can conclude that pyrene is incorporated into the micelles extremely rapidly. Thus in earlier experiments, the slow rate of exchange of hydrophobes between micelles was dominated by the rate of escape of these molecular species from the micelle. This conclusion was reached in an earlier paper because the rate of exchange was not influenced by mixing (*1*). In the next sections we examine sensitized fluorescence and try to differentiate the rate at which the captured pyrene enters the interfacial region or the core region of the micelle.

Steady-State Energy Transfer (λ_{exc} = 293 nm). For experiments on a shorter time scale after mixing (TRSS), a single wavelength, either 340 nm (naphthalene) or 393 nm (pyrene), was observed. Because these experiments are separate kinetic runs and because of the difficulties with the absolute intensity of fluorescence (*see* **Experimental Details**), the absolute loss and gain of fluorescence for each species cannot be directly compared. For TRSS experiments the intensity was normalized to the premixing intensity, with dilution effects taken into account. For longer time intervals the samples were excited at 293 nm, and fluorescence emission was monitored between 300 and 500 nm. The curves were integrated over the 315–360-nm region for naphthalene and the 388–450-nm region for pyrene. Because the optical density of the pyrene at 293 nm is very small compared to that of naphthalene, all loss of naphthalene fluorescence immediately after mixing can be assumed to be because of energy transfer to pyrene. The total fluorescence of pyrene is both directly excited and sensitized. If no sensitization of pyrene occurred, its fluorescence would be decreased by a factor of ca. 0.33 because of the naphthalene inner-filter effect. Typical spectra are shown in Figure 3.

TRSS Results. When pyrene is added to the S–N–M micelle solution, naphthalene fluorescence immediately drops (Figure 4). As mentioned in a preceding section, this drop is not a result of pyrene absorption at 293 nm. A rapid ca. 10% increase in naphthalene fluorescence (monitored at 340 nm) often occurs after the initial drop; we ascribe this increase to a mixing artifact. At pHs 7 and 11, very little further change in the naphthalene fluorescence occurs on this time scale, but at pH 3.5, the decrease in the fluorescence is more gradual. As we will see later (Figure 5), this trend is also noted on a longer time scale for pHs 7 and 11, with solutions at pH 3.5 showing no change after the first hour of mixing. The growth of pyrene fluorescence (monitored at 393 nm) at pHs 7 and 11 is easily seen. At pH 3.5 the pyrene intensity decreases after the initial mixing, as it also does on longer time scales. This decrease is consistent with the ϕ_{fl}^{Py} decrease observed in the direct excitation experiments at low pH just discussed.

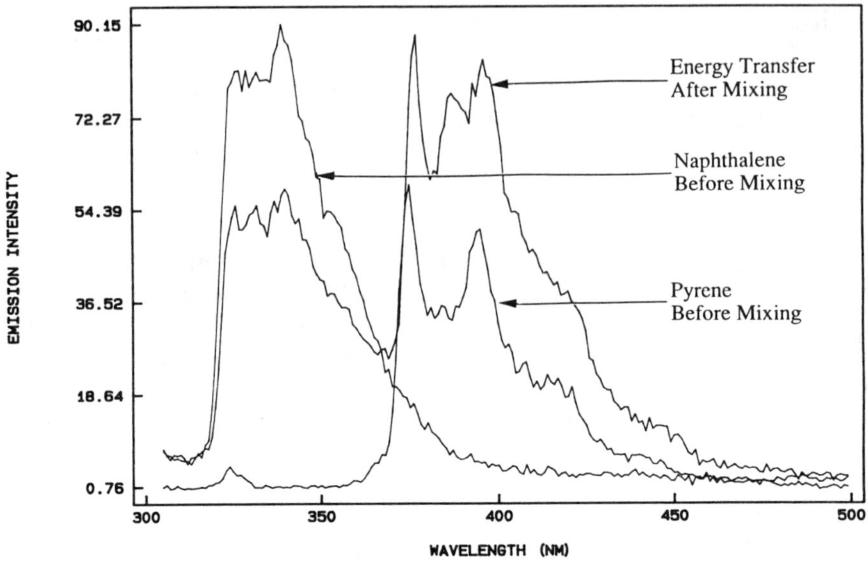

Figure 3. Steady-state fluorescence spectrum of initial pyrene solution and naphthalene-tagged micelle solution before and after (ca. 8 h) mixing (λ_{exc} = 293 nm).

The N–S–M case is much simpler. The fluorescence of naphthalene decreases steadily, after the initial dip from mixing, and that of pyrene increases steadily. The pH does not appear to modify the kinetics of this process, even at longer time scales. Once again, the pyrene fluorescence growth is more obvious than the naphthalene decrease. At pH 3.5, sensitized pyrene fluorescence shows no decrease after the initial jump, unlike directly excited pyrene or pyrene sensitized by S–N–M. We conclude that the effect that lowers ϕ_{fl}^{Py} in the direct excitation and the S–N–M TRSS experiments at pH 3.5 must involve the corona interfacial region. Thus by comparing sensitized and direct excitation, we are carrying out a kind of photoselection experiment.

Long-Time Steady-State Results. Data collected on a longer time scale reflect trends similar to those just discussed. For pHs 7 and 11 after the first 20 min, only minor changes are observed in the normalized naphthalene or pyrene emission intensity for S–N–M micelles (Figure 5). We plot the intensity of naphthalene fluorescence in terms of the efficiency of energy transfer to the pyrene, χ (25):

$$\frac{I_D{}^0 - I_D(t)}{I_D{}^0} = \chi \tag{1}$$

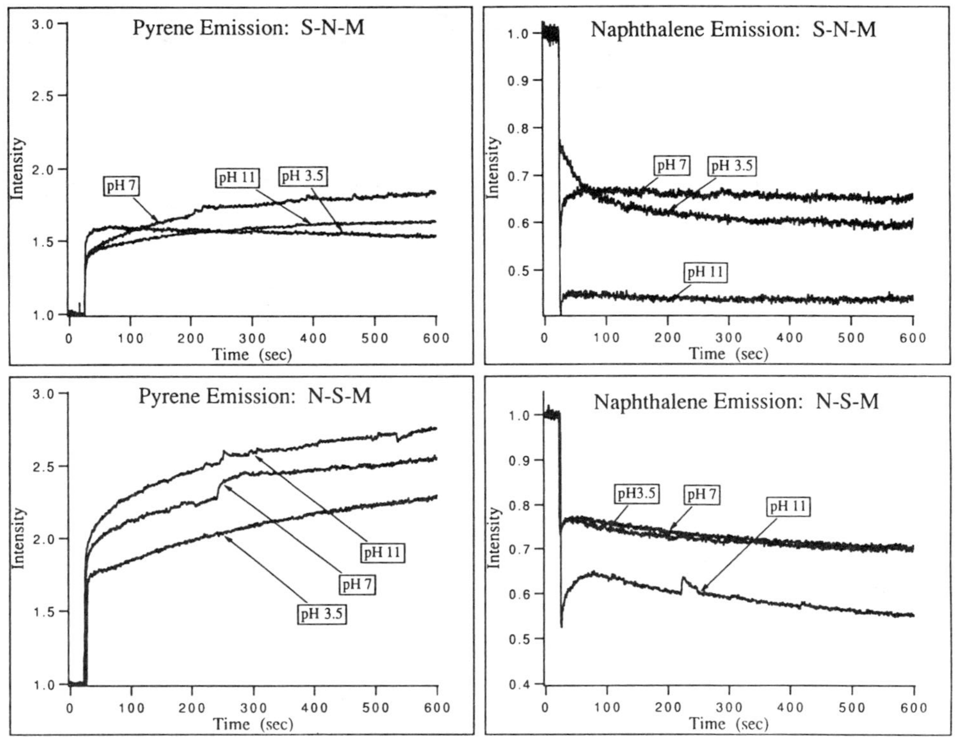

Figure 4. Change in the fluorescence of naphthalene (λ_{obs} 340 nm) and pyrene (λ_{obs} 393 nm) as a function of time after mixing for pH indicated (fluorescence was normalized to the premixing intensity for convenience).

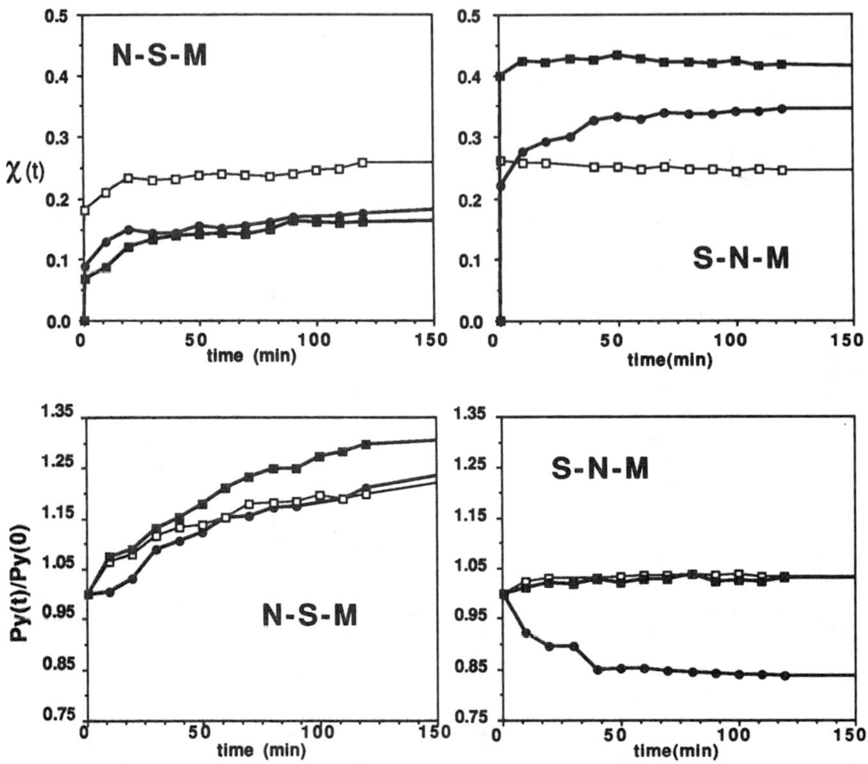

Figure 5. Like Figure 4 but collected on a longer time scale by integrating the steady-state spectrum over different wavelength regions. Pyrene is normalized to the intensity of the earliest time after mixing, and naphthalene is normalized to that of the time immediately before mixing and plotted as χ (see text). Key: ●, χ(t) at pH 3; □, χ(t) at pH 7; ■, χ(t) at pH 11.

where $I_D{}^0$ is the fluorescence of the naphthalene before mixing (corrected for dilution). For S–N–M at pH 3.5, pyrene emission shows an initial sharp increase when the micelle and pyrene solutions are mixed (*see* Figure 4), but then the pyrene begins to lose intensity (*see* Figure 5). The final equilibrium intensity is achieved on the same time scale as the other mid-tagged samples but at an intensity lower than the initial intensity. This observation is very reproducible. For S–N–M over the same time scale at pHs 7 and 11, naphthalene fluorescence decreases slightly (Figure 5). Thus very little of the pyrene is diffusing out of sensitization range of the naphthalene. This diffusion does not occur at pH 3.5, a pH level at which χ increases steadily, similar to the N–S–M case described next.

The normalized curves for N–S–M are simpler, with a slower decrease in naphthalene fluorescence and a steady increase in pyrene intensity. Because of the difficulty with the stability of the pyrene solution, the absolute value of the changes plotted in Figure 5 for different pHs should not be compared, but the kinetic behavior is reproducible.

The main conclusions are that the radial density of the corona and the overall radius, which change drastically with pH, seem to have very little effect on the rate of capture of diffusing molecules. The behavior for S–N–M emission implies that the pyrene molecule reaches the naphthalene at the interface very rapidly (<30 s). Because there is no significant increase in the naphthalene emission intensity after 8 h, the pyrene concentration appears to remain essentially constant near the interfacial region. For S–N–M at low pH, the loss of pyrene fluorescence without a corresponding increase in naphthalene fluorescence indicates that the pyrene is transferred into a region that has a lower fluorescence quantum yield. In general this case is much harder to rationalize than any of the others.

Changes in the intensity of sensitized emission occur on a much slower time scale for N–S–M. After mixing, the emission shows a more gradual change that persists for the first few hours. This gradual change implies a distinctly longer delay time before pyrene diffuses within the Förster radius of the naphthalene groups in the core of the micelle.

SPC Results. SPC data were acquired by exciting degassed samples at 293 nm and observing the fluorescence time-resolved emission decay of naphthalene at 340 nm. Typical decay curves are presented in Figure 6. The shortening of the naphthalene excited-state lifetime is quite obvious and is expected for Förster energy transfer quenching. Because acquiring a satisfactory decay curve requires ca. 15 min, the time steps between subsequent decay curves are much larger than those in the steady-state quenching data. The quenching of the naphthalene emission can be computed by integration of the naphthalene decay curves and is presented in Figure 7. In the absence of static quenching by naphthalene–pyrene interactions or by modification of the naphthalene fluorescence decay from environmental effects, these curves should be equivalent to the steady-state quenching curve. We reemphasize that the magnitude of quenching varies from experiment to experiment, so quantitative comparisons are dangerous.

The magnitude of the lifetime quenching is pH dependent for S–N–M; the least quenching occurs at pH 11. The lifetime quenching is much less efficient than steady-state quenching for pH 11 (cf. Figures 5 and 7) by a factor that is well outside the experimental uncertainty. This difference implies that static quenching is occurring in this case.

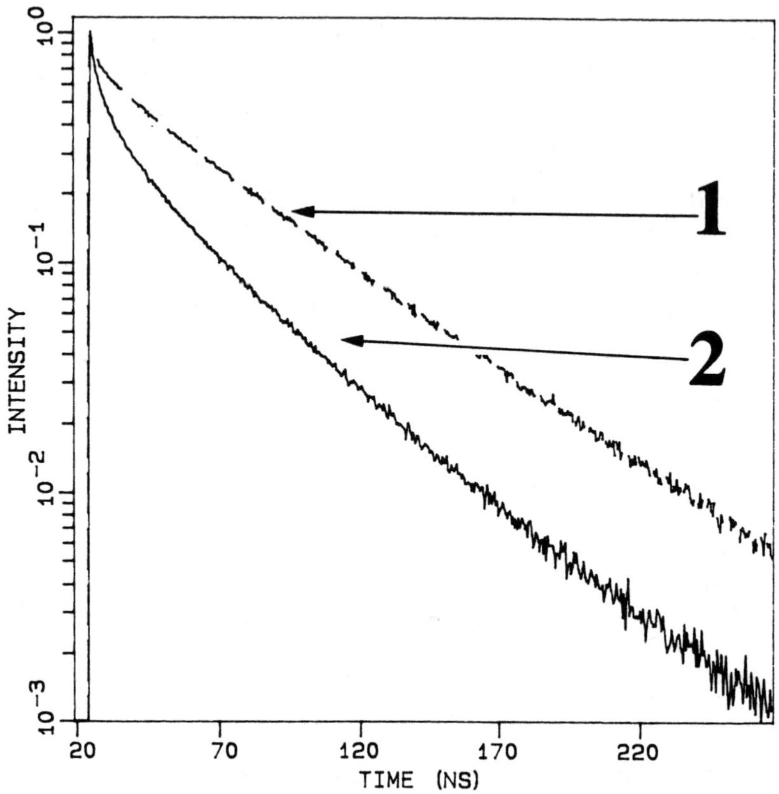

Figure 6. Time-resolved fluorescence decay of naphthalene in S–N–M micelle immediately after addition of pyrene (curve 1) and 8 h after mixing (curve 2) at pH 7 (see text).

The kinetics of the quenching of N–S–M micelles occur on a longer time scale than those for S–N–M, as was also shown for the steady-state data (Figure 5). The quenching derived from time-dependent decays is larger than that derived from steady-state (cf. Figures 5 and 7), which can arise only from differences in the amount of pyrene added to the solution or in environmental effects on the naphthalene lifetime. Within experimental error these curves can be taken to be identical. We conclude that pH has no important effect on pyrene sensitization kinetics for N–S–M, unlike S–N–M, which exhibits significant static quenching at pH 11. Also, for N–S–M, no evidence for static quenching of the naphthalene groups has been obtained, because the lifetime quenching is smaller than the steady-state quenching.

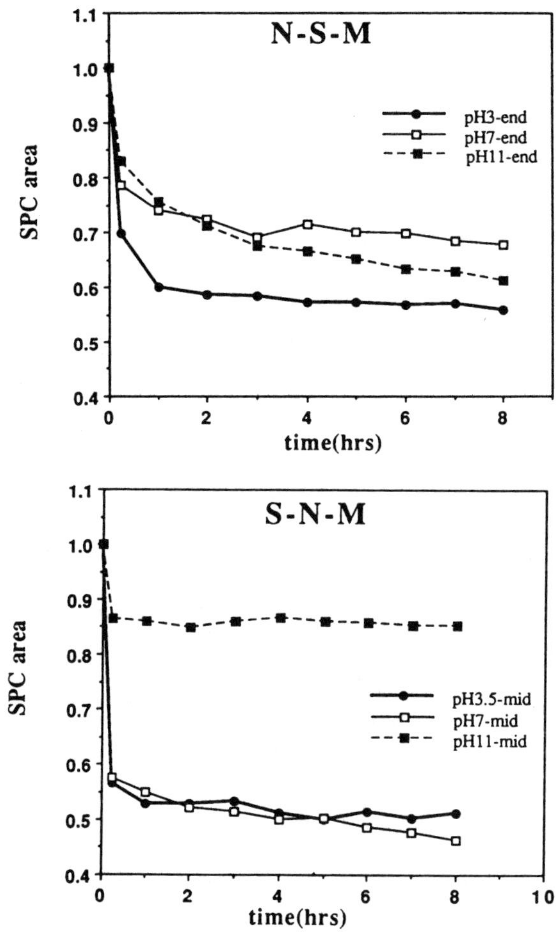

Figure 7. Quenching curves derived from integration of time-dependent naphthalene fluorescence.

Interpretation of the Kinetics of Pyrene Sensitization

In this discussion the energy transfer between the naphthalene and pyrene groups is assumed to occur by the Förster mechanism (*26, 27*):

$$k_{ET}(r) = k_D{}^0 \, (R_0{}^{DA}/r)^6 \tag{2}$$

where $k_D{}^0$ is the intrinsic decay rate of the donor excited state, $R_0{}^{DA}$ depends on the donor–acceptor pair, and r is the separation of the pair. A random orientation between the D–A pair is assumed in the values of R_0^{DA} tabulated by Berlman (*12*). The numerical differences in the appropriate R_0^{DA} for chromophores that rotate rapidly during their excited-state lifetimes versus a random ensemble of static chromophores can be ignored in this semiquantitative discussion. (For chromophores that are static during their lifetimes, R_0^{DA} would have to be multiplied by a factor of (0.475/0.667)1/6 = 0.945 (see reference 12, p 28).) The ensemble average probability of energy transfer from donor to acceptor for a viscous homogeneous solution was also obtained by Förster (*12, 26, 27*):

$$P_{DA}(x) = \pi^{1/2} \times e^{(x^2)} \{1 - \mathrm{erf}(x)\} \tag{3}$$

where $x = C_A/C_{DA}^0$, C_A is the concentration of acceptor, and C_{DA}^0 is the critical concentration of acceptor based on R_0^{DA} and equals

$$\frac{3000}{2\pi^{3/2}N_0} \frac{1}{(R_0^{DA})^3} \tag{4}$$

where N_0 is Avogadro's number. This expression ignores the effect of the finite size of these micelles, which in general tends to decrease the energy efficiency (*28*). (This tendency is illustrated in Chapter 1, section 1.21, of reference 28 in the comparison of a donor on the surface of a sphere transferring energy to acceptors in the interior of the sphere.) This expression also ignores energy transfer between donors, which tends to increase $P_{DA}(x)$. ($P_{DA}(x)$ corresponds to χ in equation 1.) When $x = 1$, then $P_{DA}(x) = 0.76$. For a donor–acceptor pair precisely separated by R_0^{DA}, $\chi = 1/2$. Therefore C_{DA}^0 and R_0^{DA} characterize the critical concentration and spatial resolution, respectively, of an energy transfer experiment. Also, for D–A pairs separated by $2R_0^{DA}$, energy transfer is negligible ($\chi = 0.015$). In a later discussion, we will use a cutoff of $\sqrt{2}R_0^{DA}$ ($\chi = 0.125$). The R_0^{DA} values for naphthalene and pyrene range between 28.6 and 30.7 Å depending on the naphthalene substitution (*12*), and in this discussion we use the value 30 Å, which corresponds to $C_{DA}^0 = 16.6$ mM in equation 4. Because of the restricted geometric aspects of these systems, this value tends to overestimate the energy transfer efficiency.

The R_0^{Np-Py} value is ca. 10% of the hydrodynamic radius of these micelles (Table I), but at higher pH when the corona is fully extended, we expect the pyrene to rapidly encounter the PS core of the micelle. The core radius (R_{core}) can be estimated by assuming that the core has the same density as bulk PS (1.1 g/cm^3) (*29*). This calculation assumes that all dioxane used in the micelle preparation has been completely removed by dialysis and ignores the effect of the 5 vol%

of dioxane used in the pyrene stock solution. To the extent that the core is swollen by dioxane, R_{core} is underestimated in this discussion. From the N_{agg} values and the molecular weights of the PS segment given in Table I, we estimate $R_{core}{}^{S-N-M} \approx 83$ Å and $R_{core}{}^{N-S-M} \approx 150$ Å. As we will see, the fact that these micelles have different R_{core} values makes their comparison less direct than desirable.

The maximum concentration of pyrene in the core is computed from the condition that 1.68 mL of 1.07×10^{-6} M pyrene is combined with 0.32 mL of a 3 mg/mL micelle solution. These concentrations work out to 7.1 and 31 pyrene molecules per micelle for S–N–M and N–S–M, respectively, or $C_{Py}{}^{S-N-M} = 4.82 \times 10^{-3}$ M and $C_{Py}{}^{N-S-M} = 3.73 \times 10^{-3}$ M if the above estimates for R_{core} are used. (The partition coefficient for pyrene is 2.2×10^5 (*1*), from which we estimate that less than 0.2% of pyrene is in the bulk phase. The slight amount of dioxane present tends to increase the pyrene content in the bulk solvent phase.) We are ignoring solubilization in the corona in these calculations.

N–S–M Micelles. As stated in the preceding section, we use equation 3 in the calculations that follow. We treat the naphthalene and pyrene concentration within the core of N–S–M micelles as homogeneously distributed and obtain $C_{Py}{}^{N-S-M}/C_{Np-Py}{}^{0} = 0.225$ and $P_{Np-Py} = 0.315$. If at the earliest time after mixing the pyrene resides in the outer portion of the micelle, then the portion of the "inner" core located ca. $\sqrt{2}R_0$ from the outer surface is quenched only slightly (i.e., $(\sqrt{2})^{-6} = 0.125$). The volume fraction of this portion of the inner core is given by

$$\phi_{inner}{}^{N-S-M} = \left(\frac{R_{core}{}^{N-S-M} - \sqrt{2}R_0}{R_{core}{}^{N-S-M}}\right)^3 = \left(1 - \frac{\sqrt{2}\,30}{150}\right)^3 = 0.369 \tag{5}$$

Thus a rough estimate of χ immediately after initial mixing and equilibration would be

$$\chi_{initial}{}^{N-S-M} = (1 - 0.369)(0.315) = 0.199$$

$$\chi_{final}{}^{N-S-M} = 0.315$$

that is, the quenching of the naphthalene should increase by a factor of ca. 1.6 over time. These values are slightly smaller than those observed for χ^{N-S-M} (*see* Figure 5; the average ratio between the first time step and the last is ca. 1.85 for pHs 7 and 11).

S–N–M Micelles. If we consider the naphthalene groups for S–N–M to be confined to a shell of thickness $\sqrt{2}R_0^{\text{Np–Py}}$ around the exterior of the PS core, then the volume fraction of this outer region is $\phi_{\text{outer}}^{\text{S–N–M}} = 1 - \phi_{\text{inner}}^{\text{S–N–M}}$, for which an equation corresponding to equation 5 is used, and we obtain $\phi_{\text{outer}}^{\text{S–N–M}} = 0.883$. Thus the initial pyrene concentration in this region is ca. $4.82 \times 10^{-3}/0.883 = 5.46 \times 10^{-3}$ M, and the ratio with respect to C_{DA}^0 is $5.46/16.6 = 0.329$, which corresponds to $\chi_{\text{initial}}^{\text{S–N–M}} = 0.416$. If the pyrene eventually diffuses uniformly throughout the core, then $\chi_{\text{final}}^{\text{S–N–M}} = (0.416)(0.883) = 0.367$; that is, a slight decrease in naphthalene quenching should occur. The pH 7 data show a hint of this behavior, but the pH 11 data do not (*see* Figure 5).

At pH 3.5, the collapsed corona probably presents a barrier to diffusion into the PS core, and hence the appropriate geometry must take into account the collapsed corona (R_{H} for N–S–M at pH 3.5 is ca. 185 Å according to QELS (*11*)). If we assume that the pyrene initially resides in a shell centered on $R_{\text{core}}^{\text{S–N–M}}$ (≈83 Å) ± $\sqrt{2}R_0^{\text{Np–Py}}$ (≈42.4 Å), then the volume in which the pyrene resides is given by

$$\frac{V_{\text{initial}}}{4\pi/3} = (R_{\text{core}}^{\text{S–N–M}} + \sqrt{2}R_0^{\text{Np–Py}})^3 - (R_{\text{core}}^{\text{S–N–M}} - \sqrt{2}R_0^{\text{Np–Py}})^3 \qquad (6)$$

(This calculation overestimates the initial concentration of pyrene in this region if significant pyrene is solubilized in the outermost part of the corona.) This calculation also yields $\chi_{\text{initial}}^{\text{S–N–M}} \approx 0.143$, which increases to a final value of 0.367 (assuming that all pyrene eventually resides in the core). This ratio of initial to final $\chi^{\text{S–N–M}}$ agrees reasonably well with the pH 3.5 data.

A full interpretation of these experiments would require a much more elaborate diffusion–reaction theory and an accounting for the effects of specific shape factors on the efficiency of Förster energy transfer. However, the preceding discussion does yield χ values that agree roughly with experimental values, and this approach probably contains the important physical ideas that underlie these experiments. Except for the prediction for S–N–M that naphthalene quenching at higher pHs should drop by ca. 12% from the time of initial pyrene capture to equilibration, most predictions for short- and long-time values of χ are qualitatively correct. The prediction for higher pHs is approximately correct for pH 7 but not for pH 11. This predicted effect would be diminished if the naphthalene groups were not located precisely at the outer edge of the PS core. A representation of how a diffuse interface might be achieved is presented in Scheme I, in which

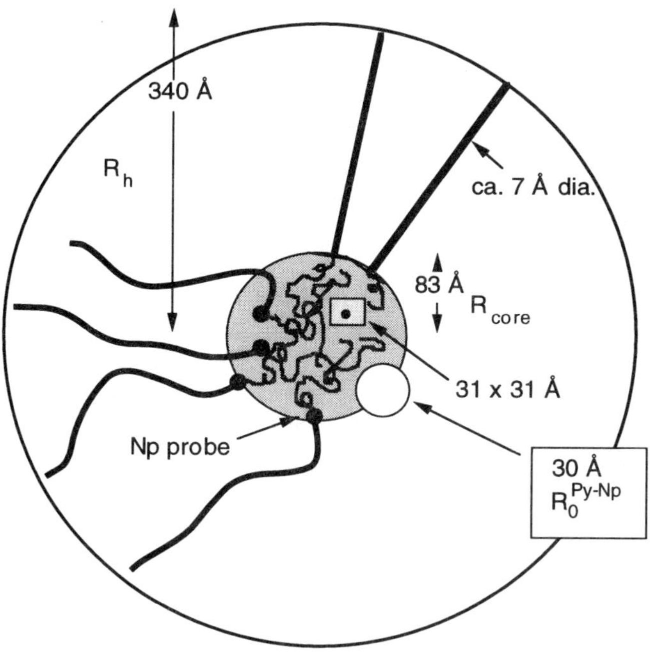

Scheme I. S–N–M at high pH. Np is naphthalene.

the micelle is depicted as being formed from a group of disordered PS coils in juxtaposition. On the basis of fluorescence depolarization experiments, we have argued that this interface is not sharp (*11*). From other ongoing work, we know that 90% of the naphthalene groups in S–N–M micelles are accessible to Tl^+ quencher at high pH, while the naphthalene groups in N–S–M are totally inaccessible (*30*). Thus the representation of S–N–M in Scheme I must include porosity and cation access to most naphthalene groups. Additionally, from the calculations provided elsewhere (*29*), the area per emerging chain corresponds to a region approximately 31–33 Å on a side; this area is large compared to the expected size of a PMA coil (ca. 6.8-Å diameter). Thus, according to this picture, also depicted in Scheme I, the open area per PS coil is considerable if PMA is deprotonated.

Apparent Pyrene Diffusion Constant in Micelles. The constant for pyrene diffusion into the glassy PS core can be roughly estimated from the present data. Figure 5 shows that the quenching of naphthalene fluorescence for N–S–M micelles equilibrates in ca. 20–30 min. We estimate $R_{core} \approx 150$ Å for these micelles, and the core region should be homogeneously occupied by pyrene on a time scale such that $(D_{Py}t_{eq})^{1/2} \approx R_{core}$. From this rough estimate we obtain $D_{Py} \approx 2 \times 10^{-15}$ cm²/s. Typically, the diffusion inside a sphere is given by a sum of exponentials of the form $\exp(-n^2\pi^2Dt/R_{core}^2)$ (*31*). Taking the slowest $n = 1$ component and setting it equal to 0.05 at 20 min, we obtain 1.4×10^{-16} cm²/s. These very small estimated values seem reasonable given the large van der Waals radius of pyrene and the values determined by Berens and Hopfenberg (*32*).

Unusual Photophysical Behavior of Pyrene. Direct excitation of pyrene demonstrated that pyrene is rapidly incorporated into a "hydrophobic environment" as defined by the I/III ratio. Thus all slower kinetic steps as just discussed must involve diffusion within the core region or, at low pH, between the corona and the core region. One of the surprising results is that at low pH, ϕ_{fl}^{Py} is diminished after the initial mixing. We argued that this effect occurs in the core–corona interfacial region, but we have no suggestion as to the origin of the effect. The PMA coil may be forced into a different conformation than would be found for the normal "hypercoiled state" of PMA at low pH because of the constraints of the micelle structure. Thus we can speculate on a quenching interaction with the –COOH groups that does not exist for pyrene solubilized in normal linear PMA at low pH.

Conclusions

This set of experiments revealed some very interesting properties of block copolymer micelle sorption of the highly hydrophobic pyrene probe from the surrounding solution.

Pyrene is sorbed into a hydrophobic region (presumed to be the corona–shell interface) very rapidly (<30 s). Diffusion into the core and away from the interface is somewhat slower, as shown by the time-dependent quenching of the N–S–M micelles. The compactness of the shell appears to have little effect on pyrene diffusion into the N–S–M core, but the situation is different for the S–N–M interfacial region. The magnitude of the naphthalene sensitization efficiency is approximately what is expected for a core–shell model, given the different locations of the naphthalene probes. However, the data hint at a broader interfacial region for S–N–M, which might be on the order of R_0^{Py-Np}. Addressing this point meaningfully requires more detailed kinetic modeling.

Future work of this type should use a hydrophobic quencher, which forms more stable aqueous solutions than pyrene, in order to further quantify the degree and kinetics of quenching experienced by the naphthalene tags in these micelles. Certainly we have encountered some unexplained effects on ϕ_{fl}^{Py} that do not enhance our understanding of these micelles except to suggest the existence of a more heterogeneous structure for the core.

Acknowledgments

This research was supported by the Department of the Army (Grant No. DAAA L03-90-G1047). S. E. Webber also acknowledges the support of the National Science Foundation Polymers Program (Grant No. DMR 9000562). S. L. Fox was a recipient of a Department of Education Fellowship during part of this research.

References

1. Cao, T.; Munk, P.; Ramireddy, C.; Tuzar, Z.; Webber, S. E. *Macromolecules* **1991**, *24*, 6300.
2. Kiserow, D.; Procházka, K.; Ramireddy, C.; Tuzar, Z.; Munk, P.; Webber, S. E. *Macromolecules* **1992**, *25*, 461.
3. Kataoka, K.; Kwon, G. S.; Yokoyama, M.; Okano, T.; Sakurai, Y. J. *Controlled Release* **1993**, *24*, 119.
4. Yokoyama, M. *Crit. Rev. Ther. Drug Carrier Syst.* **1992**, *9*(3,4), 213.
5. Rolland, A.; O'Mullane, J.; Goddard, P.; Brookman, L.; Petrak, K. *J. Appl. Polym. Sci.* **1992**, *44*, 1196.
6. Tuzar, Z.; Kratochvíl, P. *Surf. Colloid Sci.* **1993**, *15*, 1.
7. Wilhelm, M.; Zhao, C.-L.; Wang, Y.; Xu, R.; Winnik, M. A. *Macromolecules* **1991**, *24*, 1033.
8. Xu, R.; Hu, Y.; Winnik, M. A. *Langmuir* **1991**, *7*, 831.
9. Xu, R.; Winnik, M. A.; Hallett, F. R.; Riess, G.; Croucher, M. D. *Macromolecules* **1991**, *24*, 87.
10. Zhao, C.-L.; Winnik, M. A.; Riess, G.; Croucher, M. D. *Langmuir* **1990**, *6*, 514.
11. Chan, J.; Fox, S.; Kiserow, D.; Ramireddy, C.; Munk, P.; Webber, S. E. *Macromolecules* **1993**, *26*, 7016.
12. Berlman, I. B. *Energy Transfer Parameters of Aromatic Compounds;* Academic: New York, 1978.
13. Kalyanasundaram, K.; Thomas, J. K. *J. Am. Chem. Soc.* **1977**, *99*, 2039.
14. Chen, T. S.; Thomas, J. K. *J. Polym. Sci.* **1979**, *17*, 1103.
15. Olea, A. F.; Thomas, J. K. *Macromolecules* **1989**, *22*, 1165.
16. Dong, D. C.; Winnik, M. A. *Photochem. Photobiol.* **1982**, *35*, 17.
17. Kiserow, D.; Chan, J.; Ramireddy, C.; Munk, P.; Webber, S. E. *Macromolecules* **1992**, *25*, 5338.
18. Ramireddy, C.; Tuzar, Z.; Procházka, K.; Webber, S. E.; Munk, P. *Macromolecules* **1991**, *25*, 2541.
19. Tian, N.; Qin, A.; Ramireddy, C.; Webber, S. E.; Munk, P. *Langmuir* **1993**, *9*, 174.
20. Schwartz, F. P. *J. Chem. Eng. Data* **1977**, *22*, 273.
21. Procházka, K.; Kiserow, D.; Ramireddy, C.; Tuzar, Z.; Munk, P.; Webber, S. E. *Macromolecules* **1992**, *25*, 454.

22. Sturtevant, J. L.; Webber, S. E. *Macromolecules* **1989,** *22,* 3564.
23. May, W. E.; Wasik, S. P.; Freeman, D. H. *Anal. Chem.* **1978,** *50,* 175.
24. Ibid., p 997.
25. Holden, D. A.; Guillet, J. E. *Macromolecules* **1980,** *13,* 289.
26. Förster, T. *Ann. Phys. (Leipzig)* **1948,** *2,* 55.
27. Förster, T. *Discuss. Faraday Soc.* **1959,** *27,* 7.
28. *Molecular Dynamics in Restricted Geometries;* Klafter, J.; Drake, J. M., Eds.; John Wiley: New York, 1989.
29. Qin, A.; Tian, M.; Ramireddy, C.; Webber, S. E.; Munk, P. *Macromolecules* **1994,** *27,* 120.
30. Cao, T.; Yin, W.; Webber, S. E. *Macromolecules* **1994,** *27,* 7459.
31. Crank, J. *The Mathematics of Diffusion;* Oxford University: London, 1970.
32. Berens, A. R.; Hopfenberg, H. B. *J. Membr. Sci.* **1982,** *10,* 283.

RECEIVED for review November 18, 1993. ACCEPTED revised manuscript October 31, 1994.

10

Associative Thickeners

An Overview with an Emphasis on Synthetic Procedures

Wylie H. Wetzel, Mao Chen, and J. Edward Glass

Department of Polymers and Coatings, North Dakota State University, Fargo, ND 58105

The lack of concern for and understanding of the structure of surfactant-modified, water-soluble polymers is widespread. This lack is surprising, because structural variations can promote significant differences in solution behavior and application properties. In this chapter, we consider the structure of surfactant-modified, water-soluble polymers and evaluate comparative differences in the placement of hydrophobes in chemically different families.

HYDROPHOBE MODIFICATION of three chemically different families is considered in this chapter.

1. Cellulose ethers. In this family, a few large hydrophobes or many small ethyl groups are added to hydroxyethylcellulose (HEC) to form hydrophobe-modified HEC (HMHEC) or ethyl HEC (EHEC);
2. Chain growth terpolymers. Three monomers are commonly modified: acrylamide, methacrylate, and styrene. All three hydrophobe-modified monomers are included in small percentages. Methacrylate and styrene contain large amounts of ionogen monomers and have at least

0065–2393/96/0248–0163$12.00/0

20 oxyethylene units between the hydrophobes and the methacrylate or styrene group. These terpolymers (hydrophobe-modified, alkali-swellable emulsions, or HASE) also contain small amounts of difunctional monomers that form cross-linked networks.

3. Step-growth hydrophobe-modified, ethoxylated urethane (HEUR). This series is the only one in which hydrophobe placements can be quantified, but the polymers have a broad molecular weight distribution and, depending on the molecular weight, a significant amount of unreacted poly(oxyethylene) (POE).

Cellulose Derivatives

HMHEC. The reaction of alkali cellulose with ethylene oxide (discussed in Chapter 8) produces pendant nonionic oxyethylene units that promote solubility in water, in part by disruption of hydrogen bonding among the glucopyranosyl (GP) repeating units of cellulose and in part by a hydrophilic contribution arising from a unique interaction of water with the oxyethylene group (*1*). The distributions of oxyethylene chains on the segmentally rigid cellulose chain are given in Table I. HEC has been the primary thickener of latex coating formulations for 25 years. The limitations of formulations thickened with HEC are viscosities that are too high at low shear rates ($< 1\ s^{-1}$) (*2*), viscosities that are too low at high shear rates ($10^4\ s^{-1}$) (*2*), spatter and misting during roll applications (*3*), high water sensitivity (*4*), and poor gloss in applied films (*5*). These limitations are addressed in the references just cited.

The first report on the synthesis and solution behavior of HMHEC was published in 1982 (*6*). This and subsequent publications from the Aqualon laboratories (*7, 8*) discuss hydrophobic associations and viscosity maxima with increasing anionic surfactant concentrations. On average, only three hydrophobes (ranging from $C_{12}H_{25}$- to $C_{16}H_{33}$- in size) are placed on a cellulose chain (*9*). The original HMHEC studies contained only 2.5 mol of oxyethylene units (molar substitution [M.S.]) per molar GP unit. This product has limited solubility in aqueous solutions because of the size of the hydrophobes, and a second HMHEC derivative, with an M.S. of 3.3, was introduced for use in coatings formulations.

The structure HMHEC is illustrated in Figure 1. In the commercial or laboratory synthesis of HEC (*10, 11*), all of the hydroxyl groups are available, and the placement of oxyethylene units becomes a statistical matter depending on the relative reactivities of the oxyanions generated (Chapter 8). The structure of HEC has been defined by

Table I. Characteristics of Monosubstituent Cellulose Ethers

Cellulose Ether	M.S.[a]	% Unsubstituted Glycopyranose Units	Unsubstituted Units in Trains of Length n: n	%	Adduct Units: n	%	Degree of Order in Dry Powder[b]
HEC	1.0	37.0 ± 1.4	2	26.4 ± 2.9	2	32.1	—[c]
			3	14.9 ± 2.4	3	9.5	
			4	9.1 ± 0.7	4	2.5	
			5	4.9 ± 2.0	5	0.5	
	2.0	17.0 ± 1.1	2	21.9 ± 7.6	2	34.6	
			3	3.6 ± 3.5	3	18.2	10–20% crystalline
			4	1.8 ± 3.1	4	8.8	
			5	0.0	5	3.4	
	4.3	2.1 ± 0.6	2	11.1 ± 6.4	2	31.3	<5% crystalline
			3	0.0	3	26.2	
					4	14.0	
					5	3.5	
HPC	1.0	35.0 ± 0.8	2	29.1 ± 2.7	2	23.6	—[c]
			3	12.0 ± 3.4	3	1.8	
			4	12.2 ± 1.5	4	0.3	
			5	2.9 ± 1.5	5	0.2	
	2.0	11.0 ± 0.3	2	16.6 ± 6.9	2	36.2	—[c]
			3	2.8 ± 2.8	3	11.7	
			4	0.0	4	3.1	
	4.0	0.8 ± 0.3	2	0.0	2	32.9	—[c]
					3	25.3	
					4	13.6	
					5	5.1	

[a] Molar substitution.
[b] X-ray diffraction patterns from pressed plaques.
[c] Materials were not examined.

using chemical analysis with stochastic modeling techniques. Depending on the number of moles of ethylene oxide and the amounts of caustic and water used in the process, 5 to 20% of the repeating GP units remain unsubstituted. Three hydrophobes are placed on the HEC chain (*9*). Their placement geometry is unknown. If the derivatization is achieved by epoxide addition, then according to similar studies (*11, 12*), the hydrophobes are placed at the end of the oxyethylene oligomers attached to cellulose. The hydrophobe with an olefin group (*13*) could be added by a ceric ion redox reaction. However the hydrophobes are added to HEC, they are positioned close to the main chain, which is segmentally rigid because of acetal linkage through the β-1,4 positions of repeating GP rings.

In studies comparing an HMHEC that has an M.S. of 2.5 in dilute solutions with an unmodified HEC of the same molecular weight, fluorescence data indicated that above a critical polymer concentration of 500 ppm, the hydrophobic groups of HMHEC associate to form

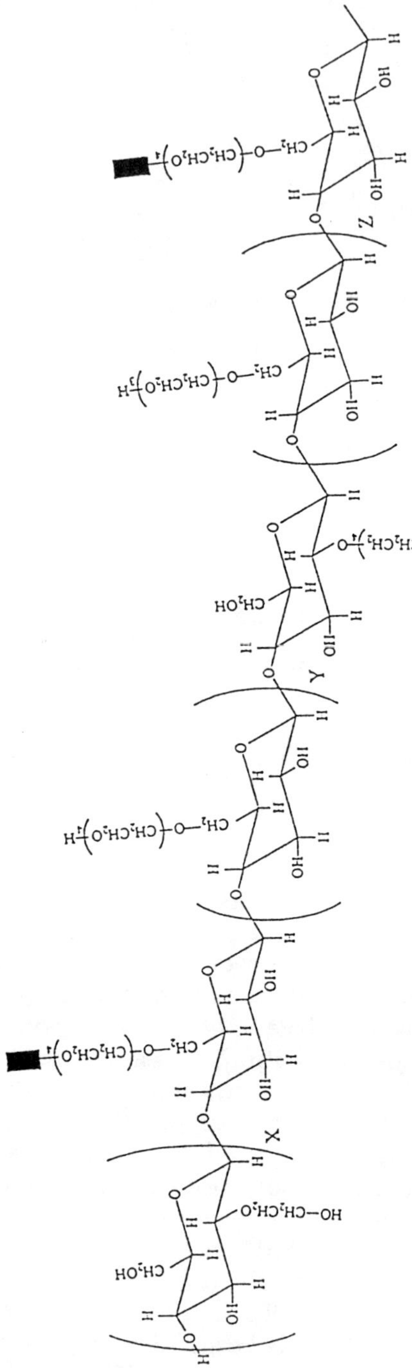

Figure 1. General structure of hydrophobe-modified hydroxyethylcellulose (HMHEC).

clusters. A significant viscosity increase (*14*) occurs above a critical polymer concentration (c_h) of approximately 0.2 wt%. The c_h is considerably lower than the predicted polymer coil overlap concentration (c^*) of approximately 1%.

A broad spectrum of surfactants and their influences on the rheological properties of aqueous solutions of HMHEC have been observed (*15*) at surfactant concentrations considerably lower than the surfactant's critical micelle concentration (CMC). The magnitude of the observed rheological effect when surfactant is added is dependent on both the alkyl chain length and the nature of the hydrophilic head group. In studies at a high HMHEC concentration (*16*), the viscosities suggest that interactions between an anionic surfactant, sodium dodecyl sulfate (SDS), and the polymer affect both interpolymer–polymer association and chain expansion because of electrostatic repulsions of SDS aggregates associated with HMHEC. In the presence of SDS, the solutions are elastic (*17*). At high surfactant concentrations, complete disruption of the polymer network is evident, with the HMHEC–surfactant systems exhibiting rheological characteristics typical of a dilute polysaccharide solution. Nonionic surfactants with adequate solubility do not promote significant viscosity increases in HMHEC solutions (*18*).

EHEC and Hydroxypropyl Cellulose (HPC). Studies of a commercial EHEC polymer were also reported during this period of HMHEC development, as was a study of HPC. An ethyl group is too small to promote micelle formation, but in EHEC, an ethyl unit occurs, on average, on most repeating GP units of the cellulose chain ($D.S._{Et} = 0.9$). Ethoxy groups are placed on the cellulose backbone via Williamson synthesis (ethyl chloride + alkali cellulose). The comparative reactivity of methyl chloride with the three oxyanions at the C_2, C_3, and C_6 positions is 5:1:2. Ethyl chloride is added at a higher reaction temperature, and its reactivity with alkali cellulose may be less selective. No experimental determination or simulation of ethyl groups distribution in EHEC has been reported. In commercial HPC, the substitution level is high (M.S. = 4.0). With less chaining of the reaction sites (Chapter 8), oxypropyl groups are more evenly substituted than the oxyethylene or ethyl groups in EHEC. The pendent oxypropylene chains are smaller than the oxyethylene units and significantly less hydrophobic (*19*) than traditional nonylphenol or dodecyl hydrophobes. Nevertheless, EHEC, HPC, and HMHEC show several common trends.

The binding of SDS to EHEC initially leads to an increase in viscosity that is due to electrostatic repulsions of the bound anionic SDS aggregates; SDS binds to the ethyl units of EHEC in a cooperative manner (*20*). In a more recent study (*21*) using EHEC at 0.5 wt%, surfactant binding to polymer was reduced, the degree of counterion binding was decreased, and the micelle aggregation number became lower as the temperature was increased. Nothing specific was observed in the premicellar range of SDS, but the role of micelles in the behavior of EHEC–surfactant combinations was important. The CMC in the presence of EHEC decreases with increasing temperature, and the difference between CMCs in the presence and absence of EHEC increases with temperature. Fluorescence studies of more hydrophilic cellulose ethers (methyl-HEC and methyl-HPC with a low M.S. of oxirane substitutents and HPC are discussed in Chapter 22).

The influence of ionic and nonionic surfactants on the cloud point and shear viscosity behavior of an aqueous 1 wt% HPC solution has been examined with a variety of surfactants (*22*). At the surfactant concentrations studied (up to 10^{-2} mol dm^{-3}), micelles of a nonionic surfactant (e.g., dodecyloxyheptaethoxyethanol) do not promote a viscosity response in HPC or HMHEC solutions (Chapter 17).

Associative Thickeners Prepared by Free-Radical-Initiated Chain Growth Syntheses

RAM. In the free-radical copolymerization of acrylamide with an alkyl-modified acrylamide (RAM), the RAM must be kept at a low concentration to minimize intrahydrophobic association that decreases the copolymer's solution viscosity (*23*). Under the conditions used in this study, RAM monomers are not soluble in aqueous media. A surfactant above its CMC (e.g., SDS) is required for solubilization. Use of such a surfactant can lead to micelle structures and the formation of contiguous RAM units (*23, 24*) in the acrylamide copolymer (Figure 2). One may postmodify the acrylamide chain (*25*); the relative placement of a low number of hydrophobes is still unknown, but they would be random with an extremely low probability of contiguity. Comparison of the solution properties of RAM copolymers (*26*) with those of the postmodified poly(acrylamide) polymer strongly suggests that copolymers made in aqueous SDS solutions have contiguous hydrophobe placement. The greatest complexity in any chain growth free-radical synthesis of associative thickeners is the drift in composition with conversion. Thus, given the possibility of contiguous groupings due to hydrophobic association in micellar aggregates and inherent differences in the reactivity ratios (*13*) of the hydrophobe and unmodified monomers, macromolecular chains formed at low conversions will have hydrophobe concentrations different from those of chains formed at intermediate and high conversions. This difference is influenced by the SDS concentration (*26*). The blockiness can also

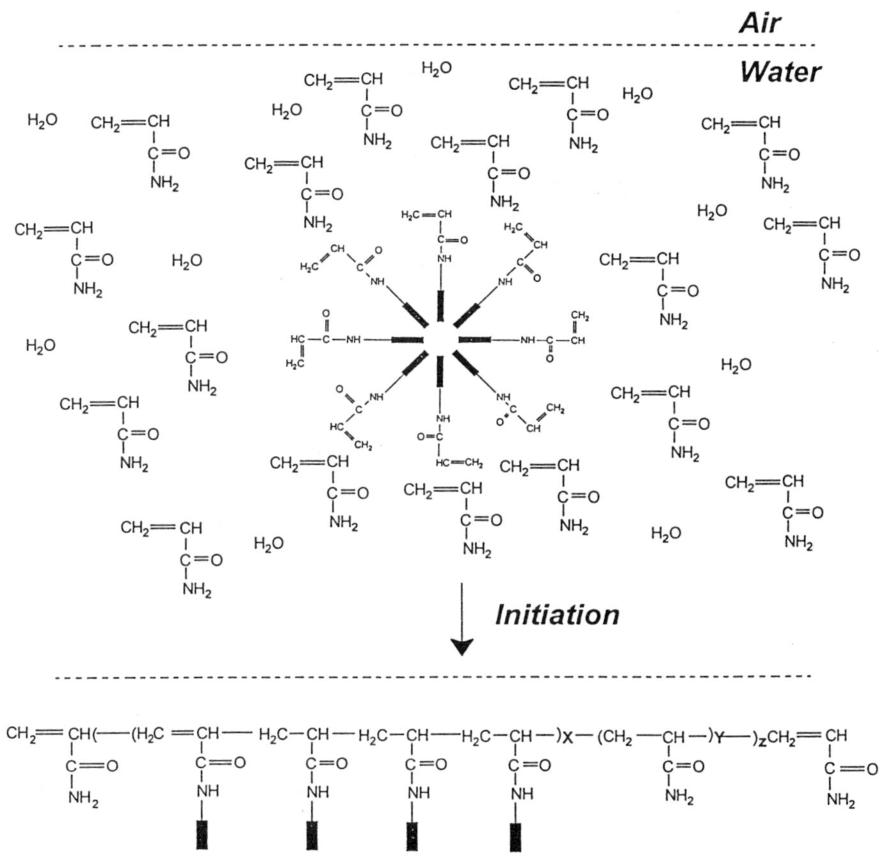

Figure 2. General structure of hydrophobe-modified polyacrylamide.

be controlled by the ratio of hydrophobe to surfactant at constant hydrophobe content (*27*). Details of the complexities and progress (*28*) in this area are highlighted in Chapter 15. It is difficult to quantify the structural influence on solution viscosities with this type of associative thickener.

HASE. Alkali swellable emulsions have been commercial products for several decades. As neutral dispersions, they are nonviscous. They contain a substantial amount of oligomeric acrylic or methacrylic acid at the latex's surface. The oligomeric acids expand upon addition of NH_4OH and increase the effective volume fraction of the dispersion and its viscosity. Positioning sequenced oligomeric acids at the surface of the disperse phase requires a balance of hydrophilicity and glass transition temperature in the final copolymer composition (*29, 30*).

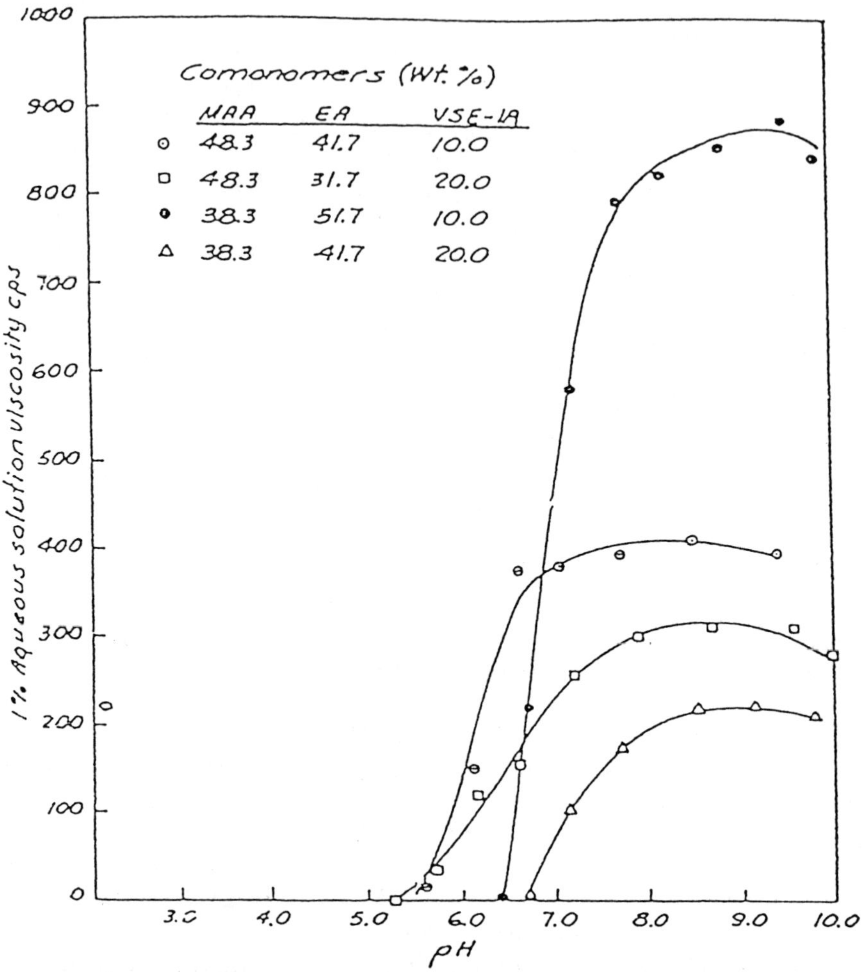

Figure 3. Aqueous solution viscosity dependence on solution pH of hydrophobe-modified alkali-swellable emulsion terpolymers. MAA is methacrylic acid, EA is ethyl acrylate, and VSE is hydrophobe-modified monomer.

The monomer that best accomplishes this balance is ethyl acrylate (*31*) (Figure 3). The hydrophilic nature of the latex may affect solubility of the dispersion when NH_4OH is added; a difunctional monomer is included in the synthesis to provide cross-links and some degree of dispersion integrity in alkaline solutions. The thickened solution is generally transparent.

More recently, a nonhydrolyzable monomer containing hydrophobes has been included in the polymerization procedure to ob-

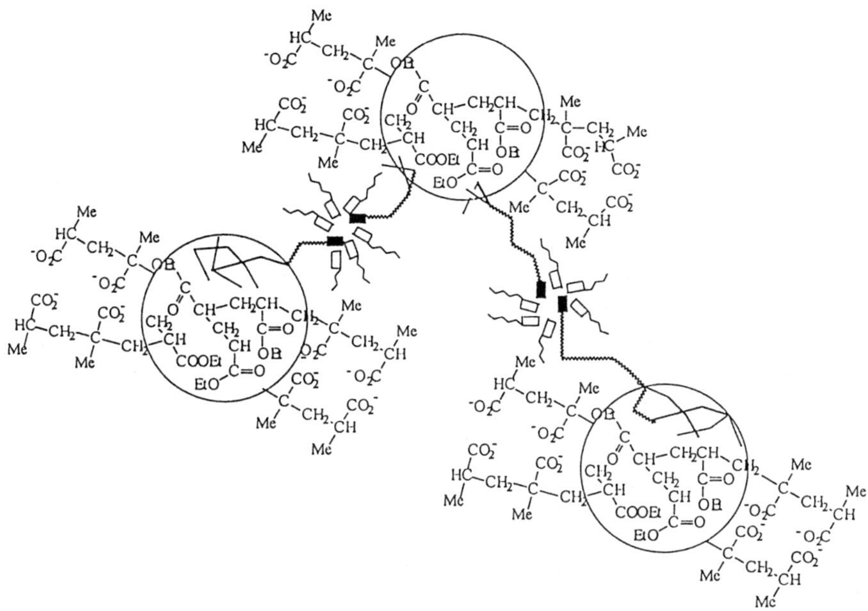

Figure 4. General structure of hydrophobe-modified alkali-swellable emulsion.

tain a hydrophobe-modified thickener (HASE). Such thickeners can be fractionated into soluble and insoluble fractions through acidification and neutralization sequences (*32*). In HASE thickeners, the hydrophobes are separated from the nonhydrolyzable monomer by at least 20 oxyethylene units. A schematic of this type of HASE thickener is shown in Figure 4. Both hydrophobic associations and electrostatic repulsions contribute to viscosity, but electrostatic repulsion is the primary contribution to viscosity build in most model and commercial HASE thickeners studied thus far. The hydrophobic contribution can be increased by increasing the number and size of the hydrophobes, but with increasing salinity, such products precipitate (*33*).

HEUR Prepared by Step-Growth Polymerization

HEUR thickeners used in coating formulations are prepared by a step growth process. The mechanism of propagation is distinctly different from that of the chain growth, free-radical process (*13*) used in the synthesis of RAM and HASE thickeners. In a step-growth process, the monomers are consumed early, but high molecular weights are not realized until high conversions are achieved and then only if stoichio-

$$2\ C_{18}H_{37}\text{—NCO} + \text{HO—(EtO)}_n\text{—H} \xrightleftharpoons[\text{Catalyst}]{\text{Tin}}$$

$$C_{18}H_{37}\text{-NH—CO—O—(EtO)}_n\text{—CO—NH—}C_{18}H_{37}$$

Scheme I. Reaction scheme for step-growth synthesis of HEUR thickener (octadecyl-modified POE).

metric ratios are used (Scheme I). A wide variation in HEUR molecular weight can be obtained by varying the stoichiometric ratio of diol to diisocyanate (Chapter 17). To achieve water solubility, a POE diol is used. A generic structure for a step-growth HEUR is illustrated in Figure 5. The primary asset of a step-growth process is the ability to place a given functional group at the terminal positions of the chains through control of reactant stoichiometry. The alkyl groups of internal isocyanate coupling units do not promote hydrophobic associations (*34, 35*), and HEUR associative thickeners therefore are the only family with known "effective" hydrophobe placement from which structure–property relationships can be defined.

POE chains are readily degraded by oxygen (*36*) and mechanical stresses. In the synthesis of HEURs, the presence of ppm amounts of oxygen can be a primary cause of oxidative degradation. Another equally damaging parameter is the reaction of the isocyanate with water (Scheme II). POE is a hygroscopic material; it is difficult to remove water from POE and the solvent commonly used in HEUR synthesis, tetrahydrofuran, without degrading POE. A substantial number of the studies reported in the literature have been done with HEUR structures not consistent with the reaction conditions used, and the so-called HEURs studied are ill (or not at all) defined. The synthetic aspects are avoided with a statement that the materials were synthesized by the "Emmons procedure." In a composition-of-matter patent, corporate lawyers claim every conceivable pathway available to obtain a material; patents for associative thickeners do not deviate from this professional standard, and citing an expired patent as a method of synthesis is providing noninformation. The term "telechelic" has also been misused as a method of hiding a step growth synthesis HEUR and, with it, the large excess of unreacted POE present.

Another misconception in the literature is that HEURs synthesized by a step growth process have narrow molecular weight distributions. The distribution in a step growth synthesis was delineated in the early work of Carothers (*37*) and Flory (*38*) as being broad, and this breadth is observed in size-exclusion chromatography of S-G HEURs

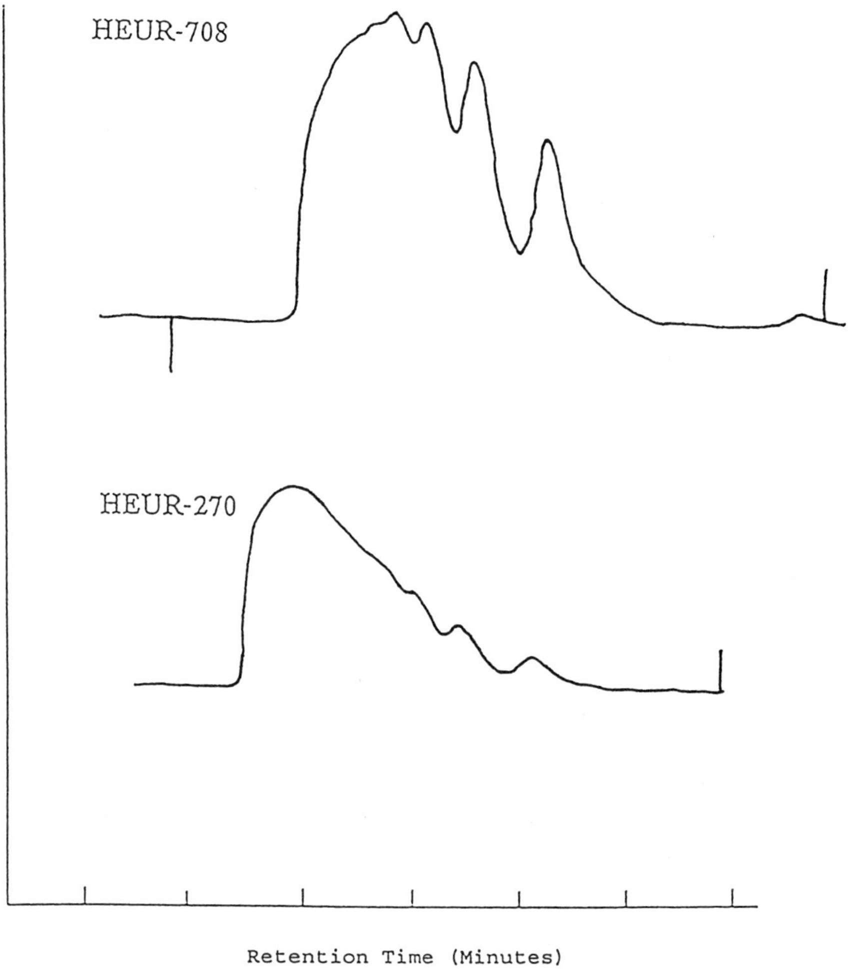

Figure 5. Size-exclusion chromatograph of two commercial HEUR thickeners (HEUR-708 and HEUR-270).

$$\mathrm{R{-}NCO} + \mathrm{H_2O} \rightleftharpoons \mathrm{R{-}NH_2} + \mathrm{CO_2}$$

$$\mathrm{R{-}NH_2} + \mathrm{R{-}NCO} \rightleftharpoons \mathrm{R{-}NH{-}CO{-}NH{-}R}$$

Scheme II. Reaction of isocyanate with water.

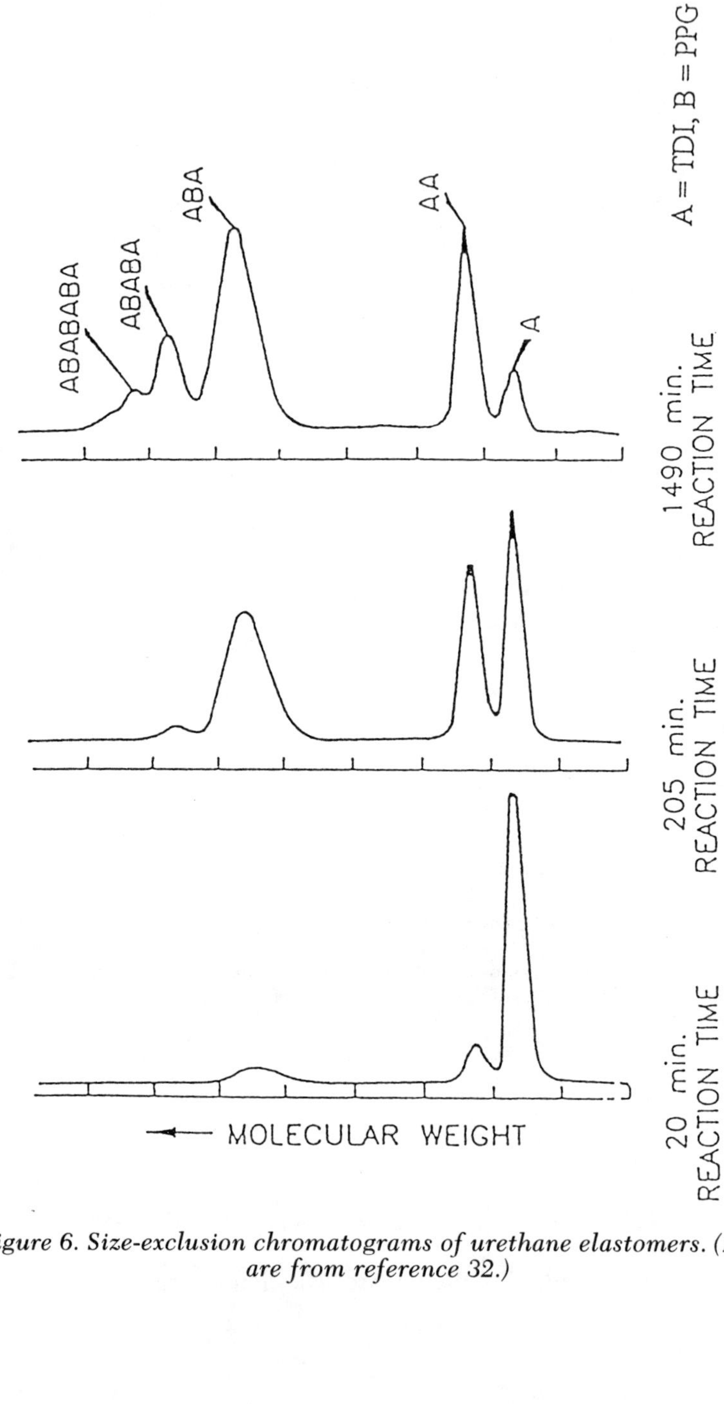

Figure 6. Size-exclusion chromatograms of urethane elastomers. (Data are from reference 32.)

using a proper column set. The size-exclusion chromatographs of commercial HEURs (*39*) are given in Figure 6. Similar data for model step-growth HEURs are given in Chapter 17. The distribution data (*40, 41*) for urethane elastomer products are similar (Figure 7).

Synthesizing HEURs with narrow molecular weight ranges is possible. The procedure involves the direct addition of a monoisocyanate to POE or the addition of a large excess of diisocyanate to POE followed by the addition of an alkyl amine to the terminal isocyanates (Chapter 17). Narrow-molecular-weight-range, surfactant-modified, water-soluble polymers similar to HEURs but without the urethane linkages were examined by Persson and coworkers (*42*). Their approach was not quantitative, and unreacted POE would be present during the procedure if it was not specifically removed. Narrow-molecular-weight-range POEs with terminal hydrophobes coupled with urethane or urea linkages maintain higher viscosities to higher shear rates before shear thinning. They also exhibit a greater tendency toward yield stress behavior than the step-growth HEURs. The narrow-molecular-weight-range POEs are referred to as "uni-HEURs." This type of associative thickener is examined in Chapters 17 and 18 and is the type of HEUR used in the most definitive rheological study of HEURs (*43*) because the classic step-growth HEURs could not be analyzed with established rheological models.

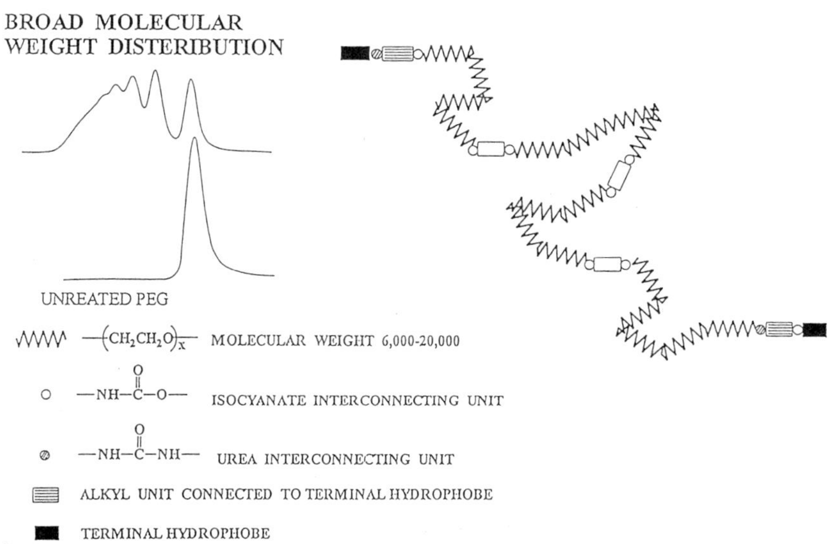

Figure 7. General structures of step-growth hydrophobe-modified ethoxylated urethane thickener.

Overview of Associative Thickener Technology

The chapters that follow discuss the mechanism by which structural variations among and within associative thickener families influence different solution behaviors and interactions with different disperse phases. The primary commercial applications of surfactant-modified polymers are in coatings, cosmetics, and detergents, and as airplane deicers.

In cosmetic and detergent applications, the viscosity at low shear rates with a moderate yield stress is desirable. These characteristics can be achieved with the uni-HEURs discussed above but not with the step-growth HEURs. HEC and its derivatives are used in cosmetic applications, but none of these derivatives, including HMHEC, generally provide the yield stress behavior desired, even when compared with the uni-HEURs that are not fully substituted (*44*) (Figure 8). These properties are discussed from a coating viewpoint in Chapters 17, 23, and 24, but they are equally applicable to other commercial areas because they also improve the environmental acceptance of commercial products.

For example, the rheology of cosmetic formulation areas related

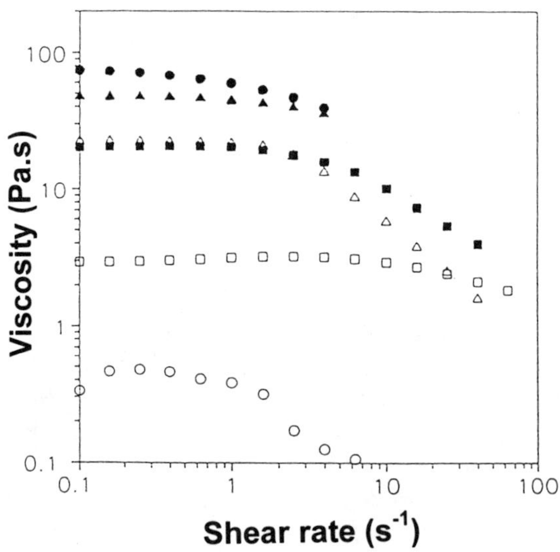

Figure 8. Viscosity dependence on shear rate of 9.4 wt% of a nonylphenol ethylated surfactant with a terminal sulfate group adjusted to an ionic strength of 0.43 M. Key: ○, no polymer added; □, $[C_{17}H_{35}C(O)O]_2[EtO]_{139}$, 1.1 mM; △, aminol A15, 50 mM; ●, $[C_{18}H_{37}NHC(O)]_{0.8}[EtO]_{182}$, 1.0 mM; ■, $[C_{18}H_{37}NHC(O)]_{0.6}[EtO]_{531}$, 0.9 mM; and ▲, $[C_{18}H_{37}NHC(O)]_{2.0}[EtO]_{182}$, 1.0 mM.

to shampoos and bath and shower products is important so that an impression of quality and substance can be obtained in a viscous product that is easy to dose, apply, and distribute. A viscous formulation also promotes phase stability and handling. Electrolytes such as NaCl transform spherical micelles of anionic surfactants into rodlike micelles (*45*). The interaction of the rods at high concentrations with one another (Chapter 14 and references therein) provides additional increases in solution viscosity. Rods are created by the addition of salts or fatty acid alkanolamides. In personal care products, however, the presence of traces of toxic nitrosamine (a carcinogen) contained in alkanolamides has caused concern about this approach to increasing cosmetic formulation viscosities (*46*). This problem is avoided through the use of surfactant-modified oligomers that do not promote stringing in formulations and the unfavorable feel of stickiness in skin applications that occur if the formulation is thickened with high-molecular-weight, water-soluble polymers. The same need for viscosity control also exists in detergent formulations.

The solution rheology of HMHEC is examined in Chapter 17. Interesting thixotropic behavior is observed in certain concentration ranges, but this difference is not observed in coatings rheology (Chapter 24), and the desirable Newtonian plateau at low shear rates observed in uni-HEURs is not evident in HMHEC solutions. HMHEC is observed to affect coating films with high gloss when oligomeric dispersants containing hydrophobes are preadsorbed on the TiO_2 pigment. In most other respects, coatings properties are similar to HEC-thickened latex coatings. This similarity is tentatively related to the closeness of the hydrophobes to the segmentally rigid cellulose backbone.

The rheology of HASE thickeners, both in solutions and in coating formulations, is discussed in Chapters 17, 23, and 24. These studies clearly demonstrate that HASE structures can be varied to obtain thickeners that mimic HMHEC and HEUR coating rheology. In many formulations, the elastic (storage modulus, G') contribution is lower than that in HEC-thickened formulations. This difference is consistent with the claim of lower spatter behavior (with smaller hydrophobes) when HASE-thickened coating formulations are applied by roll. HASE thickeners also increase applied film gloss relative to that of HEC formulations. In a general mechanism, this difference could be explained by adsorption of the unneutralized acids of HASE thickeners on the TiO_2 pigment (Chapter 24); however, increasing the number of hydrophobes in HASE thickeners decreases film gloss. This result would not have been predicted from the interaction of HEUR thickeners with preadsorbed hydrophobe-containing dispersants, in which an interaction promotes higher film gloss (*47*).

The production costs of HASE thickeners are lower than those of HEUR or HMHEC thickeners discussed in this chapter. If the reactivity (*48*) of the carboxylate groups is not detrimental to the application (e.g., if they are not a primary factor in interior architectural coatings), then HASE thickeners will be the materials of choice.

The HEUR thickeners cover a broad spectrum of applications: coatings for architecture original equipment manufactured and in cosmetics and detergents. HEUR polymers offer an area of broad potential academic interests that would be direct if only well-defined materials are included in published studies.

References

1. Bailey, F. E., Jr.; Koleske, J. V. *Alkylene Oxides and Their Polymers*; Marcel Dekker, 1991.
2. Glass, J. E. *J. Oil Colour Chemists' Assoc.* **1975,** *58,* 169; *ibid.,* **1976,** *59,* 86.
3. Glass, J. E. *J. Coat. Technol.* **1978,** *50*(640), 53, 61; *ibid. 50*(641), 56, 72.
4. Ma, Z.; Lundberg, D. J.; Roberts, S.; Glass, J. E. *J. Appl. Polym. Sci.* **1993,** *49,* 1509.
5. Lundberg, D. J.; Glass, J. E. *J. Coat. Technol.* **1992,** *64*(807), 53.
6. Landoll, L. M. *J. Polym. Sci., Polym. Chem. Ed.* **1982,** *20,* 443.
7. Gelman, R. A.; Barth, H. G. In *Water-Soluble Polymers: Beauty with Performance;* Glass, J. E., Ed.; Advances in Chemistry 213; American Chemical Society: Washington, DC, 1986; Chapter 6.
8. Sau, A. C.; Landoll, L. M. *Polymers in Aqueous Media: Performance Through Association;* Glass, J. E., Ed.; Advances in Chemistry 223; American Chemical Society: Washington, DC, 1989; Chapter 18.
9. Goodwin, J. W.; Hughes, R. W.; Lam, C. K.; Miles, J. A.; Warren, C. H. reference 8, Chapter 19.
10. Glass, J. E.; Buettner, A. M.; Lowther, R. W.; Young, C. S.; Cosby, L. A. *Carbohydr. Res.* **1980,** *84,* 245.
11. Seneker, S. D. Ph.D. Dissertation, North Dakota State University, 1986.
12. Glass, J. E.; Shah, S.; Lu, D.-L.; Seneker, S. D. In *Polymer Adsorption and Dispersion Stability;* Goddard, E. D.; Vincent, B., Eds.; ACS Symposium Series 240; American Chemical Society: Washington, DC, 1984; Chapter 7.
13. Odian, G. *Principles of Polymerization,* 3rd ed.; John Wiley & Sons: New York, 1991.
14. Tanaka, R.; Meadows, J.; Philips, G. O.; Williams, P. A. *Carbohydr. Polym.* **1990,** *12,* 443–459.
15. Tanaka, R.; Meadows, J.; Williams, P. A.; Philips, G. O. *Macromolecules* **1992,** *25,* 1304.
16. Sivadasan, A.; Somasundaran, P. *Colloids Surf.* **1990,** *49,* 229.
17. Dualeh, A. J.; Steiner, C. A. *Macromolecules* **1990,** *23,* 251.
18. Tarng, M.-R. Ph.D. Thesis, North Dakota State University, 1995.
19. Jorgensen, E. B.; Jensen, J. H.; Hvidt, S. *J. Non-Cryst. Solids* **1994,** 172, 972.
20. Carlsson, A.; Karlstrom, G.; Lindman, B.; Stenberg, O. *Colloid Polym. Sci.* **1988,** *266,* 1031.

21. Kamenka, N.; Burgaud, I.; Zana, R.; Lindman, B. *J. Phys. Chem.* **1994,** *98,* 6785.
22. Zugenmaier, P. A.; Nicolai, R. A. *Makromol. Chem., Rapid Commun.* **1990,** *11,* 95.
23. Bock, J.; Siano, D. B.; Valint, P. L.; Pace, S. J. Reference 8, Chapter 22.
24. Peer, W. J. reference 8, Chapter 20.
25. Iliopoulos, I.; Wang, T. K.; Auddebert, R. *Langmuir* **1991,** *7,* 617.
26. Valint, P. L.; Bock, J.; Ogletree, J.; Zushma, S.; Pace, S. J. *Polym. Prepr. (Am. Chem. Soc., Div. Polym. Chem.)* **1990,** *31*(2), 67.
27. Hill, A.; Candau, F.; Selb, J. *Macromolecules* **1990,** *26,* 4521.
28. Biggs, S.; Hill, A.; Selb, J.; Candau, F. *J. Phys. Chem.* **1992,** *96,* 1505.
29. Muroi, S. *J. Appl. Polym. Sci.* **1966,** *10,* 713.
30. Verbrugge, C. J. *J. Appl. Polym. Sci.* **1970,** *14,* 897.
31. Sonnabend, R. U.S. Patent 4 384 096, issued to Dow Chemical, 1983.
32. Fernando, R. H. Ph.D. Thesis, North Dakota State University, 1986.
33. Chen, M. Ph.D. Dissertation, North Dakota State University, in progress.
34. Kaczmarski, J. P.; Glass, J. E. *Polym. Mater. Sci. Eng.* **1992,** *67,* 282; in greater detail, *Langmuir* **1994,** *10*(9), 3035.
35. Fonnum, G.; Bakke, J.; Hansen, F. K. *Colloid Polym. Sci.* **1993,** *271,* 380.
36. Donbrow, M. In *Nonionic Surfactants: Physical Chemistry;* Schick, M. I., Ed.; Surfactant Series 23; Marcel Dekker: New York, 1987; p 1011.
37. Carothers, W. H. *Trans. Faraday Soc.* **1936,** *32,* 39.
38. Flory, P. J. *Principles of Polymer Chemistry,* Cornell University: Ithaca, NY, 1953, Chapters 3, 8, 9.
39. Ma, Z.; Kaczmarski, J. P.; Glass, J. E. *Prog. Org. Coat.* **1992,** *21*(1), 69.
40. Furukawa, M.; Yokoyama, T. *J. Polym. Sci.* **1986,** *24A,* 3291.
41. Thompson, C. M. *J. Polym. Sci.* **1990,** *28A,* 333.
42. Persson, K.; Wang, G.; Olofsson, G. *J. Chem. Soc., Faraday Trans.* **1994,** *90*(23), 3555.
43. Annable, T.; Buscall, R.; Ettelaie, R.; Whittlestone, D. *J. Rheol.* **1993,** *37,* 695.
44. Mast, A. P.; Prud'homme, R. K.; Glass, J. E. *Langmuir* **1993,** *9,* 708.
45. Jones, M. J. *J. Colloid Interface Sci.* **1967,** *23,* 36.
46. Behler, A.; Hensen, H.; Raths, H. C.; Tesmann, H. *Seifen, Oele, Fette, Wachse* **1990,** *116*(2), 60.
47. Lundberg, D. J.; Glass, J. E. *J. Coat. Technol.* **1992,** *64*(807), 53.
48. LeSota, S.; Lewandowski, E. W.; Schaler, E. J. reference 8, Chapter 28.

RECEIVED for review March 16, 1995. ACCEPTED revised manuscript August 22, 1995.

11

Architecture-Controlled Solution Properties of Hydrophobically Associating Copolymers

Arlette R. C. Baljon[1,2] and Thomas A. Witten[2]

[1]Department of Physics and Astronomy, Johns Hopkins University, Baltimore, MD 21218
[2]The James Franck Institute, The University of Chicago, Chicago, IL 60637

Monte Carlo computer simulations predict that dilute-solution properties of water-soluble polymers that contain a small number of hydrophobic groups are dependent on the placement of these groups. We explore a simple model in which hydrophobic associations are assumed to be strong and each associating group or sticker is assumed to be constrained to be adjacent to one other. Thus we determined how the placement of the stickers controls the mutual second virial coefficient of identical chains containing two stickers apiece in dilute solution. The net interaction passes from repulsive to attractive when the ratio of the distance between the stickers over the chain length exceeds 80%. Likewise, for chains with alternating short and long intervals between stickers, the placement asymmetry controls the swelling of the chain. Our simulations show that virtually all placements result in swelling rather than collapse behavior in the chain.

MONTE CARLO COMPUTER SIMULATIONS for studying the influence of hydrophobe placement on the statistical properties of hydrophobically modified water-soluble polymers (HMWSPs) in dilute solution were developed (*1*, *2*). Because of the tendency of hydrophobes to associate

0065–2393/96/0248–0181$12.00/0

or stick together in aqueous media, these HMWSPs exhibit interesting solution properties (*3*). For example, the intermolecular associations give rise to enhanced viscosification, so these copolymers are sometimes called associative thickeners. Their particular properties under shear make them potentially useful as aqueous viscosity modifiers. Some of the possible applications are coating and oil recovery processes (*3*). A problem, however, is that enhanced viscosity very often leads to insolubility (*4*). In this work we explore the possibility of obtaining the desirable rheological properties as well as good solubility by controlling the placement of hydrophobes. Although from what we know the chain architecture cannot yet be fully controlled experimentally, such control may soon be possible. Several experiments (*4, 5*) in which the relative composition of the copolymers is varied show that because of changes in the balance between inter- and intrachain hydrophobic associations, the rheology and phase behavior of these copolymers depend strongly on their composition.

Model

The distinctive molecular feature that appears to give rise to the aforementioned solution properties is essentially geometric: It is embodied in the strong tendency of hydrophobes to stay together. Therefore understanding the effect of this geometric constraint in isolation is important. A simplified model (*6*) for polymers can aid in understanding the effect. Each polymer chain is treated as a lattice self-avoiding random walk with a small number of hydrophobic groups or "stickers" placed alongside. Each sticker is constrained to be adjacent to one other sticker, though stickers are free to change partners. The equilibrium state of the solution consists of all configurations of chains that obey these sticker constraints as well as configurations of mutual and self-avoidance.

The aforementioned Monte Carlo simulations were employed to study these equilibrium configurations. As we shall see, both the interaction between two chains containing two stickers each and the conformation of a single chain containing many stickers depend strongly on where the stickers are placed along the chain backbone. The results presented in the next couple of paragraphs were published in an extended form in references *1* and *2*. Here we emphasize the applicability of these results to hydrophobically associating polymers. A discussion of some new results for randomly placed stickers on a polymer ring is added.

Architecture-Controlled Interaction Between Two Polymers

Consider two chains with two stickers each. First, in a good solvent the interaction between two self-paired chains (Figure 1a) is repulsive. The range of the repulsion is the size of each chain (e.g., its radius of gyration). When brought to a separation comparable to their size, the chains may exchange stickers (Figure 1b). This freedom to exchange partners results in an increased number of configurations and amounts to an entropic attraction of the order kT. Since at such separations the repulsive free energy is also of the order kT (7), the free energies of repulsion and attraction are comparable. Alterations of the sticker placements can alter the balance between repulsion and attraction. Theoretical work (6) shows that near dimension $d = 4$, the net interaction is attractive if all stickers are on the ends of the chains. Since the interaction is repulsive when the stickers are next to each other, the repulsive and attractive effects just cancel each other out at critical placements. In the limit of long chains, the critical fractional separation between the stickers is a fixed universal fraction of the chain that is independent of molecular weight and the excluded volume of two monomers (6).

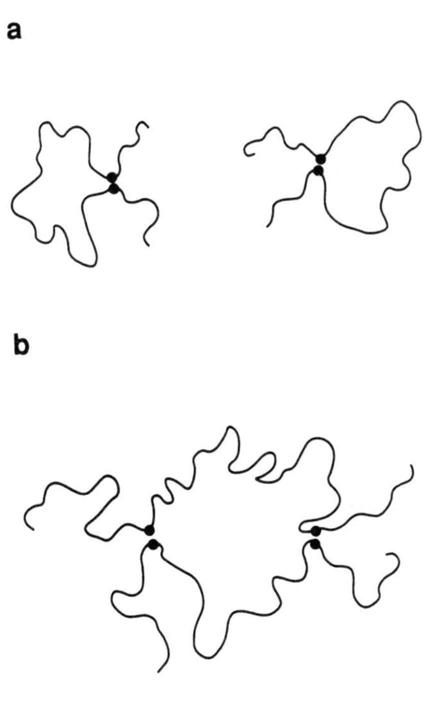

Figure 1. Two self-paired (a) and two cross-paired (b) chains. The stickers are shown as black dots. (Reproduced from reference 1. Copyright 1992 American Chemical Society.)

Simulation Strategy. Our simulations check these theoretical predictions. A configuration consists of two random-walk chains on a simple three-dimensional cubic lattice and has the following properties: (1) Each step of a random walk goes arbitrarily to a neighbor, second-neighbor, or third-neighbor site. (2) No two steps of the walks occupy the same lattice site. (3) Two particular monomers at specific points along the chain are designated as stickers. Parameter x, defined as the ratio of the sticker distance to the chain length, controls the chain architecture. Stickers are placed either symmetrically or maximally asymmetrically (Figure 2). (4) Each sticker is adjacent to one other sticker. A Monte Carlo algorithm is used to study all possible configurations of this two-chain system. The details of this algorithm can be

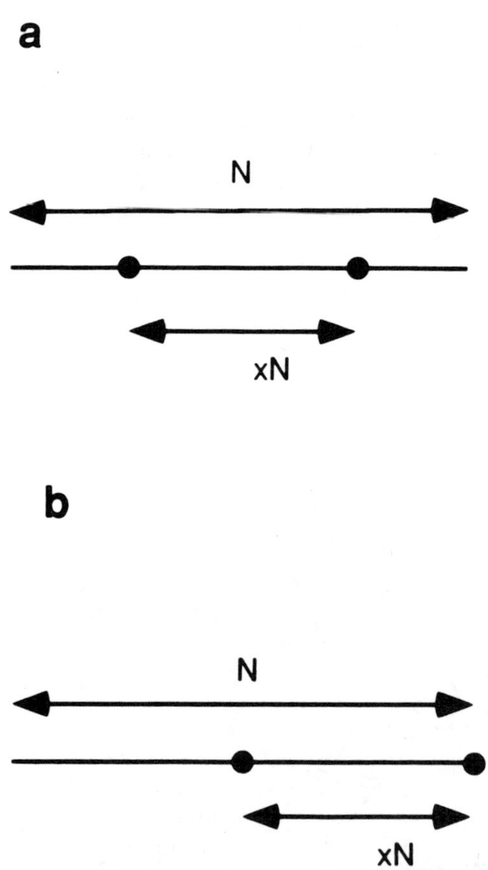

Figure 2. Symmetrical (a) and asymmetrical (b) sticker placements. The stickers are shown as black dots. (Reproduced from reference 1. Copyright 1992 American Chemical Society.)

found elsewhere (*1*). The free energies of repulsion and attraction are obtained from the number of forbidden configurations because of chain crossings and the number of additionally allowed cross-paired configurations that are due to sticker exchange.

The net interaction energy is expressed as a second virial coefficient (B_2). If the interaction between the two chains is repulsive, the second virial coefficient is positive; if the interaction is attractive, the coefficient is negative. In general, the second virial coefficient measures how the interaction between many chains in the dilute region affects thermodynamic quantities. For example, the osmotic pressure Π is given by

$$\frac{\Pi}{k_B T} = c + B_2 c^2 + O(c^3) \tag{1}$$

Here c is the number of polymers per unit volume, k_B is the Boltzmann factor, and T is the temperature. The second term on the right-hand side is a correction to the pressure to take account of interactions. In the dilute region we can neglect the $O(c^3)$ terms.

Critical Placement. Figure 3 shows the normalized second virial coefficient $\hat{B}_2$ for two polymer chains in three dimensions with symmetrically placed stickers. For reasons explained later in this chapter, we plot $\hat{B}_2$ versus $y \equiv x^{2.16}$, where x is the ratio of the sticker distance to the chain length. The data for the different chain sizes (molecular weights 34, 66, and 130) are normalized so that $\hat{B}_2(y) = 1$ at $y = 0$. We checked that the unnormalized excluded-volume data scale with chain length as $B_2 \approx N^{3\nu} \approx N^{1.8}$. At $y = 0$ we found that $B_2 \approx N^{(1.80 \pm 0.05)}$. The error bars shown in Figure 3 are obtained from the statistical error in the run: the value of the sum in a full run minus that in a half run divided by the square root of 2. The data are within error bars independent of the chain length. This superposition of the data for different chain sizes confirms the expected asymptotic scaling. Moreover, the $\hat{B}_2$ versus y data in Figure 3 show that as y increases, the chains cross from self-repelling to self-attractive behavior as anticipated. According to reference *8*, at low x, $\hat{B}_2(x) - \hat{B}_2(0) \approx x^{\nu(d+\theta_2)}$. Here θ_2 is an exponent describing certain interior correlations within a self-avoiding chain. Its value was measured as approximately 0.67 (*8*). Considering that $\nu \approx 0.588$ in $d = 3$ dimensions, we obtain $\nu(d + \theta_2) \approx 2.16$. Evidently the $\hat{B}_2$ versus y data in Figure 3 are consistent with this prediction, since $\hat{B}_2$ shows linear behavior for small y. In order to calculate the critical sticker placement y_c, for which there is no net mutual interaction between the chains ($\hat{B}_2 = 0$), we fit a poly-

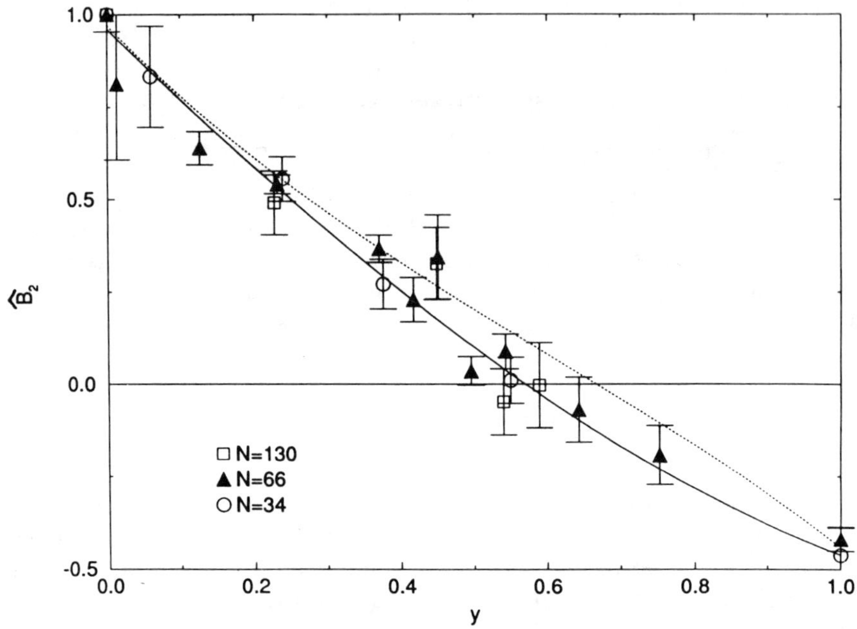

Figure 3. Normalized second virial coefficient $\hat{B}_2$ versus sticker placement for symmetrically placed stickers. $y \equiv x^{2.16}$, *where* x *is the distance between the stickers divided by the length of one chain. The values for the different chain lengths (34, 66, and 130) are scaled so that they are unity at* $x = 0$. *The solid line shows the result of a third-order polynomial fit to the data; the dotted line shows the result for maximally asymmetrically placed stickers.*

nomial through the data. A third-order polynomial fit gives a minimum value for χ^2 per degree of freedom (1.1). From the coefficients we find that $y_c = 0.58 \pm 0.02$ or $x_c = 0.776 \pm 0.012$. According to universality principles, $\hat{B}_2$, and this critical placement in particular, should be insensitive to details of the simulation, providing chains are long enough. In order to test this, we altered the $x = 1$ random walks to allow only steps of length $\leq\sqrt{2}$. The normalized second virial remained unaltered, a result that confirmed the universality of our results. The dotted curve in Figure 3 shows the results for maximally asymmetrical placed stickers. Here the critical placement is shifted to $x_c = 0.892 \pm 0.009$.

Conformations of a Single Chain Containing Many Stickers

Next we studied the equilibrium behavior of a single long chain made by joining our two sticker segments (Figure 4) head to tail. The result-

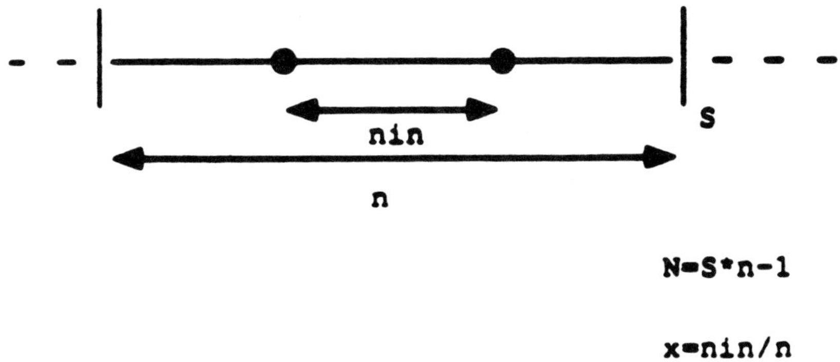

Figure 4. Chain architecture. Each chain consists of S *equal segments. Each segment contains two symmetrically placed stickers. Parameter* x, *defined as the distance between the stickers over the segment length, defines the chain architecture. The stickers are shown as black dots. Each segment consists of* n *monomers; the total number of bonds* N *equals* Sn − *1. (Reproduced from reference 2. Copyright 1993 American Chemical Society.)*

ing chain has alternating short and long intervals between the associating groups. A parameter x, defined as the length of the short interval divided by the sums of the lengths of the short and long intervals between stickers, defines the chain architecture. In the $x = 0$ limit, the chain behaves as an ordinary polymer chain and has a swollen configuration. Increasing x, however, may result in architectures for which the chain collapses. That case also shows a critical placement, x_c, with which a chain shows the statistical behavior of an ideal polymer chain in the asymptotic (large N) limit.

Influence of Sticker Placement. For details of the simulations the reader is referred to reference 2. The results presented in the next two paragraphs were published in reference 2. In three dimensions we obtain the radius of gyration R_g for $S = 2$, 4, and 8 and $n = 34$. Since the total number of monomers equals the number of segments S times the number of monomers per segment n, the total number of bonds N varies from 67 to 271. Each run lasts at least 1000 equilibration times. Figure 5 shows the results for five different chain architectures. The swelling factor α is defined as the measured radius of gyration divided by the expected radius of gyration of an ideal Gaussian chain with the same number of monomers and average squared bond length (9). For swollen chains, α increases with increasing chain length N, whereas for collapsed chains, α decreases with N in the asymptotic large-N limit. Since the swelling factor is defined as the ratio of the radius of gyration of a chain with a sticker to that of an ideal chain

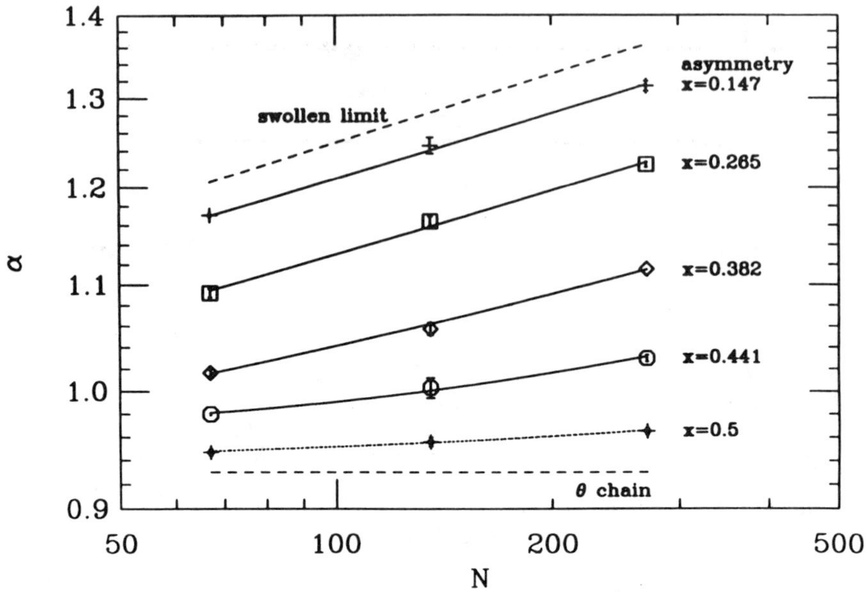

Figure 5. Log–log plot of the swelling factor α versus total chain length for different sticker placements in three dimensions. The swelling factor is defined as the average radius of gyration divided by the radius of gyration of an ideal Gaussian chain with the same length and average squared bond length. The solid lines are fits to equation 2. The dotted line is a fit to equation 5. The dashed lines indicate the expected behavior for swollen and θ chains. (Reproduced from reference 2. Copyright 1993 American Chemical Society.)

without stickers, this factor is smaller than one for a chain with critically spaced stickers would be ($x = x_c$). The error bars shown in Figure 5 are the statistical errors and are estimated by averaging over different parts of the run. The fits through the data will be explained shortly. The dashed lines indicate the expected behavior for swollen and θ chains (note the log–log scale). At first sight the data at small x seem to be in accordance with this swollen behavior, but for $x = 0.5$ the match is not as good: Although the swelling factor increases with N, the asymptotic regime is clearly not reached.

Predicting the scaling behavior of the chains at high N from the data at low N is a tricky business that is complicated by the lack of theories for associating polymers on which to rely. For small x, though, we expect the chains to behave similarly to ordinary homopolymer chains. The asymptotic N dependence of the radius of gyration of a swollen homopolymer chain is

$$R_g \approx kN^{\nu}(1 + CN^{-\nu\omega}) \tag{2}$$

Table I. Results for *k* and *C* from a Fit to Equation 2

x	k	C	χ^2
0.147	0.470 ± 0.004	0.08 ± 0.09	0.4
0.265	0.437 ± 0.003	0.13 ± 0.06	2
0.382	0.3873 ± 0.0012	0.47 ± 0.02	0.8
0.441	0.340 ± 0.002	1.22 ± 0.05	0.9
0.500	0.305 ± 0.002	1.86 ± 0.06	5

The first term on the right-hand side gives the expected scaling in the asymptotic limit. The second term is a correction for the fact that we deal with finite-size chains. ν and ω are obtained from renormalization group theory (*10*): $\nu = 0.5880 \pm 0.0010$, and $\omega = 0.850 \pm 0.015$. Fits to equation 2 are made for each architecture. They are drawn as solid lines in Figure 5. Table I shows the obtained values of k and C. The value of C increases with increasing x. In fact, $C = 0$ within error bars for the lowest x value. This means that for low x, no scaling correction is necessary and the sampled chains exhibit asymptotic behavior. One might expect this result if the excluded volume is much higher than the entropic attraction. This result is in accordance with previous results on two sticker chains (*1*), for which the entropic attraction increases with increasing x. As a goodness-of-fit measure, the values of χ^2 are given in Table I (*11*). Since we have three data points and two variables to fit, the fit is acceptable if $\chi^2 \approx 1$. Clearly, for $x = 0.5$ the data are not wholly in accordance with the swollen-chain behavior, but for all other x values they are. For other x values (i.e., $x \neq 0.5$), nothing clearly indicates that the chains will collapse on larger length scales. The sticking is predominantly local with associations between chemically nearby stickers (*2*).

Search for a Collapsed Conformation. A scaling argument is used to predict the scaling behavior for $x > 0.441$. Table I indicates that the value of k decreases with increased x. Parameter x will reach its critical value, at which the chain behaves as an ideal chain as k becomes 0. The functional dependence of k on x is obtained as follows: For homopolymers it is known from scaling theory (*12*) that

$$R_g \sim N^{0.5} \left(\beta l^{-d} \left(\frac{N}{l} \right)^{(4-d)/2} \right)^{(2\nu - 1)/(4-d)} \quad (3)$$

Here β is the excluded volume parameter, d is the dimension of space, and l is the bond length. In our case the parameter β is the effective excluded volume between two chain segments. This effective excluded volume contains the monomer repulsion as well as the sticker attraction. The parameter controlling the net effective excluded volume is x. Since β measures a local interaction, it is natural to expect it to vary smoothly with x as x passes through x_c. Thus near $x = x_c$, $\beta \sim (x - x_c)$. Then equation 3 results in

$$k \sim (x - x_c)^{2\nu - 1} \qquad (4)$$

Figure 6 shows a plot of $k^{1/(2\nu - 1)}$ versus x. For comparison, the value at $x = 0.5$ is indicated, although in this case the fit is not accepted. The other data square with our expectation: Remarkably, the anticipated linearity in x extends over the whole range of x studied. A linear least-squares fit through these data gives $x_c = 0.496 \pm 0.004$ ($\chi^2 = 0.2$). This critical point is very close to $x = 0.5$. Could the chain with equally spaced stickers be ideal in the asymptotic limit? If it is, the first thing to expect is that this ideal behavior is related to the symmetric sticker

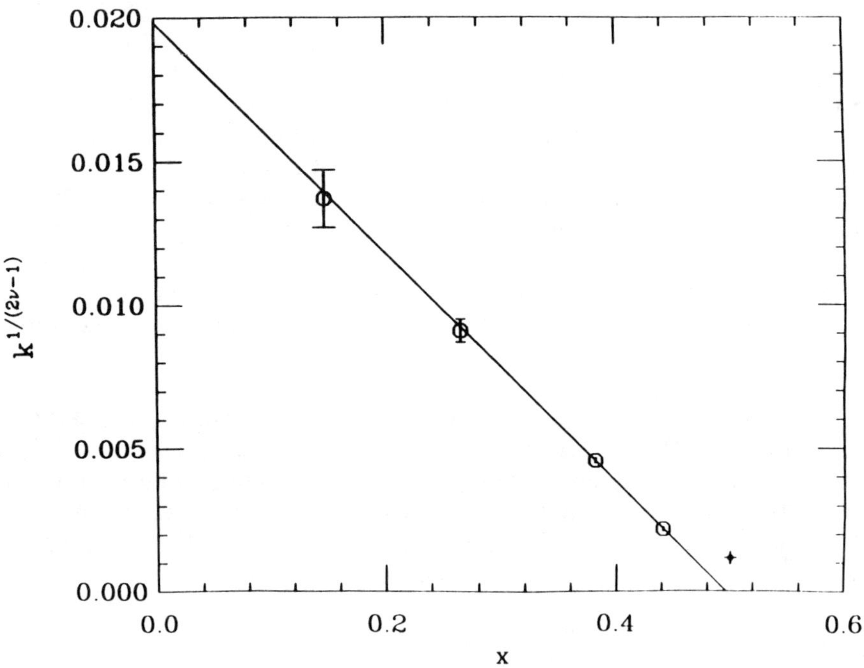

Figure 6. In order to test our scaling argument (in three dimensions), $k^{1/(2\nu-1)}$ is plotted versus x. From a linear least-squares fit we obtain $x_c = 0.496 \pm 0.004$. Although the data point at x = 0.5 was not accepted (and not used for the fit), it is indicated for comparison. (Reproduced from reference 2. Copyright 1993 American Chemical Society.)

placement, that is, to the fact that there is no distinction between long and short intervals. If this lack of distinction is the cause for the observed ideal behavior, we expect these chains to show ideal behavior in two dimensions as well. In two dimensions, swollen-chain conformations exist for all x values. In two dimensions, however, many configurations in which the sticking is not local are prohibited because the chain is not allowed to cross itself. This fact could explain the swollen behavior in two dimensions.

Our scaling argument indicates that in three dimensions the chain with equally spaced stickers is marginally collapsed. The data in Figure 5 show marginally swollen behavior, though. Indeed, if we try to treat the interaction as a small perturbation and fit the data to (*13*), then

$$R_g \approx kN^{0.5}(1 + CN^{0.5}) \tag{5}$$

The resulting fit, shown in Figure 5 as a dotted line, is quite good ($\chi^2 = 0.2$). For ordinary homopolymer chains this fit would indicate a swollen asymptotic conformation. For associating polymers, it is still possible that at high N the data will decrease or become N independent.

Randomly Placed Stickers on a Polymer Ring. Next we wanted to know what happens if stickers are placed randomly on a polymer chain. In order to circumvent end effects, we placed 4 to 16 stickers randomly on a closed loop. A loop with $2S$ stickers contains a total of $N = nS$ monomers, where $n = 34$. For two different loop sizes ($S = 2$ and 8), the radius of gyration is obtained for 16 different random sticker placements. We observed that the R_g for a specific random placement increases with an increase in the unevenness in sticker distribution. In Figure 7 this unevenness is shown for $S = 8$. We plot R_g versus the amount of randomness R. R is defined as the average variation of the distances between two stickers from the sticker distance in a chain with evenly spaced stickers:

$$R = \frac{1}{2Sn} \sum_{i=1}^{2S} \left| N_i - \frac{N}{2S} \right| \tag{6}$$

where N_i is the number of bonds in a chain segment (i) between two stickers. The increase in R_g with increasing R indicates that local sticking becomes more important at higher randomness R. Figure 8 shows the expansion factor α for chains with randomly spaced stickers. In addition, data for regularly spaced stickers ($x = 0.147$ and $x = 0.382$)

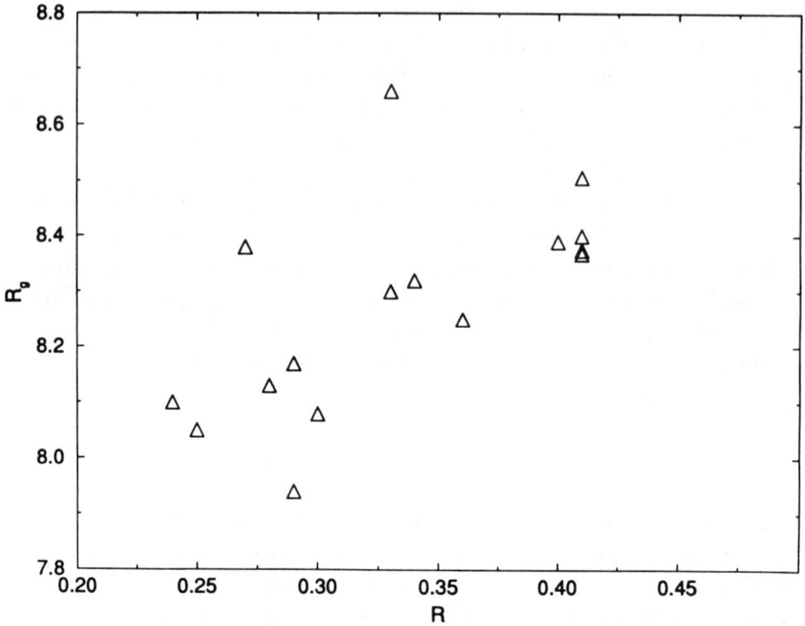

Figure 7. Dependence of the radius of gyration (R_g) on the amount of randomness R *defined in equation 6.*

are shown. Data for chains with randomly spaced stickers are not far off those obtained for a chain with $x = 0.382$. The randomness $R = 0.118$ for $x = 0.382$ is much lower than the average R for the 16 configurations of Figure 7. The fact that both chains (random stickers and $x = 0.382$) nonetheless have the same amount of swelling shows that all regularity in the sticker placement, such as a regular alternation of short and long segments, enhances swelling.

Conclusion

Using a simple topological model, we showed that the exact placement of stickers on an associative polymer chain has huge consequences for dilute-solution properties of those chains. We found that in equilibrium, the mutual interaction between two two-sticker chains passes from repulsive to attractive with increasing distance between stickers. This finding may result in an interesting new way to attain the theta state for a dilute polymer solution. The required balance is achieved by using geometrical constraints that depend only on the chain architecture. For chains with alternating long and short intervals between stickers, the amount of swelling is controlled by sticker placement. Phase separation was observed in none of the cases we studied. Al-

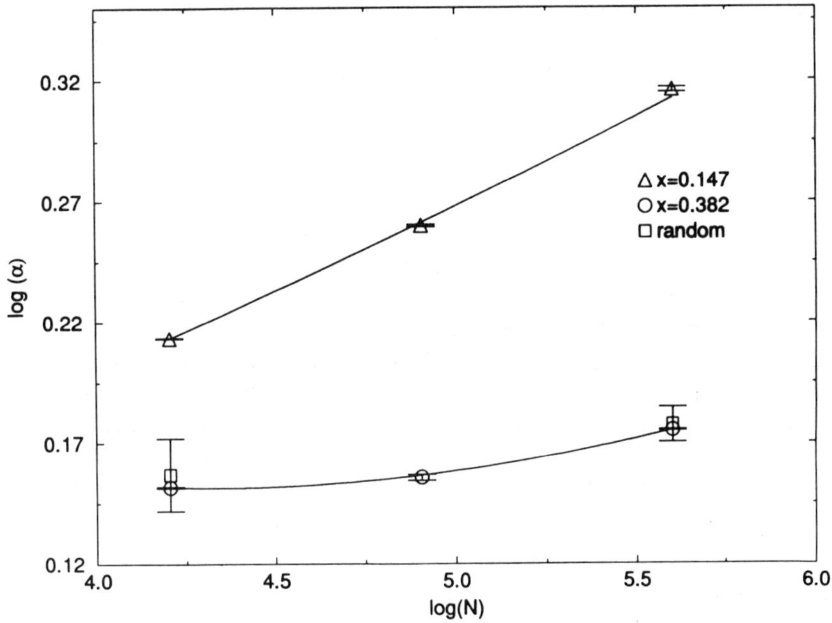

Figure 8. Swelling factor α for three-dimensional polymer loops. The squares represent the data obtained for loops on which the stickers are placed randomly.

though two two-sticker chains with stickers at the ends attract each other, we estimated that this attraction by itself is not strong enough to result in phase separation. Moreover, from the asymptotic swollen scaling behavior for single chains, we predict that for almost all architectures studied, the second virial coefficient of many chains in solution will be positive, indicating good solubility at low concentrations. We found that chains on which the stickers are placed randomly also swell. For chains with associating groups that stick in bigger multiplets, collapsed conformations and phase instability are expected for more architectures. Simulations suggest that solubility can in such cases also be controlled by sticker placement.

We expect our results to hold for real chains in a real solvent if the sticking energy is high enough, so that the hydrophobic groups are dissociated only a small fraction of the time. This fraction will depend on the binding energy of stickers as well as on sticker concentration, that is, on the average distance between two dissociated stickers. Since the interesting rheological effects are observed at high sticking energy, we studied only conformations in which all stickers are stuck. However, our simulations are easily extendable to include conformations in which some stickers are dissociated. Testing our findings experimentally would be interesting. One thing to measure would be

$\hat{B}_2$ as a function of sticker placement resulting from, for example, osmotic pressure versus concentration measurements in the dilute regime for two sticker chains. Such tests using polymers synthesized with living polymerization techniques may soon be possible (*14*).

Acknowledgments

We thank Michael Murat for his help with the initial program code. This work was supported by the Material Research Laboratory at The University of Chicago under National Science Foundation Grant DMR 88-19860 and by a grant from IBM Corporation. Part of the research was conducted using the Cornell National Supercomputer Facility, a resource of the Center for Theory and Simulations in Science and Engineering (Cornell Theory Center), which receives major funding from the National Science Foundation and IBM Corporation and additional support from New York State and members of the Corporate Research Institute.

References

1. Baljon, A. R. C.; Witten, T. A. *Macromolecules* **1992,** *25,* 2969.
2. Baljon, A. R. C. *Macromolecules* **1993,** *26,* 4339.
3. *Water-Soluble Polymers: Beauty with Performance;* Glass, J. E., Ed.; ACS Symposium Series 213; American Chemical Society: Washington, DC, 1986.
4. Hill, A.; Candau, F.; Selb, J. *Prog. Colloid Polym. Sci.* **1991,** *84,* 61.
5. Kaczmarski, J. P.; Glass, J. E. *Polym. Mater. Sci. Eng.* **1992,** *67,* 284.
6. Cates, M. E.; Witten, T. A. *Macromolecules* **1986,** *19,* 732.
7. Pincus, P. A.; Witten, T. A. *Macromolecules* **1986,** *19,* 2509.
8. des Cloizeaux, J.; Jannink, G. *Polymers in Solution: Their Modelling and Structure;* Oxford University: Oxford, England, 1990; pp 575–582.
9. Doi, M.; Edwards, S. F. *The Theory of Polymer Dynamics;* Claridon: Oxford, England, 1986.
10. Le Guillou, J. C.; Zinn-Justin, J. *Phys. Rev. Lett.* **1977,** *39,* 95.
11. Press, W. H.; Flannery, B. P.; Teukolsky, S. A.; Vetterling, W. T. *Numerical Recipes;* Cambridge University: Cambridge, England, 1988; p 502.
12. Freed, K. F. *Renormalization Group Theory of Macromolecules;* Wiley: New York, 1987; pp 81–87.
13. Yamakawa, H. *Modern Theory of Polymer Solutions;* Harper and Row: New York, 1971.
14. Storey, R. F. University of Southern Mississippi, personal communication.

RECEIVED for review February 14, 1994. ACCEPTED revised manuscript June 30, 1994.

12

Effect of Water on the Glass Transition Behavior of Hydrophilic Polyurethanes with Mixed Soft Segment

Nathaniel S. Schneider,[1] David A. Langlois,[2,3] and Cathyrine A. Byrne[2]

[1]Geo-Centers, Inc., 7 Wells Avenue, Newton, MA 02159
[2]Polymer Research Branch, AMSRL-MA-PB, Army Research Laboratory, Materials Directorate, Watertown, MA 02172

Water uptake and glass transition behavior in a series of hydrophilic polyurethane elastomers with a soft segment of poly(ethylene oxide) (PEO) alone or in a mixture with poly(tetramethylene oxide) (PTMO) were studied. In a set of samples with various PEO/PTMO ratios at fixed hard-segment content, the saturation water uptake was almost constant at 2.5 mol of water per mol of ethylene oxide. Differential scanning calorimetry measurements showed two separate glass transition temperatures (T_g) in the dry, mixed-soft-segment samples (−50 °C for PEO and −80 °C for PTMO) but only a single T_g at about −80 °C in the wet samples. The ΔC_p values for the wet samples were much higher than those for the dry samples. Measurements of the sample with pure PEO from the set just described indicated an essentially linear dependence of both T_g and ΔC_p on water content.

Hydrophilic poly(ethylene oxide) (PEO) containing polyurethanes that are based on the use of a mixed soft segment to control the extent of water swelling were first described by Tobolsky and co-

[3] Current address: Los Alamos National Laboratory, MS E549, Los Alamos, NM 87545

0065–2393/96/0248–0195$12.00/0

workers (*1*). Their study included measurements of water uptake and water and salt permeabilities and emphasized the relation of these processes to possible reverse osmosis applications. The variable and controlled degree of swelling with the related water and solute permeability is also of interest in biomedical applications of polymers as wound coverings and for controlled release. Additionally, these materials might be useful for moisture-permeable coated fabrics. Tobolsky and co-workers showed that the saturation water content was directly proportional to the PEO content in the mixed PEO–poly(propylene oxide) soft segment; obviously, then, little interaction occurs between these two components. Illinger et al. (*2, 3*) followed an alternative approach to hydrophilic polyurethanes based on the use of a block copolymer polyether consisting of a central segment of poly(propylene oxide) and terminal segments of PEO. These samples showed more complex dependence of water uptake on the soft-segment composition and the temperature. A recent reexamination of the data (*4*) suggested that the complex behavior is due to the incompatibility of the two components of the soft segment. The present study is concerned with the water uptake and glass transition behavior of a series of hydrophilic polyurethanes prepared with a mixed soft segment of PEO and poly(tetramethylene oxide) (PTMO).

Experimental Details

Polyurethane samples were formed from methylene bis(4-cyclohexylisocyanate) (H12MDI), butanediol (BD), and either a single polyol or a mixture in several mole ratios. PEO (molecular weight 1450) (Union Carbide) and PTMO (nominal molecular weight 2000) were used for the soft segment. The polyurethanes were formed by the two-stage reaction consisting of end capping with H12MDI, using dibutyl tin dilaurate as catalyst, followed by chain extension with BD. The cure was completed in a closed mold divided into equal 20- and 50-ml sections that was heated in an oven at 100 °C for 16 h.

Equilibrium sorption measurements were performed on preweighed 1-in (2.54-cm)-diameter 50-ml discs that were immersed in distilled water maintained at 30 °C. Differential scanning calorimetry (DSC) runs were made using a Perkin-Elmer DSC-2 with a liquid nitrogen reservior and helium purge gas. Samples cut from the 20-ml section of the molded sheet, weighing approximately 15 mg each, were initially quench cooled and then heated from −120 to 40 °C at 20 °C/min. A repeat run, following cooling at 40 °C/min, exhibited reproducibility within ±1.5 °C for T_g and ±5% for ΔC_p. For runs on the wet samples, the weight of the water-saturated discs was determined and the dry sample weight was calculated from the previously determined equilibrium water uptake. A different procedure was used for examining the effect of various water contents on the T_g behavior of 4PU10 (*see* "Results" for a description of sample designations). Discs of known dry weight were first saturated with water and then allowed to come to the desired final water concentration by evaporation of the excess water before they were sealed into crimped aluminum pans. The DSC scans were made at least 3 h following encapsulation to allow for uniform distribution of water throughout the sample thickness. A check one or more days later indicated that no water loss had occurred and that the DSC behavior reproduced the original result.

Results

The composition and water uptake of the set of hydrophilic polyurethanes are presented in Table I. The sample designations used in that table indicate sample compositions. The first number is the number of moles of diisocyanate used for 1 mol of total polyol. Since all the compositions were stoichiometric, this number completely specifies the molar composition. The number following PU indicates the weight percent of PEO in the mixed soft segment. Thus 4PU6 designates a sample with 4 mol of diisocyanate, 3 mol of butanediol, and 1 mol of polyol consisting of 60 wt% PEO and 40 wt% PTMO (0.678 mol of PEO/0.322 mol of PTMO).

The water uptake, recorded in Table I as grams of water per 100 g of polymer, increases progressively with increasing PEO in the first four samples, from 4PU5 to 4PU10. The next column of data, which presents water as moles of water per mole of ethylene oxide (EO), indicates that the water uptake is essentially proportional to the PEO content in these four samples of fixed hard-segment content and increases only slightly with increasing PEO content. With decreasing hard-segment content (samples 4PU10 to 2PU10), water uptake increases markedly. This increase indicates the decreasing resistance to swelling imposed by the hard-segment structure, which functions as a physical cross-link network.

In the dry samples with mixed soft segment, the glass transitions of the two individual polyethers are clearly discernible, as illustrated

Table I. Water Uptake in Hydrophilic Polyurethanes

Sample	*PTMO (g/g of polymer)*	*PEO (g/g of polymer)*	*Water Uptake*	
			%	*Mol H_2O/Mol EO*
4PU0	0.606	0.000	1	
4PU5	0.285	0.279	26	2.28
4PU6	0.221	0.332	33	2.43
4PU7	0.163	0.382	39	2.50
4PU10	0.000	0.523	57	2.67
3PU10	0.000	0.599	80	3.26
2PU10	0.000	0.702	145	5.05
2.7PU5	0.330	0.327	39	2.92

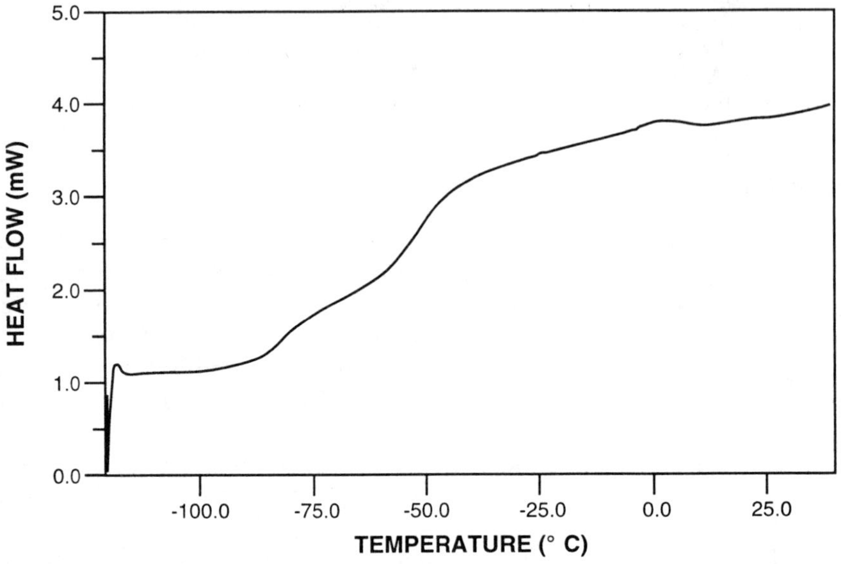

Figure 1. DSC scan of 4PU5 in the dry condition. Sample weight, 14.8 mg; scan rate, 20 °C/min for −120 to 30 °C.

by the example in Figure 1. The corresponding T_gs, which were determined as the midpoint of the increase in heat capacity, are recorded in Table II. The T_g of −78 °C for 4PU0 is typical of values reported for PTMO in MDI-based polyurethanes and is close to the value of −85 °C for the pure soft segment (MDI is diphenyl methyldiisocyanate) (5). However, uncertainty about the T_g of PEO is considerable,

Table II. Soft-Segment Thermal Transition Behavior

	T_g (°C)		ΔC_p [J/(g of SS·K)]		ΔC_p [J/(g of PEO·K)]	
Sample	*Dry*	*Wet*	*Dry*	*Wet*	*Dry*	*Wet*
4PU0	−78	−79	0.50	0.55		
4PU5	−82, −52	−75	0.62	1.20	0.74	1.92
4PU6	−83, −52	−75	0.71	1.39	0.86	1.99
4PU7	−82, −50	−78	0.76	1.44	0.87	1.84
4PU10	−51	−79	0.89	1.43	0.89	1.43
3PU10	−51	n.d.	1.03	n.d.	1.03	n.d.
2PU10	−52	−66	1.05[a]	1.72	1.05	1.72
2.7PU6	−81, −52	−78	0.79	1.40	1.09	2.31

NOTE: SS is soft segment; n.d. is not determined.
[a] Corrected for 15% crystallinity.

presumably because of the effect of the high degree of crystallinity of the pure polyether. The value of −50 °C observed here is probably elevated somewhat above the value for the pure polyether by hydrogen bonding to phase-mixed urethane segments. In the wet samples the two glass transitions coalesce, as shown by the example in Figure 2, and only a single T_g is observed at −75 °C or lower, as recorded in Table II. In addition, the water melting peak is sharp and increases in size through the set of samples with increasing PEO in the mixed-soft-segment samples. This peak represents less than 1% of the water present in these samples. Thus essentially all of the water is present as nonfreezing bound water.

The samples with a pure PEO soft segment exhibit more complicated behavior in the wet state, which, although not illustrated by additional figures, is described in what follows. In 3PU10 the T_g is obscured by soft-segment recrystallization, and additional endothermal activity occurs below and above the water melting peak. The lower endothermal region can be attributed to bound freezing water and amounts to about 4.5% of the added water. This amount is about 4 times more water than appears in the sharp melting peak at 0 °C. A peak at 19 °C represents the melting of water-induced crystallinity in the PEO segment. These complications make it difficult to quantify

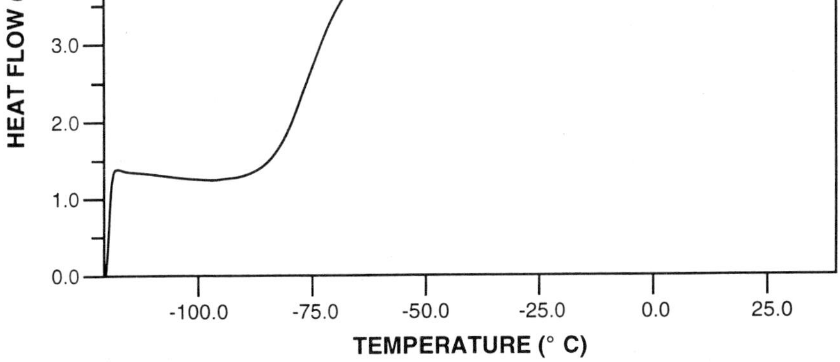

Figure 2. DSC scan of 4PU5 with saturation water uptake. Sample weight, including dissolved water, 15.9 mg; scan rate, 20 °C/min for −120 to 30 °C.

the behavior at T_g in this sample. 2PU10 shows 15% crystallinity in the dry sample, but in the wet sample the melting peak at 18 °C is erased, indicating that PEO is soluble in the higher-saturation water content of this sample. A large water peak, amounting to less than 5% of the added water, is centered at −2 °C. Otherwise, the region between T_g and the melting peak of water is relatively clear, and, therefore, the T_g behavior can be analyzed.

Some further indication of the effect of water on soft-segment behavior can be gained by a quantitative examination of the heat capacity change at the soft-segment glass transition. The results are recorded in Table II in joules per gram of soft segment per degree Kelvin. For the mixed-soft-segment samples, the recorded ΔC_ps were based on the temperature range, which included the transitions due to both types of polyol (*see* Figure 1). In comparing the ΔC_ps for these samples in the dry and wet states, significantly higher values in the wet state are immediately evident. An additional important observation is that ΔC_p increases progressively with increasing proportion of PEO in the soft segment through the set of samples from 4PU0 to 4PU10. This indicates that ΔC_p is larger for PEO than for PTMO. In 4PU10 the estimated ΔC_p in the wet state is lower than that of the mixed-soft-segment samples. This result might have been affected by the broad low-temperature endotherm, which probably represents bound freezing water. As noted in the previous paragraph, interfering thermal processes also complicated the interpretation of the T_g behavior in the other wet samples with pure PEO soft segment.

In the phase-segregated polyether polyurethanes, separation of soft- and hard-segment components is usually incomplete. Short hard-segment units can mix with the soft-segment phase and raise the T_g. In addition, a portion of the soft segment can mix with the hard-segment phase or might be immobilized at the interphase and will not contribute to the behavior at T_g. One possible explanation of the increase in ΔC_p is that a change in morphology has occurred with the addition of water and that the fraction of free soft segment has accordingly increased. An estimate of the fraction of free soft segment, F_{ss}, is given by the ratio (5)

$$F_{ss} = \frac{\Delta C_p}{\Delta C_p{}^0} \tag{1}$$

where ΔC_p is the measured value, and $\Delta C_p{}^0$ is the value for the pure soft segment. The value of $\Delta C_p{}^0$ for pure PTMO is 0.815 J/(g·K) (5). The fraction of free soft segment for PTMO, determined as the ratio of the measured value (Table II) and the value for the pure sample, is 0.61, which is close to the value reported for PTMO in an MDI–BD polyurethane.

As noted earlier in this chapter, there is no reliable $\Delta C_p{}^0$ for PEO because it is not possible to measure ΔC_p directly due to the high level of crystallinity in PEO. A value for $\Delta C_p{}^0$ can be estimated from the set of samples in the dry state with fixed hard-segment contents, that is, 4PU0 to 4PU10, by subtracting the PTMO contribution. The PTMO contribution in the mixed-soft-segment samples, ΔC_p(PTMO), is given by the following relation:

$$\Delta C_p(\text{PTMO}) = wF_{\text{ss}}\Delta C_p{}^0 \tag{2}$$

where all quantities on the right-hand side refer to PTMO, and w is the weight fraction of PTMO in the soft segment (Table I). The resulting ΔC_ps for PEO for the three samples 4PU6 to 4PU10 (next-to-last column in Table II) are almost constant and equal 0.87 J/(g·K). Thus, the results support the assumptions underlying the calculation. Why the value for 4PU5 is distinctly lower is not known. If a $\Delta C_p{}^0$ for PEO is estimated by assuming that $F_{\text{ss}} = 0.61$ (the same as that for PTMO), the resulting value is 1.43 J/(g·K), which is far larger than that of PTMO. However, even this value is below the ΔC_p results of nearly 2 J/(g·K) for the wet samples (last column of Table II). Furthermore, given the nonpolar nature of H12MDI compared to MDI, it is likely that F_{ss} is greater for PEO than for PTMO, with the result that $\Delta C_p{}^0$ for PEO would be lower than the value just indicated. Thus the higher values for the wet samples cannot be attributed to an increase of the free PEO to 100% as a result of water uptake. In fact, the comparison suggests that water contributes directly to ΔC_p.

To determine the contribution of water to T_g behavior, DSC measurements were conducted with another preparation of 4PU10 at four water contents from 15.8 to 50% (0.302 to 0.96 g of H_2O/g of PEO); the highest value corresponded to the saturation concentration in this sample. This sample differed from that on which the earlier measurements of Tables I and II were made, and some of the properties differed from those of the earlier sample. For all water concentrations below saturation, the DSC traces were free of complicating features that would interfere with a reliable measurement of ΔC_p. However, at saturation, the DSC scan was complicated by a broad peak at about −25 °C, which represented bound freezing water, and a moderately sharp peak at about 2.5 °C. These features precluded the determination of an accurate value of ΔC_p in this sample, although it was still possible to determine T_g. The T_g results, plotted as a function of water concentration in Figure 3, follow a surprisingly linear dependence, with a slope of −33.5 K/g of H_2O. ΔC_p also exhibits a clear-cut dependence

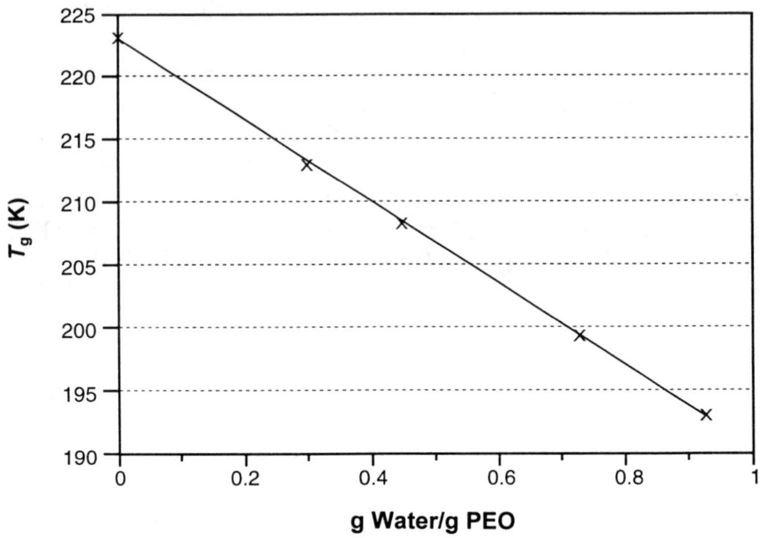

Figure 3. Plot of polymer T_g *for 4PU10 as a function of water uptake in units of grams of* H_2O *per gram of PEO.*

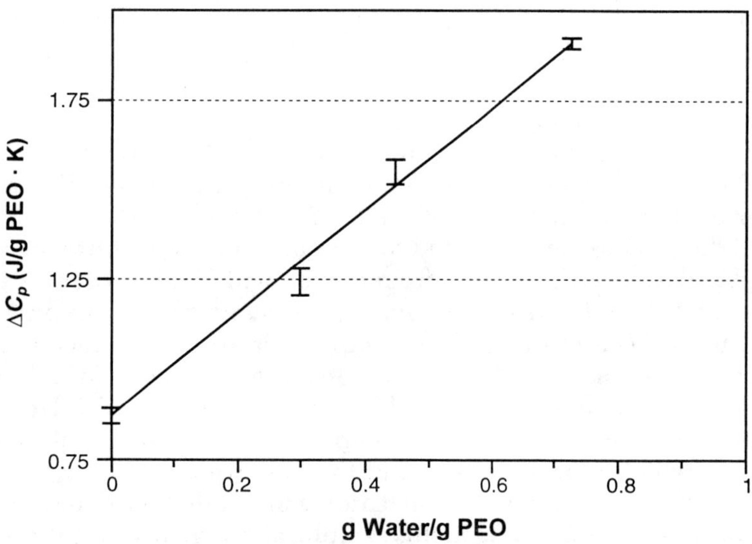

Figure 4. Plot of ΔC_p *for 4PU10 as a function of water uptake in units of grams of* H_2O *per gram of PEO. Error bars represent standard deviations for repeat scans on a single sample.*

on water content. The data are plotted in Figure 4, where the vertical segments indicate the standard deviation from four or more repeat runs. The dependence on water content is almost linear, and the slope of the line leads to ΔC_p = 1.45 J/(g of H_2O·K).

Conclusions

In the set of samples with fixed hard-segment content (4PU6 to 4PU10), the water uptake in the samples with mixed soft segment is approximately proportional to the EO content, that is, a ratio of 2.5 mol of H_2O/mol of EO. The DSC trace shows that essentially all the water is bound nonfreezing water. If all the PEO in the soft segment is assumed to be accessible to water, then this value should represent the water–EO stochiometry. Although this value is in conflict with the results of deuterium NMR data on water in a highly crystalline PEO sample (6), which indicated a one-to-one stoichiometry, it is in agreement with the results of chemical shift NMR results (7), which suggested a molar stoichiometry of 3 H_2O/EO. The ratio is even higher in the samples of lower hard-segment content, but the increasing complexity of the DSC trace makes it difficult to determine quantitatively the fraction of added water that is not freezing.

Water has a profound effect on the glass transition behavior, including a depression of the soft-segment T_g and an increase in ΔC_p. The T_g depression exhibits a linear dependence on water content. The increase in ΔC_p also appears to be proportional to water content. The nature of the underlying physical process responsible for the increase in ΔC_p is not known but is probably related to the interactions with PEO represented by the bound nonfreezing state of all or a large fraction of the water content. One possible interpretation of these results is that a loss of water mobility accompanies the freezing out of the polymer segmental mobility at T_g. Deuterium NMR experiments could be helpful in providing a direct determination of the state of water and the changes in water mobility that might be occurring at the polymer T_g.

References

1. Chen, C. T.; Eaton, R. F.; Chang, Y. S.; Tobolsky, A. V. *J. Appl. Polym. Sci.* **1962,** *16,* 2195.
2. Illinger, J. L.; Schneider, N. S.; Karasz, F. E. In *Permeability of Plastic Films and Coatings to Gases, Vapors and Liquids;* Hopfenberg, H. D.; Ed.; Plenum: New York, 1975; pp 183–196.
3. Illinger, J. L. In *Polymer Alloys;* Klempner, D.; Frisch, K. C., Eds.; Plenum: New York, 1977; pp 313–325.

4. Schneider, N. S.; Illinger, J. L.; Karasz, F. E. *J. Appl. Polym. Sci.* **1993,** *47,* 1419.
5. Camberlin, Y.; Pascault, J. P. *J. Polym. Sci., Polym. Phys. Ed.* **1984,** *22,* 1835.
6. Hey, M. J.; Ilett, S. M.; Mortimer, M.; Oates, G. *J. Chem. Soc., Faraday Trans.* **1990,** *86,* 2673.
7. Liu, K. J.; Parsons, J. L. *Macromolecules* **1969,** *2,* 529.

RECEIVED for review November 18, 1993. ACCEPTED revised manuscript July 26, 1994.

13

Rheological Properties of Poly(ethylene oxide) in Anionic Surfactant Solutions

K. C. Tam

Advanced Materials Research Center, School of Mechanical and Production Engineering, Nanyang Technological University, Nanyang Avenue, Singapore 2263, Republic of Singapore

Poly(ethylene oxide) (PEO), being a nonionic polymer, exhibits some unique properties when dissolved in water. It interacts with surfactants and other amphiphilic molecules in solutions. The effects of sodium n*-alkyl sulfates on the rheological properties of solutions of PEOs with three different molecular weights (0.6 million, 1 million, and 2 million) were examined. The viscosity and shear-thinning behavior are dependent on the surfactant chain length and concentration. The zero-shear viscosity increases with surfactant concentration up to a critical surfactant level, beyond which the viscosity decreases. The amount of surfactant required to impart maximum viscosity is dependent on the polymer concentration and the surfactant chain length. Increases in viscosity can be attributed to the formation of polymer–micelle complexes that result in a larger polymer coil. Viscosity reduction beyond the critical surfactant concentration is caused by the charge-shielding effects of excess positive sodium ions in solution. Long-chain surfactant is more efficient than short-chain surfactant, in modifying the rheological properties of PEO solutions. Graessley's reduced-variable approach can be used to correlate the shear-dependent viscosity.*

0065–2393/96/0248–0205$12.00/0

WATER-SOLUBLE POLYMERS have acquired increasing importance because of the demand for water-based instead of traditional solvent-based products. One example of such demands is for paint and coating formulations. This trend is projected to increase as environmental pressure mounts in the light of growing evidence of the danger of many organic solvents currently in use by industries. Water, being easily available and environmentally friendly, will in increasing measure be the solvent of choice for a wide range of products.

Water-soluble polymers are broadly divided into three categories depending on the charge present on the polymer chains. They are anionic, cationic, or nonionic. The behavior of the polymers dissolved in water depends greatly on the charge characteristics of the polymers.

Poly(ethylene oxide) (PEO) is a nonionic polymer, and its behavior in water differs from that of ionic polymers. It is used in many different applications, for example, as a friction reducer in turbulent flow, a thickening agent, a lubricant, and a flocculant and dispersant in food and pharmaceutical products. Being a polyether, it forms hydrogen bonds with water and is completely soluble to just below the boiling point of water. The strong hydrogen bonding affinity accounts for its associations with various polar compounds such as phenolic resins, mineral acids, urea, and surface-active agents.

PEO can associate with some small amphiphilic molecules and thereby modify its solution properties. Much work has been devoted to thermodynamic studies of these interactions. Techniques such as conductometry (*1–4*), surface tension (*2, 3, 5, 6*), dye solubilization (*1–3, 5*), and equilibrium dialysis (*7, 8*) have been used. Formation of complexes of water-soluble polymers (such as PEO) and surfactant molecules was adequately reviewed by Goddard (*9*).

As mentioned earlier, most of the work on polymer–surfactant interactions has been devoted to thermodynamic studies. Only a few studies have reported on the viscometric behavior of PEO in the presence of surfactant molecules. However, nearly all of these studies were performed with a capillary (Ubbelohde) viscometer (*10–13*), which is typically used for measuring the viscosity of very dilute solutions under Newtonian flow. The only exception is the study by Brackman (*14*), which examined the first normal stress of 0.25 wt% high-molecular-weight PEO in different proportions of sodium dodecyl sulfate (SDS).

Recently, Ortiz et al. (*15*) conducted a comprehensive study of concentrated PEO solutions. The molecular weight of each PEO sample was determined by the intrinsic-viscosity method, and the weights were comparable to the values given by the manufacturer. Different solvent types were used, and a mathematical model to predict the steady shear and dynamic behavior was developed. However, they did not examine the behavior of PEO in the presence of surfactants, which is the focus of this chapter.

In this chapter the rheological properties of PEO in sodium *n*-alkyl sulfates are reported. The interactions between PEO and surfactant molecules of different chain lengths are discussed with reference to steady-shear viscosity data.

Experimental Details

Tables I and II summarize the characteristics of the PEO samples and surfactants, respectively, used in this study. Three different grades of PEO were supplied by Union Carbide; their molecular weights were 600,000, 1 million, and 2 million. These weights are comparable to the molecular weights determined from intrinsic-viscosity measurements obtained by Ortiz et al. (*15*) as shown in Table I. The surfactants employed were sodium *n*-alkyl sulfates with chain lengths of 8, 10, and 12. The chemical structures of PEO and the anionic surfactant are shown in Figure 1.

The PEO powders were dried in an oven at 40 °C for 24 h to determine the moisture content. Stock solutions of 2 wt% of each of the three grades of PEO were made by dissolving the PEO flakes in MilliQ water and gently mixing the solution with a magnetic stirrer for 24 h. From this stock solution, 0.25, 0.5, and 1 wt% PEOs were prepared in various surfactant concentrations ranging from 0 to 0.1 M.

Shear-dependent viscosities were measured using a Bohlin VOR controlled-rate rheometer fitted with a double-gap concentric cylinder. The test samples were kept at a constant temperature of 25 °C by using a temperature-controlled bath. Depending on the viscosity of the solution, the shear rate imposed on the sample ranged from 0.1 to 500 s^{-1}.

Table I. PEO Grades as Supplied by Union Carbide Corporation

Polymer Grade	M_w[a] *(kg/kmol)*	M_v[b] *(kg/kmol)*
Polyox WSR-205	6×10^5	5.8×10^5
Polyox N-12K	1×10^6	1.0×10^6
Polyox N-60K	2×10^6	1.8×10^6

[a] Supplied by Union Carbide.
[b] Determined by Ortiz et al. (*15*) using the intrinsic-viscosity technique.

Table II. Critical Aggregation and Critical Micelle Concentrations (10^3 M) of Sodium *n*-Alkyl Sulfates

Chain Length (n)	*Aggregation*	*Micelle*
8	64	130
10	7.5	33
12	0.79	8.3

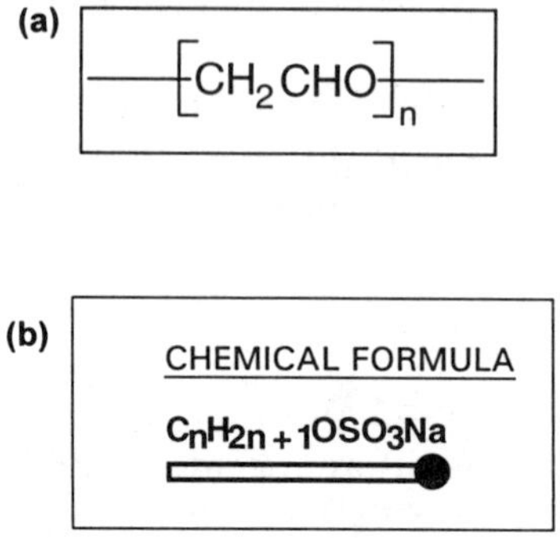

Figure 1. Structures of polymer and anionic surfactant: (a) PEO; (b) sodium n-alkyl sulfate.

Results and Discussion

The steady-shear viscosities of three grades of PEO (WSR-205, N-12K, and N-60K) at 0.5 wt% are shown in Figure 2. In the shear rate range examined, both of the lower-molecular-weight samples (WSR-205 and N-12K) exhibit Newtonian behavior. The higher-molecular-weight sample (N-60K) starts to shear thin at a shear rate of about 30 s^{-1}. The Newtonian viscosities are also consistent with the direct dependence on molecular weight.

Effects of SDS Concentrations. PEO solutions are known to interact with surfactants. The magnitude of the shear viscosity and the

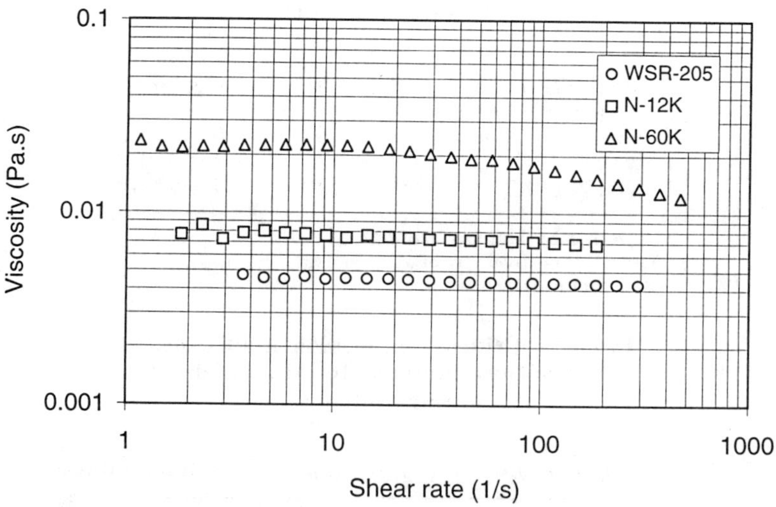

Figure 2. Steady-shear viscosity of 0.5 wt% aqueous solutions of WSR-205, N-12K, and N-60K.

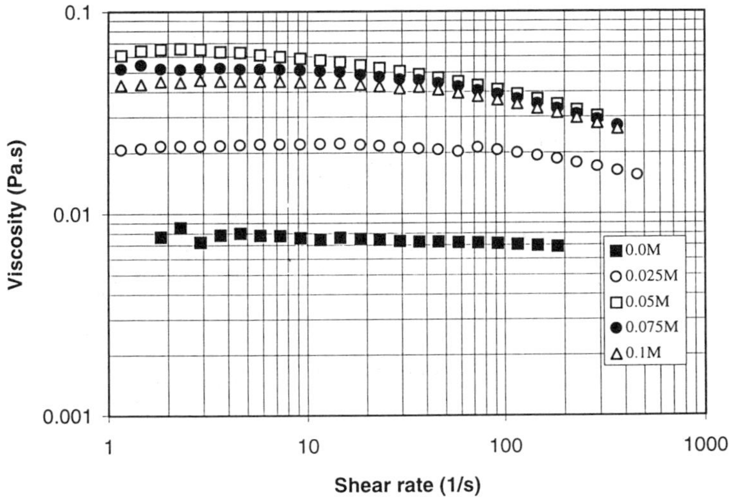

Figure 3. Steady-shear data for 0.5 wt% N-12K in SDS solutions with different concentrations.

shear-thinning characteristics of PEO solutions depend on the concentrations and types of surfactant present. The addition of SDS to PEO solution alters the polymer conformation and hence its response to applied shear gradient. The changes in PEO chain conformation are discussed more completely later in this chapter.

The effects of SDS concentrations on the rheological properties of PEO solutions were studied. Figures 3 and 4 show the viscosity curves of 0.5 wt% N-12K and N-60K in different concentrations of SDS solutions. The SDS concentrations used ranged from 0 to 0.1 M. The viscosity reaches a maximum at an SDS concentration of about 0.05 M. Any further increase in SDS concentration results in a decrease in viscosity. The same trend was observed for N-12K (molecular weight = 1 million) and N-60K (molecular weight = 2 million). However, the higher-molecular-weight PEO samples exhibit marked shear-thinning properties in solutions with various concentrations of SDS. The effect of SDS concentration on the zero-shear viscosity of PEO solutions is best summarized in Figure 5. A number of features can be concluded from the figure, namely:

1. The viscosity starts to increase at an SDS concentration of more than 0.01 M. This level represents a concentration at which sufficient SDS micelles have formed and bound onto the polymer chains to cause an increase in the radius of gyration of the polymer coil. This increase

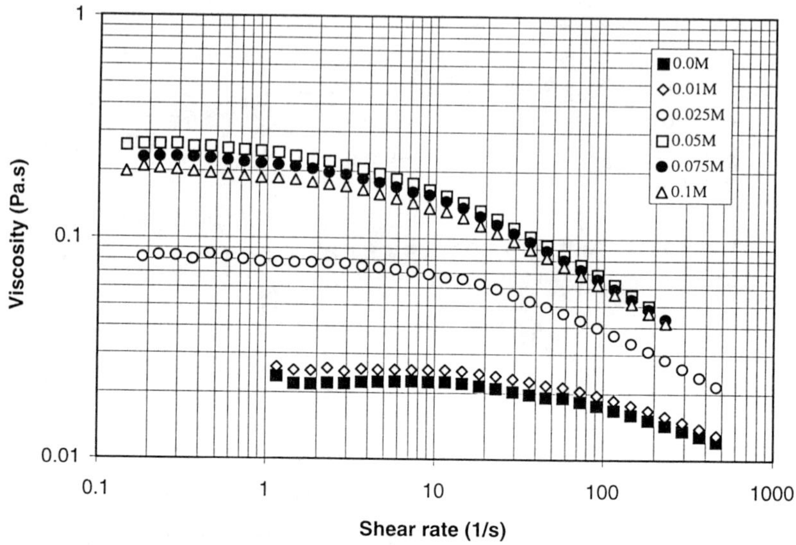

Figure 4. Steady-shear data for 0.5 wt% N-60K in SDS solutions with different concentrations.

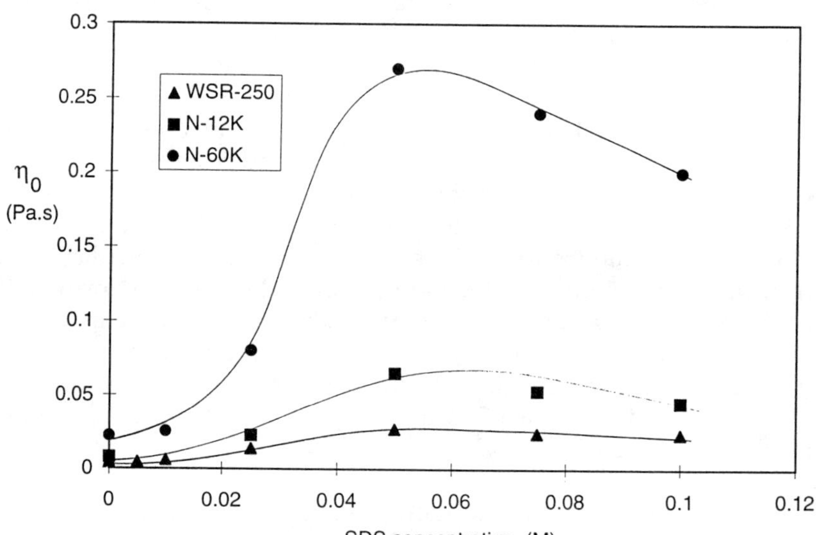

Figure 5. Effect of SDS concentrations on the zero-shear viscosity of different PEO solutions.

is due to repulsion between negative charges on the polymer chains.

2. Maximum viscosity is reached at an SDS concentration of about 0.05 M. This concentration represents a saturation of the SDS micelles along the polymer chains that corresponds to the maximum extent to which the polymer coils can expand. Any further addition of SDS to the solutions will not result in coil expansion. However, added SDS has the negative effect of reducing the size of the polymer coils because of charge shielding by the free sodium ions. Hence a reduction in viscosity is observed.
3. The maximum viscosity observed is proportional to the molecular weight of the PEO. For example, for the same polymer concentration, the maximum viscosity of N-60K is about 5 times that of N-12K. Entanglement of polymer chains due to polymer coil expansion is the most probable cause for the large increase in the viscosity of the higher-molecular-weight polymer compared to that of the lower-molecular-weight polymer.

The behavior outlined above can be best illustrated by the mechanistic model shown in Figure 6. This model is based on earlier work of Cabane and Duplessix (*16*), who proposed the PEO/SDS complex on the basis of a neutron-scattering study.

The point at which maximum viscosity occurs depends on the

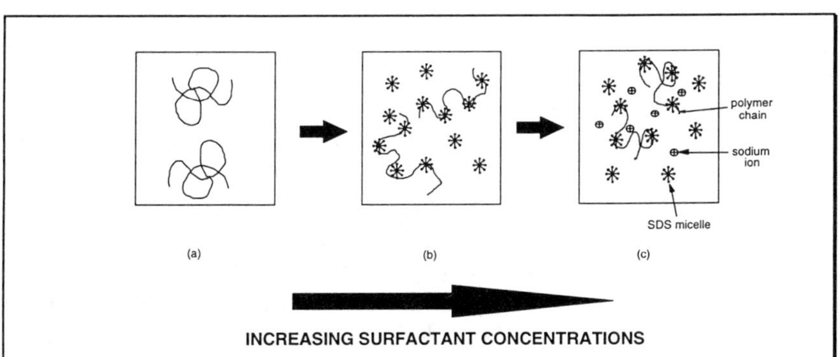

Figure 6. Model of PEO/SDS complex at [SDS] of 0 M (a), 0.05 M (maximum coil size is due to intra- and intermolecular repulsion of charges on polymer chains) (b), and >0.05 M (reduction in polymer coil is due to charge shielding by sodium ions) (c).

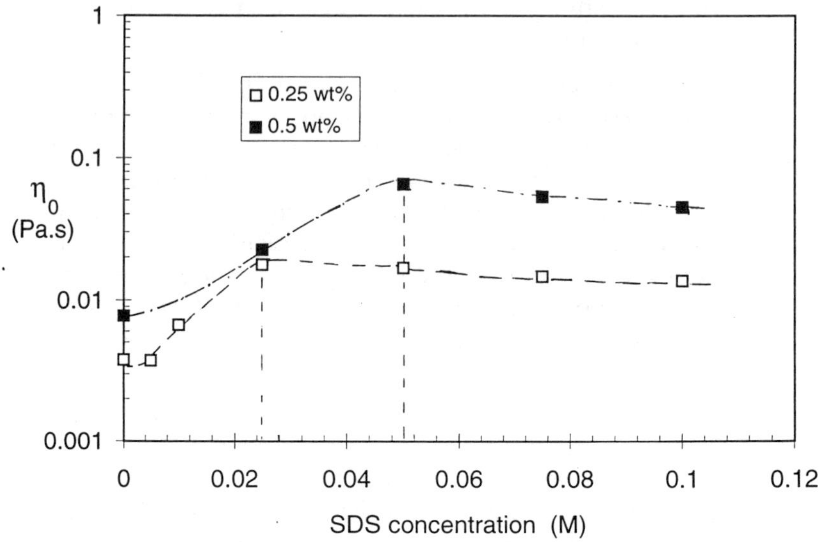

Figure 7. Effect of SDS concentration on 0.25 and 0.5 wt% N-12K solutions.

polymer concentration. As shown in Figure 7, the maximum viscosities of 0.25 and 0.5 wt% N-12K PEO solutions occur at SDS concentrations of 0.025 and 0.05 M, respectively. In order to obtain a similar density of bound SDS micelles on polymer coils, a twofold increase in polymer concentrations will similarly require a twofold increase in SDS concentration.

Effects of Chain Length of Sodium *n*-Alkyl Sulfates. For most surfactants, thermodynamic behavior and hence binding behavior are largely dependent on the surfactant chain length in addition to other parameters such as temperature, solvent type, etc. As the chain length increases, the hydrophobicity of the surfactant molecule increases. This increase is reflected in the critical micelle and critical aggregation concentrations as outlined in Table II. Thus the viscosity behavior of PEO solutions is also very probably affected by the surfactant chain length. Figure 8 depicts the effect of surfactant chain length on the shear viscosity of 0.5 wt% N-12K in 0.05 M surfactant solution. For a fixed surfactant concentration, the viscosity of the PEO solution is proportional to the surfactant chain length. The effect of surfactant concentration on the zero-shear viscosity of 0.5 wt% N-12K PEO solutions is summarized in Figure 9. A number of trends can be observed.

1. The onset of viscosity increase is dependent on the surfactant chain length. Viscosity increase commences at

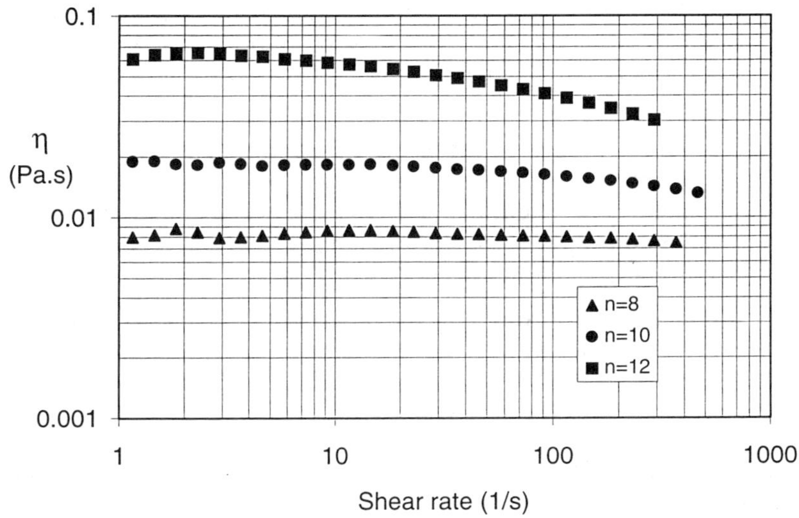

Figure 8. Steady-shear viscosity of 0.5 wt% N-12K in 0.05 M surfactant with chain length of 8, 10, or 12.

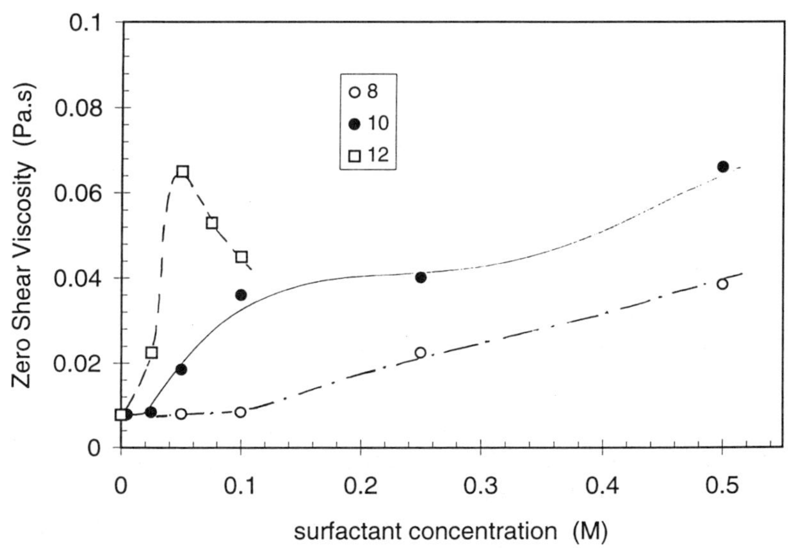

Figure 9. Effect of chain length of sodium n*-alkyl sulfates on zero-shear viscosity of 0.5 wt% N-12K solutions.*

0.1, 0.03, and 0.01 M for chain lengths of 8, 10, and 12, respectively. This increase is directly related to the critical micelle and critical aggregation concentrations of each surfactant.

2. The surfactant concentration at which maximum viscosity occurs is a function of the chain length. For $n = 12$, the maximum occurs at an SDS concentration of 0.05 M. No maximum is observed for chain lengths of 8 and 10 at concentrations as high as 0.5 M.
3. Different initial slopes of the viscosity/surfactant concentration are observed. These differences indicate that the nature and rate of binding of surfactant onto polymer coils is affected by chain length.

Master Plots of Viscosity Curves. Viscosity curves can be reduced and correlated by the reduced-variable approach proposed by Graessley (*17*). This approach has the form

$$\frac{\eta - \eta_s}{\eta_0 - \eta_s} = fn(\beta) \tag{1}$$

and

$$\beta = \frac{(\eta_0 - \eta_s)M}{cRT}\dot{\gamma} = \lambda\dot{\gamma} \tag{2}$$

where η_s is the solvent viscosity, M is the molecular weight, c is the polymer concentration, R is the gas constant, T is the temperature, and λ is the molecular relaxation time.

For most high-molecular-weight polymer solutions, $\eta_0 >> \eta_s$; hence equation 1 can be simplified to

$$\frac{\eta}{\eta_0} = fn(\beta) \tag{3}$$

The appropriate choice for the reduced shear rate (equation 2) is guided by the observation that the longest relaxation time of the polymer governs the onset of shear rate dependence. Such a dependence of the viscosity can be represented by the characteristic relaxation time in spring-bead models: $\lambda = (\eta_0 - \eta_s)M/cRT$. Such an approach to correlating viscosity data removes most of the observed variation that in organic polymer systems is due to molecular weight, polymer concentration, and temperature. This method has been applied suc-

(a)

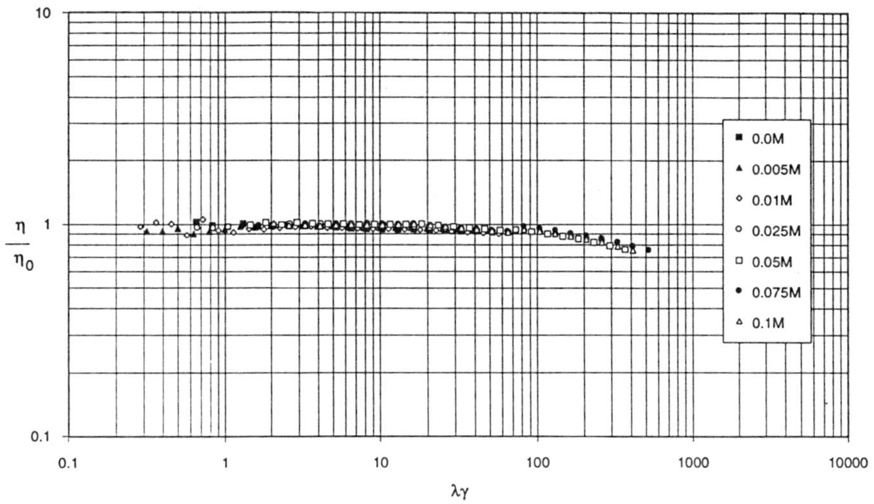

Figure 10. Reduced-viscosity plots of 0.5 wt% WSR-205 (a) in SDS solutions. Continued on next page.

cessfully to most monodispersed polymeric systems. However, when different solvents were used in the preparation of the test solutions, greater variations and spread of the reduced data were observed. Such differences are related to changes in the polymer structure in the presence of these solvents.

The applicability of the approach described above is tested with the present system, in which polymer chains and solvents, which in this case are water and surfactant molecules, interact. Figures 10a to 10c show the reduced-viscosity data for 0.5 wt% WSR-205, N-12K, and N-60K in different SDS concentrations. The master plots are not perfect, and deviations are apparent at higher β values, especially for the higher-molecular-weight PEO solutions (such as N-12K and N-60K), in which the shear-thinning behavior is more pronounced. This pronounced shear thinning indicates that the shear-dependent viscosity of high-molecular-weight PEO in surfactant solutions cannot be absorbed into the nondimensioning parameters of the master curve expression of equations 2 and 3. However, better reduced master plots are obtained for N-12K in surfactant solutions with shorter chains. All the data for PEO solutions are representative of a unique master curve when the data are superimposed onto a single plot.

Conclusions

The rheological properties of solutions of three different-molecular-weight PEOs are influenced by the addition of anionic surfactant. Vis-

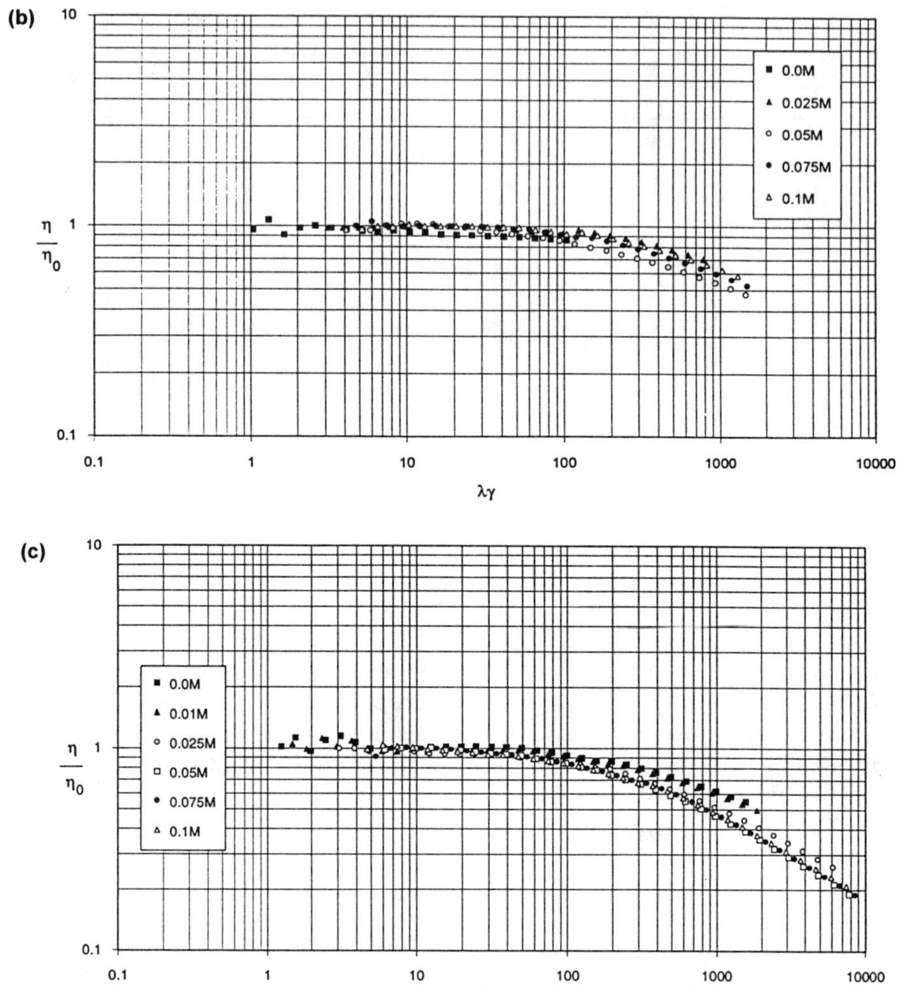

Figure 10. Continued. *Reduced-viscosity plots of 0.5 wt% N-12K (b) and N-60K (c) in SDS solutions.*

cosity increases with surfactant concentration up to a critical level, beyond which viscosity decreases. Long-chain surfactants are more efficient in altering the rheological properties than short-chain surfactants.

References

1. Saito, S. J. *J. Colloid Interface Sci.* **1967,** *24,* 227.
2. Schwuger, M. J. *J. Colloid Interface Sci.* **1973,** *43,* 491.

3. Jones, M. N. *J. Colloid Interface Sci.* **1967,** *23,* 36.
4. Botre, C.; De Martiis, F.; Solinas, M. *J. Phys. Chem.* **1964,** *68,* 3623.
5. Arai, H.; Horin, S. *J. Colloid Interface Sci.* **1969,** *30,* 372.
6. Williams, J.; Knox, J. R.; Parshall, T. O. *J. Colloid Interface Sci.* **1970,** *33,* 16.
7. Shirahama, K. *Colloid Polym. Sci.* **1974,** *252,* 978.
8. Shirahama, K.; Ide, N. *J. Colloid Interface Sci.* **1976,** *54,* 450.
9. Goddard, E. D. *Colloids Surf.* **1986,** *19,* 255.
10. Nagarajan, R. J. *Chem. Phys.* **1989,** *90,* 1980.
11. Saito, S. *J. Biochem.* **1957,** *154,* 19.
12. Lance-Gomez, E. T. *J. Appl. Polym. Sci.* **1986,** *31,* 333.
13. Francois, J.; Dayantis, J.; Sabbadin, J. *Eur. Polym. J.* **1985,** *21,* 165.
14. Brackman, J. C. *Langmuir* **1991,** *7,* 469.
15. Ortiz, M.; De Kee, D.; Carreau, P. J. *J. Rheol.* **1994,** *38,* 519.
16. Cabane, B.; Duplessix, R. *J. Phys. (Paris)* **1982,** *43,* 1529.
17. Graessley, W. W. *Adv. Polym. Sci.* **1974,** *16,* 1.

RECEIVED for review April 16, 1994. ACCEPTED revised manuscript May 10, 1995.

14

Viscoelastic Surfactant Systems Under Shear

H. Hoffmann, S. Hofmann, and U. Kästner

Physikalische Chemie I, Universität Bayreuth, D-95440 Bayreuth, Germany

Four viscoelastic aqueous fluids that each contain 1–3 wt% surfactant or water-soluble polymers are described. The viscoelastic properties are a result of three-dimensional networks that are already present at rest or are formed during flow. Two of the described systems have a yield stress that is large enough to suspend small particles from sedimentation but low enough to allow the fluids to flow. The first system consists of entangled threadlike micelles. It behaves like a simple Maxwellian fluid. Small, charged, rodlike micelles are present in the second system. Under shear, large supramolecular structures that are completely aligned in the direction of flow are formed. The third system consists of densely packed multilamellar vesicles. The yield stress of this system is a result of dense packing. The fourth system consists of hydrophobically and cationically modified hydroxyethylcellulose that is physically cross-linked by surfactant molecules.

CONCENTRATED SOLUTIONS OF SURFACTANTS can still have viscosities in the range of the solvent viscosity. On the other hand, dilute solutions with only a few percent surfactant can be highly viscous or can even have a yield value. Such solutions usually also have strong viscoelastic properties. The different flow behaviors in these solutions are consequences of the different micellar structures that are formed by the surfactants. In concentrated low-viscosity solutions, the surfactant is organized into globular micelles, and the flow behaviors of such

0065–2393/96/0248–0219$14.50/0

systems are thus similar to those of any other concentrated dispersion in which the particles are not connected. Viscoelasticity in dilute solutions is usually a result of an interconnected network of micellar aggregates. In this chapter, four systems are discussed. They all show viscoelastic properties during flow, but the viscoelastic behaviors in the solutions are the results of different structures. All four systems are aqueous solutions with 1–3 wt% surfactant. The solutions are completely transparent and at rest are optically isotropic. The solutions show the recoil phenomenon.

The casual observer could assume that the solutions behave very similarly and that the micellar structures in the solutions might be the same. However, detailed investigations using different techniques show that the microstructures in the systems are very different and that the flow behaviors of the systems under shear show marked and important differences. The four systems that will be discussed in this chapter are as follows: (1) viscoelastic surfactant solutions in which the network, which is the origin of the viscoelastic properties, is composed of entangled threadlike micelles; the network is already present in the solution at rest; (2) surfactant solutions that behave as Newtonian fluids at small shear rates but become viscoelastic when exposed to greater shear forces; (3) surfactant solutions in which the viscoelasticity originates from densely packed multilamellar vesicles; and (4) viscoelastic solutions that consist of modified water-soluble hydroxyethylcellulose (HEC) that is cross-linked by surfactants.

Experimental Details

Materials. Cetylpyridinium chloride was bought from Merck, fractionated in cold and hot water, and recrystallized several times in acetic acid–methanol. Sodium salicylate was bought from Merck and used without further purification. Hexaethyleneglycol–monododecyl ether ($C_{12}E_6$) was bought from NIKKO Chemical Co. Ltd., Tokyo, as a pure substance (GC). The sodium alkylether sulfate $C_nE_{2.5}SO_4Na$ was a gift of Hoechst AG, Gendorf (trade name GENAPOL LRO), and was received as an aqueous solution containing 27 wt% active surfactant with $\langle M_w \rangle = 382$ g/mol. The technical product consists of a mixture of surfactants with 0–8 ethylene oxide units (number average, 2.5) and a chain length distribution n of 75% C_{12} and 25% C_{14}. It was used without further purification. The zwitterionic surfactant tetradecyldimethylamine oxide (C_{14}DMAO) was also a gift of Hoechst AG, Gendorf. It was purified by freeze-drying and two recrystallizations from acetone, p.A. Sodium dodecyl sulfate (SDS) was bought from Serva (Heidelberg) as "SDS cryst. reinst" and was used without further purification. Tetradecyltrimethylammonium bromide (C_{14}TAB) was bought from Aldrich and recrystalized in diethyl ether–ethanol (1:1). The sample of HEC, the hydrophobically modified samples (F-HMHEC), and the cationically modified samples (cat-HEC) were kindly supplied by Hoechst AG (Kalle Albert, Wiesbaden). The structure of the hydrophobic substituent (F-HMHEC) is

$$-CH_2-CHOH-CH_2-O-CH(CH_2Cl)-CH_2-O-(CH_2)_2-(CF_2)_{7-11}-CF_3$$

resulting from the corresponding glycidyl ether used in synthesis. The cationic substituent (cat-HEC) is a glycidyltrimethylammonium chloride. Lithium perfluorononanoate ($C_8F_{17}CO_2Li$) was bought from Aldrich and used without further purification.

Methods. All rheological measurements were recorded with a Bohlin CS rheometer.

The static light-scattering experiments were carried out on a KMX 6 light-scattering photometer (Chromatix) by using a small angle of 6°–7°. The differential refractive index was measured on a KMX 16 differential refractometer (Chromatix). The light source was a He–Ne laser (632.8 nm).

For electric birefringence measurements, the solution was placed in a cell between crossed polarizers, and the light intensity of a small He–Ne laser was recorded when an electric field was applied to the solution (for details, *see* reference 1).

A cell of the Couette type with a rotating inner cylinder was used for flow birefringence. The experimental arrangement is described in detail in reference 2 and references cited therein.

Neutron-scattering investigations were carried out with the D11 and D17 instruments of the Institute Laue-Langevin in Grenoble, France.

The expected network structures were visualized with a CEM 902 transmission electron microscope (TEM) (Zeiss) and a JEM 840-A scanning electron microscope (SEM) (Jeol). For the different preparations, see references *3* and *4*, respectively.

Results and Discussion

Viscoelastic Surfactant Solutions from Entangled Threadlike Micelles. Viscoelastic surfactant solutions from entangled threadlike micelles have been intensively studied over the last few years and are understood best with respect to the four systems discussed in this chapter (*5a–5d*). During this conference a whole symposium was devoted to these systems (*6*). Since the contributions to this symposium are being published in a separate volume, the properties of these systems are discussed here only very briefly, and the interested reader is referred to the proceedings of the symposium (*6*).

Solutions of this type belong to the micellar L_1 phase. The aggregation behavior of practically any surfactant that is soluble in water can be influenced in such a way that the surfactant forms long cylindrical micelles. To control the aggregation behavior, various parameters such as temperature, salinity, cosurfactant concentration, mixing ratio of surfactants, and addition of strongly binding counterions can be manipulated. The threadlike micelles in these solutions can be as long as several micrometers, and even if the threads are somewhat flexible and form coils in the solution, the coils will overlap at a rather low

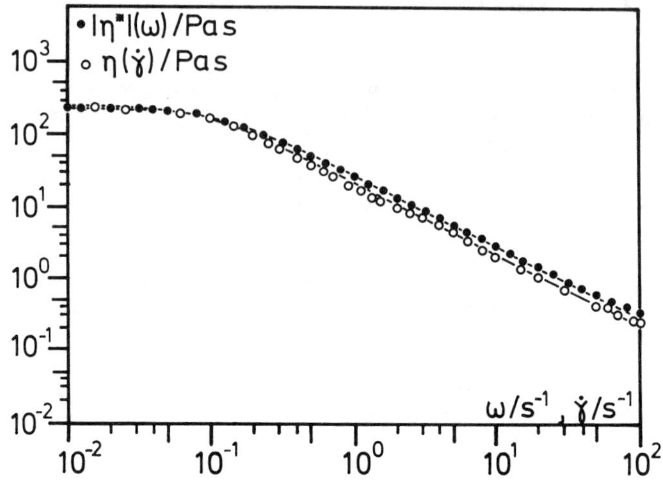

Figure 1. Steady-state shear viscosity $\eta(\dot{\gamma})$ *as a function of shear rate* $(\dot{\gamma})$ *and complex viscosity* η^* (ω) *as a function of angular frequency* (ω) *for a solution of 100 mmol cetylpyridinium chloride and 60 mmol sodium salicylate at* T = *20 °C.*

concentration of surfactant. In the overlap region of the coils ($c > c^*$), the solutions usually have strong viscoelastic properties. In spite of the seemingly very complicated situation in these solutions, the rheological behavior can be very simple. Many solutions show simple Maxwellian behavior. The zero-shear viscosity is often given by the product of a single shear modulus and a single structural relaxation time ($\eta^\circ = G^\circ \cdot \tau_S$) (*7*). Under these conditions the structural relaxation times are controlled by various kinetic processes in which the cylindrical micelles can break and re-form (*8*). A deformed entanglement network can thus relax by a combination of reptation and kinetic processes. The viscosity usually shows a power law behavior with respect to the surfactant concentration ($\eta^\circ \sim (c/c^*)^x$). The exponent x depends, however, on the special conditions, and its value can vary between 8.5 and 1.3. For different exponents of x, the modulus scales always with the same power law ($G^\circ \sim (c/c^*)^{2.3}$). Under shear the systems are shear-thinning. Some results are shown in Figure 1, where shear viscosity is simultaneously plotted against shear rate and radial frequency. The two curves are practically identical. The results show that the systems obey the Cox–Mertz rule; that is, the dynamic and shear viscosities at a given frequency or shear rate are the same.

Viscoelastic Surfactants from Shear-Induced Structures. In recent years the flow behavior of solutions of ionic surfactants that contain small rodlike micelles at well below the overlap concentration has been studied in detail. Some surfactant aggregates are able to form shear-induced structures (SIS) that are supramolecular. These structures, which are formed in the laminar flow after a critical shear rate ($\dot{\gamma}_c$) is exceeded, are responsible for new rheological properties such as an increase in viscosity (*9*), the occurrence of a normal stress difference, and flow birefringence. For $\dot{\gamma} > \dot{\gamma}_c$, the solutions also show rheopectic behavior. The SIS may also be responsible for the friction reduction effect, which is observed if these SIS-forming surfactant solutions are studied in pipe flow systems. This phenomenon, which is often called drag reduction, is also found in aqueous or nonaqueous polymer solutions or dispersions of elongated particles (e.g., asbestos fibers). The occurrence of drag reduction in surfactant solutions implies the possibility of using this energy-saving effect in closed pipe flow circuits. In this case polymer-based systems are not suitable because the long exposure to shear forces in circuit systems causes decomposition of the polymer chains.

Early investigations were performed with binary surfactant solutions of cetylpyridiniumsalicylate, tetradecyltrimethylammonium salicylate, and hexadecyloctyldimethylammonium bromide. Flow birefringence (*10–12*) and small-angle neutron scattering (SANS) experiments (*13–16*) showed that as soon as the shear rate is larger than the threshold value $\dot{\gamma}_c$, a fraction of the micelles is completely aligned in the direction of flow. This alignment cannot be the result of simple hydrodynamic orientation, because the product of $\dot{\gamma}_c$ and the rotational diffusion constant τ_{rot} of the rodlike micelles is always much less than 1. After it was realized that the charge of the micelles plays an important role in the formation of SIS, experiments with mixed nonionic–ionic surfactant systems were started. The two surfactants were used to vary the lengths of the rodlike micelles and their surface charges by altering the mixing ratio and to study the effect of this variation on SIS formation. The aim was to find out which conditions a surfactant solution must fulfill in order to be able to form SIS and, in consequence, to cause drag reduction.

Earlier investigations of the ternary system C_{14}DMAO–SDS–H_2O focused on flow birefringence and rheological experiments with solutions having different mixing ratios, concentrations, and salt contents (*17*). Some typical experimental results are shown in Figures 2 and 3. Figure 2 shows the flow birefringence data for 100 mM solutions with different C_{14}DMAO/SDS mixing ratios. The corresponding data on mean micellar distance and a rough estimation of the lengths of the rodlike micelles are given in Table I. The values were obtained by a combination of both SANS and electric birefringence measurements and by assuming that the micelles are rodlike in shape (*18*).

In Figure 3 the flow birefringence is plotted against the shear rate

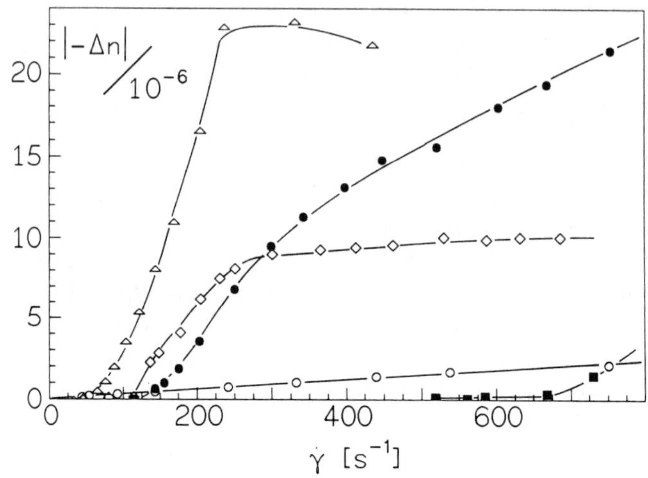

Figure 2. Plot of flow birefringence against shear rate for 100 mM solutions with different molar mixing ratios of C_{14}DMAO to SDS. Key: ○, 9:1; ●, 8:2; △, 7:3; ◇, 6:4; ■, 5:5. T = 25 °C.

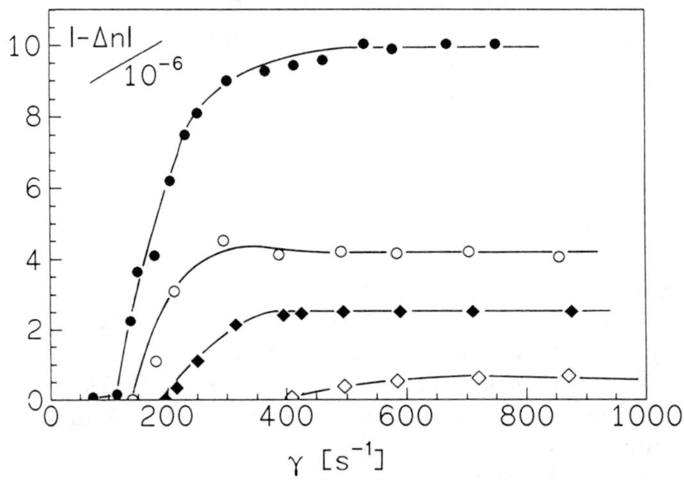

Figure 3. Plot of flow birefringence against shear rate for solutions with a constant molar mixing ratio of C_{14} DMAO to SDS (6:4) but different concentrations. Key: ◇, 30 mmol/L; ◆, 60 mmol/L; ○, 80 mmol/L; ●, 100 mmol/L. T = 25 °C.

Table I. Length of Micelles Compared with Mean Distances for 100-mM Solutions of Different Molar Mixing Ratios at $T = 25$ °C

Mixing Ratio, C_{14}DMAO/SDS	*Mean Distance Apart* d *(Å)*	*Micellar Length* L *(Å)*
9:1	240 ± 20	>300
8:2	190 ± 15	176 ± 40
7:3	170 ± 10	135 ± 20
6:4	160 ± 10	116 ± 20
5:5	150 ± 10	103 ± 20
4:6	145 ± 10	95 ± 15

for solutions with a constant mixing ratio (6:4) but different concentrations. The data in Figure 2 and Table I make it evident that SIS are preferably observed in the concentration regime below the overlap concentration c^* (*18*). The solution with a mixing ratio of 9:1, in which the micelles are longer than the mean distance from each other, shows a linear increase of Δn versus $\dot{\gamma}$ that indicates viscoelastic properties, while all other solutions give rise to the characteristic S-shaped curve of the Δn $(\dot{\gamma})$ plots, which is typical for shear-induced behavior. The critical shear rate γ_c has a minimum value at a ratio of 7:3, and the plateau value Δn_{max} becomes smaller with increasing content of SDS. The solution with a ratio of 5:5 exhibits a second critical shear rate $\dot{\gamma}_{c2}$ at 3200 L/s (not shown in Figure 2). At shear rates larger than $\dot{\gamma}_{c2}$, the SIS is destroyed and the solution becomes isotropic again. Furthermore, investigations into the dependence of $\dot{\gamma}_c$ on the ionic strength I of the solutions revealed that $\ln \dot{\gamma}_c \sim 1/\sqrt{I}$, a result that indicates the important role that surface charge and properties of the electric double layers play in the formation of SIS (*17*). It is also well known that the binding of hydrophobic organic counterions to the micellar surface promotes the ability to form SIS. When these results were combined, it was concluded that the two important parameters, $\dot{\gamma}_c$ and Δn_{max}, are determined by a combination of micellar surface charge effects (which include the effect of counterions and the influence of added salt) and structure parameters such as length and mean distance apart.

Two models were proposed. The first model was mainly based on the rheopexy of the solutions and the fact that the viscosity of the solutions increases distinctly if $\dot{\gamma}_c$ is exceeded (shear thickening) and decreases again if $\dot{\gamma}$ is increased further (shear thinning). According to this first model, at $\dot{\gamma} > \dot{\gamma}_c$, the kinetic energy of the micelles is large enough to overcome their repulsive electrostatic interaction. The micelles aggregate, and very long necklace-type aggregates that become completely aligned in the direction of flow are formed (pearl-string model) (*12*). This model is also supported by new experimental results and theoretical considerations of other groups that predict that shear-induced growth of the micelles takes place (*19–21*). The second model was based on the observation that the position of the sharp correlation peak in the SANS experiments, which is caused by the aligned fraction of the micelles (q_{max}), does not shift in comparison to the scattering maximum caused by the micelles in the quiescent solution or by the unaligned material that is still present at $\dot{\gamma} > \dot{\gamma}_c$ (*13–15, 17*). The fact that the mean distance d of the micelles can be derived from the value of the scattering vector q_{max} ($|q_{max}| = 2\pi/d$) means that d is kept constant when the SIS is formed and that therefore no aggregation takes place. According to this model, the micelles simply switch into a position in which their main axes are oriented parallel to the direction of flow. This switching clearly resembles a shear-induced transition into a nematic phase (phase–transition model).

To clarify the situation, new experiments were performed in which a third component was added to the micelles. First, C_{14}DMAO was partially replaced by $C_{12}E_6$, and in a second experiment, SDS was replaced by $C_{12}E_{2.5}SO_4Na$ (GEN). Solutions (100 mM) with a C_{14}DMAO/SDS mixing ratio of 7:3 were used in the experiments. The total concentration and the overall mixing ratio of nonionic to ionic surfactants and thus the micellar surface charge were kept constant. Figures 4 and 5 show the results of flow birefringence experiments.

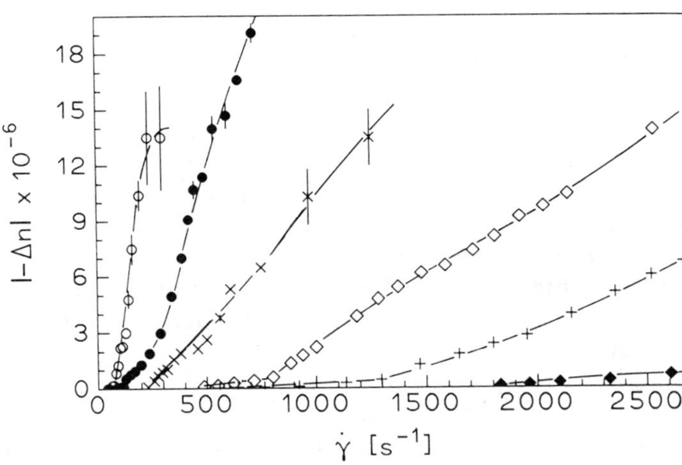

Figure 4. Flow birefringence versus shear rate for 100-mM solutions with a molar mixing ratio for (C_{14}DMAO + $C_{12}E_6$)/SDS of 7:3 and increasing replacement of C_{14}DMAO by $C_{12}E_6$. x($C_{12}E_6$) = n($C_{12}E_6$)/[n($C_{12}E_6$) + n(C_{14}DMAO)]. x($C_{12}E_6$) = 0 (○), 0.025 (●), 0.5 (×), 0.1 (◇), 0.15 (+), and 0.2 (◆). T = 25 °C.

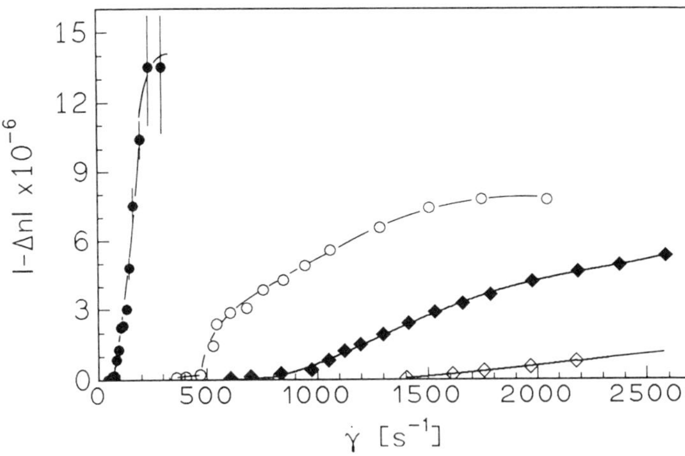

Figure 5. Flow birefringence versus shear rate for 100-mM solutions with a molar mixing ratio for C_{14}DMAO/(SDS + GEN) of 7:3 and different molar fractions of the alkylethersulfate GEN. x(*GEN*) = n(*GEN*)/[n(*GEN*) + n(*SDS*) + n(C_{14}*DMAO*)]. x(*GEN*) = *0 (●), 0.1 (○), 0.15 (◆), and 0.2 (◇).* T = *25 °C.*

Figure 4 shows that $\dot{\gamma}_c$ shifts to larger values with increasing replacement. With a 30% substitution of C_{14}DMAO, the ability to form SIS was completely lost, and the solutions remained isotropic. The same result was obtained when the SDS was replaced by GEN (Figure 5). Total replacement of SDS makes the formation of SIS impossible. Figures 6 and 7 reveal the change in the electric birefringence signal when the amine oxide (Figure 6) and SDS (Figure 7) were increasingly replaced by $C_{12}E_6$ and GEN, respectively.

The anomalous signal in the electric birefringence experiment indicates a strong interaction between the micelles. The second effect is due to the overlap of the electric double layers of the rodlike micelles. The typical shape is shown by the pure 100 mM 7:3 C_{14}DMAO/SDS solutions (Figures 6 and 7, top left). The progressive disappearance of the anomaly with an increasing share of $C_{12}E_6$ or GEN indicates that the rodlike micelles become shorter and their electrostatic interaction is diminished. At x($C_{12}E_6$ or GEN) = 0.3, the micelles are too short to maintain a strong interaction between their double layers. However, the rectangular shape of the signal proves that anisometric rodlike or spheroidlike particles are still present.

The conclusion that the incorporation of (ethylene oxide)$_x$ groups into the micellar interface is responsible for the suppression of the SIS was confirmed by another experiment. Readjustment to the original intensity of electrostatic interaction by increasing the total concentra-

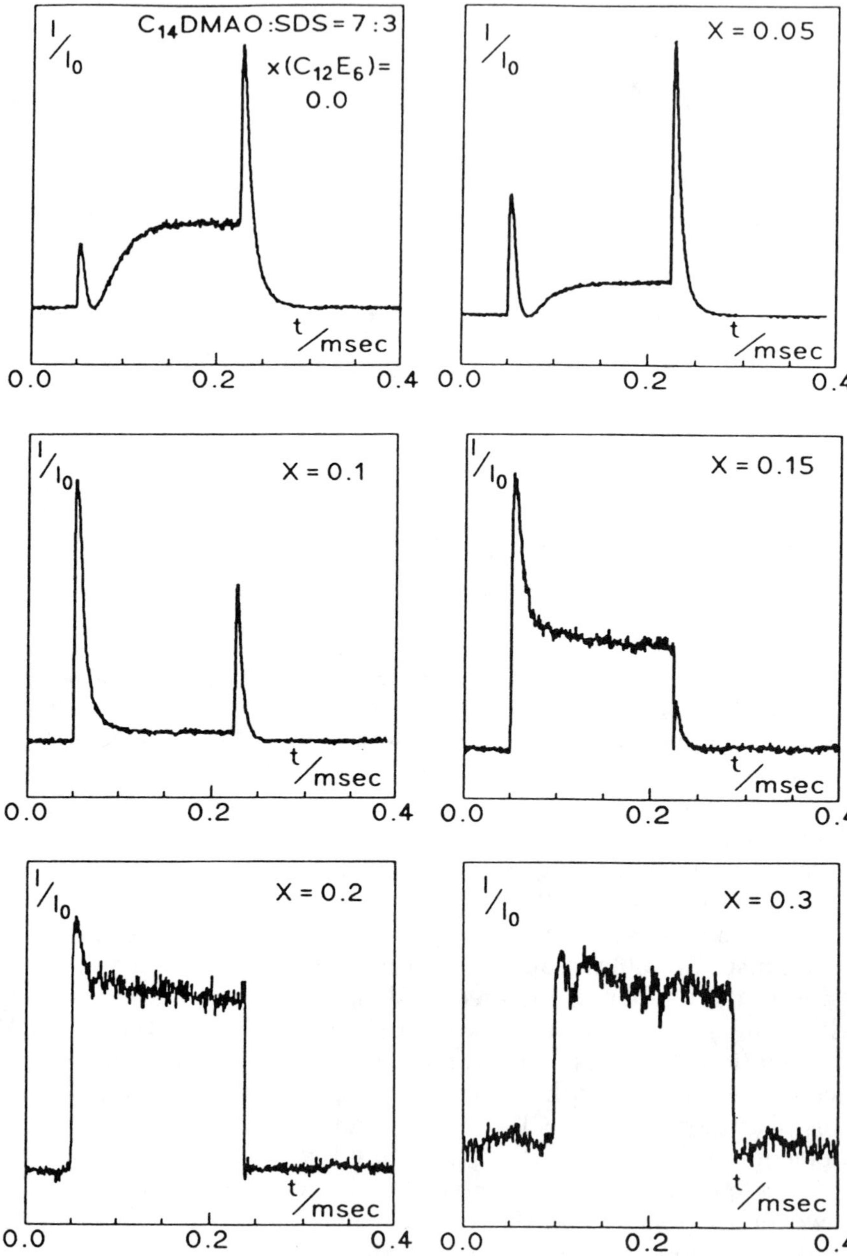

Figure 6. Disappearance of the anomalous signal with increasing replacement of $C_{14}DMAO$ by $C_{12}E_6$. ($C_{14}DMAO$ + $C_{12}E_6$)/SDS = 7:3; c = 100 mmol/L; $x(C_{12}E_6) = n(C_{12}E_6)/[n(C_{12}E_6) + n(C_{14}DMAO)]$; E = 6×10^5 V/m; T = 25 °C.

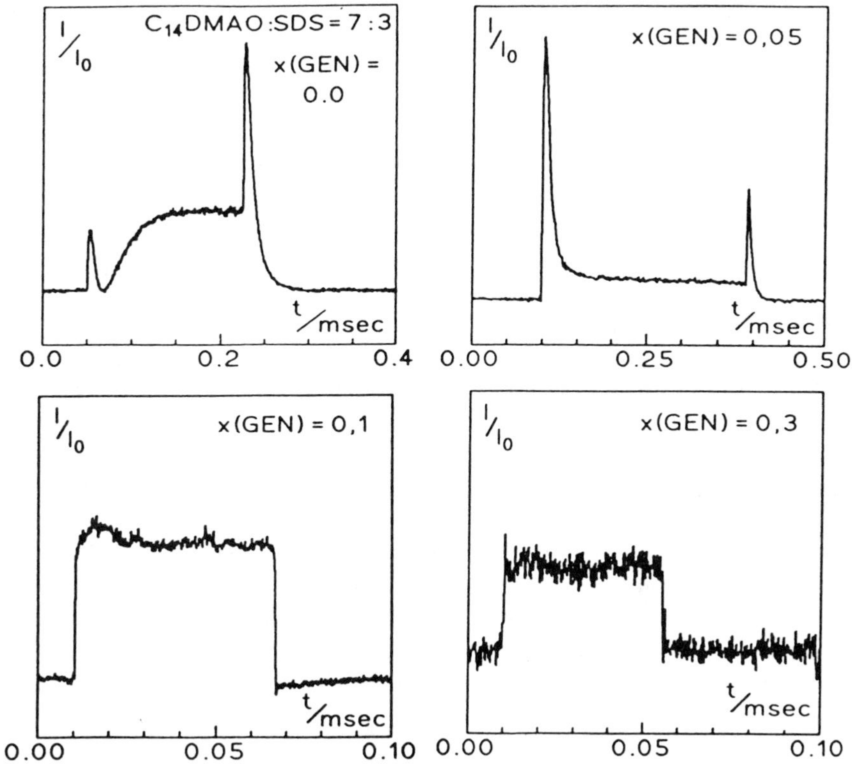

Figure 7. Disappearance of the anomalous signal with increasing replacement of SDS by the alkylethersulfate GEN. $C_{14}DMAO/(GEN + SDS) = 7{:}3$; $c = 100$ *mmol/L*; $x(GEN) = n(GEN)/[n(GEN) + n(SDS) + n(C_{14}DMAO)]$; $E = 6 \times 10^5$ *V/m*; $T = 25$ *°C.*

tion of solutions with $x(C_{12}E_6) = 0.3$ and $x(GEN) = 0.3$ (neither of which showed SIS overall) did not reestablish the ability to form SIS but resulted in viscoelastic behavior only. Only a linear increase of the birefringence in a $\Delta n/\dot{\gamma}$ plot could be observed. The readjustment of interaction was again monitored by electric birefringence. The original shape of the signal was obtained again at $c = 175$ mM (for $x(C_{12}E_6) = 0.3$) and at $c = 225$ mM (for $x(GEN) = 0.3$).

The important conclusion that must be drawn from these experiments is that rodlike micelles that carry a proper surface charge are a necessary but not sufficient condition for the formation of SIS. The occurrence of the supramolecular structure depends not only on long-range electrostatic interaction but also on short-range steric interaction. Modification of the micellar surface structure by incorporation of ethoxy groups has a marked (and in this case devastating) effect on the ability of the micelles to form SIS. Although no information about the exact conformation of the $(EO)_x$ groups at the micellar surface is available, the protrusion of these groups from the micellar interface obviously prevents the SIS from being formed. This prevention is possible only if the micelles that take part in the shear-induced superstructure have to come very close to each other in order to form the SIS. Not only the long-range electrostatic interaction but also the microscopic properties of the micellar interface structure are important for the SIS. This fact strongly supports the assumption underlying the first model that shear-induced coagulation of individual micelles is responsible for the SIS and precedes the alignment of the superstructures in the direction of flow. This finding can also provide a better understanding of why organic counterions that bind to the micellar surface promote SIS formation. Attached naphthoate or salicylate ions not only cause the micelles to grow but also give rise to a short-distance attractive hydrophobic interaction that could make conditions more favorable for micellar aggregation.

However, this explanation leaves us in need of an explanation for the nonshift of the correlation peak in the SANS experiments. We carried out some simple calculations on the basis of data in the literature (*13–15, 17*) and believe now that the limit of resolution ($\Delta\Theta/\Theta \approx 8\%$, D11 instrument) (*22*) is simply not high enough to reveal the small changes in the average mean distance of all the micellar aggregates that occur if only a fraction of the micelles aggregate and take part in the SIS (*18*).

Viscoelastic Solutions from Multilamellar Vesicles Under Shear. Phospholipids can form uni- and multilamellar vesicles. The size of these vesicles depends on the method of preparation. One way of preparing vesicles is by sonication of dispersions of phospholipids. Vesicles can also be formed by surfactants. Several systems in which vesicles form are known. Kaler et al. (*23*) showed that vesicles are formed when some cationic and anionic surfactants are mixed together. Dubois and Zemb (*24*) observed vesicles in dilute L_α-phases of didodecyldimethylammonium bromide. Hoffmann and co-workers (*25*) finally showed very recently that vesicles can be prepared from zwitterionic surfactants and cosurfactants. These vesicular solutions can have very interesting rheological properties if the vesicles are charged with some ionic surfactants and the total concentration of surfactant is high enough. Unlike the vesicles that consist of phospholipids, the vesicles from surfactants form spontaneously in aqueous solution when the components are mixed in the right proportion. The systems may, however, take several days to reach equilibrium. We

Figure 8. FF-TEM micrograph in the L_α phase with a total surfactant concentration of $C_{14}DMAO$ + $C_{14}TAB$ of 100 mM (ratio of 9:1) with 200 mM hexanol. Bar equals 1 μm.

include these systems in this summary of viscoelastic fluids under shear and discuss some of the properties of these systems.

Figure 8 is a TEM micrograph of a vesicular system that was prepared by the freeze–fracture method. The system consists of 90 mM $C_{14}DMAO$, 200 mM hexanol, and 10 mM cationic surfactant $C_{14}TAB$. The cationic surfactant in the system can also be replaced by the anionic surfactant SDS without changing the rheological properties. The systems are transparent and have a weak birefringence if the solutions are not under strain. Under small deformations or strains, the viscoelastic solutions become strongly birefringent. This birefringence is demonstrated in Figure 9, which shows that samples tilted between crossed polarizers become highly birefringent.

The electron micrograph in Figure 8 has several features that are relevant to the properties of the systems. The vesicles show a rather large polydispersity. Some vesicles seem to consist of a single shell, while others have up to at least 10 bilayers. The interlamellar spacing between the bilayers is rather uniform and is in the range of 800 Å. The vesicles are very densely packed, and the whole volume of the system is filled with the vesicles. The system is obviously in a single-phase situation. The vesicles are spherical even though the outermost shells can have a radius of several thousand angstroms. One multilamellar vesicle can thus be considered a small-volume element of an L_α-phase. Some vesicles do not consist of concentric shells but do have defects. Typical larger vesicles are about 1 μm in diameter, and the empty spaces that would result from the packing of dense large vesicles are filled with smaller vesicles so that the whole volume of the samples is completely used up. The micrograph shows clearly that

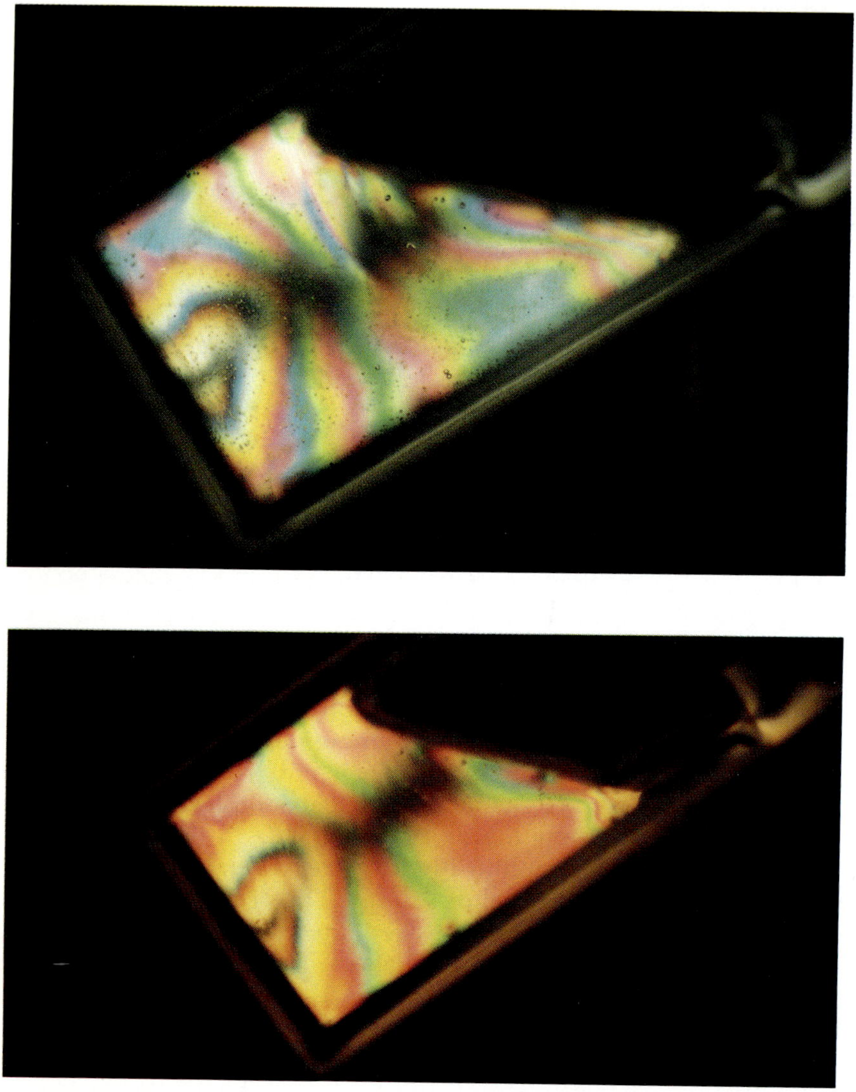

Figure 9. Two tilted samples of a viscoelastic fluid with a yield value held between crossed polarizers. The meniscus of the solution adjusts to the horizontal because the yield cannot withstand macroscopic gravity forces of the sample. The stress birefringence in the solution does not relax. Each sample is 2 cm thick. The same colors span the whole width of the samples. The colors reflect the stress and hence the deformation.

each vesicle is completely surrounded by other vesicles and is sitting in a cage from which it cannot escape by a simple diffusion process without deformation of the shells. The systems must therefore have viscoelastic properties under deformation or shear. So far we have made rheological measurements only in the single-phase region. By diluting the solutions, we would reach a two-phase region. We have not carried out rheological investigations in the two-phase area of the phase diagram yet.

The viscoelastic properties of the systems are demonstrated in Figures 10 and 11. In Figure 10, the storage modulus, loss modulus, and complex viscosity are plotted against the oscillating frequency in a double log plot. The storage modulus is much larger than the loss modulus over the whole frequency range and is not very dependent on frequency. The systems thus behave like a soft solid material. The systems actually have a real yield stress value (shown in Figure 11). To determine the yield stress value, we measured the deformation at two different times when the systems were exposed to increasing shear stresses. If the system responds only elastically, the deformation should depend only on the shear stress and not on the time the stress

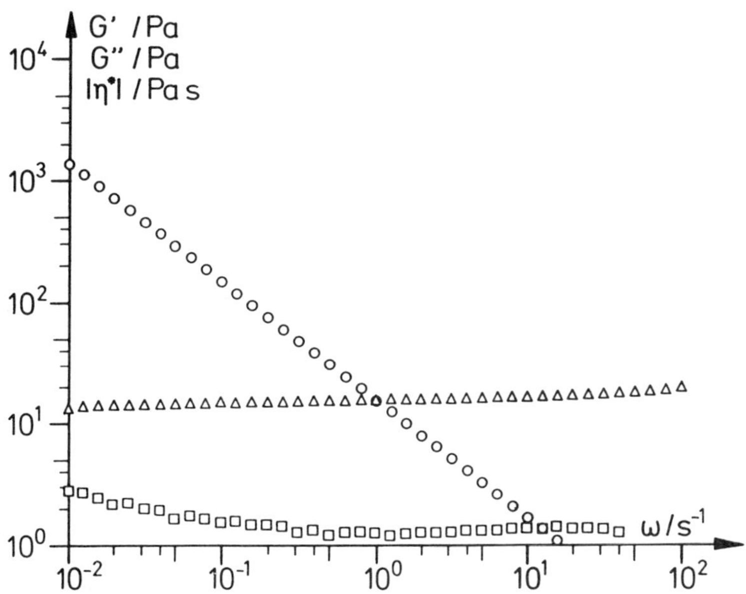

Figure 10. Plot of the storage modulus G' (△), loss modulus G" (□), and complex viscosity (η) (○) as a function of the angular frequency at a total surfactant concentration ($C_{14}DMAO$ + $C_{14}TAB$) of 100 mM with 200 mM hexanol.*

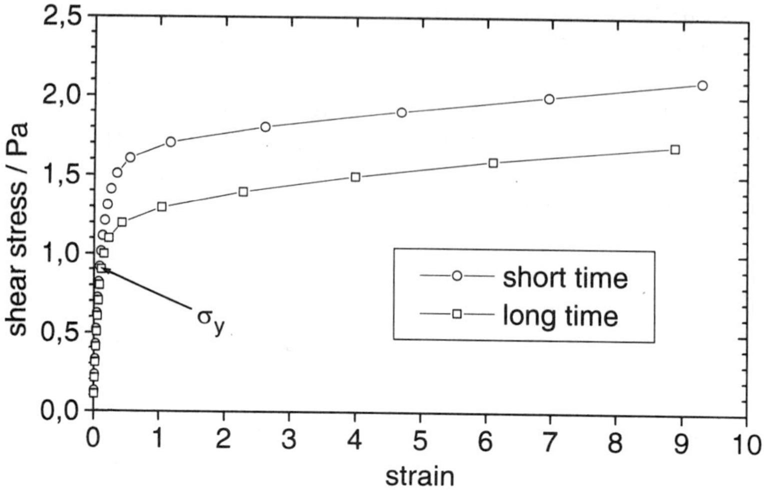

Figure 11. Demonstration of the yield stress behavior of the L_α phase with a total surfactant concentration ($C_{14}DMAO$ + $C_{14}TAB$) of 100 mM with 200 mM hexanol.

is applied. Figure 11 shows that the systems do indeed have a yield value.

The shear modulus and the yield value depend on the total surfactant concentration, as is demonstrated in Figure 12, which shows that the modulus and the yield value disappear for concentrations of less than 1% surfactant. For these concentrations the vesicles are no longer densely packed, and they can easily pass each other under shear flow. The yield value varies linearly with the modulus and is always about 10% of the modulus. Therefore when the vesicles are deformed about 10%, they can pass each other under shear.

In the densely packed region, the modulus depends strongly on the charge density of the vesicles, as shown in Figure 13. With increasing charge density, the modulus is S-shaped and saturates with a charge density of less than 10%. If the charge density on the vesicles is shielded by excess salt, the density decreases rapidly with ionic strength, as shown in Figure 14 in a plot of the modulus against the square root of the ionic strength.

Figures 13 and 14 seem to indicate that the shear modulus is determined by electrostatic repulsion between the bilayers. It could be argued that as for other colloidal systems, the shear modulus depends linearly on the compression modulus and that the compression modulus would be given by the osmotic pressure of the charged bilayers. If this scenario is the case and the modulus is determined only by the charge density of the system, then the shear modulus is independent

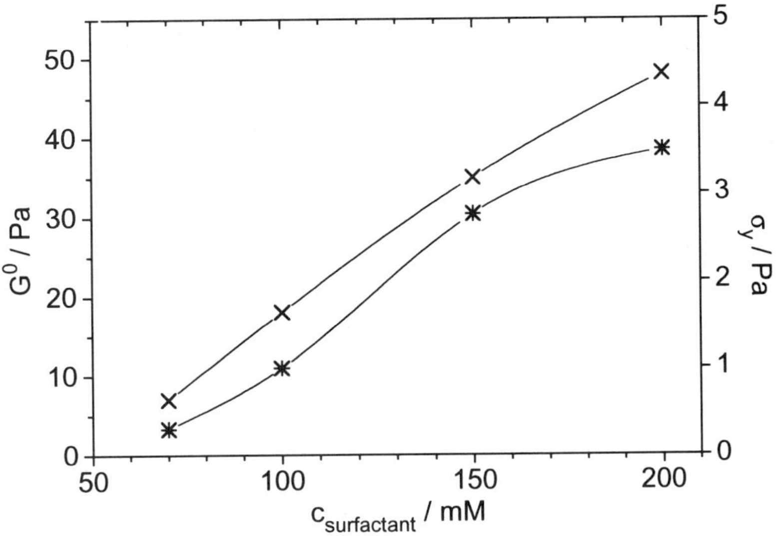

Figure 12. The yield stress value σ_y (*) and the modulus G° (×) as a function of the total surfactant concentration ($C_{14}DMAO$ + $C_{14}TAB$).

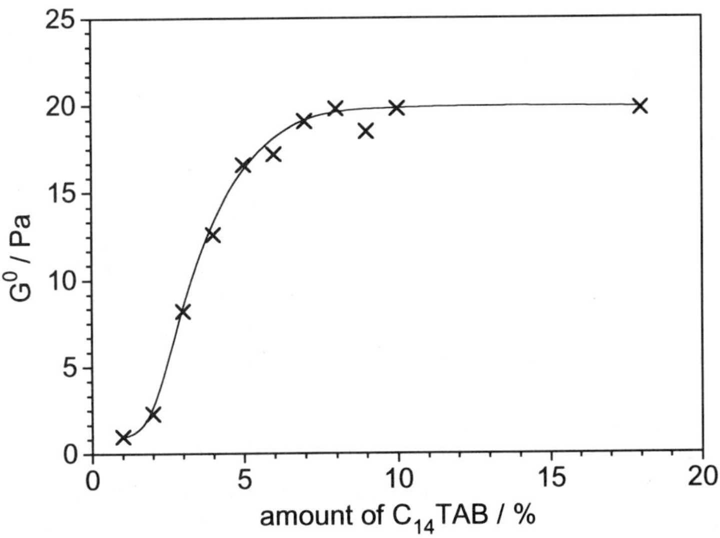

Figure 13. Modulus G° of the L_α phase with a total surfactant concentration ($C_{14}DMAO$ + $C_{14}TAB$) of 100 mM with 220 mM hexanol as a function of $C_{14}TAB$ concentration.

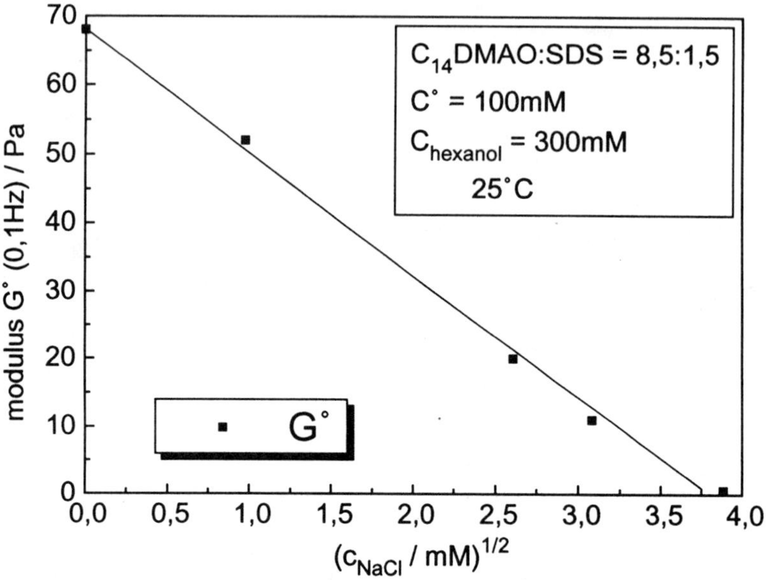

Figure 14. Modulus G° plotted against the square root of the NaCl concentration of a sample of C_{14}DMAO + SDS (C_{total} = 100 mM) with 300 mM hexanol.

of the chain length of the surfactant. This is, however, not the case, as is shown in Figure 15, where the moduli for five different systems in which the chain length was varied are plotted against the frequency. The modulus is different by about a factor of 10 for the C_{16} and the C_{12} surfactants if all other parameters are kept the same. Thus the modulus clearly depends on the thickness of the bilayers also. One possible reason for this dependence could be that the modulus also depends on the bending constant of the bilayer. Surfactants in a bilayer aggregated state can usually solubilize considerable amounts of hydrocarbon, and the vesicle state can, too. The hydrocarbon is expected to solubilize in the interior of the vesicles between the monolayers. If it does, one would expect the bilayers to become thicker with the solubilized amount of hydrocarbon, but the interlamellar distance should not change very much, and the modulus could therefore be expected not to change very much. These results, however, did not occur, as shown in Figure 16 in a plot of the modulus against the solubilization of hydrocarbon. Actually, the data show an increase of the modulus with solubilization. This result probably indicates that the total bilayer area is increasing and the interlamellar spacing is decreasing with solubilization. The hydrocarbon is therefore probably

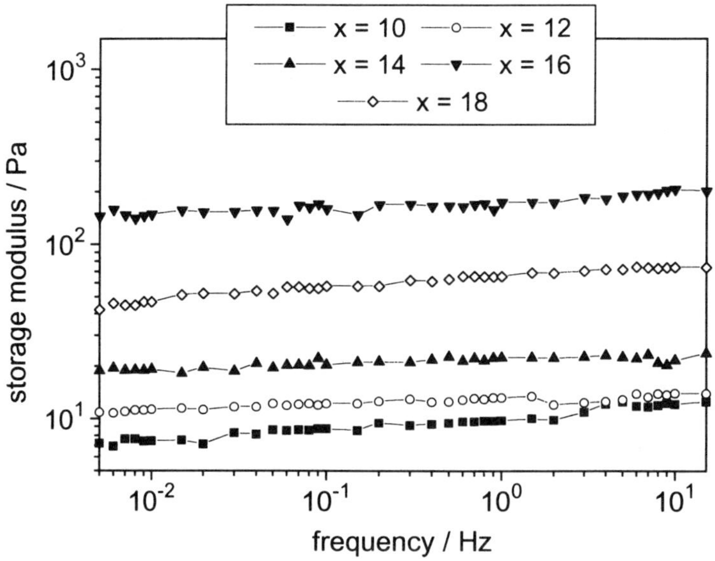

Figure 15. Storage modulus G′ as a function of the frequency (T = 25 °C) *of the system $C_xDMAO + C_{14}TAB$ (C_{total} = 100 mM) with 220 mM hexanol. The chain length of C_xDMAO varies between C_{10} and C_{18}.*

solubilized into the palisade layer and is oriented parallel to the surfactant chains.

The viscosities of many viscoelastic solutions follow the Cox–Mertz rule that shear viscosities as a function of shear rate and complex viscosities as a function of frequency have the same value in the shear-thinning region. This rule holds true for viscoelastic solutions of entangled threadlike micelles, as discussed earlier in this chapter. The vesicle solutions do not show this simple behavior, as is shown in Figure 17, where the shear viscosities and complex viscosities for two different systems are plotted against shear rate and angular frequency, respectively. The viscosities are the same for low frequencies. For higher shear rates $\eta, (\dot{\gamma}) > \eta^*(\omega)$.

The $C_{14}DMAO–C_{14}TAB$ system shows two breaks in the plot. For shear rates that are larger than the critical values, the multilamellar vesicles probably undergo transformations to new structures. Such transitions have been proposed by Roux (*26*) for vesicular systems of different compositions.

Viscoelastic Systems from Modified HEC. Water-soluble cellulose derivatives are large-scale commercial products. Because of their compatibility, HEC and its modified derivatives are used in

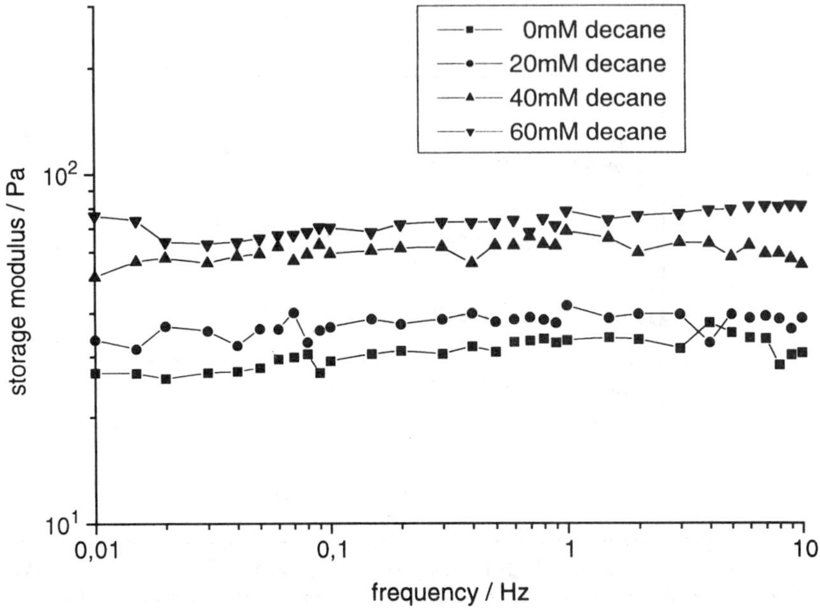

Figure 16. Storage modulus G′ as a function of the frequency of the system 100 mM C_{14}DMAO + 10 mM HCl + 220 mM hexanol with increasing solubilization of decane.

many industrial applications as thickeners. A solution of 1% HEC (M_W = 5 × 10^5 g/mol) has a viscosity of about 100 mPa·s. Chemical modification of HEC causes an increase in viscosity. During the last few years, several groups (*27–29*) have shown that the efficiency of HEC as a thickener can be considerably improved by hydrophobic modification of the HEC. The degree of hydrophobic substitution has to be kept low. Otherwise the polymers become insoluble in water. The thickening behavior of these polymers is due to intra- and intermolecular cross-linking of their hydrophobic side chains. The unmodified HEC and the four hydrophobically modified samples (F-HMHEC) investigated have the same molecular weights (M_W). The degree of perfluoro substitution is very low (*see* Table II). The number of substituents (z_s) varies between 1 and 8 per polymer chain.

The addition of small amounts of surfactant increases and strengthens the cross-link points and optimizes the macroscopic properties of the compounds. Figure 18 shows the plot of the viscosity of a sample of HEC and two samples of F-HMHEC against the concentration of an anionic perfluoro surfactant. The unmodified HEC sample shows

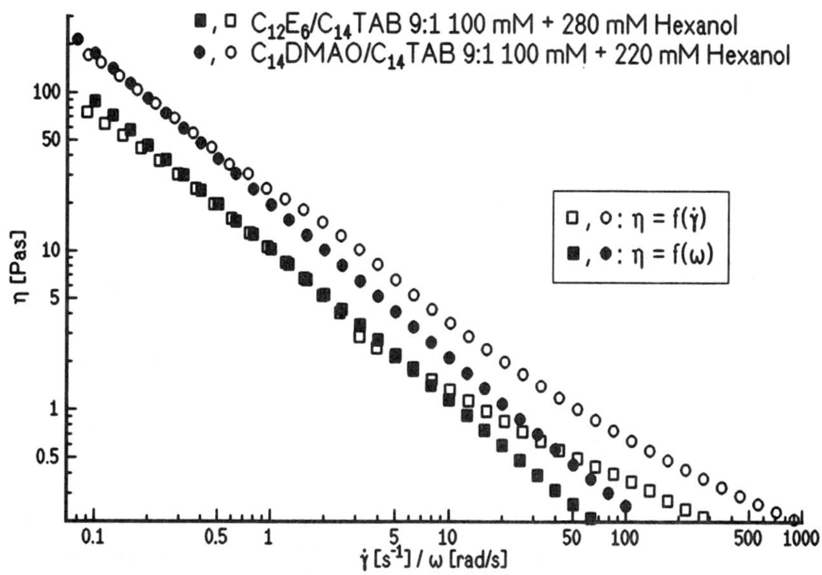

Figure 17. A comparison of dynamic viscosity versus angular frequency (closed symbols) and shear viscosity versus shear rate (open symbols) in two different systems.

Table II. Some Characteristic Parameters of Modified HEC Samples

Sample	M_w[a] (g/mol)	Molar Substitution: Hydroxyethyl	Molar Substitution: Substituent	z_s[b]
HEC	500,000	2.3–2.6	—	—
F-HMHEC 1	500,000	2.3–2.6	0.0007	1
F-HMHEC 2	500,000	2.3–2.6	0.0027	4
F-HMHEC 3	500,000	2.3–2.6	0.0031	6
F-HMHEC 4	500,000	2.3–2.6	0.0043	8
cat-HEC 1	33,000	1.90	0.27	37
cat-HEC 2	43,000	1.38	0.21	41
cat-HEC 3	120,000	2.09	0.10	48
cat-HEC 4	150,000	2.36	0.11	63

[a] Determined from light-scattering measurements.
[b] Number of substituents per polymer chain.

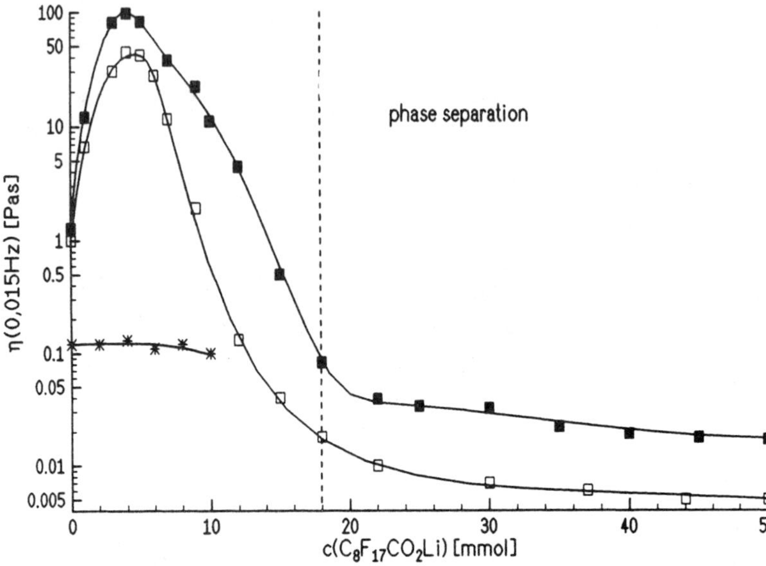

Figure 18. Viscosity of 1% solutions of HEC (*), F-HMHEC sample 2 (□), and F-HMHEC sample 4 (■) as a function of the concentration of lithium perfluorononanoate.

no increase in viscosity, while the viscosity of the F-HMHEC sample exceeds the maximum. This maximum characterizes the maximum number of cross-link points. After passing the maximum, the junction points separate with increasing surfactant concentration until the network structure is finally destroyed. Phase separation occurs by the addition of excess surfactant concentration because of the different osmotic pressures between the polymeric coils and the solution (depletion flocculation [*30, 31*]).

Solutions with properties similar to those just described can be obtained with cat-HEC by the addition of anionic surfactants, a process that has been investigated during recent years (*32–34*). The anionic surfactants bind to the cationic parts of the polymer, and the tails of the surfactants act again as cross-links between different polymer backbones. The four cat-HEC compounds that were used for the experiments have different molecular weights (M_W) and different degrees of molar substitution of hydroxyethyl and of the cationic substituent glycidyltrimethylammonium chloride (*see* Table II).

Figure 19 shows the viscosity of two samples of cat-HEC as affected by the addition of SDS. The viscosity again exceeds the maximum. This maximum corresponds to a charge compensation of 90%.

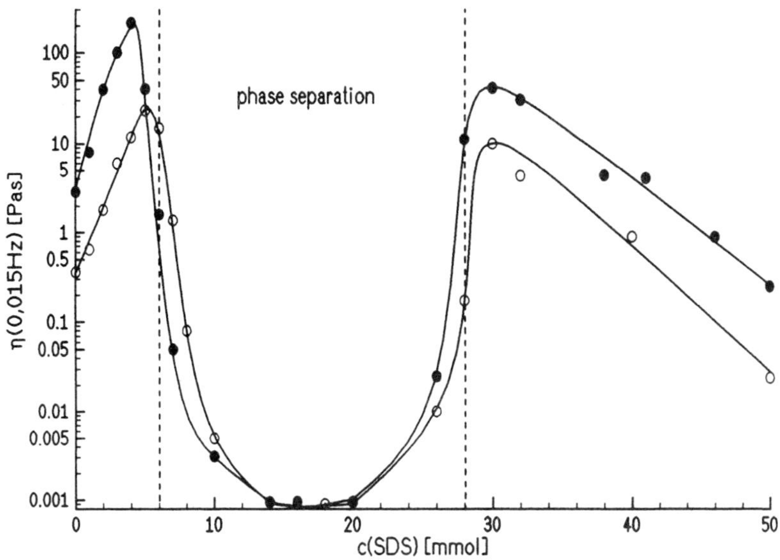

Figure 19. Viscosity of 1% solutions of cat-HEC samples 2 (○) and 4 (●) as a function of SDS concentration.

At a charge compensation of 100%, phase separation occurs. When excess surfactant concentration is added, the solution becomes single phase again and the viscosity exceeds a second maximum. In the upper single phase, the polymer is completely recharged.

At the first maximum, cat-HEC sample 4 possesses a yield stress. From strain-dependent measurements, a yield stress value of about 10 Pa is determined (*3*). Figure 20 is a plot of the storage and loss moduli of this sample versus frequency. The storage modulus (G′) is larger than the loss modulus (G″) over the whole frequency range.

At a characteristic polymer concentration c^*, the zero-shear viscosity of the HEC sample shows a strong increase in viscosity. This coil overlap concentration corresponds to a situation in which the single polymer coils are in contact with each other. After exceeding c^*, the viscosity can be represented by power law behavior with an exponent x of approximately 4.5 ($\eta \sim (c/c^*)^x$). The F-HMHEC samples show a similar increase in viscosity at lower polymer concentrations. This situation corresponds to the beginning of intermolecular association of the hydrophobic side chains.

The cat-HEC samples show an increase in viscosity at very low polymer concentrations. This increase marks the point in which the electric double layers of the extended polymeric chains start to interact

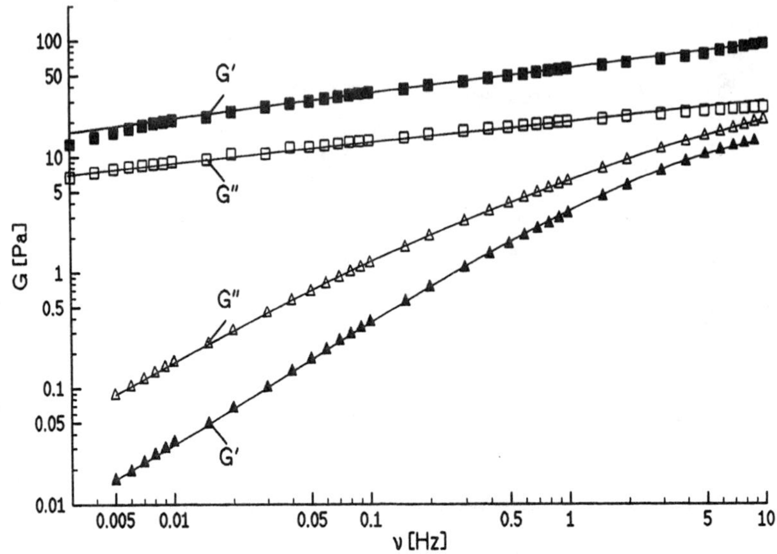

Figure 20. Storage (G′) and loss (G″) moduli as a function of the frequency of a 1% cat-HEC sample 4 solution (▲ and △) and a 1% cat-HEC sample 4 solution with 4 mmol of SDS added (maximum viscosity) (■ and □).

and form polymeric domains. At higher concentrations the polymers are coiled and the polyelectrolyte behaves like a nonionic polymer (*35, 36*). The viscosity then follows the scaling law of polymer solutions (*see* Figure 21) (*37*).

In the presence of surfactant, the two polymer types behave very similarly. The viscosity at the first (and second) maximum increases with increasing molecular weight and increasing number of cationic substituents for the cat-HEC samples and with increasing number of perfluoro side chains for the F-HMHEC samples. Both compounds at their maxima form strong transient network structures. These three-dimensional network structures can be made visible by freeze-fracture SEM. Figures 22a and 22b show cat-HEC sample 4 and F-HMHEC sample 2, respectively, at their maximum viscosities. Both network structures are formed from connected spherical particles. The diameter of the F-HMHEC coils is about 100 nm and corresponds to the size of the coils without surfactant, which was determined from light scattering and electric birefringence measurements. The diameter of a cat-HEC globule is about 50 nm. The length of each cat-HEC molecule, which was determined from electric birefringence measurements when no surfactant was added, was about 185 nm. The increas-

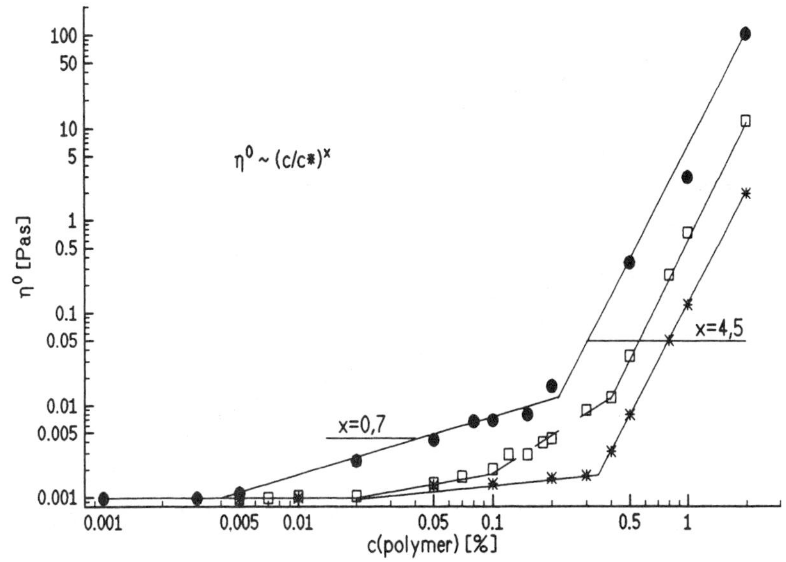

Figure 21. Zero shear viscosity versus polymer concentration of HEC (*), F-HMHEC sample 2 (□), and cat-HEC sample 4 (●).

ing charge compensation by the oppositely charged surfactant thus seems to lead to a more coiled conformation of these compounds.

Polymers in aqueous solutions show a characteristic flow behavior under shear. At very low shear rates the viscosity of entangled polymer solutions is constant and high (first Newtonian plateau). With increasing shear rate, the entanglements dissolve and the viscosity decreases.

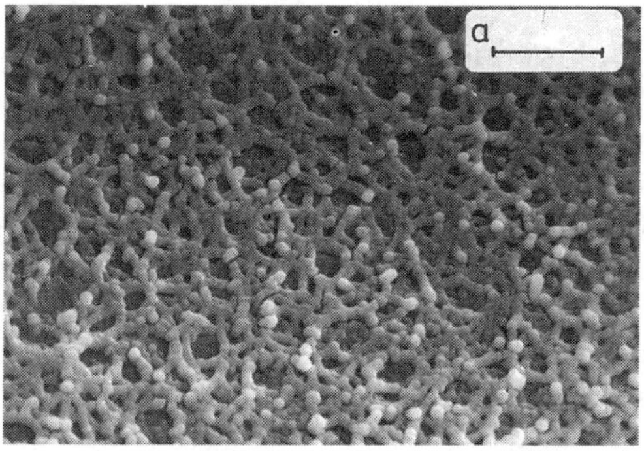

Figure 22. FF-SEM pictures of 1% cat-HEC sample 4 with 4 mmol of SDS (a) and 1% F-HMHEC sample 2 with 4 mmol of lithium perfluorononanoate (b). Bars equal 1 μm.

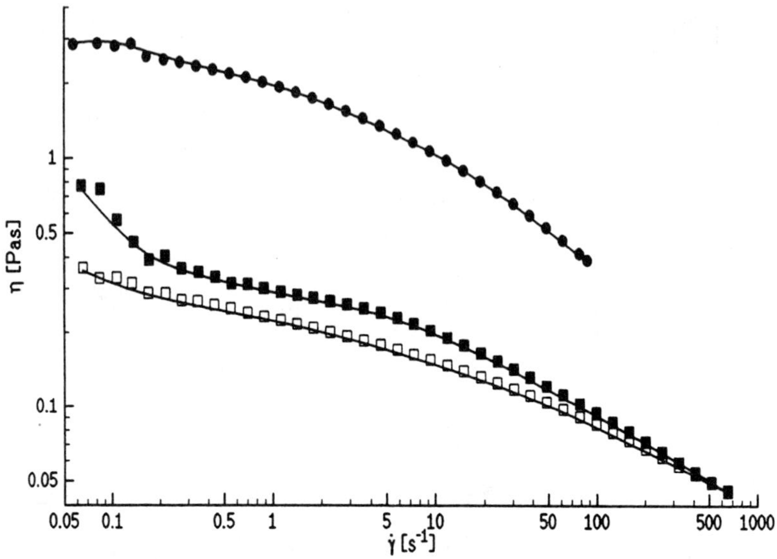

Figure 23. Viscosity as a function of the shear rate of 1% cat-HEC sample 4 (●) and 1% F-HMHEC sample 2 with increasing (■) and decreasing (□) shear rates.

Under special conditions the viscosity reaches a second constant value at very high shear rates. This situation corresponds to the completely untied polymeric chains. This behavior is called shear-thinning behavior, and we found such behavior for both compounds (*see* Figure 23). The F-HMHEC samples also show thixotropic behavior, which is seen in the difference in the viscosity curves with increasing and decreasing shear rates.

When surfactant is added, polymer solutions keep their shear-thinning behavior. Figure 24 shows the complex and shear viscosities of both polymer compounds with added surfactant. The slopes of the decreasing-viscosity curves in the middle region are about 0.8 (cat-HEC; Figure 24a) and 0.65 (F-HMHEC; Figure 24b). The F-HMHEC sample therefore behaves like polymeric melts, while the cat-HEC sample behaves like systems with yield stress (*37*). The viscosity of the cat-HEC sample shows no constant value at very low shear rates (or frequency) (*see* Figure 24a). This behavior is characteristic for gels with a yield stress. The difference between the shear viscosity and the complex viscosity in the two polymer–surfactant systems shows that the systems do not behave like simple threadlike networks.

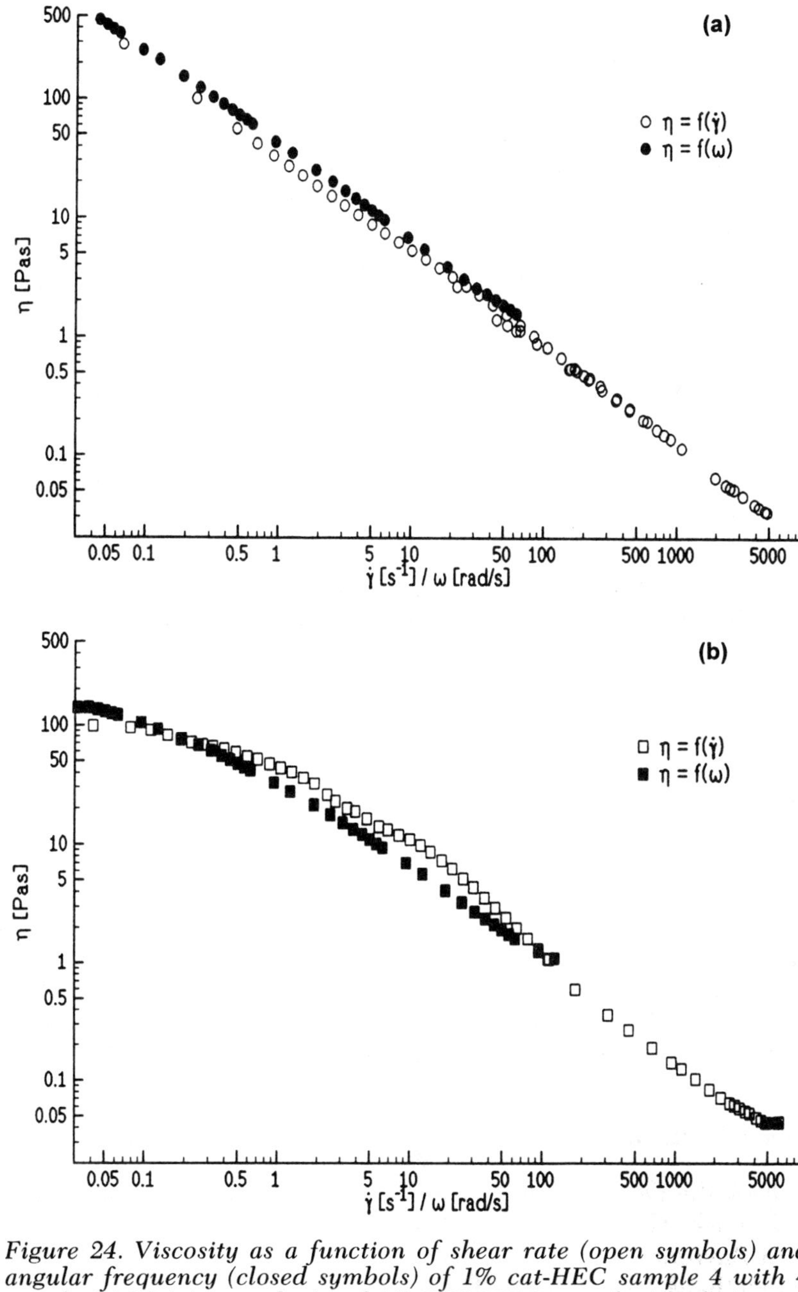

Figure 24. Viscosity as a function of shear rate (open symbols) and angular frequency (closed symbols) of 1% cat-HEC sample 4 with 4 mmol of SDS (a) (○ and ●) and 1% F-HMHEC sample 4 with 4 mmol of lithium perfluorononanoate (b) (□ and ■).

Conclusions

Viscoelastic surfactant solutions under shear show different flow behavior depending on the microstructure in solution. Systems with networks that are composed of entangled threadlike micelles often have a simple rheological behavior. Oscillatory experiments show a Maxwell-type behavior. The systems show shear-thinning behavior. The dynamic and shear viscosities at a given frequency and shear rate, respectively, are identical (Cox–Mertz rule).

Solutions with small, charged, rodlike micelles show very different behavior under shear. Simple Newtonian behavior is observed at small shear rates, and systems show the recoil phenomenon at higher shear rates. Above a critical shear rate, supramolecular structures that seem to consist of connected rodlike micelles are formed in the solution.

Viscoelastic solutions of densely packed multilamellar vesicles have a yield stress when the vesicles are charged and the concentration of surfactant is large enough. At lower concentrations the vesicles can pass each other under shear, and the yield stress disappears. For higher concentrations the value of the modulus depends on the charge density and the chain length of the surfactant. The dynamic and shear viscosities of these vesicular solutions do not follow the Cox–Mertz rule. We concluded that for shear rates larger than a critical value, the multilamellar vesicles change into new structures.

Water-soluble polymers that are modified by hydrophobic or cationic groups show viscoelastic properties and shear-thinning behavior. Both dynamic and shear viscosity are the same. When surfactant is added, the viscosity and the modulus exceed a maximum. Under some conditions, yield stress is observed at the maximum viscosity. The dynamic and shear viscosities of these polymer–surfactant systems are not the same. The network is built up of coiled polymeric chains that are connected through hydrophobic interactions of the surfactant molecules. The microstructure of the network is more complex than the network of simple threadlike micelles.

Acknowledgments

We thank the Deutsche Forschungsgemeinschaft for financial support of this work. The project was sponsored by Sonderforschungsbereich 213 in Bayreuth. One of us (U. Kästner) is thankful for a fellowship. We also thank Ute Munkert for the preparation of the freeze–fracture TEM micrograph of the multilamellar vesicles, Gisela Singer for carrying out some of the rheological measurements, and Christine Thunig for preparing the samples with vesicles.

References

1. Schorr, W.; Hoffmann, H. *J. Phys. Chem.* **1981,** *85*(21), 3160–3167.
2. Hofmann, S.; Hoffmann, H.; Rauscher, A. *Ber. Bunsenges. Phys. Chem.* **1991,** *95*(2); 153–164.
3. Hoffmann, H.; Munkert, U.; Thunig, C.; Valiente, M. *J. Colloid Interface Sci.* **1994,** *163*(1), 217–228.
4. Kästner, U.; Hoffmann, H.; Dönges, R.; Ehrler, R. *Colloids Surf. A* **1994,** *82*, 279–297.
5. (a) Rehage, H.; Hoffmann, H. *Mol. Phys.* **1991,** *74*(5), 933–973. (b) Clausen, T. M.; Vinson, P. K.; Minter, J. R.; Davis, H. T.; Talmon, Y.; Miller, W. G. *J. Phys. Chem.* **1992,** *96*(1), 474–484. (c) Toshiyuki, S.; Hirotaka, H.; Tadao, K. *Langmuir* **1987,** *3,* 1081–1086. (d) Candau, S. J.; Hirsch, E.; Zana, R.; Delsanti, M. *Langmuir* **1989,** *5,* 1225–1229.
6. Hoffmann, H. *Abstracts of Papers,* 206th National Meeting of the American Chemical Society, Chicago, IL; American Chemical Society: Washington, DC, 1993; PMSE 94.
7. Khatory, A.; Lequeux, F.; Candau, S. J. *Langmuir* **1993,** *9,* 1456–1464.
8. Turner, M. S.; Marques, C.; Cates, M. E. *Langmuir* **1993,** *9,* 695–697.
9. Gravsholt, S. *J. Colloid Interface Sci.* **1976,** *57,* 575–577.
10. Rehage, H.; Hoffmann, H. *Rheol. Acta* **1982,** *21,* 561–563.
11. Rehage, H.; Wunderlich, I.; Hoffmann, H. *Prog. Colloid Polym. Sci.* **1986,** *72,* 51–59.
12. Wunderlich, I.; Hoffmann, H.; Rehage, H. *Rheol. Acta* **1987,** *26,* 532–542.
13. Kalus, J.; Hoffmann, H.; Ibel, K. *Colloid Polym. Sci.* **1989,** *267,* 818–824.
14. Kalus, J.; Hoffmann, H.; Lindner, P. *Prog. Colloid Polym. Sci.* **1989,** *79,* 233–238.
15. Jindal, V. K.; Kalus, J.; Pilsl, H.; Hoffmann, H.; Lindner, P. *J. Phys. Chem.* **1990,** *94,* 3129–3138.
16. Pilsl, H.; Hoffmann, H.; Hoffmann, S.; Kalus, J.; Kencono, A. W.; Lindner, P.; Ulbricht, W. *J. Phys. Chem.* **1993,** *97,* 2745–2754.
17. Hoffmann, H. S.; Rauscher, A.; Hoffmann, H. *Ber. Bunsenges. Phys. Chem.* **1991,** *2,* 153–164.
18. Hofmann, S. Ph.D. Thesis, Universität Bayreuth, Bayreuth, Germany, 1994.
19. Bruisma, R.; Gelbart, W. M.; Ben-Shaul, A. *J. Phys. Chem.* **1992,** *96,* 7710–7727.
20. (a) Wang, S.-Q. *J. Phys. Chem.* **1990,** *94,* 8381–8384. (b) Wang, S.-Q. *Colloid Polym. Sci.* **1992,** *270,* 1130–1134.
21. Hu, Y.; Wang, S.-Q.; Jamieson, A. M. *J. Colloid Interface Sci.* **1993,** *156,* 31–37.
22. Ibel, K. *J. Appl. Crystallogr.* **1976,** *6,* 296–309.
23. Kaler, E.; Herrington, K. L.; Murthy, A. K.; Zasadzinski, J. *J. Phys. Chem.* **1992,** *96,* 6698–6707.
24. Dubois, M.; Zemb, T. *Langmuir* **1991,** *7,* 1352–1360.
25. Munkert, U.; Hoffmann, H.; Thunig, C.; Meyer, H. W.; Richter, W. *Prog. Colloid Polym. Sci.* **1993,** *93,* 137–145.
26. Roux, D.; Nallet, F.; Diat, O. *Europhys. Lett.* **1993,** *24*(1), 53–58.
27. Sau, A. C.; Landoll, L. M. In *Polymers in Aqueous Media;* Glass, J. E., Ed.; Advances in Chemistry 223; American Chemical Society: Washington, DC, 1989; pp 343–364.
28. Tanaka, R.; Meadows, J.; Phillips, G. O.; William, P. A. *Macromolecules* **1992,** *25,* 1304–1310.

29. Dualeh, A. J.; Steiner, C. A. *Macromolecules* **1991,** *24,* 112–116.
30. Vincent, B.; Edwards, J.; Emmett, S.; Jones, A. *Colloids Surf.* **1986,** *18,* 261–281.
31. van de Pas, J. C.; Buytenhek, C. *J. Colloids Surf.* **1992,** *68,* 127–139.
32. Dualeh, A. J.; Steiner, C. A. In *Polyelectrolyte Gels;* Harland, R. S.; Prud'homme, R. K., Eds.; Advances in Chemistry 480; American Chemical Society: Washington, DC, 1992; pp 42–52.
33. Goddard, E. D.; Leung, P. S.; Padmanabham, K. P. A. *J. Soc. Cosmet. Chem.* **1991,** *42,* 19–34.
34. Goddard, E. D.; Leung, P. S. *Colloids Surf.* **1992,** *65,* 211–219.
35. Odijk, T. *Macromolecules* **1979,** *12,* 688–693.
36. Skolnick, J.; Fixman, M. *Macromolecules* **1977,** *10,* 944–948.
37. Barnes, A.; Hutton, J. F.; Walters, K. *An Introduction to Rheology;* Elsevier: Amsterdam, 1989; Vol. 15, pp 71–77.

RECEIVED for review March 7, 1994. ACCEPTED revised manuscript August 17, 1994.

Associating Polymers

B. Performance in Solution

1. Hydrophobe-Modified Acrylamide Polymers

15

Hydrophobic and Electrostatic Interactions in Water-Soluble Associating Copolymers

Joseph Selb, Simon Biggs, Delphine Renoux, and Françoise Candau

Institut Charles Sadron (Centre de Recherches sur les Macromolécules–Ecole d'Application des Hauts Polymères), 6 rue Boussingault, 67083 Strasbourg Cédex, France

*Associative polyacrylamide derivatives containing both ionic sites and small numbers of hydrophobic groups were prepared, and their thickening properties in aqueous solution were investigated. Two different radical micellar copolymerization processes in aqueous media were used: The comonomer of acrylamide was either a hydrophobic monomer (*N*-ethylphenylacrylamide) solubilized within surfactant micelles (sodium dodecyl sulfate) or a micelle-forming cationic polymerizable surfactant (*n*-hexadecyldimethyl-4-vinylbenzylammonium chloride). Relationships between the copolymerization mechanism and the copolymer microstructure are proposed. Owing to the competition between attractive hydrophobic interactions and repulsive electrostatic interactions, such hydrophobically modified polyacrylamides exhibit various rheological behaviors in aqueous solution depending on shear rate, shear time, ionic strength, and copolymer characteristics.*

WATER-SOLUBLE POLYMERS modified with relatively low amounts of a hydrophobic comonomer (<5 mol%) have been the subject of extensive research efforts during the past decade because of their interesting rheological behavior in aqueous solution (*1–4*). Intermolecular hydrophobic associations give rise to enhanced viscosification proper-

0065–2393/96/0248–0251$14.00/0

ties. Under shear, the interchain physical links are disrupted, and this disruption leads to a strong shear-thinning effect. However, this dissociation phenomenon is reversible, and interchain liaisons re-form as shear decreases. Additionally, kinetic effects on the association–dissociation processes are responsible for thixotropic properties (*5, 6*). Shear thickening and rheopectic behavior may also occur in the low-shear region (*7–10*) because of the balance between intra- and intermolecular interactions. Finally, interactions of classical surfactants with this type of amphiphilic polymer strongly affect the association process and thus provide another way of controlling the thickening properties (*10–14*). All of these features make it possible to use hydrophobically modified water-soluble polymers as thickening agents in many applications involving viscosity control of various aqueous systems (*2–4, 15–17*).

On the other hand, polyelectrolytes have been known for a very long time as efficient thickeners, especially in salt-free solution, because of intermolecular charge–charge repulsions that lead to coil expansion.

For the production of thickeners with improved properties, we therefore considered water-soluble polymers containing both hydrophobic and ionic groups in order to combine the advantages of polymolecular association and the polyelectrolyte effect (*7, 18–31*). The so-called polysoaps (*32–35*) are related compounds that also consist of polyelectrolytes bearing hydrophobic side chains but at a much higher level (>10 mol%). However, such compounds are generally not efficient thickeners because the intramolecular hydrophobe associations are responsible for a hypercoiled conformation and intermolecular associations are not favored even in the semidilute range.

We report here some results of an investigation into the synthesis, characterization, and solution properties of two classes of ion-containing hydrophobically modified polyacrylamides. The specific features of the copolymerization methods used as well as their influence on the copolymer microstructure were examined. The aqueous solution properties of the copolymers were studied in the absence and the presence of salt. We compared the thickening abilities of the various samples, giving special attention to the effect of introducing charges onto the polymer chain; the nature of the hydrophobe groups, their number, and their distribution along the polymer backbone were taken into account. Shear rate and shear time effects were also investigated. Additional information on the association process was provided by fluorescence studies.

Experimental Methods

Materials. Acrylamide (Merck) was recrystallized twice from chloroform. *N*-4-Ethylphenylacrylamide (eΦAM) was synthesized by the reaction of 4-ethylaniline with acryloyl chloride according to a procedure described in the literature (*36*). Both reagents (Aldrich) were distilled just prior to use. *n*-Hexadecyldimethyl-4-vinylbenzylammonium chloride (N16) was prepared by quaternization of hexadecyldimethylamine (Hoechst, distilled) with 4-vinylbenzyl chloride (Kodak, used as received) in ethanol at 50 °C for 1 day and in the presence of *tert*-butylcatechol as inhibitor of polymerization. N16 was purified by recrystallization from ethyl acetate.

Copolymer Synthesis. ***Acrylamide–Sodium Acrylate–Ethylphenylacrylamide Series.*** Terpolymers of acrylamide (AM), sodium acrylate (NaA), and ethylphenylacrylamide (e*Φ*AM) were prepared in a two-step procedure. AM–e*Φ*AM copolymers were first obtained by radical copolymerization using an aqueous micellar method well detailed elsewhere (*6*, *36*). Typical experimental conditions were as follows: The total concentration of monomers in water was 3 wt%; the proportion of e*Φ*AM in the monomer feed varied from 1 to 3 mol% (i.e., 2.4–7 wt%); sodium dodecyl sulfate (SDS) was used as surfactant at concentrations ranging from 1 to 5 wt% of the total recipe; the concentration of $K_2S_2O_8$ used as water-soluble initiator was 0.3 wt% based on monomers. The reaction was allowed to proceed at 50 °C for 7 h under nitrogen. About 90% conversion occurred. The copolymer was recovered by precipitation in methanol and purified by repeated filtrations and washings in methanol before being dried under vacuum at 50 °C. Subsequent partial hydrolysis (10–40% of AM units) to create carboxylate groups was carried out at a polymer concentration of 0.7 wt% in water under alkaline conditions ([NaOH] = 0.25 M); the reaction times and temperatures were 1 h at 30 °C, 2 h at 30 °C, and 2.5 h at 50 °C for 10, 20, and 40% hydrolysis, respectively. The recovery and purification of hydrolyzed samples were like those just described. Some AM–NaA–e*Φ*AM samples were prepared in a single step by direct copolymerization of the three monomers under similar micellar conditions.

For some polymerization kinetics studies, the reaction solution was divided into several bottles by a double-needle transfer technique before $K_2S_2O_8$ was added. The separate reactions were terminated at various reaction times by addition of hydroquinone and cooling in an ice-water bath (*37*).

AM–N16 Series. Typical experimental conditions for AM–N16 copolymerization were as follows: The concentration of monomers in water was 3 wt%; the proportion of N16 in the monomer feed varied from 0.5 to 5 mol% (i.e., 2.9–27 wt%); and 4,4′-azobis(4-cyanovaleric acid) (ACVA) (Aldrich), at 3 wt% based on monomers, was used as water-soluble initiator instead of $K_2S_2O_8$ because the latter strongly reduces the solubility of N16. Other experimental conditions (temperature, reaction time, and recovery procedure of the polymer) were like those for the process just described.

Copolymer Characterization. The hydrophobe content in AM–NaA–e*Φ*AM copolymers was determined by UV spectrophotometry using the calibration established with ethylphenylpropionamide (*36*, *37*). The level of carboxylate groups was determined by potentiometry and sodium elemental analysis. The copolymer molecular weights before hydrolysis were measured in formamide by static light scattering and were in the range 1.6–2.6 × 10^6. The composition of AM–N16 copolymers was determined from the chlorine content.

Rheological Measurements. The viscosity of aqueous polymer solutions was measured at 25 °C with either a controlled stress rheometer (CarriMed) using cone plan geometry or a Contraves Low-Shear 30 instrument. The polymer concentration ranged from 0.25 to 3% wt/wt, and the shear rate ranged from 0.01–10 to 1000–2500 s^{-1} depending on the viscosity of the sample.

Other Methods. Conductometry was used for determining the critical micellar concentration (cmc), as described elsewhere (*38*). Details on the experimental procedure for fluorescence measurements were given in a previous paper (*6*).

Copolymer Synthesis

Two classes of associating polyacrylamide derivatives were synthesized: one contains anionic-type derivatives and one contains cationic-type derivatives. The former class (structure **1**) contains terpolymers of AM with NaA as the ionic component and e*Φ*AM as the hydrophobe (AM–NaA–e*Φ*AM series). The second class (structure **2**) consists of copolymers of AM with a comonomer containing both hydrophobic and ionic groups, the N16 (AM–N16 series). Samples of various hydrophobicities and ionic levels were prepared by varying the copolymer composition. Two particular "micellar" radical copolymerization processes were used. In both cases the kinetics of copolymerization and hydrophobe incorporation were investigated, allowing us to propose the most probable reaction mechanism. Such information was necessary for a clear understanding of the solution properties, because varia-

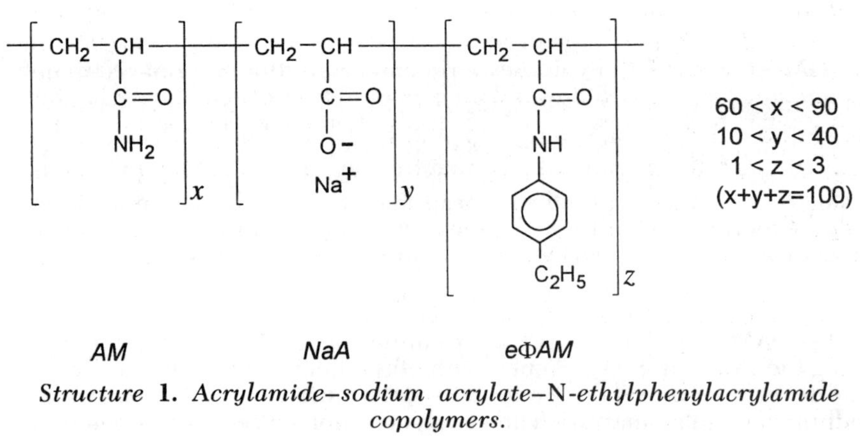

Structure **1.** *Acrylamide–sodium acrylate–*N*-ethylphenylacrylamide copolymers.*

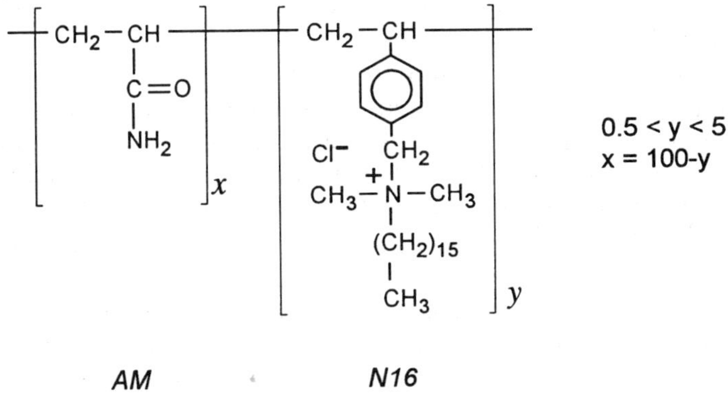

Structure 2. *Acrylamide–*n*-hexadecyldimethyl-4-vinylbenzylammonium chloride copolymers.*

tions in the number and distribution of hydrophobes along the polymer backbone are of major importance in the association phenomenon.

AM–NaA–eΦAM Series. ***Principle of the Synthesis.*** Two different methods were used to prepare AM–NaA–eΦAM copolymers.

Method 1. In a two-step procedure, AM–eΦAM copolymers were first obtained by radical copolymerization, and then chemical modification introduced NaA units (partial hydrolysis of AM functions).

As in the micellar process initially developed by Evani (*39*), Valint et al. (*36*), and Turner et al. (*40*), copolymerization was carried out in an aqueous medium containing the hydrophilic monomer (AM), while the presence of a surfactant at a concentration above its cmc ensured solubilization of the hydrophobic monomer (eΦAM) within micelles. The striking feature of this process is that it leads to copolymers with somewhat blocky structures as a consequence of the microheterogeneous nature of the reaction medium (*5, 6, 24, 41–44*). As shown in previous studies, the average length of the hydrophobic sequences is directly related to the micelle concentration (*5, 6, 24, 41, 44*). Thus, by varying the [hydrophobe]/[micelle] ratio at a constant hydrophobe content, the blockiness of the copolymer can be modified. In this way it is possible to control the thickening efficiency, because longer hydrophobic sequences are more favorable than shorter ones to the formation of interchain liaisons (*5, 6*).

The subsequent partial hydrolysis to convert AM–eΦAM copolymers into AM–NaA–eΦAM copolymers was carried out under alkaline conditions. The level of hydrolysis was controlled by varying the temperatures, reaction times, and concentrations of the reagents (*30*) (*see* ***Experimental Methods***).

The advantage of this method is that it provides a homologous series of copolymers with variable ionic contents but constant hydrophobe content, hydrophobe sequence distribution, and degree of polymerization. The poor solubility in water of some AM–eΦAM intermediates (hydrophobe level, ≳2 mol%) (*6*) is, however, a limiting factor, because it prevented us from performing hydrolysis in a homogeneous medium.

Method 2. In an alternative method, AM–NaA–eΦAM copolymers were prepared in a single step by direct polymerization of the three monomers under micellar conditions similar to those just described. Terpolymerization is faster than the previously described method, and control of the ionic unit content is easier, being accomplished by adjusting the composition of the initial monomer feed. Furthermore, because of the stronger hydrophilicity provided by the carboxylic groups, more hydrophobes can be incorporated. On the other hand, this single-step procedure is not well suited to the preparation of homologous series of copolymers that differ only by ionic-site content. It is not easy to prepare samples with the same molecular weights and the same molecular-weight distributions when the formulation of the reaction mixture is changed to obtain different copolymer compositions. Furthermore, the reaction mechanism discussed in the next section indicates that the copolymer microstructure should depend on the NaA level.

Copolymerization Kinetics. The kinetics of polymerization and hydrophobe incorporation were investigated only for AM–eΦAM copolymers. The mechanism for the micellar terpolymerization of AM, NaA, and eΦAM is more complicated. Even for conventional solution polymerization, the presence of a third monomer is obviously a complicating factor. The situation is still more complex in micellar media for the following reasons: (1) because of its ionic nature, NaA acts as a salt and thus should strongly modify the micellization process of the ionic surfactant (cmc and aggregation number); (2) conversely, the presence of an ionic surfactant must affect the copolymerization of AM and NaA, which is known to be very sensitive to ionic strength (*45–47*); and (3) electrostatic repulsions between anionic acrylate groups of the growing polymeric chain and anionic micelles containing hydrophobic monomers should alter the hydrophobe incorporation process. In conclusion, even if the [hydrophobe]/[surfactant] ratio is kept constant, variations in NaA content will undoubtedly affect the hydrophobe distribution along the polymer backbone.

In the binary AM–eΦAM system, monomer-to-polymer conver-

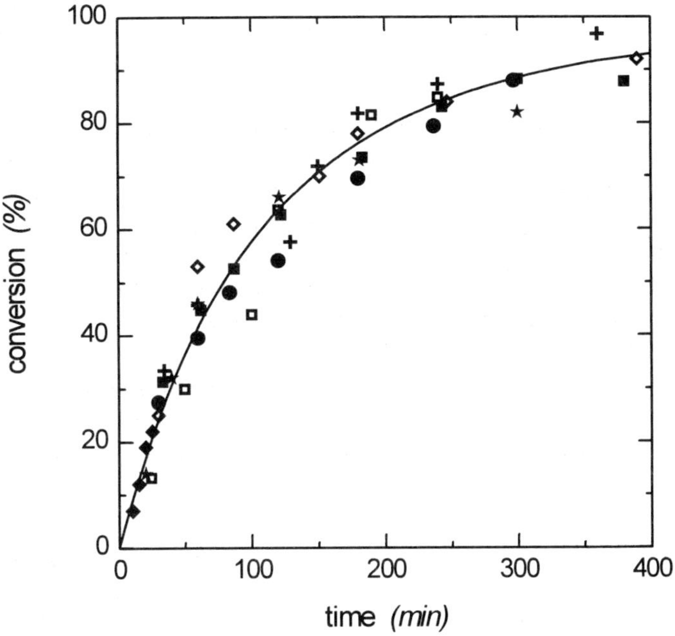

Figure 1. Conversion–time data for various AM–eΦAM micellar copolymerizations at various eΦAM and SDS concentrations. Key: eΦAM (mol% in monomer feed)/SDS (wt% in aqueous solution): 0/3 (+); 1/3 (◆, ●, ■); 1/7.5 (★); and 3/3 (□, ◇). The line is the theoretical curve predicted for AM in the absence of surfactant and hydrophobe (see *ref. 48).*

sion was investigated as a function of time for various hydrophobic monomer contents (0–3 mol% based on the total monomer feed) and various surfactant contents (3–7.5 wt% based on total recipe), with all other experimental parameters kept constant (*see* **Experimental Methods**). The conversion–time data reported in Figure 1 for different reaction mixtures reveal no noticeable change in the rate of monomer conversion with time. All experimental data can be fitted to the theoretical curve calculated for the homopolymerization of AM in water with $K_2S_2O_8$ as the initiator and in the absence of surfactant and comonomer (*48*). These results show that neither the hydrophobic monomer nor the surfactant exerts any significant influence on the rate of polymerization (at least in the concentration ranges of both additives investigated).

Copolymer Composition. The influence of the surfactant on the copolymerization mechanism and the rate of hydrophobe incorporation was investigated by comparing the micellar process with a homogeneous process. In the homogeneous process, the use of a cosolvent (formamide) instead of a surfactant ensures the solubility of the hydrophobic monomer within the aqueous reaction medium (6). Figure 2 shows that for a given initial hydrophobe level in the feed (1 mol%), the variation of copolymer composition as a function of conversion depends strongly on the conditions of the synthesis. A more or less marked drift in copolymer composition is observed in the micellar process, but the composition does not vary when copolymerization is carried out in a homogeneous solvent mixture without surfactant. In this case, the constancy of copolymer composition with conversion suggests that monomers exhibit similar reactivity; this result was expected, because the two monomers bear identical acrylamide functional groups. In contrast, the presence of surfactant leads to an apparent higher reactivity for the hydrophobic monomer: At the beginning

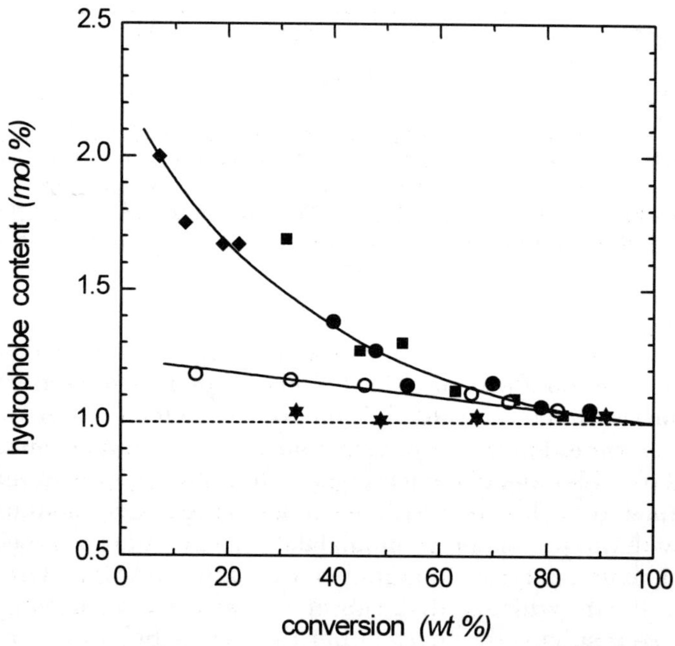

Figure 2. Variation in copolymer composition as a function of conversion for different AM–eΦAM copolymerizations at constant eΦAM concentration (1 mol% of the monomer feed). Key: ◆, ●, ■, micellar copolymerization, [SDS] = 3 wt%; ○, micellar copolymerization, [SDS] = 7.5 wt%; and ★, homogeneous copolymerization (formamide–water mixture, 10/90 vol/vol). Other experimental conditions: [total monomer] = 3 wt%; $[K_2S_2O_8] = 3.3 \times 10^{-4}$ M; and T = 50 °C.

of polymerization the copolymer samples are richer than the feed in eΦAM. As the copolymerization reaction proceeds, the average hydrophobe content decreases toward that of the feed. In agreement with results reported by other authors (*49*), the initial deviation from feed composition increases when the surfactant concentration is reduced, that is, when the [eΦAM]/[SDS] ratio is increased (Figure 2). This effect is even stronger for higher eΦAM concentrations (*37*). As a consequence, when the micellar process is used, the copolymers recovered at high levels of conversion exhibit an inherent compositional heterogeneity that increases as the initial number of hydrophobes per micelle increases.

The faster incorporation of the hydrophobe in the early stages of copolymerization can be explained by taking into account the dynamics of the micellar system, namely, the intermicellar exchange of solubilized material. The rate of transfer of one hydrophobe between two micelles is 3 orders of magnitude greater than the rate of addition of a monomer onto a growing radical (*37*). For an unnucleated micelle, the exit and entry processes are equilibrated. When a macroradical head group enters a micelle, hydrophobic monomers polymerize and the entry process therefore overcomes the exit process. Consequently, during the residence time of the macroradical head group within a micelle, there is a net flow of hydrophobic monomers into this micelle. When the growing radical head group leaves the micelle to add AM units in the bulk solution, the number of hydrophobes constituting the sequence is larger than that initially solubilized within the micelle. This feature causes a concomitant lowering of the average number of hydrophobes per micelle as a function of reaction time, and the hydrophobic sequences that form subsequently are thus shorter. For systems initially containing a larger number of hydrophobes per micelle, the time required to consume all of the monomers inside the micelle is increased, and the transfer of hydrophobes from other micelles is therefore favored (*50*). This situation leads to a still faster hydrophobe incorporation. The possible variation in hydrophobe sequence length as a function of conversion is a complicating factor in the association process. Actually, the situation is similar to that of a sample functionalized with hydrophobe groups of variable size. This means that for a given sample, all macromolecular chains do not exhibit equivalent ability for hydrophobic interchain liaisons.

In conclusion, by using a micellar copolymerization process, the average length of the hydrophobe sequences can be modified at a constant hydrophobe level by varying the [hydrophobe]/[micelle] ratio, but the resulting samples present some heterogeneity in composition.

AM–N16 Series. ***Principle of the Synthesis.*** Previous studies (*26, 51, 52*) have shown that hydrophobically modified water-soluble polymers can be obtained by copolymerization of AM with a polymerizable surfactant. In this case no additional surfactant is required to solubilize the hydrophobic monomer, because the hydrophobic monomer itself forms micelles in an aqueous medium owing to its amphiphilic character. Compared to the composition of the reaction mixture in the previously described micellar process, the composition in this procedure is simplified, because the hydrophobic monomer–surfactant pair is replaced by a micelle-forming single component.

According to this concept, cationic hydrophobically associating polyacrylamides were synthesized by using N16 as the comonomer. This cationic micelle-forming monomer was chosen because its properties in aqueous solution and its homopolymerization in the micellar state have been studied in detail in our laboratory (*38, 53, 54*). Under the experimental conditions of the copolymerization (i.e., at 50 °C and in the presence of AM), the cmc was slightly modified ($\sim 3.8 \times 10^{-4}$ instead of 2.5×10^{-4} mol/L at 25 °C in pure water). In the range of monomer feed compositions investigated, the N16 concentration (c) was always well above the cmc ($c \simeq$ 5–50 cmc). The micelle aggregation number did not vary significantly (~60) in the range of N16 concentrations used in our experiments.

Although the copolymerization reaction mixture is quite homogeneous (optically transparent), it is actually a microheterogeneous system, as in the previously described micellar process. Micelles behave as microdomains with very high local N16 concentrations. Therefore a blocky structure is also expected here, because when a growing polyacrylamide radical encounters a micelle, all the monomers forming the micelle should be consumed.

Copolymerization Kinetics. Figure 3 shows conversion–time data corresponding to various N16 contents. From these results it is difficult to draw definitive conclusions about the effect of the N16 level on the kinetics of copolymerization. In the early stages of polymerization, the differences in polymerization rates for various systems are within the range of experimental errors. Indeed, because of the very high initial rate of polymerization, slight differences in a possible inhibition period could be responsible for the observed variations. However, the lack of experimental data at very low conversion degrees does not allow us to support this assumption.

In contrast with AM–eΦAM copolymerization, the conversion does not progress significantly beyond ~3 h of reaction time, and the final degree of conversion is lower; the higher the N16 concentration,

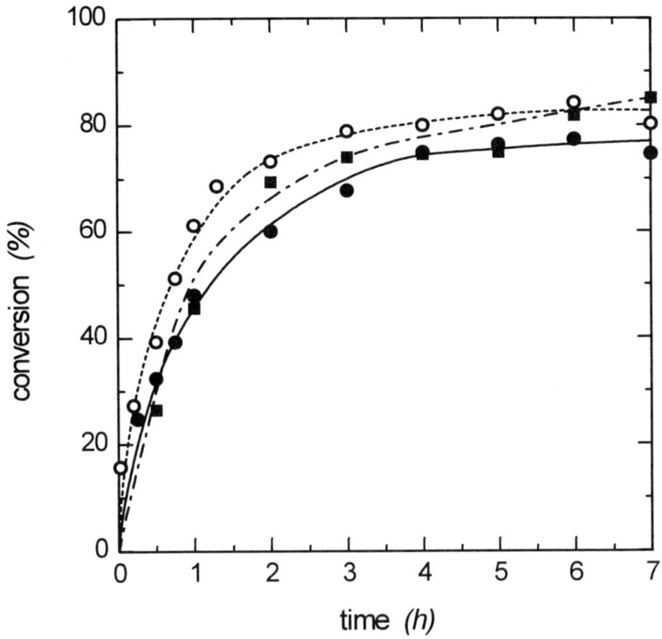

Figure 3. Conversion–time data for three AM–N16 copolymerizations at N16 contents (mol% of monomer feed) of 3% (■), 4% (○), and 5% (●).

the stronger the effect (final weight conversions of ~85, 80, and 75% for 3, 4, and 5 mol% of N16, respectively). As shown in the next section, this behavior is related to an incomplete consumption of the polymerizable surfactant.

Copolymer Composition. Figure 4 shows the variation in polymer composition as a function of conversion for two different initial N16 concentrations in the monomer feed. Examination of the data leads to the following comments.

Unlike the compounds in the AM–eΦAM series, the composition of the final compounds does not correspond to the initial monomer composition. This difference reveals that incorporation of the polymerizable surfactant is far from being complete (~50–60%), a situation that leads to the observed lowering in the global monomer conversion already discussed (*see* Figure 3). (The low molar levels in N16 actually represent a much higher weight proportion in the feed: 16.3–23.8 wt% for 3–5 mol%.) Such behavior can be ascribed to polymer–surfactant interactions. Many studies (*10–14*) have shown that surfactants interact strongly with hydrophobically modified water-soluble polymers via hydrophobic associations. Such a process is likely to occur during

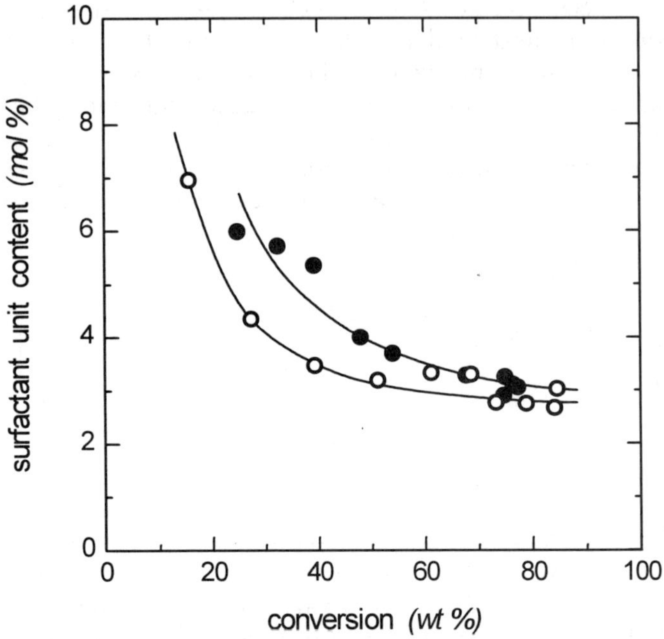

Figure 4. Variation in copolymer composition as a function of conversion for AM–N16 copolymerizations at N16 contents (mol% of monomer feed) of 4% (○) and 5% (●).

the polymerization reaction, when the surfactant monomers adsorb onto the previously formed hydrophobic sequences. In other words, mixed micelles composed of polymerized and nonpolymerized surfactant species are formed. This trapping effect of the polymerizable surfactant molecules hinders their further consumption by the polymerization reaction.

As is observed for AM–eΦAM samples prepared in micellar solution (Figure 2), an important drift in copolymer composition is observed for AM–N16 copolymers (Figure 4), and the hydrophobe content exceeds that of the feed at the early stage of the reaction. This drift could be attributed to differences in reactivity parameters, because the two monomers have quite different chemical structures. However, from our knowledge of the polymerization mechanism for micelle-forming monomers (*38, 53, 54*), we can infer that the drift in copolymer composition is mainly related to the dynamics of the micellar system. In a previous study from our laboratory on the homopolymerization of the same polymerizable surfactant, we showed that the degree of polymerization largely exceeds the micellar aggregation number, because the polymerizing micelles are fed surfactant molecules coming from unnucleated micelles (*38, 53, 54*). The same feeding process should occur here and lead to a faster incorporation of N16 at the beginning of the copolymerization. That also means that the number of N16 units constituting one sequence is larger than the initial aggregation number of the micelle.

It can be concluded that the samples taken at high degrees of conversion exhibit some compositional heterogeneity, as was the case in the first micellar process.

Rheological Properties in Aqueous Solution

Before we compare the behaviors of AM–NaA–eΦAM and AM–N16 copolymers, it is worthwhile to examine the similarities and differences in their chemical structures.

1. Both series exhibit a somewhat blocky structure.
2. The C_{16} alkyl chains of AM–N16 samples constitute larger hydrophobic groups than the ethylphenyl groups of AM–NaA–eΦAM samples (roughly equivalent to C_6 alkyl chains). This factor favors stronger hydrophobic interactions. However, the presence of the quaternary ammonium group close to the C_{16} chain may electrostatically hinder the hydrophobic attractions. It must also be kept in mind that the effective hydrophobicity depends on the lengths of the sequences (*5, 6, 24, 42–44*). Thus several adjoining hydrophobic units form a large hydrophobic group, which can be more favorable to intermolecular interactions than the same number of units singly and randomly distributed along the chain.
3. If one disregards the differences in their chemical nature, AM–NaA–eΦAM and AM–N16 copolymers differ mainly by the relative locations of the hydrophobic and ionic groups along the polyacrylamide backbone.

   In the AM–NaA–eΦAM series, the hydrophobic and ionic groups are on different monomer units. The level of charges can be varied while the hydrophobe level and hydrophobe distribution are held constant and vice versa. In this study, a large range of ionic-group contents was covered (0–40 mol%); the hydrophobic character was varied essentially by modifying the blockiness of the copolymer at a constant hydrophobe content (1 mol%) (i.e., by using different surfactant concentrations in the synthesis).

> In contrast, for the AM–N16 series, the hydrophobic and ionic groups are on the same monomer unit (N16). Obviously, in this case, the levels of these two functional groups vary concomitantly. In this series the hydrophobic character was controlled by modifying the surfactant unit content (~0.5–5 mol%). All the samples therefore have a low charge density imposed by the number of N16 units.

It can be predicted that ion-containing associating copolymers will exhibit a more complicated behavior than that observed for the corresponding neutral copolymers: In addition to the attractive hydrophobic interactions, repulsive electrostatic interactions occur.

The presence of ionic sites should have two opposite effects on the viscosifying properties: (1) a factor favorable to an increase in viscosity is the well-known polyelectrolyte effect, that is, coil expansion resulting from intramolecular charge–charge repulsions; and (2) on the other hand, a lowering of the degree of association with a concomitant decrease in viscosity should result from the intermolecular electrostatic repulsions, which hinder some interpenetration of the macromolecular chains. In addition, the hydrophilic–hydrophobic balance of the samples is shifted toward an increased polymer solubility, which also contributes to reducing the hydrophobic interactions.

The screening of the charges when salt is added should also have two opposite effects: (1) at the intramolecular level, the chain contraction favors a decrease in viscosity; and (2) at the intermolecular level, the macromolecular interpenetration is facilitated and thus promotes enhanced polymolecular associations.

As a consequence, depending on which of these effects is dominant, a great variety of behaviors can occur as a function of the molecular characteristics of the samples, polymer concentration, and ionic strength (*7, 20, 21, 25, 26, 30*).

AM–NaA–eΦAM Series: Behavior in Pure Water. The results reported here deal essentially with AM–NaA–eΦAM samples prepared according to the two-step procedure (i.e., hydrolysis of AM–eΦAM copolymers). A typical example of the variation in viscosity as a function of polymer concentration before and after partial hydrolysis is shown in Figure 5. At all polymer concentrations examined, the charged AM–NaA–eΦAM copolymer exhibits a higher viscosity than the noncharged parent AM–eΦAM copolymer. Similar behavior was observed for other degrees of hydrolysis (10–40%) (*30*). The shape of the curve for viscosity versus concentration is different, depending on whether the sample is charged or not. For the neutral sample (i.e.,

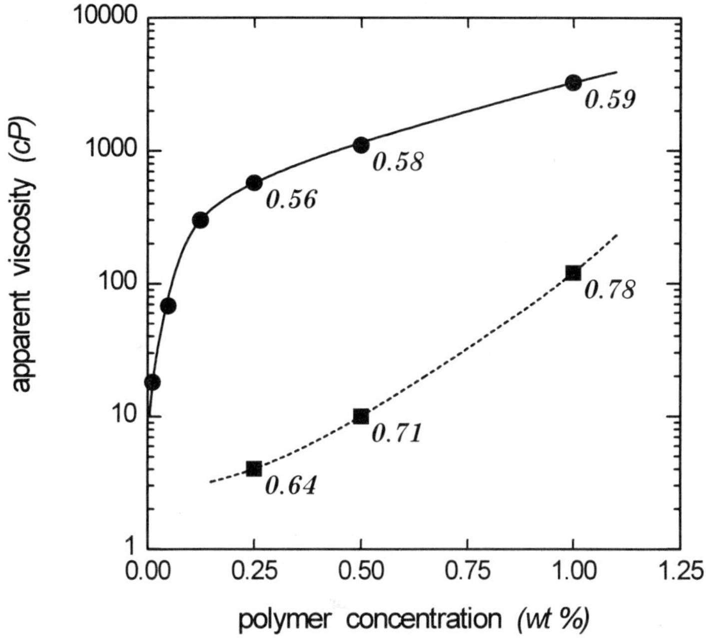

Figure 5. Variation in viscosity as a function of copolymer concentration for an AM–eΦAM sample in pure water before (■) and after (●) hydrolysis. Data point labels indicate I_3/I_1 pyrene fluorescence ratios. [eΦAM] = 1 mol%; hydrolysis degree = 18.5 mol%; and shear rate = 0.5 s^{-1}.

before hydrolysis), an "exponential-like" curve is observed, as is usual for associating polymers. For the charged polymer (i.e., after hydrolysis), however, the curvature is reversed: Viscosity increases rapidly at low concentrations and tends toward a plateau value at higher concentrations. Such behavior is typical of polyelectrolytes, for which chains are highly expanded at low concentrations owing to intramolecular charge–charge repulsions, while coil dimensions are reduced at higher concentrations owing to autoscreening of the charges.

The presence of charges also affects the flow properties as a function of shear rate. Figure 6 gives viscosity data obtained first from scanning of increasing shear rates and then from scanning of decreasing shear rates. For the uncharged AM–eΦAM copolymer, the down-ramp curve lies down the up-ramp curve, which is characteristic of thixotropic behavior. In contrast, for the charged AM–NaA–eΦAM copolymer, no significant thixotropic effect appears, because the two curves are nearly superimposed. This thixotropic behavior is a time-dependent effect that can be related to the kinetics of the associa-

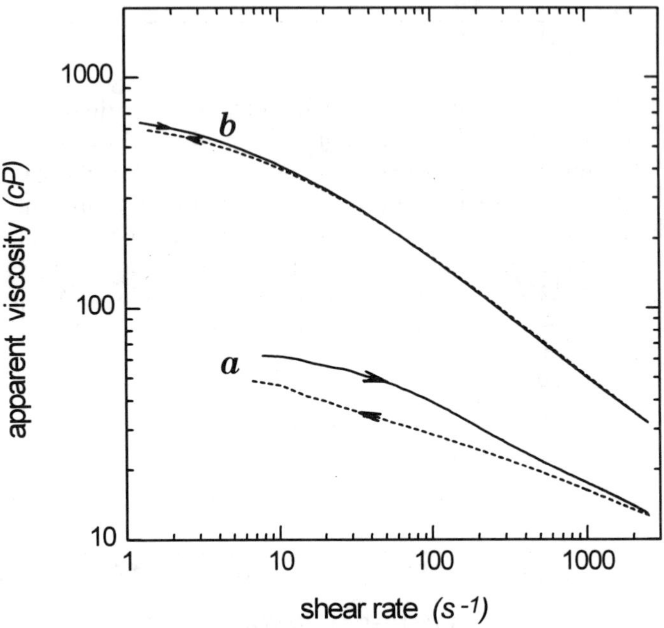

Figure 6. Variation in viscosity as a function of shear rate for an AM–eΦAM sample in pure water before (a) *and after* (b) *hydrolysis. Key: →—, increasing shear; and ----←----, decreasing shear. Polymer concentration = 3 wt%; [eΦAM] = 1 mol%; and hydrolysis degree = 18.5 mol%.*

tion–dissociation phenomenon (*5, 6*). Thixotropic character markedly increases with the extent of the association process (*6*). The lack of thixotropy for the hydrolyzed sample therefore indicates a less marked degree of association. This behavior is confirmed by fluorescence studies using pyrene as a probe. The ratio of the intensities of the third and first peaks of the pyrene emission spectrum (I_3/I_1) is sensitive to the local environment of the probe. Then the I_3/I_1 ratio can be used as a hydrophobicity scale (the higher the I_3/I_1 ratio, the stronger the hydrophobic character) to show the formation of hydrophobic microdomains in an aqueous solution (*55*). Fluorescence data reported in Figure 5 show that the uncharged copolymer leads to rather high values of I_3/I_1 (0.64–0.78), and these values increase as a function of polymer concentration. (For homopolyacrylamide, i.e., for a hydrophobe-free sample, I_3/I_1 is 0.54–0.58 in the same polymer concentration range.) This result reveals the formation of hydrophobic microdomains whose numbers or hydrophobicity increases with copolymer concentration owing to an increasing effectiveness of interchain liaisons. On the other hand, for the hydrolyzed sample, the values of I_3/I_1 are much lower, and this result confirms a lower degree of association. In conclusion, in salt-free solution, the increase in viscosity due to the polyelectrolyte effect prevails over the decrease in intermolecular interactions.

The occurrence of these two opposing effects is evidenced by looking at the influence of ionic-site content. Figure 7 shows viscosity variation for a series of samples of increasing hydrolysis levels prepared from the same AM–eΦAM. The viscosity passes through a maximum while the simultaneous and progressive disappearance of the hydrophobic microdomains is depicted by a continuous decrease in I_3/I_1. Because of the competition between the loss of interchain associations and the greater chain expansion upon increasing charges, there is a certain level of charge for which the two effects combine and lead to a maximum thickening efficiency. This maximum as a function of the hydrolysis level also was observed by Peiffer (*26*) for another type of hydrophobically associating polyacrylamide.

The level of ionic charge at which hydrophobic interactions are lost depends on the initial hydrophobicity of the parent AM–eΦAM sample. Thus at constant hydrophobe content, copolymers prepared

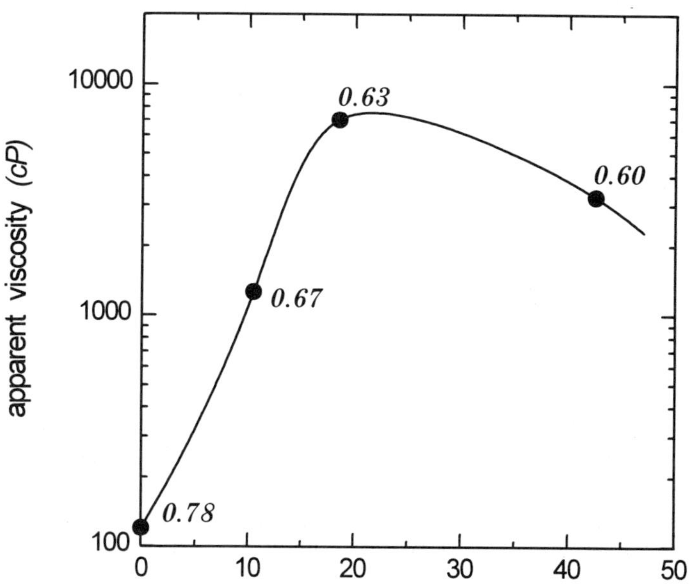

Figure 7. Variation in viscosity as a function of hydrolysis level (AM–NaA–eΦAM copolymers). Data point labels indicate I_3/I_1 pyrene fluorescence ratios. Polymer concentration = 1 wt%; [eΦAM] = 1 mol%; and shear rate = 0.5 s^{-1}.

with a higher surfactant concentration (i.e., with a lower thickening efficiency due to shorter hydrophobe sequences) are no more associated even for a low NaA content (<20 mol%) (*30*).

AM–NaA–eΦAM Series: Behavior in Salt Solution. Figure 8 gives an example of the viscosity behavior of a partially hydrolyzed copolymer (AM–NaA–eΦAM) as a function of salt content (NaCl). For reference, viscosity data for the corresponding uncharged copolymer and the polyacrylamide analogs, both nonhydrolyzed (PAM) and hydrolyzed (HPAM), are also given. Figure 9 shows the corresponding fluorescence results. Examination of the data in Figures 8 and 9 leads to the following comments.

The viscosity of the homopolyacrylamide solution does not vary as a function of salt content as would be expected. (However, the absence of the salt effect is not the rule for all neutral water-soluble polymers. In some cases, depending on the chemical nature of the polymer, the viscosity of the aqueous solution is modified by adding

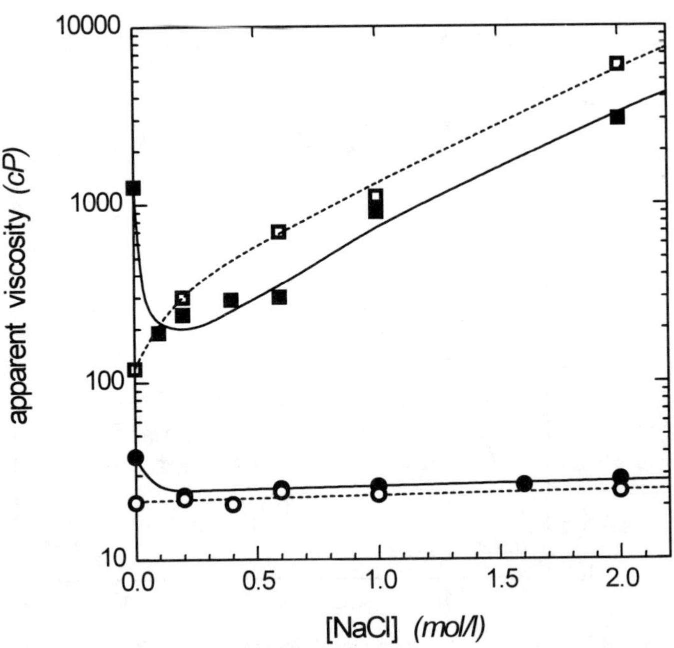

Figure 8. Variation in viscosity as a function of salt content for solutions of various samples. Key: ○, PAM; ●, HPAM; □, AM–eΦAM; and ■, AM–NaA–eΦAM. Polymer concentration = 1 wt% and shear rate = 0.5 s^{-1}. For the copolymer, [eΦAM] = 1 mol% and hydrolysis degree = 10.5 mol%.

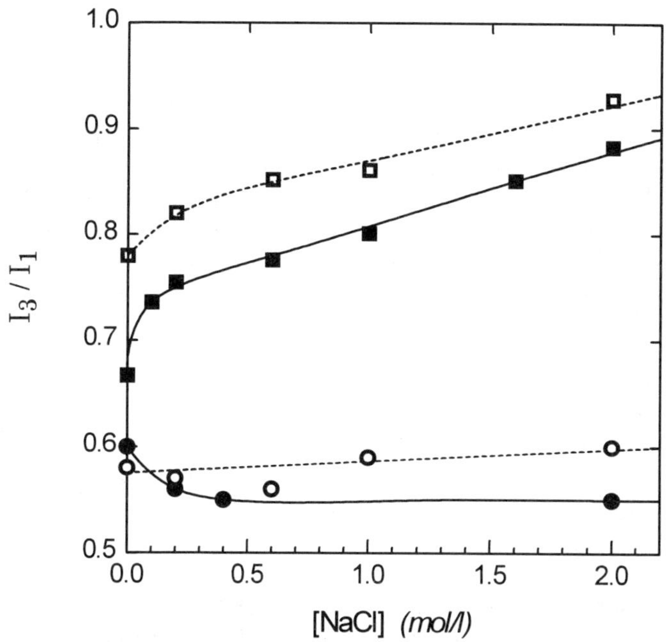

Figure 9. Variation in I_3/I_1 pyrene fluorescence ratio as a function of salt content for solutions of various samples. Key: ○, PAM; ●, HPAM; □, AM–eΦAM; and ■, AM–NaA–eΦAM. Polymer concentration = 1 wt% and shear rate = 0.5 s^{-1}. For the copolymer, [eΦAM] = 1 mol% and hydrolysis degree = 10.5 mol%.

salt (*56*). Such an effect may be related to the hydrophobic character of the polymer.)

The partially hydrolyzed homopolyacrylamide exhibits a higher viscosity in the absence of salt owing to the well-known polyelectrolyte effect, and when salt is added, the viscosity decreases because of the charge-screening effect. However, after complete charge screening, that is, for [NaCl] > 0.1 M, the HPAM behaves like neutral PAM. The slightly higher viscosities observed for the charged sample reveal its increased solubility.

In contrast, for both charged and uncharged hydrophobe-containing copolymers, very large variations in viscosity are observed over the range of salt contents investigated, a result indicating that the association degree is modified for both systems. This result is confirmed by fluorescence data. Although for PAM and HPAM samples the I_3/I_1 pyrene ratio is low and does not vary significantly as a function of salt concentration, for the copolymers, either neutral or charged, the I_3/I_1 values increase strongly and this increase clearly reveals a stronger association process.

For the neutral copolymer, the influence of the salt does not imply electrostatic effects but a salting-out effect, that is, a lowering of the solubility of the hydrophobic moieties (*57*). This assumption is supported by the fact that for ethylphenylpropionamide, the model molecule of the hydrophobic units, the solubility is ~3.5 times lower when the NaCl concentration is increased from 0 to 2 M. Consequently, the hydrophilic–hydrophobic balance of the copolymer is shifted toward decreased solubility. This shift leads to a continuous increase in the degree of association and therefore to an increase in viscosity as a function of the salt content. Similar behavior has been observed for other neutral polyacrylamide derivatives bearing hydrophobes of a different nature (*7, 8, 27, 58*).

For the charged copolymer, the initial sharp loss in viscosity is due to charge screening by the salt and hence to chain contraction, as already observed for the HPAM. According to the previously discussed features, the further increase in viscosity upon increase in salt content can be primarily attributed to a salt effect on the hydrophobe solubility and not to electrostatic screening effects between charged chains.

After charge screening by the salt (>0.1 M), the hydrolyzed copolymer sample exhibits lower viscosities than the nonhydrolyzed sample does. Once again, fluorescence data are in agreement and show a lower hydrophobicity for the hydrolyzed copolymer. That means that the degree of association of the charged copolymer is always lower than that observed for the corresponding noncharged sample, owing to the stronger hydrophilic character provided by the ionic NaA units.

The screening of the polyelectrolyte effect when salt is added is also apparent in the viscosity variation as a function of polymer concentration (Figure 10): in the presence of salt, the shape of the curve becomes similar to that observed for the neutral associating polymer. The higher I_3/I_1 pyrene fluorescence ratios in salt solutions than in pure water (Figure 10) reveal a stronger degree of association in the presence of salt: The higher the salt concentration, the stronger the effect. However, in the range of copolymer concentrations investigated (≤1 wt%), the polyelectrolyte coil contraction effect in the presence of salt is predominant, and the viscosities are lower than in pure water.

From the study of AM–NaA–e*Φ*AM copolymers we can draw the following conclusions. For homopolyacrylamide, the introduction of ionic sites by partial hydrolysis always improves the thickening properties in both pure water and salt solution, but this improvement is not true for AM–e*Φ*AM associating copolymers. For AM–e*Φ*AM copolymers, the main consequence of the hydrolysis is to lower the degree of association. For this reason, the effect of partial hydrolysis is

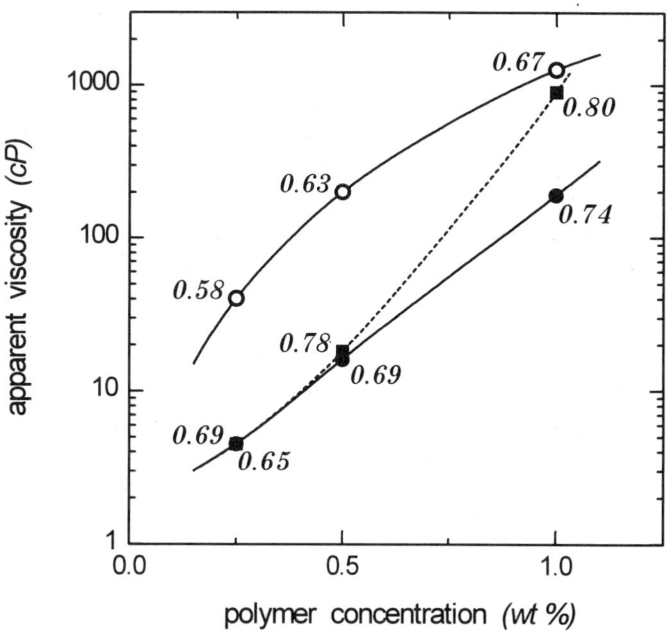

Figure 10. Variation in viscosity as a function of polymer concentration for an AM–NaA–eΦAM sample in pure water (○) or in 0.1 M (●) or 1 M (■) aqueous NaCl solution. Data point labels indicate I_3/I_1 pyrene fluorescence ratios. [eΦAM] = 1 mol%; hydrolysis level = 10.5 mol%; and shear rate = 0.5 s^{-1}.

beneficial only in salt-free solution or at low salt content (≲0.1 M), that is, when polyelectrolyte coil expansion is prevailing. In contrast, at a higher salt concentration, that is, when coil expansion is reduced, the benefit of introducing ionic sites is lost and the unfavorable effect of the lowered association degree is predominant. To keep the advantages of both polyelectrolyte and association effects, it is therefore necessary to prepare copolymers containing more hydrophobes. However, as discussed in the section on synthesis, this is not possible if the two-step procedure is used, owing to the poor solubility in water of AM–eΦAM intermediates. The one-step terpolymerization procedure allows us to overcome this problem. For example, a sample (M_w = 3 × 10^6) prepared under micellar conditions ([SDS] = 3 wt%) with a monomer molar ratio of 89/9/2 for AM–NaA–eΦAM in the feed exhibits very interesting thickening properties; at a copolymer concentration of 0.25 wt% in an aqueous 0.1 M NaCl solution, the viscosity is ~80 cP, and at 1 wt%, the solution behaves like a gel.

AM–N16 Series. Unlike the structures of the AM–NaA–eΦAM series, the chemical structures of the AM–N16 copolymers preclude a direct comparison with the corresponding uncharged samples. The general behavior of AM–N16 copolymers in aqueous solution in the absence or presence of salt is roughly similar to that of AM–NaA–eΦAM copolymers. Figure 11 shows the variation in viscosity of an AM–N16 sample as a function of polymer concentration. In salt-free solution the curve is typical of polyelectrolytes. However, compared to the AM–NaA–eΦAM samples (*see* Figure 5), the tendency toward a plateau value is less marked, a difference that must be attributed to a much lower ionic-site content. As already observed for the previous copolymer series (Figure 10), the shape of the curve is modified by adding salt (0.1 M NaCl), and the viscosity increases faster as a function of polymer concentration (Figure 11).

For a fixed copolymer concentration (1 wt%), the viscosity passes through a minimum as increasing amounts of NaCl are added (Figure 12), as already observed for AM–NaA–eΦAM copolymers (Figure 8). As already discussed, the opposite effects of coil contraction and enhanced degree of association upon increasing ionic strength account for this behavior.

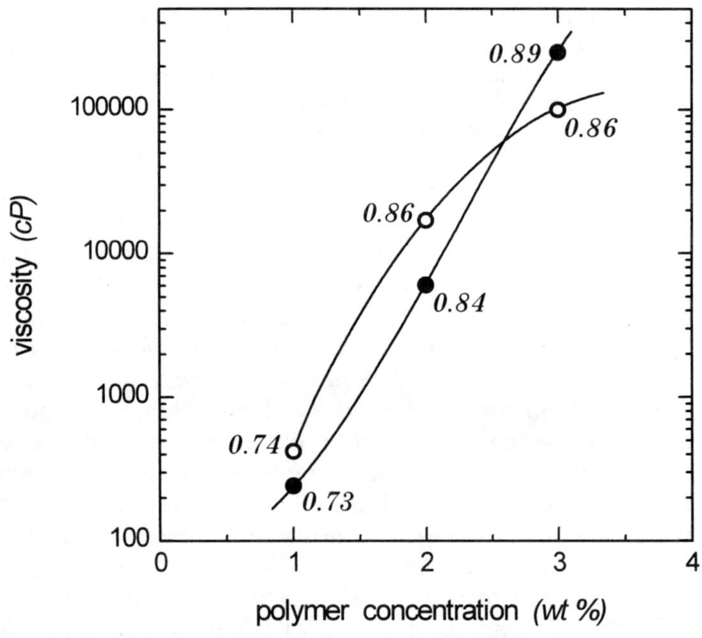

Figure 11. Variation in viscosity as a function of copolymer concentration for an AM–N16 sample in pure water (○) or 0.1 M NaCl aqueous solution (●). Data point labels indicate I_3/I_1 *pyrene fluorescence ratios. [N16] = 1 mol%; viscosities are extrapolated to zero shear.*

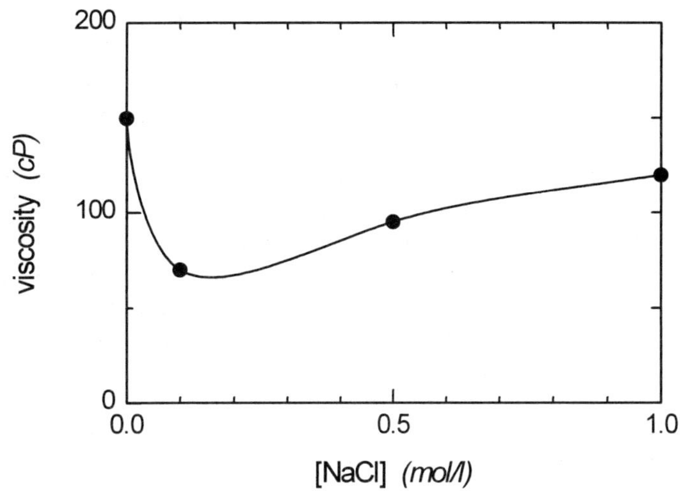

Figure 12. Viscosity as a function of salt content for an AM–N16 sample. Polymer concentration = 1 wt% and [N16] = 1 mol%; viscosities are extrapolated to zero shear.

We observed that the minimum of the curve disappears when the polymer concentration is increased (3 wt%): The viscosity is always higher in the presence of salt than in pure water. As the interpenetration of macromolecular chains is enhanced at a high polymer concentration, intermolecular hydrophobic associations are favored, and this effect becomes predominant whatever the ionic strength. In a previous work on ionic hydrophobically associating acrylamide-based copolymers (*26*), it was noted that the hydrophobic effect dominates either when the ionic site content is low or when the polymer concentration is increased.

The results of the investigation into the thickening properties of AM–N16 copolymers call for the following comments. Many previous studies on hydrophobically associating water-soluble copolymers have stressed that the association process is very sensitive to the numbers or sizes of hydrophobic groups. Thus slight variations in the hydrophobe level within a narrow range (typically ~0.5–2 mol%) or in the length of a hydrophobic alkyl chain (for example, C_8 to C_{16}) induce large modifications of viscosification efficiency. In the AM–N16 copolymers, the C_{16} alkyl chain can be considered a relatively large hydrophobe group, and the hydrophobe content is comparatively high (1–3 mol%). When the number and size of hydrophobes, the molecular weights of the samples (in the range of 1.5–3 million), the polymer concentrations (1–3 wt%), as well as data in the literature on related compounds are taken into account, stronger thickening properties could reasonably have been expected. It is significant that for an identical hydrophobe level (1 mol%), the AM–eΦAM and AM–NaA–eΦAM copolymers discussed here exhibit an equivalent if not better thickening ability than that of AM–N16 copolymers, despite smaller hydrophobe groups. In addition, even if an increase in the N16 content (up 3 mol%) significantly improves the viscosification properties, we did not observe the drastic effect generally found for other types of associating copolymers. Finally, the rise in viscosity with increasing AM–N16 concentration was not as steep as expected.

From all these observations it can be inferred that only a very low fraction of N16 units contributes to the intermolecular association process. A possible explanation could be electrostatic repulsions between quaternary ammonium groups that prevent C_{16} chains from associating. However, such an explanation is not quite satisfying, because fluorescence measurements show that hydrophobe interactions are actually operative. The high I_3/I_1 pyrene fluorescence ratios in both the absence and the presence of salt (Figure 11) unambiguously reveal the presence of hydrophobic microdomains within the aqueous solution. Their presence implies that associations between the C_{16} alkyl chains are essentially intramolecular. The blocky structure of AM–N16 copolymers and the specific conformation of the N16 sequences account for such behavior. Viscometric and fluorescence experiments (*38, 53*) have shown that the aqueous homopolymerization of N16 in the micellar state leads to a system that exhibits a structure similar to that of a polysoap with intramolecular hydrophobic microdomains. In the AM–N16 copolymers, it can be assumed that each N16 sequence also adopts a compact conformation that is due to hydrophobic interactions between neighboring side chains. Such a micellelike conformation, in which C_{16} alkyl chains form a hydrophobic core surrounded by cationic charges of the quaternary ammonium groups, does not favor intermolecular hydrophobe interactions.

We conclude that the more suitable copolymer architecture depends on the nature of the hydrophobe units. For AM–eΦAM and AM–NaA–eΦAM copolymers, the blocky structure favors the intermolecular association process because single ethylphenyl groups are not hydrophobic enough, while several adjoining eΦAM units form larger and more hydrophobic groups that are able to promote interchain liaisons. In contrast, for AM–N16 copolymers, long sequences of surfactant units lead locally along the backbone to an unwanted polysoap conformation that reduces the possibilities of interchain liaisons. Further investigations of the properties of copolymers in which surfactant units are randomly distributed along the backbone are in progress.

Shear Time Effect. The rheological properties of AM–NaA–eΦAM and AM–N16 associative thickeners are various and complex.

Depending on the sample and the experimental conditions, all types of behaviors are observed: shear thinning, shear thickening, thixotropy, and rheopexy.

In the low-shear region ($\lesssim 10\ s^{-1}$), behavior is generally rheopectic, and the viscosity at a constant shear rate increases with time to a plateau value. An example of such behavior is given in Figure 13 for an AM–N16 copolymer: under very low shear, the viscosity increases by 1 order of magnitude within 5 min, whereas at a higher shear rate, the equilibrium viscosity is reached almost instantaneously. The magnitude of the viscosity increase and the time required to obtain the plateau value depend also on the nature of the sample, the polymer concentration, and the salt content. These results suggest that new interchain links are formed when the sample is subjected to a low shear stress. A possible explanation is that under low shear, the chain extends slightly and thus some intramolecular connections rupture, because their formation requires a more compact coil conformation. The newly liberated hydrophobic units then are able to associate intermolecularly with other hydrophobic groups. For certain samples, the rheopectic effect is followed by a smaller thixotropic effect (i.e., de-

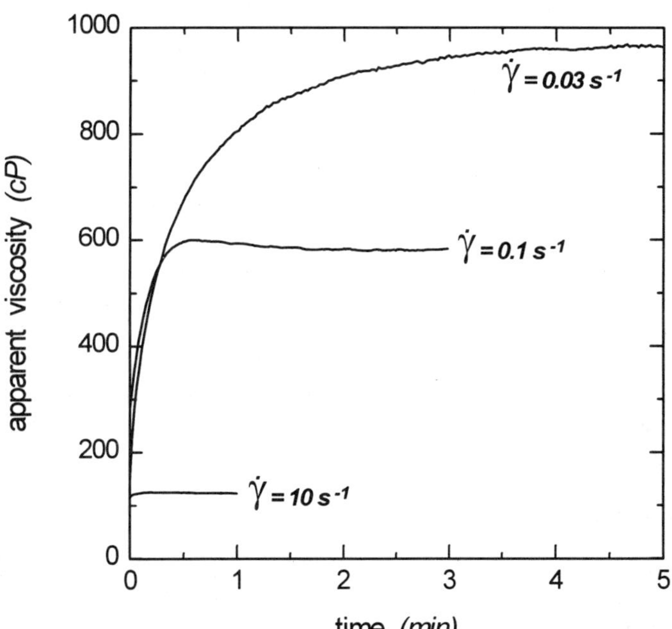

Figure 13. Viscosity as a function of shear time at three different shear rates ($\dot{\gamma}$) for an AM–N16 sample in 0.5 M NaCl aqueous solution. Polymer concentration = 1 wt% and [N16] = 3 mol%.

creasing viscosity with time at a constant shear rate) before the plateau value is attained.

A consequence of this behavior is that the variation in viscosity as a function of shear rate cannot be correctly investigated by using a conventional scanning procedure in which shear rate (or shear stress) is continuously increased, because time effect and shear effect act simultaneously. Reliable viscosity data can be obtained only in the stationary state, that is, when the viscosity becomes constant as a function of time under a fixed shear rate or a fixed shear stress. All viscosity data reported in this chapter were obtained under such conditions.

Conclusion

The radical copolymerization of acrylamide in an aqueous medium with either a micelle-forming polymerizable surfactant or a hydrophobic monomer solubilized within surfactant micelles yields copolymers with somewhat blocky structures as a consequence of the microheterogeneous nature of the reaction medium. Both micellar processes also induce some compositional heterogeneity because the monomer that is located within the dispersed micellar phase is more rapidly consumed at the beginning of the reaction than the acrylamide solubilized within the aqueous continuous phase. The dynamics of the micellar systems, that is, material exchanges between micelles, account for this result.

For polyacrylamides containing both hydrophobic groups and ionic sites, the competition between hydrophobic attractions and electrostatic repulsions allows us to control the thickening efficiency in aqueous solution by varying the composition and microstructure of the copolymer. Compared to neutral hydrophobically associating polymers, one of the main effects of introducing charges onto the polymer backbone is a lowering of the degree of association. In addition to the well-known charge-screening effect, another important result is the decrease in solubility of the hydrophobe as observed in brine solution. Both contributions are responsible for the associative behavior.

Acknowledgments

The financial assistance of ARTEP (Association de Recherches sur les Techniques d'Exploitation du Pétrole) and PIRSEM (Programme Interdisciplinaire de Recherches sur les Sciences pour l'Energie et les Matières Premières) is gratefully acknowledged. S. Biggs is grateful to SERC/NATO for the provision of a European postdoctoral fellowship. D. Renoux thanks PPG Industries France for financial support.

References

1. Landoll, L. M. *J. Polym. Sci., Polym. Chem. Ed.* **1982,** *20,* 443.
2. Evani, S.; Rose, G. D. *Polym. Mater. Sci. Eng.* **1987,** *57,* 477.
3. *Polymers in Aqueous Media: Performance Through Association;* Glass, J. E., Ed.; Advances in Chemistry 223; American Chemical Society: Washington, DC, 1989.
4. McCormick, C. L.; Bock, J.; Schulz, D. N. In *Encyclopedia of Polymer Science and Engineering,* 2nd ed.; Mark, H. F.; Bikales, N. M.; Overberger, C. G.; Menges, G., Eds.; Wiley-Interscience: New York, 1989; Vol. 17, p 730.
5. Hill, A.; Candau, F.; Selb, J. *Prog. Colloid Polym. Sci.* **1991,** *84,* 61.
6. Hill, A.; Candau, F.; Selb, J. *Macromolecules* **1993,** *26,* 4521.
7. Bock, J.; Siano, D. B.; Valint, P. L., Jr.; Pace, S. J. *Polym. Mater. Sci. Eng.* **1987,** *57,* 487.
8. McCormick, C. L.; Nonaka, T.; Johnson, C. B. *Polymer* **1988,** *29,* 731.
9. Jenkins, R. D.; Silebi, C. A.; El-Aasser, M. S. In ref. 17, Chapter 13, p 222.
10. Biggs, S.; Selb, J.; Candau, F. *Langmuir* **1992,** *8,* 838.
11. Sau, A. C.; Landoll, L. M. In ref. 3, Chapter 18, p 343.
12. Karunasena, A.; Brown, R. G.; Glass, J. E. In ref. 3, Chapter 26, p 495.
13. Magny, B.; Iliopoulos, I.; Audebert, R.; Piculell, L.; Lindman, B. *Prog. Colloid Polym. Sci.* **1992,** *89,* 118.
14. Dualeh, A. J.; Steiner, C. A. *Macromolecules* **1990,** *23,* 251.
15. *Water-Soluble Polymers: Beauty with Performance;* Glass, J. E., Ed.; Advances in Chemistry 213; American Chemical Society: Washington, DC, 1986.
16. *Water-Soluble Polymers: Synthesis, Solution Properties, and Applications;* Shalaby, S. W.; McCormick, C. L.; Buttler, G. B., Eds.; ACS Symposium Series 467; American Chemical Society: Washington, DC, 1991.
17. *Polymers as Rheology Modifiers;* Schulz, D. N.; Glass, J. E., Eds.; ACS Symposium Series 462; American Chemical Society: Washington, DC, 1991.
18. Bock, J.; Valint, P. L., Jr.; U.S. Patent 4 730 028, 1988.
19. Ezzell, S. A.; Middleton, J. C.; McCormick, C. L. *Polym. Prepr. (Am. Chem. Soc., Div. Polym. Chem.)* **1991,** *32*(1), 573.
20. McCormick, C. L.; Middleton, J. C.; Cummins, D. F. *Macromolecules* **1992,** *25,* 1201.
21. Wang, K. T.; Iliopoulos, I.; Audebert, R. *Polym. Bull.* **1988,** *20,* 577.
22. Magny, B.; Iliopoulos, I.; Audebert, R. *Polym. Commun.* **1991,** *32,* 456.
23. Magny, B.; Iliopoulos, I.; Audebert, R. *Polym. Prepr. (Am. Chem. Soc., Div. Polym. Chem.)* **1991,** *32*(1), 577.
24. Dowling, K. C.; Thomas, J. K. *Macromolecules* **1990,** *23,* 1059.
25. Watterson, A. C.; Hunter, C. R.; Thompson, A. M.; Salamone, J. C. *Polym. Prepr. (Am. Chem. Soc., Div. Polym. Chem.)* **1991,** *32*(1), 126.
26. Peiffer, D. G. *Polymer* **1990,** *31,* 2353.
27. Zhang, Y.-X.; Da, A.-H.; Butler, G. B.; Hogen-Esch, T. E. *J. Polym. Sci., Polym. Chem. Ed.* **1992,** *30,* 1383.
28. Shay, G. D. In ref. 3, Chapter 25, p 457.
29. Rauscher, A.; Hoffmann, H.; Rehage, H.; Fock, J. *Tenside, Surfactants, Deterg.* **1992,** *29,* 101.
30. Biggs, S.; Selb, J.; Candau, F. *Polymer* **1993,** *34,* 580.

31. Kuo, P. L.; Hung, M. N.; Lin, Y. H. *J. Appl. Polym. Sci.* **1993,** *47,* 1295.
32. Strauss, U. P. In ref. 3, Chapter 16, p 317.
33. Chu, D. Y.; Thomas, J. K. *Macromolecules* **1987,** *20,* 2133.
34. Binana-Limbelé, W.; Zana, R. *Macromolecules* **1990,** *23,* 2731.
35. McCormick, C. L.; Hoyle, C. E.; Clark, M. D. *Macromolecules* **1990,** *23,* 3124.
36. Valint, P. L., Jr.; Bock, J.; Schulz, D. N. *Polym. Mater. Sci. Eng.* **1987,** *57,* 482.
37. Biggs, S.; Hill, A.; Selb, J.; Candau, F. *J. Phys. Chem.* **1992,** *96,* 1505.
38. Cochin, D.; Zana, R.; Candau, F. *Polymer Int.* **1991,** *30,* 491.
39. Evani, S. U.S. Patent 4 432 881, 1984.
40. Turner, S. R.; Siano, D. B.; Bock, J. U.S. Patents 4 520 182, 1985; 4 521 580, 1985; and 4 528 348, 1985.
41. Peer, W. J. In ref. 3, Chapter 20, p 381.
42. Ezzell, S. A.; McCormick, C. L. *Macromolecules* **1992,** *25,* 1881.
43. Ezzel, S. A.; Hoyle, C. E.; Creed, D.; McCormick, C. L. *Macromolecules* **1992,** *25,* 1887.
44. Branham, K. D.; Middleton, J. C.; McCormick, C. L. *Polym. Prepr. (Am. Chem. Soc., Div. Polym. Chem.)* **1991,** *32*(1), 106.
45. Plochocka, K. *J. Macromol. Sci., Rev. Macromol. Chem.* **1981,** *C20,* 67.
46. Topchiev, D. A.; Shakirov, V. Z.; Kalinina, L. P.; Karaputadze, T. M.; Kabanov, V. A. *Polym. Sci. USSR* **1972,** *14,* 652.
47. Ponratnam, S.; Kapur, S. L. *Makromol. Chem.* **1977,** *178,* 1029.
48. Shawki, S. M.; Hamielec, A. E. *J. Appl. Polym. Sci.* **1979,** *23,* 3341.
49. Valint, P. L., Jr.; Bock, J.; Ogletree, J.; Zushma, S.; Pace, S. J. *Polym. Prepr. (Am. Chem. Soc., Div. Polym. Chem.)* **1990,** *31*(2), 67.
50. Harkins, W. D. *J. Polym. Sci.* **1950,** *5,* 217.
51. Schulz, D. N.; Kaladas, J. J.; Maurer, J. J.; Bock, J.; Pace, S. J.; Schulz, W. W. *Polymer* **1987,** *28,* 2110.
52. Van Phung, K.; Evani, S. U.S. Patent 4 728 696, 1988.
53. Cochin, D.; Candau, F.; Zana, R. *Macromolecules* **1993,** *26,* 5755.
54. Cochin, D.; Zana, R.; Candau, F. *Macromolecules* **1993,** *26,* 5765.
55. Kalyanasundaram, K. *Photochemistry in Microheterogeneous Systems;* Academic: New York, 1987.
56. Saito, S. *J. Polym. Sci. Part A1* **1969,** *7,* 1789.
57. Long, F. A.; McDevit, W. F. *Chem. Rev.* **1952,** *51,* 119.
58. Bock, J.; Siano, D. B.; Schulz, D. N.; Turner, S. R.; Valint, P. L., Jr.; Pace, S. J. *Polym. Mater. Sci. Eng.* **1986,** *55,* 355.

RECEIVED for review November 10, 1993. ACCEPTED revised manuscript July 26, 1994.

16

Associating Polymers Containing Fluorocarbon Hydrophobic Units

Eric J. Amis,[1] Ning Hu, Thomas A. P. Seery,[2] Thieo E. Hogen-Esch, Mariam Yassini, and Frank Hwang

Department of Chemistry, University of Southern California, Los Angeles, CA 90089–0482

Associating polymers with hydrophilic backbones and with perfluoroalkane comonomers as the hydrophobic units were synthesized, and their properties were investigated. The viscosity of aqueous solutions at low concentration can be increased by more than 4 orders of magnitude. This effect far exceeds that of comparable hydrocarbon hydrophobic copolymers. For random copolymers, optimum comonomer contents for viscosity enhancement reflect the balance of intramolecular and intermolecular associations. Studies of model telechelic and comb-type copolymers with variable lengths and spacing of hydrophobes demonstrate the optimization of intermolecular associations and strong (long-lifetime) associations. For the model systems, Arrhenius activation energies are 16 to 29 kcal/mol for viscous flow and 5 kcal/mol for unentangled homopolymer chains. The results are interpreted in terms of models for transient networks with modifications to account for the postulate that neither loops nor chain extensions produce active network strands.

POLYMERS THAT ASSOCIATE VIA PHYSICAL INTERACTIONS in solutions have received much attention as a replacement for high-molecular-

[1]Current address: Polymers Division, National Institute of Standards and Technology, 224/B210, Gaithersburg, MD 20899
[2]Current address: Department of Chemistry, University of Connecticut, Storrs, CT 06269

0065–2393/96/0248–0279$13.00/0

weight polymeric viscosifiers. Such associating polymers are now used in a variety of applications because of their unique rheological properties (*1, 2*). Water-soluble associating polymers are especially interesting because of their roles in food thickeners, coatings, paints, enhanced oil recovery, and water treatment (*3, 4*). Many of these associating polymers are amphiphilic: They contain a hydrophilic main chain with hydrophobic side chains. For example, polyacrylamide (PAM) and cellulose can be chemically modified by attachment of hydrophobic units along the chain or at the chain ends. Strong associations between these hydrophobic units (sometimes refered to as "stickers") lead to the formation of transient networks that greatly enhance solution viscosity and viscoelasticity. The properties of these polymer solutions depend not only on the structure of the polymer chain, that is, molecular weight of the main chain and structure, number density, size, and distribution of the stickers, but also on external factors such as added surfactants, cosolvents, and salts.

Most theoretical efforts to understand the mechanisms of associations have focused on the rheological behavior (*5–9*). The classic explanation for the driving force of hydrophobic associations is the entropic gain accompanying the association. The removal of a hydrophobe from aqueous solution into a micellar aggregate is accompanied by the breakdown of a structure of highly ordered water molecules around the hydrophobe, leading to a positive change in entropy. The contribution to free energy from the enthalpy by the transfer of a hydrophobe from water to a micelle is usually much smaller than the entropic term. One difference between polymeric hydrophobic associations and small-molecule micelles is that hydrophobic associations usually have aggregation numbers of less than 10 (*10*), and small-molecule micelles have aggregation numbers of several tens to more than 100 (*11*).

Associating polymers containing fluorocarbon stickers form much stronger associations than the corresponding hydrocarbons, even when the polymers have much lower hydrophobe contents (*12*). This phenomenon is consistent with the fact that the micellization of fluorocarbon surfactants occurs at much lower concentrations than that of hydrocarbon surfactants (*13*). This condition is believed to reflect a more hydrophobic character of the fluorocarbon groups. The entropy increase upon micellization is higher. Furthermore, the solubility of CF_4 is one-seventh of that of CH_4 in water. The solubility of water is almost 25 times less in perfluoroheptane than in heptane. On the basis of critical micelle concentration data, Mathis et al. (*13*) suggested that fluorocarbon surfactants are about 1.5–1.8 times more effective than their hydrocarbon analogs.

The dramatic effect of the greater hydrophobicity of fluorocarbons

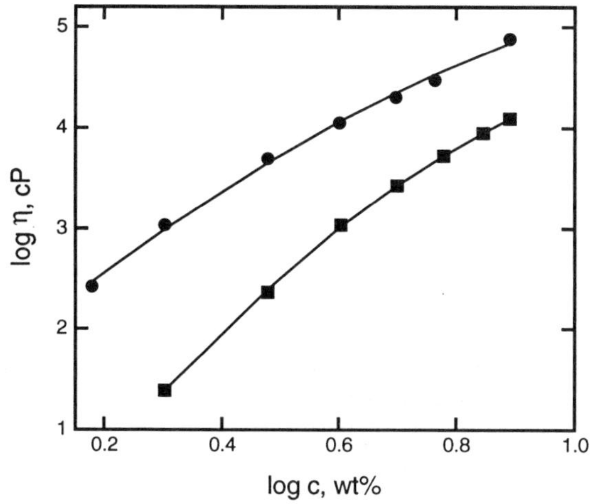

Figure 1. Viscosity (η) versus concentration (c) for telechelic polymers with molecular weight 35,000 and hydrophobic unit $-C_8F_{17}$ (●) *or* $-C_{16}H_{33}$ (■).

is reflected in the viscosity data for hydrocarbon- and fluorocarbon-containing associating polymers, as shown in Figure 1 (*14*). The viscosity of a model telechelic associating polymer of polyethylene oxide (PEO) with fluorocarbon ($-C_8F_{17}$) hydrophobic end units is about 10 times higher than that of one with hydrocarbon ($-C_{16}H_{33}$) hydrophobic units. Even though the hydrocarbon is twice as long, the fluorocarbon is more effective. The two polymers have the same mole contents of hydrophobic units.

Thus our focus is on the similarities and differences imparted to associating polymers by fluorocarbon hydrophobes. The first part of this chapter reviews experimental results obtained with fluorocarbon-containing, acrylamide-based random copolymers and hydrophobically modified cellulose derivatives. The second section focuses on characterization of polymers with regular structures, telechelic and comb types, and comparisons with theoretical models.

Random Copolymers

Fluorocarbon-containing, water-soluble associating polymers were synthesized by chemical modification of a preformed polymer and by copolymerization of the appropriate monomers. Cellulose derivatives, hydroxyethylcelluloses (HEC), were chemically modified by linking HEC with fluorocarbon-containing modifiers through an ether linkage (*15*). The degree of substitution is controlled by reaction extent. The

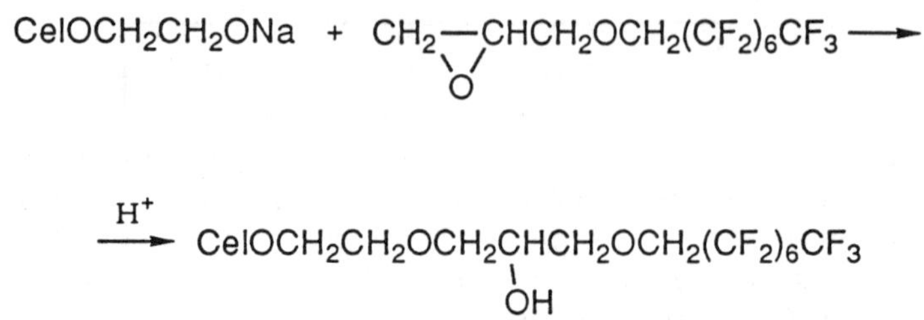

Scheme 1.

synthetic route is shown in Scheme 1. PAM-based copolymers were synthesized by copolymerizing acrylamide monomers with fluorocarbon-containing acrylate (e.g., 3M monomer FX13) or methacrylate comonomers (*12*). Typical fluorocarbon comonomers are shown as structures **1** and **2**. The hydrophobe content incorporated into the chain was controlled by the ratio of the comonomers. One difficulty common to both routes is that the fluorocarbon-containing monomers usually have very poor solubility in water because of their extreme hydrophobicity. The hydrophobic comonomer can, however, be dispersed by adding a fluorocarbon surfactant or by using a mixed-solvent system such as water–acetone.

A major difficulty associated with these materials is ascertaining the degree of hydrophobic comonomer incorporated into the polymer after synthesis. Because the content of the hydrophobic comonomer is exceedingly low, no method has been completely successful in determining the hydrophobe content in the polymer or the hydrophobe distribution in the polymer chain. Schulz et al. (*16*) addressed this problem in their studies of hydrocarbon associating polymers by incorporating a phenyl ring in their hydrophobic unit to allow direct measurement of comonomer content by UV spectroscopy. Fluorescence techniques using pyrene tags have also been used to probe the microstructures of hydrophobic associations, especially at low concentrations (*17*).

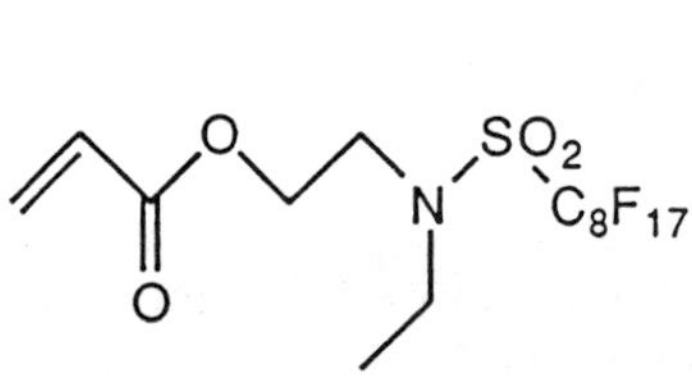

Structure **1.** *FX13, 2-(N-ethylperfluorooctanesulfoamido)ethyl acrylate.*

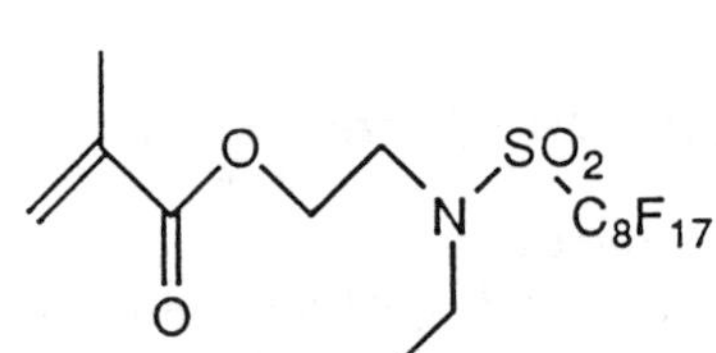

Structure **2.** *FX14, 2-(N-ethylperfluorooctanesulfoamido)ethyl methacrylate.*

With fluorine-containing polymers, ^{19}F NMR spectroscopy is a natural possibility for determining the hydrophobe content because of the natural abundance (100%) and relative sensitivity (83.3) of fluorine. However, although ^{19}F NMR spectroscopy shows spectra of comonomers with well-resolved peaks, ^{19}F NMR absorption for the copolymer is extremely broadened (*18*). Although the direct determination of hydrophobe content is prohibited, monitoring the loss of the sharp monomer peaks during the reaction allows us to measure the incorporation rate and thus the hydrophobe content. The broadening reflects changes in the environment surrounding the fluorine hydrophobic units, and thus we anticipate that future work will use NMR spectroscopy to probe the microstructures of associating junctions.

Viscoelasticity. ***Shear Rate.*** A typical feature of associating polymers is a strong shear dependence of viscosity. Figure 2 shows the effect of shear rate on the measured viscosity for an associating copolymer of acrylamide and fluorocarbon-containing acrylate with a molecular weight on the order of 10^6. A very strong shear-thinning effect is observed as the viscosity of a 300-ppm solution drops from 2500 to 400 cP over a shear rate range of only one decade. This effect is also polymer concentration dependent. At lower concentrations, shear thinning is much less prominent. Similar results are seen for associating copolymers of *N,N*-dimethylacrylamide and fluoroacrylate (*19*) and for copolymers with hydrocarbon associating units (*20*). The strong shear thinning is attributed to the destruction of intermolecular

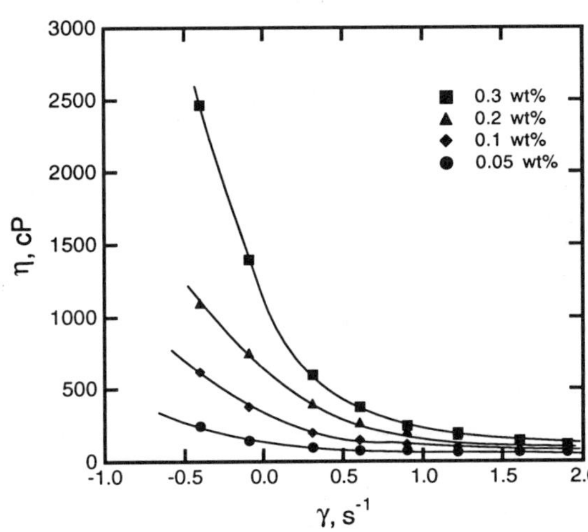

Figure 2. Effect of shear rate ($\dot{\gamma}$) on solution viscosity of copolymer of acrylamide and 0.07 mol% FX13 at various solution concentrations.

association junctions. The suppression of shear thinning at low concentrations may therefore reflect both the overall decrease in associations and a decrease in the proportion of intermolecular versus intramolecular associations. For such a complex system as these random copolymers, it is difficult to separate these effects.

Comonomer Content. The comonomer content of associating polymers plays an important role in their viscosifying ability. For all our systems, if solution viscosity at constant polymer concentration is plotted versus the comonomer content for polymers synthesized under conditions varying only in comonomer addition, a maximum is observed, as seen in Figure 3 (*21*). A complete incorporation of comonomers has been confirmed by the ^{19}F NMR technique, and viscosities are thus plotted against the contents of fluorocarbon acrylate or fluorocarbon methacrylate comonomers. Curves with maxima at intermediate comonomer contents are obtained for these and other fluorocarbon copolymers. This maximum is consistent with the formation of an extended network of intermolecular associations favored by the increasing comonomer content. The network is eventually overcome by a preference for intramolecular associations when the comonomer content becomes very high. Chains or clusters of chains collapse under the effect of the association junctions.

Several other important points can be illustrated by Figure 3. First, the viscosity varies as much as 2 orders of magnitude, meaning that

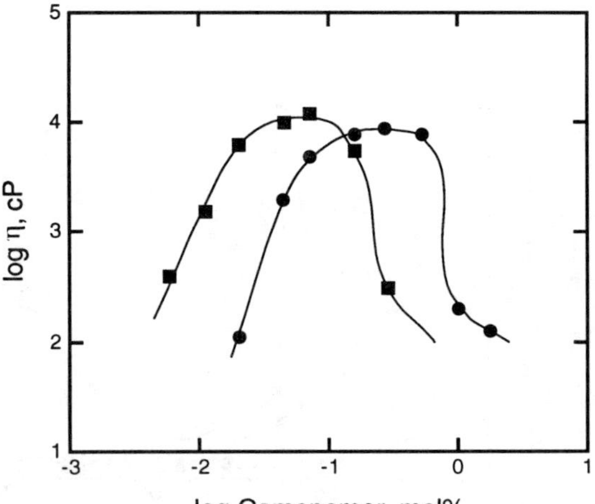

Figure 3. Viscosity of PAM copolymers versus content of fluorocarbon-containing acrylate (■) or methacrylate (●) comonomers at 25 °C. Measurements were made at 0.5 wt% concentration with 0.4-s^{-1} shear rate.

care must be taken to synthesize a copolymer with maximum viscosity. Second, comonomer content at the viscosity maximum is as little as 0.1 mol%, showing the astonishing associative power of fluorocarbon-containing polymers. Third, the peak position changes as different comonomers are used. These changes imply that the interplay of inter- and intramolecular associations depends on the type of hydrophobic comonomer.

Spacer. One structural parameter that alters the association is a variable-length hydrophilic spacer placed between the hydrophobic unit and the hydrophilic main chain. A series of comonomers with nonionic hydrophilic spacers of ethylene oxide units ($-CH_2CH_2O-$) placed between the fluorocarbon chain and the acrylate were synthesized. The comonomers $CH_2=CHCO(OCH_2CH_2)_nCH_2C_7F_{15}$ (n = 1, 2, or 3) were prepared by reacting $CF_3(CF_2)_6CH_2ONa$ with $Cl(CH_2CH_2O)_n-H$ in a mixture of toluene–diglyme (70/30, vol/vol) to give $CF_3(CF_2)_6CH_2(CH_2CH_2O)_n-H$. This compound was then reacted with acryloylchloride to give the corresponding acrylates. These were copolymerized with acrylamide (22). Figure 4 shows a plot of viscosity versus comonomer content for several different spacer lengths (one, two, or three). The viscosity increases dramatically as the hydrophilic spacer length increases, especially with the introduction of the first unit. Similar results were obtained on hydrocarbon associating polymers, and these results have been interpreted as dem-

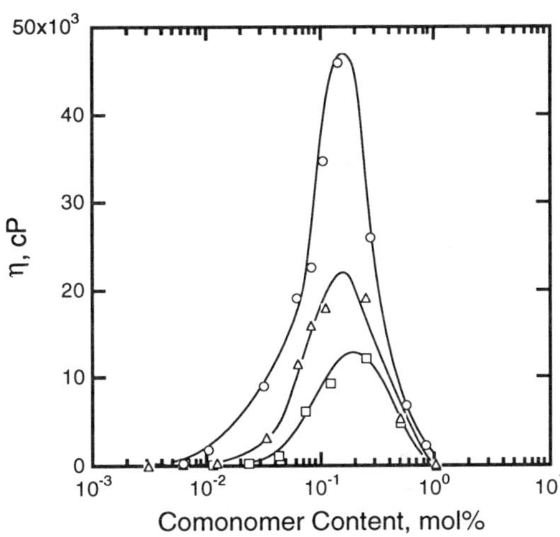

Figure 4. Viscosity versus hydrophobe content for PAM copolymers with one (□), two (△), or three (○) ethylene oxide spacer units. Measurements were made at 0.5 wt% concentration with 0.4-s^{-1} shear rate.

onstrating enhanced formation of micellar structures (*23*). In that case, a charged moiety in the spacer group was used with the goal of increasing the driving force toward hydrophobic association.

Several factors may be responsible for the viscosity enhancement. First, the spacer may increase accessibility of the hydrophobes by decreasing the steric hindrance to association. This increased accessibility would allow denser, more compact, micellelike associations that are stronger and lead to higher viscosities. A second factor could be a decoupling of the motions of associating junctions from the main chain dynamics. To the extent that the solution viscosity of associating polymers is determined by the long lifetime of the junctions, this decoupling will reduce the influence of the faster chain dynamics on junction dynamics. There may also be questions regarding the distribution of hydrophobes along the polymer chain. This distribution can also lead to changes in viscosity, and because changing the spacer length may change the kinetics during synthesis, it will be important to characterize the distribution of the hydrophobes. Alternatively, preparation of regular combs with these spaced hydrophobe comonomers would address this question.

Salt. The addition of salt steadily increases the viscosity of fluorocarbon associating polymer solutions. This same effect is seen with hydrocarbon counterparts. A typical plot is shown in Figure 5 (*12*). Here the viscosity of a fluorocarbon copolymer is plotted against the concentration of added NaCl for several polymer concentrations. The

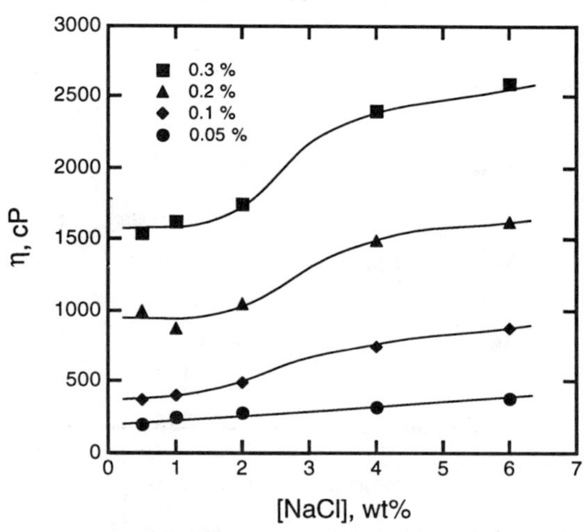

Figure 5. Viscosity versus salt concentration for PAM copolymers in aqueous NaCl solutions at different polymer concentrations.

salt effect increases as the polymer concentration increases, although it also appears that until the salt concentration reaches about 2%, it has little effect on the viscosity of the solution. It has been argued that salt enhances hydrophobic interactions by making the solvent more polar (*11*). This action is consistent with the effect of salts on the solubility of hydrocarbon molecules in water, where the solubility decreases noticeably when the salt concentration exceeds a certain value (*11*).

Surfactant. Upon the addition of surfactants, the solution viscosity of fluorocarbon associating polymers initially increases, passes through a maximum, and then decreases to low levels, as shown in Figure 6. This particular example shows the effect of fluorocarbon surfactant on the viscosity of a solution of an associating copolymer with fluorocarbon hydrophobes (*22*). The viscosity at a high surfactant concentration is actually lower than the initial viscosity with no surfactant present, and this observation is consistent with results for hydrocarbon hydrophobes (*24*). The effect of surfactants arises from interactions between the hydrophobic unit of the polymer and the hydrophobic part of the surfactant molecule. As a result, the associating polymer interactions are altered by the surfactant (*24*).

At small doses, surfactant molecules are incorporated into existing polymeric micelles. Whether surfactants promote the formation of additional associating junctions is not clear. However, the incorporation of surfactants into the junctions stabilizes the associations and thus

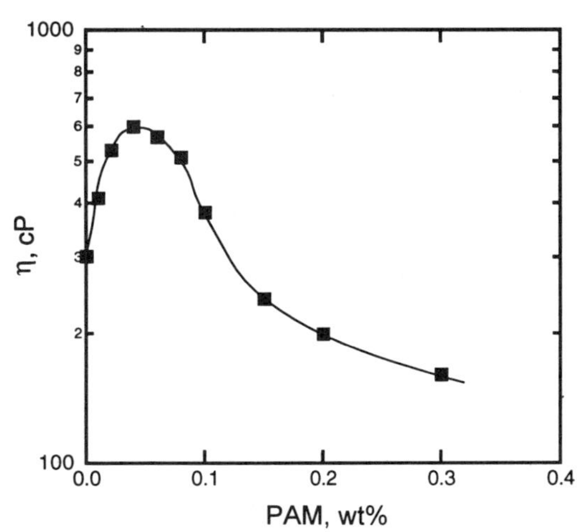

Figure 6. Viscosity versus added surfactant for 0.25 wt% PAM copolymer solution.

increases their lifetimes. With the formation of more stable micelles, an increase in solution viscosity is expected even if the number of associating junctions does not increase (*6*). As the concentration of surfactant increases to a point at which there are sufficient surfactant molecules to form stable micelle structures with individual hydrophobic units, the hydrophobes separate and exit surrounded by surfactant molecules. As a result, the polymer associations break up, and solution viscosity decreases. In the extreme case when all associating junctions are broken by a high concentration of surfactant, the solution viscosity would be similar to that of polymers without any association. With this as the basis for a technique, single-chain molecular weights have been measured by light scattering or intrinsic viscosity for polymers that would otherwise be aggregated by their strongly associating fluorocarbon or hydrocarbon units (*25, 26*). Nevertheless, the surfactants change the nature of the system and may even interact with the polymer chain itself (*27, 28*). This method should be applied with caution.

Solvent. The addition of an appropriate organic solvent to the aqueous solutions of fluorocarbon associating polymers causes a dramatic decrease in solution viscosity. Figure 7 shows how the viscosity of a fluorocarbon associating polymer solution changes as a function of acetone content in a water–acetone mixed solvent (*29*). Initially, the viscosity decreases slowly, but it shows a rapid downturn as the acetone content increases further. An increase in acetone content from 4 to 14% decreases the solution viscosity by almost 2 orders of magni-

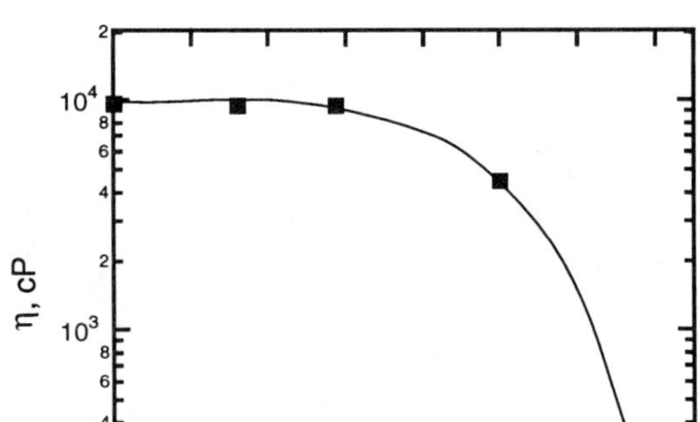

Figure 7. Effect of acetone on solution viscosity of hydrophobically modified associating PAM copolymer.

tude. Mixed solvents of water and dimethyl sulfoxide can also break up associations of acrylamide-based fluorocarbon copolymers. In the presence of 50% dimethylformamide, the reduced viscosity of the solution drops dramatically, a result indicating that few associations exist at such levels of added solvent (*22*). The organic solvent probably acts similarly to such hydrophobic bond breakers as urea to break up the association junctions and dramatically decrease the viscosity. Similar experimental results have been obtained on hydrocarbon-containing associating polymers (*26*) and have been used to determine the molecular weights of associating polymers. As long as the cosolvent is a good solvent for the chain backbone, use of the cosolvent is preferable to use of small-molecule surfactants for measuring isolated chain properties. In addition to maintaining the chain solvation, the cosolvent does not introduce the complication of small surfactant micelles.

Temperature. The effect of temperature on viscosity is not straightforward for fluorocarbon associating polymers, as Figure 8 (*22*) shows. Similar behavior is seen in other fluorocarbon associating polymer systems (*29*) and is reproducible. The viscosity first decreases with increasing temperature and then, around 60–80 °C, increases, only to decrease again as the temperature goes higher. An increase in viscosity with temperature may be viewed as consistent with entropically driven hydrophobic bonding, but a decrease in viscosity with temperature is usually associated with an enthalpic effect. To separate

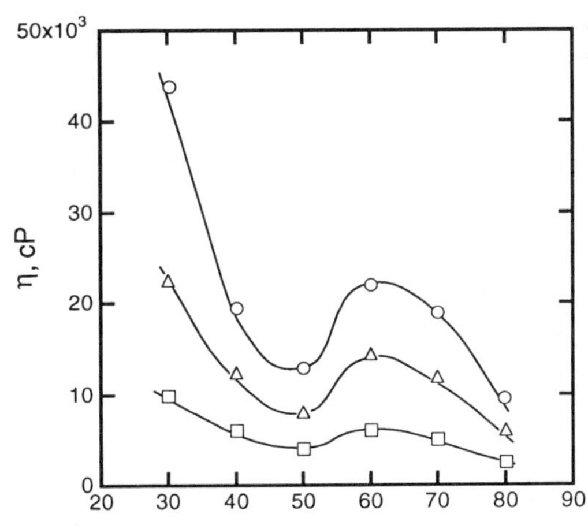

Figure 8. Viscosity versus temperature for copolymers with one (□), two (△), or three (○) ethylene oxide spacer units in the hydrophobic comonomer at 0.5 wt% concentration with 0.1-s^{-1} shear rate.

two regions of decrease by a region of increase, another explanation must be invoked. Lundberg and Makowski (*30*) obtained similar results on the temperature dependence of viscosity for ionomers in a mixed-solvent system. Their explanation was that the first region shows a viscosity decrease due simply to the viscosity of the solvent. The decrease in the second region is primarily due to cosolvent–polymer interactions. In the third region, the ionic interactions weaken as the temperature increases sufficiently. This unusual temperature dependence is seen only for mixed-solvent systems and random fluorocarbon copolymers. Studies of hydrocarbon associating polymers show a consistent decrease in viscosity with increasing temperature (*10, 14*). A suitable explanation for the difference is still lacking.

The characterization of hydrophobic associations as entropically driven leads to an expectation that as temperature is increased, associations will be favored, and if viscosity is characterized by the number of associations, this favoring will result in an expected increase in viscosity. In this model, viscosity would increase as temperature increases, contrary to the usual result and the Arrhenius description. In fact, the viscosity decrease with temperature increase presents a problem for the hydrophobic model. We can resolve this inconsistency by considering that although the number of associations may increase by entropic forces as temperature increases, the strength of these associations will nevertheless decrease. According to the transient-network theories of Green and Tobolsky (*5*) extended to associating polymers by Tanaka and Edwards (*6*), the lifetime of the associations determines viscosity as long as chain dynamics are rapid enough to reequilibrate chains as stickers migrate from one association junction to another. Lifetime is evidently the overriding factor suppressing the effect of the number of associations. One can also imagine these two influences leading to complex temperature dependencies in some cases. The random copolymers are probably not the appropriate model system for testing these ideas.

Light Scattering. Light scattering is a natural choice for characterizing polymers in general, but for the associating polymers, this method is quite difficult to employ. The high viscosities make it difficult to clean the solutions either by filtration or by centrifugation. For the fluorocarbon hydrophobes, these problems are compounded by the strength of the interactions and the fact that even for the best cosolvent and the lowest concentrations, significant intermolecular associations persist. In a recent paper, Seery et al. (*25*) investigated three such copolymers with hydrophobic contents of 0.007, 0.07, and 0.7 mol% by using a combination of static and dynamic light scattering. For the lowest hydrophobic content (0.007%), dynamic light scattering is essentially the same as for a homopolymer with no hydrophobes. Static scattering, however, shows a negative osmotic second virial coefficient, A_2, indicative of attractive interactions between these chains that have on average only 1.3 hydrophobes per chain.

At a hydrophobe level of 0.07%, solutions at typical 0.1% concentrations are extremely viscous and the dynamic scattering at concentrations down to 10 ppm shows large aggregates together with a few small species that are apparently collapsed single chains. The aggregates have 300-nm hydrodynamic radii and do not dissociate upon addition of cosolvents or increase in temperature. Addition of a fluorocarbon surfactant disperses the aggregates and makes it possible to characterize the polymer chains and determine their apparent molecular weight as 6.5×10^6 and their hydrophobe content as 68 per chain. One interesting observation is that the apparent A_2 for the aggregated species is positive, a result suggesting a structure of hydrophobes buried within the hydrophilic chains.

Increasing the hydrophobe content to 0.7% did not change the scattering behavior much from that of the 0.07% samples. At high concentrations (greater than 1000 ppm), the dynamic scattering collapses to a single mode, and a comparison of the radius of gyration to the hydrodynamic radius indicates that the clusters become increasingly dense. Newly added chains apparently fill in around aggregated ones and thereby cause the volume of the cluster to grow more slowly than its mass. The semidilute overlap concentration is never attained.

Polymers with Regular Structures

As mentioned, the synthesis of random associating copolymers is straightforward, but the distribution of hydrophobic units and the hydrophobe content in polymer chains remain unresolved. Many studies of associating polymers have avoided the problem and implicitly assumed a random incorporation of hydrophobic units into the polymer chain. Such an assumption, however, is not always valid. For hydrocarbon-associating polymers, McCormick and coworkers (*31*) argued that hydrophobic units form blocky structures and that the length of the blocky structure is related to the surfactant used during the emulsion polymerization. The blocky structures of the hydrophobe that arise by different synthetic routes in turn affect its rheological behavior (*20*). Because of the uncertainty regarding the distribution and content of hydrophobes in the random copolymers, a systematic study is extremely difficult, and the results may not be consistent between studies.

To circumvent this difficulty, we studied associating polymers with two types of well-defined structures: telechelic and comb. Telechelic polymers have a hydrophobic unit attached at each end of a PEO hydrophilic chain. Combs are composed of fluorocarbon stickers spaced regularly along a chain of coupled PEO units. For telechelic polymers, monodispersed PEO chains are end-capped by a fluorocarbon hydrophobic unit, and for comb polymers, the hydrophobe becomes part of a linking unit that is used in a condensation reaction of the PEO prepolymer chains. The telechelic polymers simply have two hydrophobes per chain on the ends. The combs have a distribution in overall chain molecular weight, but on every chain the hydrophobe spacing is well defined by the length of the PEO prepolymer. Telechelic polymers have often been used as model associating polymers, as illustrated in the rheological studies of Jenkins (*32*). Care must always be taken to ensure that both ends are capped with hydrophobic units.

The focus of our study is the temperature dependence of viscosity for solutions of each type of polymer structure. We are also interested in how viscosity and network-forming ability differ between telechelic and comb polymers and between polymers with similar structures but different hydrophobe lengths and spacing.

Theory. Theoretical studies of the effects of reversible associations on rheological properties were pioneered by Green and Tobolsky (*5*). Their theory is based on an extension of classical rubber elasticity theories to transient networks formed by entanglements or breakable physical bonds. It predicts that the steady shear viscosity $\eta(\dot{\gamma})$ equals the zero shear viscosity $\eta(0)$

$$\eta(\dot{\gamma}) = \eta(0) = \tau G_{\infty}$$

where the relaxation time τ is the reciprocal of the bond-breaking and bond-reformation rate, and G_{∞} is the high-frequency storage modulus given by

$$G_{\infty} = \nu kT$$

where ν is the number density of elastic chains and kT is thermal energy.

Tanaka and Edwards (*6*) expanded the Green–Tobolsky theory. They assumed that the dissociation of an end group from a junction can be approximated as an activation process characterized by a bonding potential with well depth E_m. In the Green–Tobolsky limit, that is, when the total number of junctions is unaffected by the shear, the association breakage rate β_0 can be expressed by

$$\beta_0 = \omega_0 e^{-E_m/kT}$$

where ω_0 is a characteristic frequency of thermal motion that is estimated to have a typical value on the order of 10^8–10^9 Hz. The lifetime of association τ_x is the reciprocal of β_0:

$$\tau_x = \omega_0^{-1} e^{E_m/kT}$$

From this theory the zero shear viscosity is

$$\eta_0 = f_N \frac{\nu_0 kT}{\beta_0}$$

where f_N is an unessential prefactor on the order of unity. Thus the static viscosity η_0 takes the Green–Tobolsky form, but relaxation time is controlled by the lifetime of the association τ_x, which depends exponentially on temperature. The important conclusion is that viscosity will decrease exponentially as the temperature increases because the bond breakage frequency β_0 depends exponentially on temperature.

Viscoelasticity. ***Shear.*** As already stated, solutions of associating polymers are characterized by strong shear dependence. In this study we have eliminated this complication by measuring zero shear viscosity in order to study the dynamics and make comparisons with theory. Low shear rate viscosities are measured with a magnetic sphere viscometer built in this laboratory. In this instrument, a 1-mm magnetic sphere is floated in the sample solution by using a magnetic field from an external field coil. The current to the coil is controlled by a computer that is also interfaced to a quadrant photodetector that images the position of the sphere. As the sample cell containing the solution and the floating sphere is translated vertically, the additional current to the field coil supplied to keep the sphere in position reflects the force of the solution on the sphere. Shear rates are easily controlled by the velocity of the sample cell translation. The working shear rate of the magnetic sphere viscometer is usually in the range of 3×10^{-3} to 4 s^{-1}, with higher shear rates for low-viscosity samples and lower shear rates for high-viscosity samples. The working range was sufficient to ensure that no shear rate dependence of viscosity was observed for any of the solutions studied. Thus all viscosity data that follow are zero shear values.

Surfactant. In methods described in the previous section, fluorocarbon surfactants were used to break associations. In the method de-

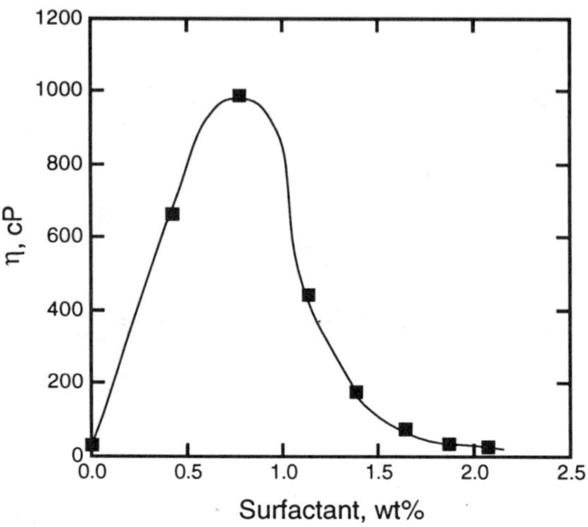

Figure 9. Viscosity versus concentration of added SDS surfactant for 2.15 wt% solution of comb polymer with $-C_4F_9$ hydrophobes.

scribed here, a hydrocarbon surfactant, sodium dodecyl sulfate (SDS), is used to alter the association. Upon the addition of SDS to a solution of our comb polymers, the viscosity initially increases, passes through a maximum, and then decreases to low levels, as shown in Figure 9. The results are very similar to the effects of the fluorocarbon surfactant as discussed. The surfactant is apparently building stronger associations, which are then dispersed by excess surfactant.

Activation Energy. Viscosities of comb and telechelic polymer solutions decrease dramatically as the temperature is increased from 10 to 60 °C. The natural logarithms of viscosities of several comb polymers with different lengths of hydrophobic units and different polymer molecular weights are plotted against inverse absolute temperature in Figure 10. The polymer concentrations are fixed at 3.7 wt%. In contrast to the temperature dependence of viscosity for the random associating polymers discussed in the first section, good linear relationships are observed for all the comb samples studied. From the slopes of these lines, apparent activation energies for viscous flow can be derived by using the Arrhenius relationship. The obviously steeper slopes for the larger hydrophobes demonstrate their higher activation energies. The viscosities of telechelic polymer solutions also were measured as a function of temperature. As for the comb polymers, the Arrhenius plots show good linear relationships (Figure 11).

The fact that both comb and telechelic polymers show an Arrhenius relationship of temperature and viscosity is evidence that a single

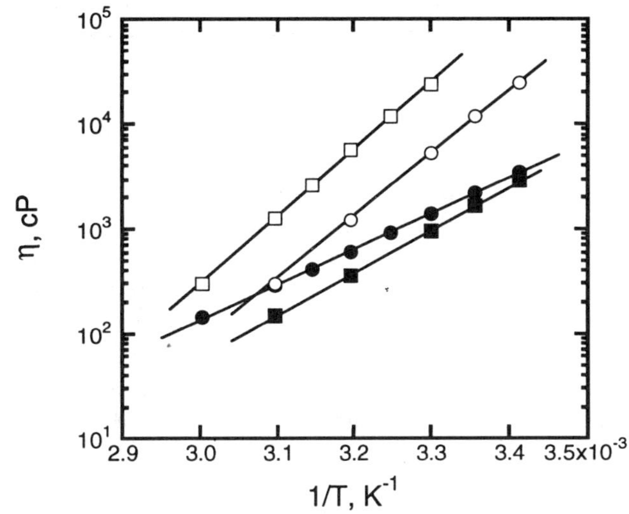

Figure 10. Viscosity versus inverse temperature for comb polymers at 3.7 wt% concentration with different spacing molecular weights and hydrophobic units. Solid symbols are for $-C_4F_9$ hydrophobe, and open symbols are for $-C_8F_{17}$ hydrophobe.

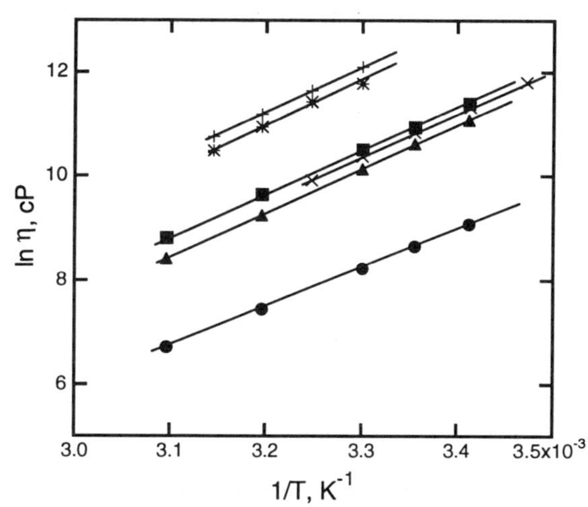

Figure 11. Viscosity versus inverse temperature for several telechelic polymers with different molecular weights (4,600 (top) to 15,000 (bottom)) and hydrophobic units ($-C_7F_{15}$ (solid symbols) and $-C_8F_{17}$) at 7.8 wt% concentration.

process underlies the relaxation kinetics of these fluorocarbon associating polymers. In other words, the long lifetime of association bonding determines the dynamics of the solutions that lead to the high viscosity of the associating polymers. Furthermore, the barrier potential model of Tanaka and Edwards seems to apply to both comb and telechelic polymers, even though the theory was originally developed to explain the behavior of telechelic polymers.

Data for apparent activation energies E_m for several comb polymers with different length hydrophobes are plotted in Figure 12 as a function of polymer concentration. For comparison, the activation energies are shown on the same graph for solutions of two PEO without hydrophobes. In the low-concentration range, E_m for associating polymers initially increases as concentration increases. When the polymer concentration becomes greater than about 3 wt%, E_m reaches a plateau. Clearly, the activation energies of comb polymers with $-C_8F_{17}$ hydrophobic units are much greater than those of comb polymers with $-C_4F_9$ hydrophobic units. Activation energies in the plateau region mainly depend on hydrophobe size. Neither the total polymer molecular weight nor the molecular-weight spacing between the hydrophobes significantly affects these energies. Along the plateau, the activation energy increases by about 13 kcal/mol from the homopolymer to the $-C_4F_9$ hydrophobe and again by the same amount for the

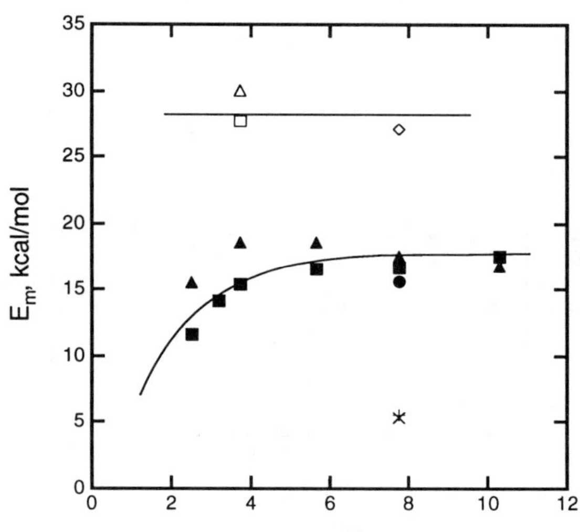

Figure 12. Activation energy versus concentration for comb polymers of differing molecular weight with $-C_4F_9$ hydrophobe (solid symbols), $-C_8F_{17}$ hydrophobe (open symbols), or no hydrophobe (cross symbols).

$-C_8F_{17}$ hydrophobe. Each $-CF_2-$ group apparently contributes approximately 3 kcal/mol to the activation energy for viscous flow.

Telechelic polymers with $-C_8F_{17}$ hydrophobic units have activation energies of about 17 kcal/mol, which is equivalent to about $29kT$ at 298 K. This value is comparable to the activation energy value of 16 kcal/mol measured by Annable et al. (*10*) and Hu (*14*) for telechelic polymers with $-C_{16}H_{33}$ hydrophobic units. This result shows again that the associations formed by fluorocarbon hydrophobic units are much stronger than those formed by hydrocarbons of the same lengths.

The behavior of the activation energy of comb polymer solutions in relation to concentration includes reaching a plateau at high concentrations and then gradually decreasing upon dilution. The plateau value of the activation energy for the comb polymers at high concentration can be explained according to the Tanaka–Edwards theory. The activation energy, in contrast to viscosity itself, is an intensive quantity and should not depend on the amount or density of hydrophobe units. Changing the spacing of hydrophobic units along the polymer backbone or changing the total molecular weight affects the viscosity of the solution (*14*). However, the chain dynamics do not alter the strength of the association, and hence the activation energy of the viscous flow in this region remains controlled by the association lifetime. E_m thus characterizes the strength of associations.

As the concentration decreases, the data show that E_m also decreases gradually. We can understand this decrease in the effective interaction strength by considering that below some concentration, it will be more probable that multiple stickers from the same chain will meet in the same association junction. These looped structures do not contribute to the transient network, and their dissociation rate does not affect the solution viscosity to the same extent as do the active chain junctions. In systems in which the solvent is of marginal quality for the chain backbone, one could even suspect that raising the temperature might expand the chains and enhance the probability of effective chain junctions. This behavior would yield a larger apparent E_m.

Although self-looping of associating polymers has been demonstrated in very dilute solutions (*17*), no direct evidence of loop formation at relatively higher concentrations has been reported. The practical difficulties of such experiments are enormous; however, simple entropic considerations argue for a transition from a network composed predominantly of loops at low concentration to one that is mainly links at high concentration. This transformation is driven with the entropy loss penalty being less severe for associations formed through intermolecular junctions than for those formed through intramolecular looping. A similar entropic loss penalty has been discussed (*33, 34*) in the context of the micellization of ABA triblock copolymers in a solvent selective for the center block. The comb polymers are actually more likely to form intramolecular loops, because a single comb polymer with multiple stickers has multiple opportunities to form thermodynamically stable intramolecular junctions.

This role of self-looping is further evidenced in the basic viscosity–concentration studies. One prediction from the Tanaka model is $\eta \sim c$. Figure 13 shows an example of the concentration dependence of viscosity for a telechelic polymer. The concentrations of the solutions were considerably below the entanglement regime for linear chains; nevertheless, the viscosity increased enormously with concentration. If one were to describe the viscosity with a power law, it would be nearly c^6 at low concentration and would decrease to about c^3 as concentration increased. The same is true for comb polymers, although the slopes are as high as c^9. Annable et al. (*10*) pointed out that this deficiency in the model could be corrected by considering intramolecular loops leading to inactive chains. Loops are favored at low concentrations, and thus the results in Figure 13 can be interpreted as demonstrating the conversion of inactive loop chains to active network chains as well as the overall addition of more chains to the network as concentration increases.

However, several observations indicate that self-looping is not sufficient to explain the data. The activation energies for telechelic polymers were calculated and are plotted as a function of molecular weight in Figure 14. The concentrations of the polymers were 7.8%. The plot shows that the apparent activation energies depend not only on hydrophobe length but also on molecular weight for the telechelic-type

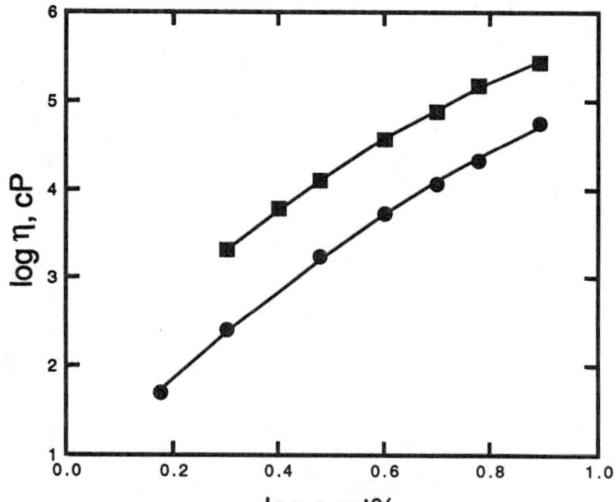

Figure 13. Viscosity versus concentration for telechelic polymers with M_w *= 8000 and hydrophobic unit* $-C_7F_{15}$ *(●) or* $-C_8F_{17}$ *(■).*

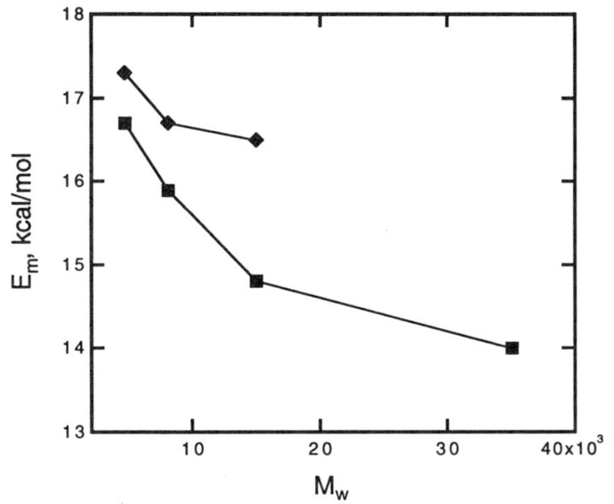

Figure 14. Activation energy versus M_w *for telechelic polymers with hydrophobic unit* $-C_8F_{17}$ *(◆) or* $-C_7F_{15}$ *(■) at 7.75 wt% concentration.*

polymers. Longer hydrophobic units are characterized by larger activation energies for the same polymer molecular weight, a trend that is consistent with that for the comb polymers. However, for telechelic polymers having the same hydrophobic units, longer chains have lower activation energies. Furthermore, a comparison between the activation energies of telechelic polymers and of comb polymers carrying the same hydrophobic unit ($-C_8F_{17}$) is shown in Figure 15. Clearly, the apparent activation energy for viscous flow of telechelic polymers is substantially lower than that for comb polymers.

It has been assumed that the activation energy reflects the strength of association regardless of the molecular structure, that is, either comb or telechelic. The activation energies of telechelic polymers and those of the comb polymers should be the same when the same hydrophobic unit is used and the concentrations are sufficiently high. What is found is that the activation energies of the telechelic polymers with $-C_8F_{17}$ hydrophobes are comparable to those of the comb polymers with $-C_4F_9$ hydrophobes. From the assumptions of the model, polymer molecular weight should not affect the strength of telechelic chain associations, but instead we see in Figure 14 a systematic decrease as the telechelic molecular weight increases.

Two simple arguments for the effect of molecular weight on association strength lead to opposite expectations. If looping were a factor, the shorter chains would be expected to loop more and thus show a lower apparent E_m. Likewise, one could imagine chains being stretched to form associations and the restoring force from the chain

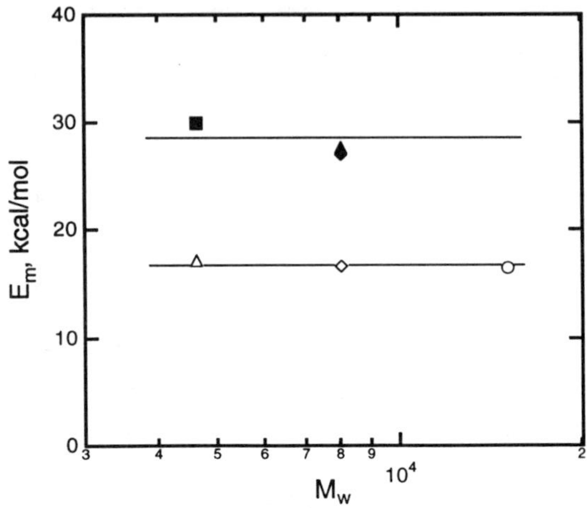

Figure 15. Activation energy versus spacing molecular weight for comb polymers (top line) or telechelic polymers (bottom line). Both have a $-C_8F_{17}$ hydrophobic unit.

extension weakening the association. The effect would also be more substantial for shorter chains.

A different explanation is necessary, and here we focus on the fundamental difference between telechelic and comb or random copolymers. If associations occur by strictly pairwise interaction, then telechelic polymers can never form a network. Only when the two ends of a telechelic chain are associated into different junctions of three or more units from different chains does the chain become an active chain that can contribute to the formation of the network. Since telechelic polymers have only two associations, many of their junctions may simply extend the chain length. This extension leads to an increase in viscosity that is much less than that produced by networked chains. As the temperature increases, the enhanced hydrophobe dissociation rate appears as a decrease in chain molecular weight. This decrease in turn decreases the viscosity but to a much smaller extent than does the breakup of a network. For comb or random polymers, on the other hand, even pairwise interactions lead to networking of chains. The main result is that associating units of comb polymer chains are much more efficient than those of telechelic chains in terms of networking. This difference will be especially important when a substantial fraction of the associations are pairwise.

As associations increase to more than pairwise, the linearly associated telechelic polymers become equivalent to comb polymers, in which the internal junctions are "preassociated" and each additional association produces active strands. Thus the hydrophobes of low-molecular-weight telechelic polymers become relatively more important because a larger fraction of their interactions are likely to produce active network junctions. This model also explains Annable's observations from measurements of storage moduli that the active chain fraction is apparently higher for low-molecular-weight telechelic polymers than for high-molecular-weight ones (*10*).

Conclusion

In general, the properties of fluorocarbon-containing associating polymers are similar to those of associating polymers with hydrocarbon hydrophobes. The main difference is that the fluorocarbon associations are much stronger than the hydrocarbon ones, and thus the effects of the associations are magnified. For random copolymers of high molecular weight, the influences of hydrophobe linking to the backbone and of additives such as salt, cosolvent, and surfactants were considered. Studies of comb and telechelic polymers have given us access to variables of chain length, spacing of hydrophobes along the chain, and size of hydrophobic unit. In general, the effectiveness of the hydrophobic associations in increasing viscosity is reduced by intramolecular loops that do not yield network strands, or, in the telechelic polymers, by limited pairwise interactions that serve only to extend chains. The data for concentration, molecular weight, and temperature dependence on the regular copolymers all support the model proposed by Annable et al. that includes the effect of loops as an extension of the Tanaka and Edwards theory for transient associating networks. One additional effect that should be considered in the models is the distinction that telechelic chains require more than pairwise association junctions in order to participate fully in network formation. Comb or random copolymers are inherently more effective.

Acknowledgments

This work was supported by the U.S. Department of Energy Office of Basic Energy Sciences and by Applied Biosystems, Inc. We are extremely grateful to Steve Menchen of Applied Biosystems for providing the telechelic and comb polymers.

References

1. Broze, G.; Jerome, R.; Teyssie, P.; Marco, G. *Macromolecules* **1983,** *16,* 996.

2. Agarwal, P.; Garner, R. T.; Lundberg, R. D. *Macromolecules* **1984,** *17,* 2794.
3. Shaw, K. G.; Leipold, D. P. J. *J. Coatings Technol.* **1985,** *57,* 63.
4. Evani, S.; Rose, G. D. *Polym. Mater. Sci. Eng.* **1987,** *57,* 477.
5. Green, M. S.; Tobolsky, A. V. *J. Chem. Phys.* **1946,** *14,* 80.
6. Tanaka, F.; Edwards, S. F. *J. Non-Newtonian Fluid Mech.* **1992,** *43,* 247.
7. Baljon-Haakman, A.; Witten, T. A. *Macromolecules* **1992,** *25,* 2969.
8. Brown, G.; Chakrabarti, A. *J. Chem. Phys.* **1991,** *96,* 3251.
9. Wang, S. Q. *Macromolecules* **1992,** *25,* 7003.
10. Annable, T.; Buscall, R.; Ettelaie, R.; Whittlestone, D. *J. Rheol.* **1993,** *32,* 695.
11. Tanford, C. *The Hydrophobic Effect: Formation of Micelles and Biological Membranes;* John Wiley: New York, 1973.
12. Zhang, Y.-X.; Da, A.-H.; Hogen-Esch, T. E.; Butler, G. B. *J. Polym. Sci., Polym. Lett.* **1990,** *28,* 213.
13. Mathis, D.; Leempoel, P.; Ravey, J.-C.; Selve, C.; Delpuelch, J.-J. *J. Am. Chem. Soc.* **1984,** *106,* 6162.
14. Hu, N. Ph.D. Thesis; University of Southern California, 1994.
15. Hwang, F. S.; Hogen-Esch, T. E. *Macromolecules* **1993,** *26,* 3156.
16. Schulz, D. N.; Kaladas, J. J.; Maurer, J. J.; Bock, J.; Pace, S. J.; Schulz, W. W. *Polymer* **1987,** *28,* 2110.
17. Duhamel, J.; Yekta, A.; Hu, Y. Z.; Winnik, M. A. *Macromolecules* **1992,** *25,* 7024.
18. Yassini, M.; Hogen-Esch, T. E. *Polym. Mater. Sci. Eng.* **1992,** *33,* 933.
19. Xie, X.; Hogan-Esch, T. E. *Polym. Mater. Sci. Eng.* **1994,** *35*(1), 498.
20. Hill, A.; Candau, F.; Selb, J. *Macromolecules* **1993,** *26,* 4521.
21. Hogen-Esch, T. E.; Yassini, M.; Zhang, Y.-X.; Hwang, F.; Amis, E. J.; Seery, T. *Polym. Mater. Sci. Eng.* **1990,** *31,* 460.
22. Hwang, F. Ph.D. Thesis; University of Southern California, 1993.
23. Peiffer, D. G. *Polymer* **1990,** *31,* 2353.
24. Biggs, S.; Selb, J.; Candau, F. *Langmuir* **1992,** *8,* 838.
25. Seery, T.; Yassini, M.; Hogen-Esch, T. E.; Amis, E. J. *Macromolecules* **1992,** *25,* 4784.
26. Gelman, R. A.; Barth, H. G. In *Water Soluble Polymers: Beauty with Performance;* Glass, J. E., Ed.; Advances in Chemistry 213; American Chemical Society: Washington, DC, 1986; pp 101–110.
27. Goddard, E. D. *Colloids Surf.* **1986,** *19,* 255.
28. Tadros, Th. F. *J. Colloid Interface Sci.* **1980,** *74,* 196.
29. Zhang, Y.-X.; Da, A.-H.; Butler, G. B.; Hogen-Esch, T. E. *J. Polym. Sci., Polym. Chem.* **1992,** *30,* 1383.
30. Lundberg, R. D.; Makowski, H. S. *J. Polym. Sci., Polym. Phys. Ed.* **1980,** *18,* 1821.
31. Ezzel, S. A.; Hoyle, C. E.; Creed, D.; McCormick, C. L. *Macromolecules* **1992,** *25,* 1887.
32. Jenkins, R. Ph.D. Thesis; Lehigh University, 1986.
33. tenBrinke, G.; Hadziioonnou, G. *Macromolecules* **1987,** *20,* 486.
34. Balsara, N. P.; Tirrell, M.; Lodge, T. P. *Macromolecules* **1991,** *24,* 1975.

RECEIVED for review April 16, 1994. ACCEPTED revised manuscript July 19, 1994.

Associating Polymers

B. Performance in Solution

2. Hydrophobe-Modified, Ethoxylated Urethane Polymers

17

Comparative Flow Properties of Model Associative Thickener Aqueous Solutions

Ming-Ren Tarng,[1] J. Philip Kaczmarski,[2] David J. Lundberg,[3] and J. Edward Glass

Department of Polymers and Coatings, North Dakota State University, Fargo, ND 58105

In this chapter the flow behavior of model hydrophobe-modified, water-soluble polymers of the type used in coating formulations (hydrophobe-modified, ethoxylated urethane [HEUR], hydrophobe-modified hydroxyethyl cellulose [HMHEC], and hydrophobe-modified alkali-swellable [HASE] polymers) alone and in the presence of surfactants is examined. The synthesis, characterization, and solution rheology of well-characterized HEUR water-soluble polymers are described for three types of model HEURs. The first type consists of linear poly(oxyethylene)s with terminal hydrophobes of different sizes and with variable oxyethylene spacer lengths between termini. The second type is synthesized by a one-step addition of an ethoxylated nonylphenol surfactant to a diisocyanate or an isocyanato functional biuret or isocyanurate. In both types, swollen gels are realized with full terminal modification. With the linear HEURs at moderate concentrations, the phase separation of HEUR thickeners is dependent on the size of the hydrophobe and the number of oxyethylene units between the hydrophobes. The critical aggregation concentration for viscosity increase is

[1] Current address: Dunn-Edwards Corp., 4885 East 52nd Place, Los Angeles, CA 90040
[2] Current address: GE Plastics, 1 Lexan Lane, Mount Vernon, IN 47620
[3] Current address: 3M Center, Building 236-1N-05, St. Paul, MN 55144

0065–2393/96/0248–0305$16.25/0

lowest when 500 oxyethylene units separate the terminal hydrophobes. In the presence of surfactant, either anionic or nonionic, the fully modified HEURs also exhibit a maximum in viscosity when the spacer is ~500 oxyethylene units. The viscosities realized with the nonionic surfactant are equal to or greater than (with smaller HEUR oxyethylene spacings) those with the anionic surfactant sodium dodecyl sulfate. The third type of HEUR synthesized is by the traditional step-growth process, which is characterized by broad molecular weight distributions. This compositionally defined step-growth HEUR and HMHEC and a chain-growth HASE thickener containing 35 mol% of a hydrophobe-containing monomer are also examined. Their flow profiles and oscillatory responses alone and in the presence of anionic and nonionic surfactants are studied.

STRUCTURE–PROPERTY RELATIONSHIPS among associative thickeners are discussed in this chapter. In the first section, telechelic and direct-addition methods of synthesizing hydrophobe-modified ethoxylated urethane (HEUR) polymers with narrower molecular weight ranges and with hydrophobes only in the terminal positions (as in Schemes I and II) are explained. The viscosity of these polymers builds in aqueous solutions. We also examine their thixotropy. In the second section, the solution properties of HEURs prepared by step-growth (S-G) polymerizations (described in Chapter 10 of this book, along with the synthesis of hydrophobically modified water-soluble polymers and the complexity in placement of the hydrophobes by conventional procedures) are examined, and their solution viscosities are compared with those of the narrower-molecular-weight-range terminal-hydrophobe HEURs.

In the third section, the influence of both anionic and nonionic surfactants on the solution rheology of the terminal models and S-G HEURs is compared with that of two other hydrophobically modified water-soluble polymer types: hydrophobe-modified hydroxyethyl cellulose (HMHEC) and hydrophobe-modified alkali-swellable emulsions (HASE). The structural aspects of the HMHEC and HASE families are discussed in Chapter 10. The contributions of these materials to the elastic behavior of aqueous surfactant solutions are considered in this chapter. These observations in model systems are extended to encompass the influence of these polymers in coating formulations in Chapter 24 of this text, where they are compared with commercial associative thickeners.

Narrow-Molecular-Weight-Range "uni-HEURs" with Terminal Hydrophobe Placement

Hydrophobe and Spacer Size Effects. As noted in Chapter 10, the most economic methods for the synthesis of HMHEC, HASE, and HEUR polymers produce complex product mixtures that do not lend themselves to a fundamental understanding of how hydrophobic interactions influence the responses of thickened aqueous dispersions. The first HEUR studied in this chapter is synthesized by the procedure shown in Scheme I. This sequence provides a means of placing the hydrophobes only at the terminal positions and permits the study of hydrophobe size with a relatively narrow HEUR molecular weight distribution. A small, rigid diisocyanate (isophorone) can be added in a 200-fold molar excess to poly(oxyethylene) (POE) of any molecular weight to produce a telechelic isocyanate precursor for reaction with alkyl amines or alcohols. This method of HEUR synthesis uses a large excess of diisocyanate to inhibit the chain extension observed in conventional S-G polymerization (Chapter 10). The telechelic isocyanato prepolymer is reacted with methanol for molecular weight (size-exclusion chromatography) analysis, which indicates the absence of chain extension. When an excess of alkylamine or alcohol is added to the solution containing the diisocyanate, a variety of by-

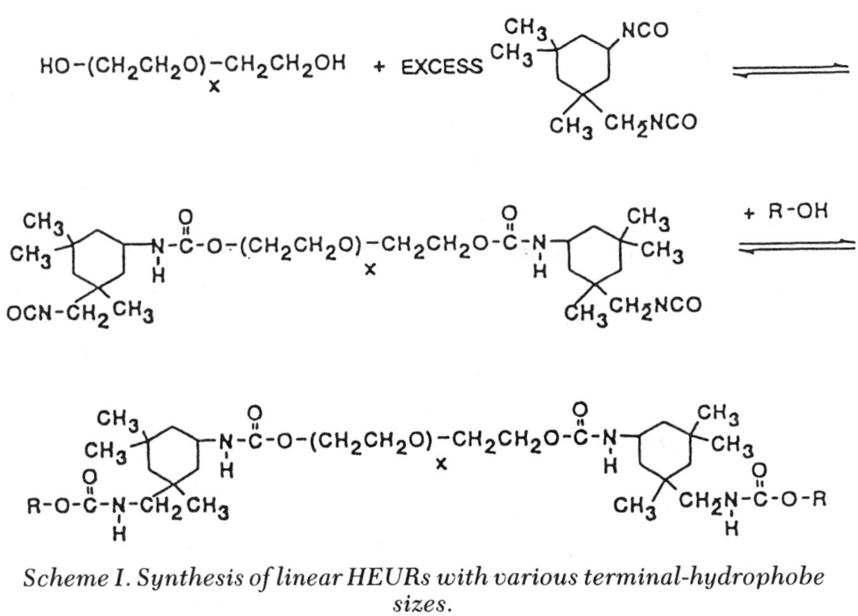

Scheme I. Synthesis of linear HEURs with various terminal-hydrophobe sizes.

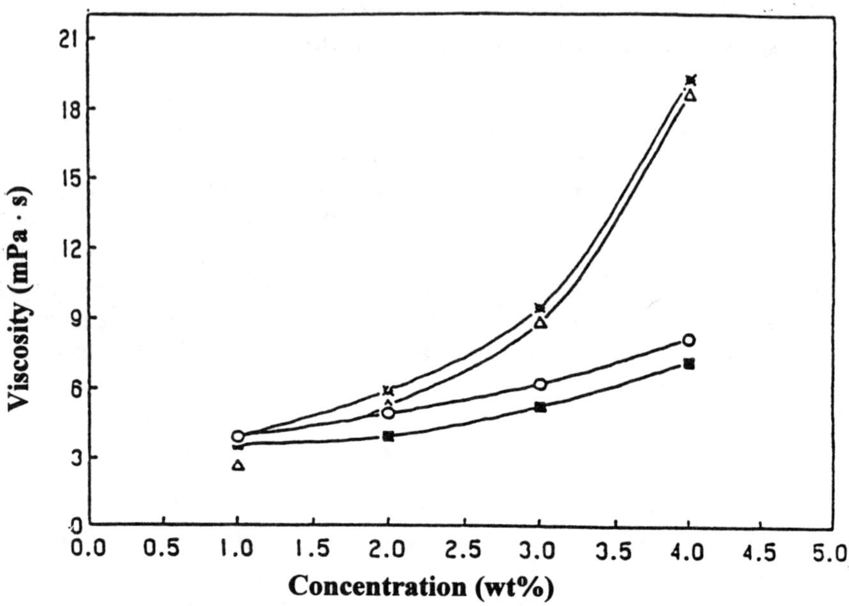

Figure 1. Low-shear-rate (2 s^{-1}) viscosity as a function of the aqueous-solution concentration of linear HEUR polymer (Scheme I). Key: ■, *$HO(EtO)_{595}H$;* ○, *$CH_3OIP(EtO)_{595}IPOCH_3$;* △, *$C_4H_9OIP(EtO)_{595}$-$IPOC_4H_9$; and *, branched $C_8H_{17}OIP(EtO)_{595}IP$-branched-$OC_8H_{17}$.*

products are formed (*1*) that should be removed if definitive structure–property relationships are to be defined.

When the isocyanato prepolymers are reacted with alcohols or amines, the viscosities increase (Figure 1) in aqueous solution with increasing hydrophobe size (*2*). When the terminal groups are small, viscosity among the telechelic HEURs increases only at relatively high concentrations. The rigid isophorone unit contributes relatively little to the size of the terminal hydrophobe with small hydrophobe adducts. With increasing hydrophobe size, the viscosities increase, a situation consistent with other studies of HEUR (*3, 4*), hydrophobe-modified R-acrylamide (*5, 6*), and HMHEC (*7*) polymers.

A second procedure for the synthesis of a terminal-hydrophobe HEUR with a relatively narrow molecular weight distribution is the direct addition of a monoisocyanate to POE (Scheme II). In this study, POEs of different molecular weights are also used to evaluate the effect of spacer length between hydrophobes on the solution behavior of associative thickeners. The quantitative reaction of stoichiometric amounts of POE with octadecyl isocyanate is difficult to achieve because of the large difference in molecular weights of the two reactants

$$H_{37}C_{18}NCO + HO\text{—}(CH_2CH_2O)_n\text{—}H \xrightleftharpoons[\text{catalyst}]{\text{tin}}$$

$$H_{37}C_{18}\text{—}NHCOO\text{—}(CH_2CH_2O)_{n-1}\text{—}CH_2CH_2O\text{—}CONH\text{—}C_{18}H_{37}$$

Scheme II. Synthesis of octadecyl-modified POE.

and the reaction of water with the isocyanate (*1*). POEs are hygroscopic and contain small amounts of water. In theory, the water can be removed by several procedures: by azeotropic distillation with toluene, by melting of polyols under vacuum with stirring at 80 °C, or by freeze-drying (*8*). Complete removal of the water was difficult, and stoichiometric ratios led to incomplete hydrophobe modification. Complete modification of the POE was accomplished by using a fourfold equivalent excess of isocyanate. After removal of the *N,N'*-di-$C_{18}H_{37}$ urea (*1*), the extent of modification was estimated from the ^1H NMR spectroscopic spectra (δ = 1.3 ppm, hydrophobe methylene group; δ = 3.6 ppm, oxyethylene (EtO) methylene group [9]) (Figure 2).

Viscosity dependence on the concentration of fully substituted octadecyl- and nonylphenol-isophorone-modified POE thickeners (*1*) was studied with 0.002 M sodium dodecyl sulfate (SDS), which was added to disrupt the gel-like phase formed with fully substituted HEUR polymers. The influence of EtO spacer length and size among larger terminal hydrophobe groups is evident in Figure 3. The $[C_{18}H_{37}NHC(O)O]_2(EtO)_{531}$ thickener exhibits a greater increase in viscosity at lower concentrations than does a fully modified $C_{18}H_{37}$ HEUR with a shorter or a longer EtO spacer. The same trend is exhibited by $(NPIP)_2(EtO)_x$ model thickeners (NPIP is nonylphenol isophorone). In a prior study, pyrene-modified POE exhibited more intramolecular associations when separated by 100 EtO units than when separated by 200 EtO units (*10*). This fluorescence study along with the low-shear-rate viscosity data in this study suggest that the shorter EtO chains promote intramolecular associations in aqueous media. The viscosity increases with increasing spacer length until the concentration of hydrophobes falls below an effective level: $-EtO_{>500}-$ spacing for maximum domain participation.

Shear Thickening and Thixotropy Among Model Terminal "uni-HEUR" Thickeners. The shear viscosity of surfactant-modified, water-soluble polymers can be very sensitive to the sample's shear history. Figures 4–6 illustrate this sensitivity with the $[C_{18}H_{37}NHC(O)O]_2(EtO)_x$ series. In a previous examination of com-

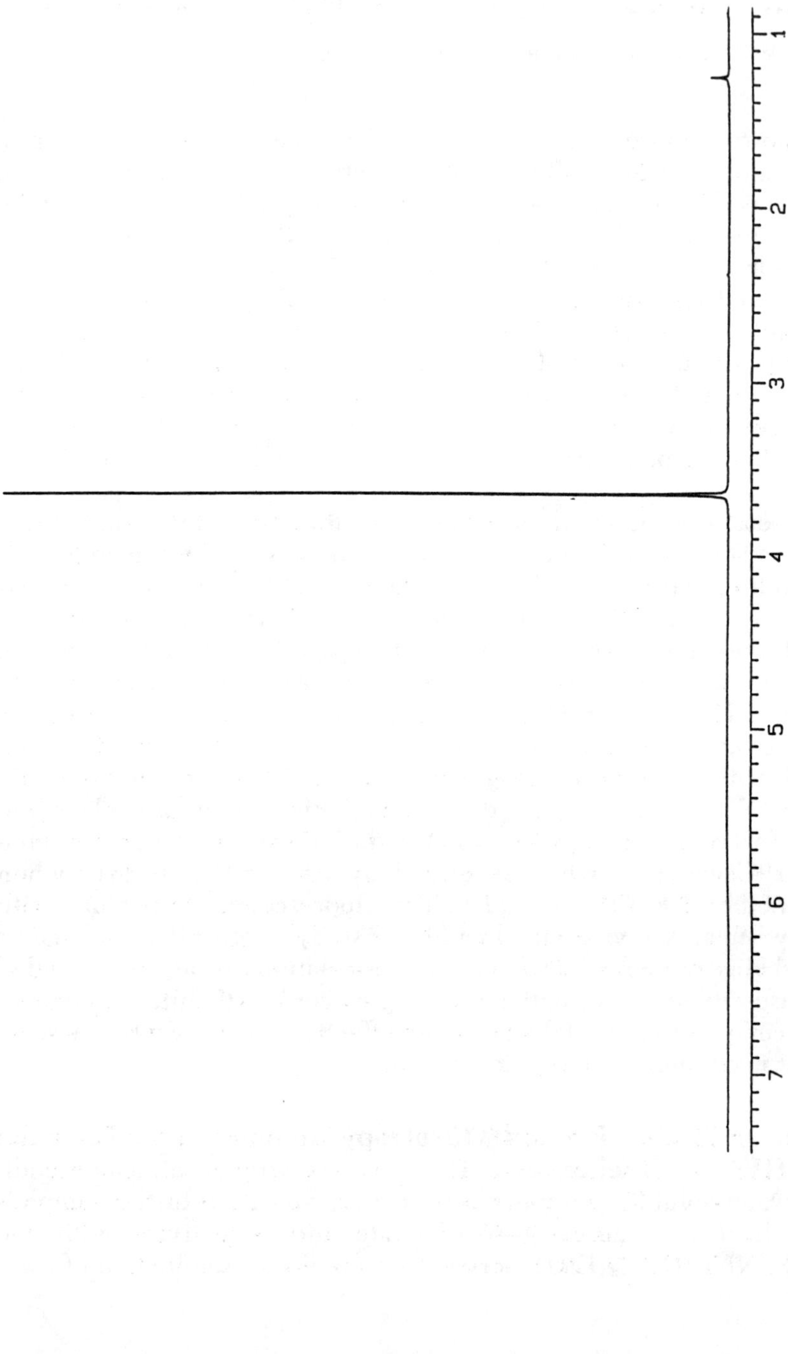

Figure 2. ^{1}H NMR spectrum of a $(C_{18}H_{37}NHCOO)_2$-$(EtO)_{182}$ HEUR.

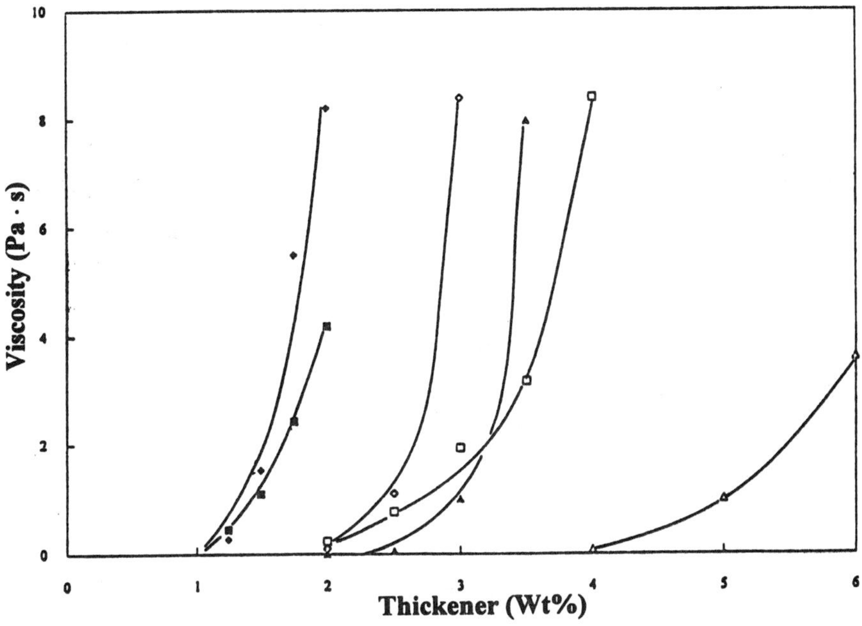

Figure 3. Low-shear-rate viscosity (2 s^{-1}) of $C_{18}H_{37}$-modified (closed symbols) and NPIP-modified (open symbols) POE as a function of thickener concentration. Key: △, $(C_{18}H_{37}NHCOO)_2(EtO)_{182}$ with 0.002 M SDS; ◇, $(C_{18}H_{37}NHCOO)_2(EtO)_{531}$ with 0.002 M SDS; and □, $(C_{18}H_{37}NHCOO)_2(EtO)_{663}$ with 0.002 M SDS.

mercial thickeners, shear thickening was observed among certain HEURs (*11*). According to a number of observations reviewed in that study, shear thickening is a transition from intra- to intermolecular hydrophobic bonding. One would therefore expect shear thickening in these model terminal HEURs to occur with short EtO spacings and to become less important with increasing EtO separations. In Figures 4–6 the shear rate is ramped from 0.1 to 500 s^{-1} within 3 min in most of our studies, and then the rate is decreased over the same period; the sample then remained undisturbed for 5 min before the cycle was repeated with a 3-min and then a 1-min gestation period after completion of each cycle. The HEUR with only 200-EtO-unit oxyethylene spacing is notably shear thickening in each cycle, a result consistent with a transition from intra- to interhydrophobic associations. The shear thinning at low shear rates (<2 s^{-1}) of the initially unperturbed sample is due to the disruption of aggregates.

The solubilization of water-soluble polymers has always been difficult because of the solvation of poorly dispersed powders. Aggregation of these particles with viscous outer surfaces results in poor solvation of the complete powder particles; this poor solvation is not

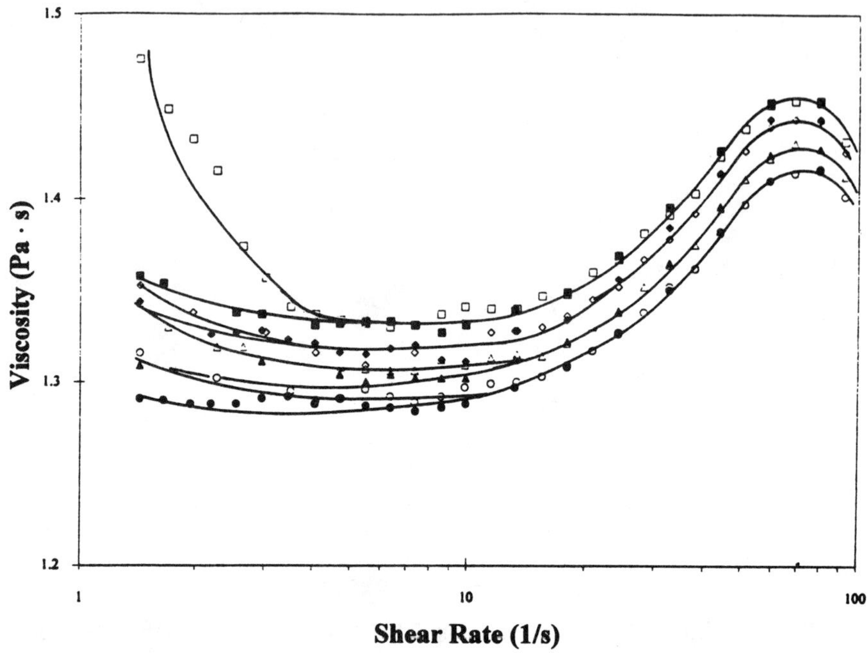

Figure 4. Hysteresis loop of 4-wt% $(C_{18}H_{37}NHCOO)_2(EtO)_{182}$ aqueous phase. Key: □ and ■, initial loop; ◇ and ◆, 5-min elapse; △ and ▲, 3-min elapse; and ○ and ●, 1-min elapse. Open symbols, ascending time was 3 min; closed symbols, descending time was 3 min.

visually evident because the refractive index of the swollen powder is similar to that of water. When improper dispersion is carefully avoided, the viscosity/shear rate profile of the HEUR with the optimal EtO spacing, $[C_{18}H_{37}NHC(O)O]_2(EtO)_{531}$, provides a very unusual deformation response. The shear thickening is minimal, but the apparent thixotropy at an intermediate deformation range is very unusual. In additional studies, the HEUR with the smaller –EtO– spacing exhibits a similar viscosity shear rate profile in nonionic surfactant solutions. These differences will be discussed in detail in a future publication. The data indicate the importance of the associative aggregate's configuration in the viscosity shear rate profile.

S-G HEURs with Defined Composition and Broad Molecular Weight Distributions

As noted in Chapter 10, S-G polymerizations lead to broad distributions in molecular weights (*12, 13*), and intermediates in S-G propaga-

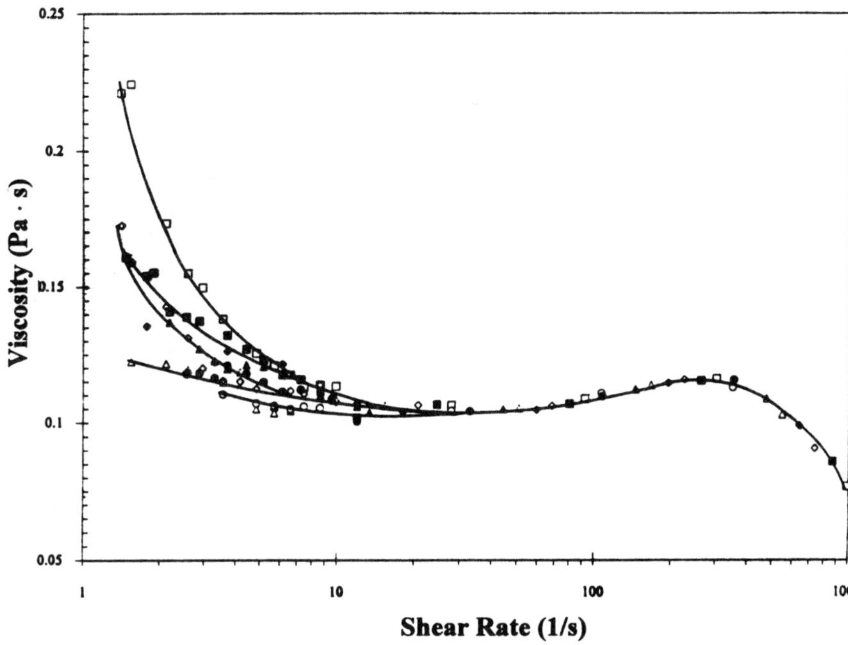

Figure 5. Hysteresis loop of 3-wt% $(C_{18}H_{37}NHCOO)_2(EtO)_{331}$ aqueous phase. Key: □ and ■, initial loop; ◇ and ◆, 5-min elapse; △ and ▲, 3-min elapse; and ○ and ●, 1-min elapse. Open symbols, ascending time was 3 min; closed symbols, descending time was 3 min.

tion may be terminated by hydrophobic moieties or may remain at that molecular weight but unreacted (*14*). The extent of modification among the different, distinct molecular weights is unknown, and two types of potentially hydrophobic alkyl groups are present: internal alkyl groups within the POE backbone and external alkyl groups at the termini of the HEUR chain (*15*).

Reaction of a slight excess of POE with the diisocyanate in a S-G polymerization provides HEURs with terminal hydroxyl groups. This synthetic procedure allows evaluation of the relative contributions of the internal alkyl groups. HEUR polymers prepared with well-defined components vary in molecular weights (17,000 to 54,000, in relation to the reaction ratios used), molecular weight distributions, and aqueous solution behavior (Table I).

To minimize the amount of unreacted POE present and to optimize the possible aggregation of internal alkyl groups, a 7:6 mole ratio of a POE (with a lower degree of polymerization [n = 139]) to diisocyanate was used to obtain HEURs of 31,000–35,000 M_n. The lower POE molecular weight was used to enhance hydrophobic differ-

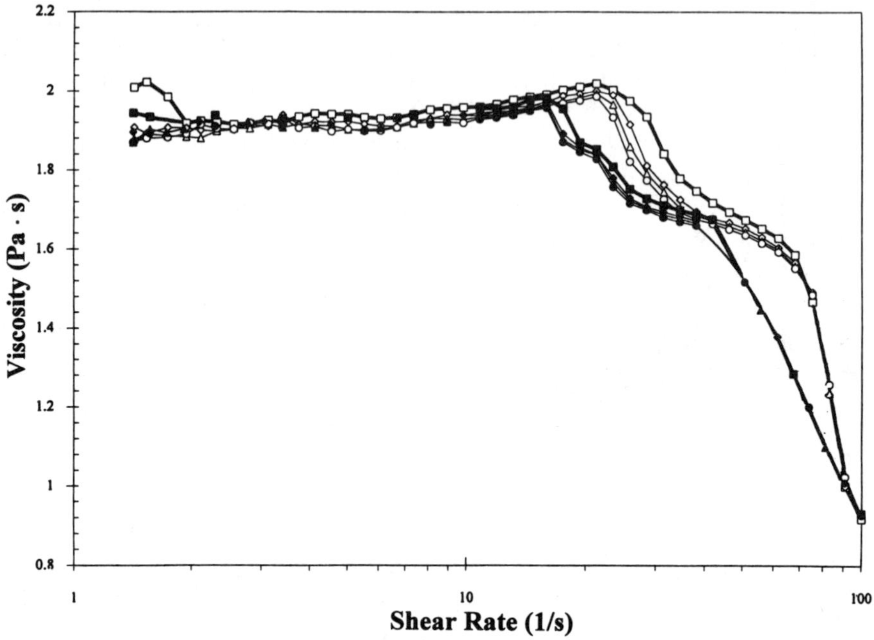

Figure 6. Hysteresis loop of 2-wt% $(C_{18}H_{37}NHCOO)_2(EtO)_{531}$ aqueous phase. Key: □ and ■, initial loop; ◇ and ◆, 5-min elapse; △ and ▲, 3-min elapse; and ○ and ●, 1-min elapse. Open symbols, ascending time was 3 min; closed symbols, descending time was 3 min.

Table I. Molecular Weights of $HO(EtO)_nH$/Diisocyanate HEURs with Terminal Hydroxyl Groups

R[a]	n	X	M_n *(theoretical)*	M_n	M_w	*PDI*[b]
$H_{12}MDI$	182	1	16,500	20,500	27,500	1.34
$H_{12}MDI$		2	25,000	25,000	34,700	1.38
$H_{12}MDI$		3	33,000	28,600	40,800	1.43
IPDI	182	1	16,500	17,400	24,600	1.42
IPDI		2	24,900	24,500	38,700	1.58
IPDI		3	33,200	27,800	45,000	1.62
HDI	182	1	16,500	18,400	26,500	1.44
HDI		2	24,700	24,200	37,400	1.55
HDI		3	33,100	31,400	49,800	1.58
TMXDI	182	1	16,500	18,700	25,000	1.33
TMXDI		2	24,900	24,000	33,100	1.38
TMXDI		3	33,300	27,100	38,100	1.40
IPDI	139	6		31,200	49,500	1.59
HDI	139	6		35,300	56,500	1.60
TMXDI	139	6		35,100	55,890	1.59
$H_{12}MDI$	139	6		33,400	53,000	1.59

NOTE: General HEUR structure: $HO(EtO)_{182}–(CONH–R–NHCOO–(EtO)_{182}–)_xH$.
[a] IPDI is isophorone diisocyanate; TMXDI is *m*-tetramethylxylene diisocyanate.
[b] PDI is polydispersability index (M_w/M_n).

ences that may arise from the internal groups with variations in the alkyl group size of the diisocyanate. The four types of diisocyanates used are discussed in Chapter 10.

A fluorescent probe, 8-anilino naphthalene sulfonic acid, is sensitive to the local environment. The wavelength of the maximum emission intensity will shift and the intensity of the emission will increase when this probe enters a hydrophobic domain (*16–18*). The intensity with the unmodified $HO(EtO)_{663}H$ does not change sharply with increasing concentration, and no sharp intensity increase is observed with the HEURs that contain only internal alkyl groups. These HEURs also do not exhibit viscosity increases over moderate concentrations (Figure 7).

HEURs with terminal hydrophobes were prepared directly by reversing the stoichiometry to an excess of diisocyanate compared to POE in the S-G polymerization (Scheme III in Chapter 10). The isocyanato intermediates were reacted with an alkyl amine without isolation of the intermediate. To obtain a range of molecular weights, the mole ratios of hexamethylene diisocyanate (HDI) were 3:2, 5:4, and

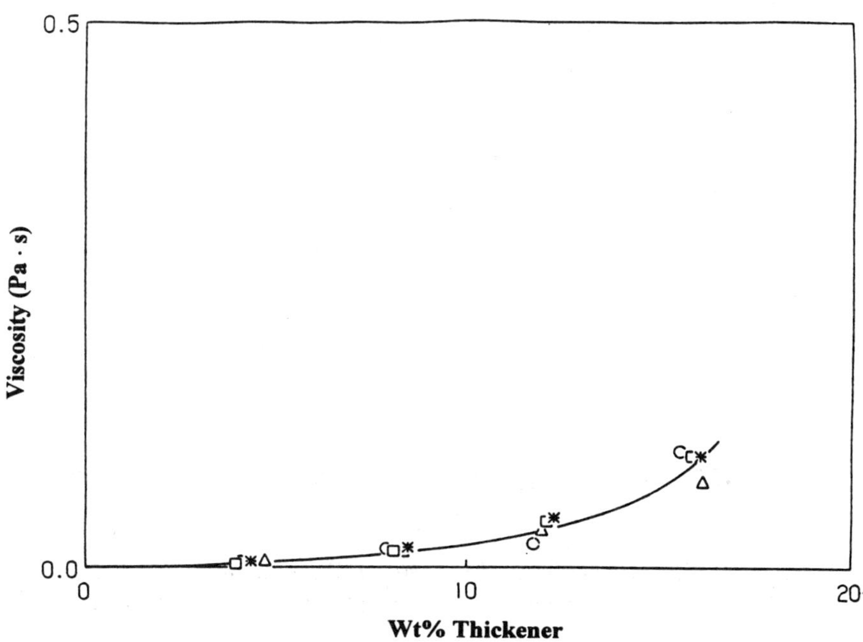

Figure 7. Newtonian shear viscosity as a function of HEUR concentration. Key: ○, $HO-[(EtO)_{139}HDI]_6-O(EtO)_{139}H$; △, $HO-[(EtO)_{139}IPDI]_6-O(EtO)_{139}H$; *, $HO-[(EtO)_{139}TMXDI]_6-O(EtO)_{139}H$; *and* □, $HO-[(EtO)_{139}H_{12}MDI]_6-O(EtO)_{139}H$.

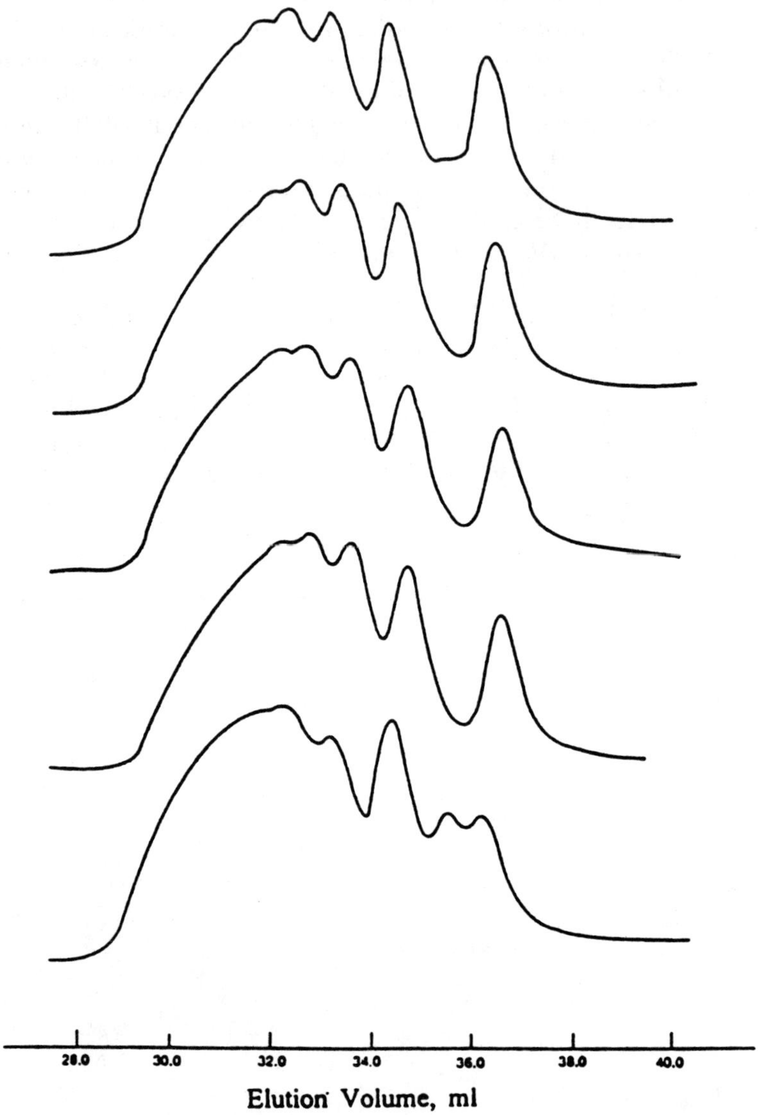

Figure 8. SEC chromatographs of R–[HDI–(EtO)$_{182}$]$_4$–HDI–R thickeners. From top to bottom, R = hexyl, octyl, dodecyl, tetradecyl, and octadecyl.

7:6. The terminal isocyanates were reacted with amines varying in alkyl size from hexyl to octadecyl. The influence of the diisocyanate coupler was examined by changing the diisocyanate from HDI to H_{12}MDI (dicyclohexylmethane diisocyanate; structures are given in Chapter 10) while keeping the mole ratio of diisocyanate to HO(EtO)$_{182}$H constant at 3:2. The molecular weight distribution is broad (Figure 8), but it does not change during the final modification step.

Influence of Effective Terminal-Hydrophobe Size. When large alkyl groups are positioned at the termini (e.g., $C_{12}H_{25}$–NHCONH – [HDI–NHCOO – (EtO)$_{182}$ – CONH –]$_4$ – HDI – NHCO NH – $C_{12}H_{25}$), a sharp upturn in fluorescence intensity is observed (Figure 9) with increasing concentration. This change indicates that the external alkyl groups are forming hydrophobic aggregates. The upturn occurs at lower concentrations in the R–NHCONH–[HDI–NHCOO–(EtO)$_{182}$–CONH–]$_2$–HDI–NHCONH–R series with an in-

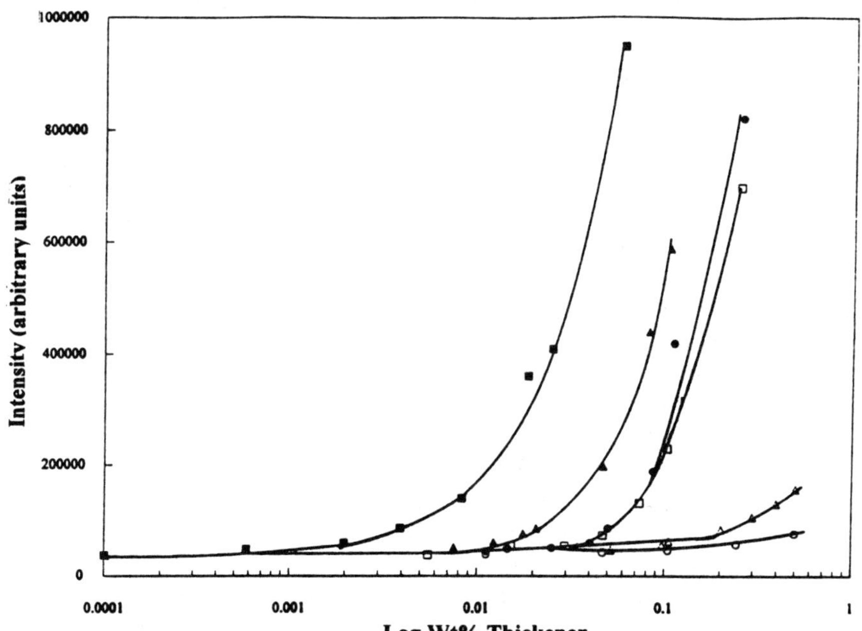

Figure 9. Intensity of emission of 8-anilinonaphthalene sulfonic acid as a function of R–[R′–(EtO)$_{182}$–]$_2$–R′–R concentration. Key: □ and ■, R = $C_{12}H_{25}$; △ and ▲, R = C_8H_{17}; and ○ and ●, R = C_6H_{13}. Open symbols, R′ = HDI; closed symbols, R′ = H_{12}MDI.

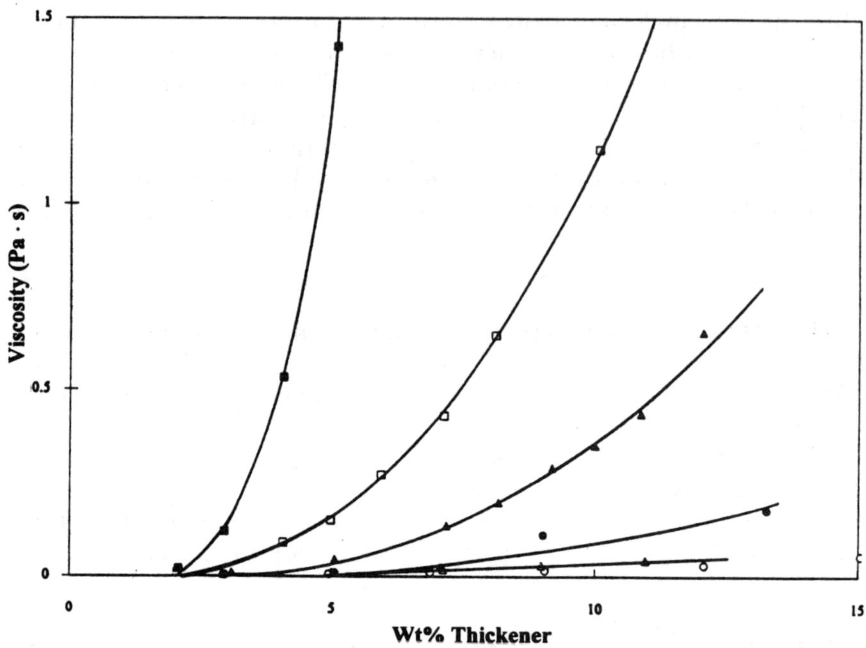

Figure 10. Newtonian viscosity as a function of R–[R″–(EtO)$_{182}$–]$_2$–R′–R concentration. Key: □ *and* ■, *R = $C_{12}H_{25}$;* △ *and* ▲, *R = C_8H_{17}; and* ○ *and* ●, *R = C_6H_{13}. Open symbols, R′ = HDI; closed symbols, R′ = H_{12}MDI.*

crease in the size of the alkyl amine (i.e., R is varied from C_6H_{13}– to $C_{12}H_{25}$–). Similar results are observed with the R–NHCONH–[H_{12}-MDI–NHCOO–(EtO)$_{182}$–CONH–]$_2$–H_{12}MDI–NHCONH–R series; the alkyl amine size is varied from C_6H_{13}– to $C_{12}H_{25}$–. The viscosities follow this trend (Figure 10).

Intrinsic viscosities of these two series were measured in *N,N*-dimethylformamide (DMF), a solvent that disrupts micelle formation. The data (Table II) demonstrate the reproducibility of the synthetic procedure. The intrinsic viscosities in DMF increase as the molecular weights are increased, but within a given molecular weight series, no increases in intrinsic viscosity are observed as the size of the external hydrophobe is varied.

Intrinsic viscosities are typically measured at dilute concentrations at which polymer–polymer interactions can be neglected (*19*). If the terminal alkyl groups associate in aqueous media at very low concentrations, polymer–polymer interactions through hydrophobic aggregation negate the dilute concentration conditions. Fluorescence measurements (Figure 9) indicate that such associations occur in water

Table II. Reduced (η) and Intrinsic (η_{sp}/c) Viscosities of Direct-Addition Thickeners

	η_{sp}/c			
Thickener	*DMF*	*H_2O*	*[η] DMF*	*η_{sp}/c Ratio (H_2O/DMF)*
R–NHCONH–(H_{12}MDI–NHCOO–$(EtO)_{182}$–CONH–$)_2$–H_{12}MDI–NHCONH–R				
R = C_6H_{13}	0.47	0.47	0.40	1.0
R = C_8H_{17}	0.45	0.64	0.40	1.4
R = $C_{12}H_{25}$	0.46	1.50	0.40	3.3
R = $C_{14}H_{29}$	0.42	1.10	0.36	2.6
R–NHCONH–(HDI–NHCOO–$(EtO)_{182}$–CONH–$)_2$–HDI–NHCONH–R				
R = C_6H_{13}	0.52	0.51	0.40	1.0
R = C_8H_{17}, run 1	0.69	0.70	0.52	1.0
run 2	0.51	0.51	0.45	1.0
R = $C_{12}H_{25}$, run 1	0.62	1.15	0.53	1.8
run 2	0.52	0.85	0.40	1.6
R = $C_{14}H_{29}$	0.54	1.67	0.41	3.1
R = $C_{18}H_{37}$	0.49[a]	1.74[a]	0.46[b]	3.5
R–NHCONH–(HDI–NHCOO–$(EtO)_{182}$–CONH–$)_4$–HDI–NHCONH–R				
R = C_6H_{13}	0.64	0.66	0.53	1.0
R = C_8H_{17}	0.62	0.64	0.51	1.0
R = $C_{12}H_{25}$	0.65	0.80	0.52	1.2
R = $C_{14}H_{29}$	0.65	1.75	0.53	2.7
R = $C_{18}H_{37}$	0.70	15.90	0.55	22.7
	0.62[a]	1.12[a]		1.8
R–NHCONH–(HDI–NHCOO–$(EtO)_{182}$–CONH–$)_6$–HDI–NHCONH–R				
R = C_6H_{13}	0.80	0.82	0.67	1.0
R = C_8H_{17}	0.78	0.80	0.62	1.0
R = $C_{12}H_{25}$	0.72	0.80	0.58	1.1
R = $C_{14}H_{29}$	0.74	1.86	0.58	2.5

SOURCE: Reproduced from reference 15. Copyright 1994 American Chemical Society.
[a] η_{sp}/c at 0.5 g/dL.
[b] [η] at 0.5 g/dL and lower concentrations.

at very low concentrations for HEURs with terminal alkyl groups. In view of these observations, specific viscosities at concentrations of 1.0 g/dL in water rather than intrinsic viscosities were measured and are compared with specific viscosities in DMF. Specific viscosity increased in aqueous media relative to that measured in organic media. In the R–NHCONH–[HDI–NHCOO–$(EtO)_{182}$–CONH–$]_2$–HDI–NHCONH–R series (Table II), the HEURs with terminal alkyl groups R = C_6H_{13} and R = C_8H_{17} (12 and 14 carbons, respectively, if the alkyl group of the HDI is considered) do not show increased viscosity at the concentrations studied. As the terminal alkyl groups (R) are enlarged to $C_{12}H_{25}$–, $C_{14}H_{29}$–, and $C_{18}H_{27}$– in this series, a deviation from unity is observed.

The low-shear-rate viscosity data as a function of HEUR concentration for the R–NHCONH–[HDI–NHCOO–$(EtO)_{182}$–CONH–$]_2$–HDI–NHCONH–R and the $H_{12}MDI$ series parallel the specific-viscosity data. When the interconnecting unit is changed from HDI to $H_{12}MDI$ and the alkyl amine is kept constant, the influence of the diisocyanate on the hydrophobicity of the external alkyl unit is evident. The importance of the "effective" size of the external alkyl group (defined as the combination of the long-chain alkyl amine and the alkyl group of the diisocyanate used to couple the amine to the POE) is evident when the viscosity data are compared with the fluorescence data. The aggregation phenomenon occurs at lower concentrations with HEURs containing the larger external "effective" alkyl size.

Molecular Weight Influence. A molecular weight influence is also observed in the data on specific viscosity in water. As the stoichiometric ratio of HDI to POE is changed from 3:2 to 5:4 to 7:6 (Table II), the specific-viscosity ratio decreases toward unity if the terminal alkyl size is constant. The low-shear-rate viscosities parallel the specific-viscosity data. With increasing HEUR molecular weight, the concentration of the external alkyl groups decreases, the extent of association is lower, and higher concentrations of the thickener are required before the aggregation phenomenon and the viscosity increase occur.

Fully modified HEURs with $C_{18}H_{37}$–NHCOO hydrophobes and narrow molecular weight distributions (*1*) exhibit phase-separated solutions in the absence of surfactant. This phenomenon was also observed in star-type HEURs previously studied (*2*). None of the S-G HEURs in this study exhibited phase-separated solutions, even though the HEUR $C_{18}H_{37}$–NHCONH–[HDI–NHCOO–$(EtO)_{200}$–CONH–$]_2$–HDI–NHCONH–$C_{18}H_{37}$ possesses a significantly greater effective terminal-hydrophobe size. According to an analysis of the size of external alkyl groups, this HEUR is expected to exhibit phase separation. The differences observed arise from differences in molecular weight distribution, and the low molecular weight substituted products in S-G HEURs could act as in situ surfactants. The extent of hydrophobe modification of the S-G HEURs presented in this study is not easily measured, and this difficulty may also contribute to the lack of phase separation. Our studies in this area will be reported in the near future.

Activation Energy for Flow. The viscosities of both the terminal-hydrophobe and the S-G HEURs decrease with increasing temperature (20 to 50 °C), indicating a lower residence contact time of the

Table III. Activation Energy of Flow (E_m) from Arrhenius Plots

Polymer	E_m *(kJ/mol)*[a]
HMHEC	49
HEC	43
$C_{18}H_{37}$–NHCOO–$(EtO)_{531}$–CONH–$C_{18}H_{37}$	85
$C_{18}H_{37}$–NHCOO–$(EtO)_{331}$–CONH–$C_{18}H_{37}$	85
$C_{18}H_{37}$–NHCOO–$(EtO)_{182}$–CONH–$C_{18}H_{37}$	90

[a] Standard deviation is ±4 kJ/mol.

hydrophobes. The activation energies (E_m) for viscous flow were obtained from the Arrhenius equation:

$$\log \eta = \log k - (E_m/2.303RT)$$

where η is low-shear-rate Newtonian viscosity. The activation energy is dependent upon hydrophobe size and not on the separation distance between terminal hydrophobes (Table III); thus the transfer in hydrophobic associations in concentrated solutions is rapid (*20*).

Surfactant Influences and Elasticity Contributions

Viscosity increases with associative thickeners below their critical aggregation concentrations can be realized with the addition of conventional anionic (*21*) and nonionic (*22*) surfactants. A review of prior surfactant studies is helpful in understanding the nature of the HEUR–surfactant interactions.

Mixed solutions of homologous surfactants exhibit ideal mixing behavior in micelles and monolayers; the homolog with the larger hydrophobe is the most interfacially active. Mixed micelles and monolayers of anionic and nonionic surfactants, however, exhibit a strong deviation from ideal behavior in enhanced association and surface activities (*23*). At low surfactant concentrations, micelles of the nonionic component are dominant. With increasing concentration, incorporation of the anionic surfactant is enhanced below the critical micelle concentration (cmc) of the pure anionic surfactant. The difference arises from a decrease in electrostatic repulsions among the ionic groups that is facilitated by the EtO chains of the nonionic surfactant (*24*). The interactions are attributed to a partial charge transfer to the ether oxygen of the EtO groups and, consequently, an increased attraction of the hydrophobic group of the nonionic surfactant to the anionic surfactant.

POE–SDS mixtures have been the most extensively investigated among polymer–surfactant mixtures and are the most pertinent to this study; HEUR thickeners are generally >95% EtO units by chemical composition. In one of the first studies of POE and SDS solutions, which used conductance, surface tension, and viscosity measurements, two transition points were observed and interpreted in terms of a polymer–surfactant complex or micelle (*25*). In dilute POE–SDS solutions, a critical SDS concentration is noted for the formation of the aggregates. The SDS micelles bound (*26*) to POE are smaller than the micelles formed in the absence of POE. NMR data indicate that the surfactant–polymer–water interface tends to retain a certain stoichiometric composition. The POE–SDS aggregate is a mixed micelle in which some of the polymer units are wrapped around a given micelle (*27*); the free energy for the initially bound micelles is 10–20 kT. Other EtO units are randomly spaced above the micelle interface (*28*). The gain in free energy, *dG*, per additional bound micelle decreases as the repulsions between micelles within a POE–SDS aggregate become more important. As long as the free energy is greater than 1 kT, all the available micelles bound to POE are equally distributed. Beyond a stoichiometric number of micelles per aggregate, the micelles remain free or are bound weakly to the aggregates (*29*). The influence of such interactions on solution viscosities is discussed in Chapter 13 for very high molecular weight POEs. In comparison to the viscosity effect noted for surfactant-capped POEs (given below), the SDS–POE interaction influence is low.

Surfactant Influences. ***Maxima in Terminal Hydrophobe HEURs.*** For a given associative thickener structure below its critical aggregation concentration, the viscosity increases with increasing surfactant concentration (*21, 22*), achieves a maximum, and then decreases. The concentration of surfactant required to achieve the maximum, the viscoelasticity of the solution in the area of the viscosity maximum, and factors influencing the magnitude of the viscosity increase in model hydrophobically modified, water-soluble polymers are addressed below.

The viscosity of the linear HEURs without an internal hydrophobe and a high-molecular-weight POE spacer link (Scheme I) was examined in both anionic (SDS) and nonionic (branched $C_{13}H_{27}(OEt)_9OH$) surfactant solutions. In the sections to follow, the number of cmcs noted refer to the properties of SDS in the absence of an associative thickener. In the figures, SDS is illustrated in molar concentrations. Nonionic surfactants are reported in weight percentages because they are mixtures of products (*30*). The anionic–nonionic mixed-surfactant interactions noted in previous figures are reflected in the viscosity maxima near the cmc (0.8 ± 0.2, 0.008 M; Figure 11a) of SDS. The

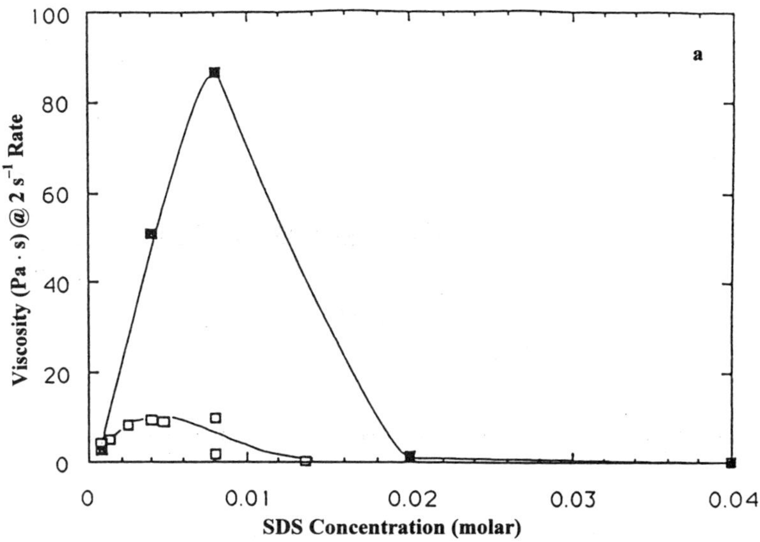

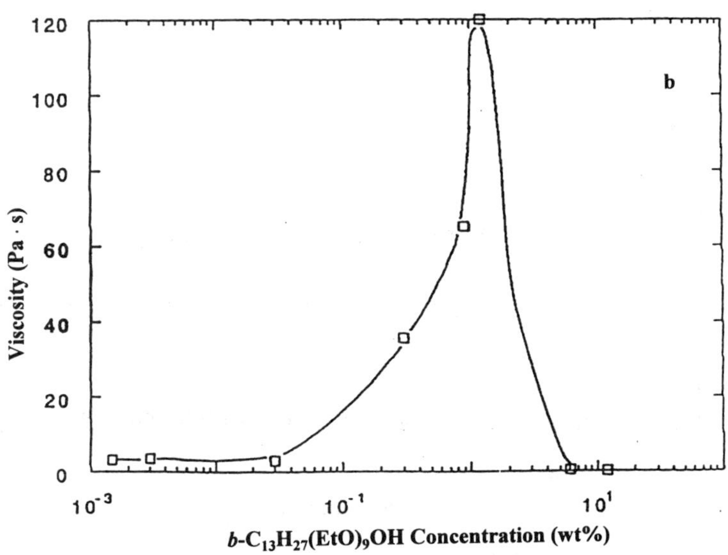

Figure 11. a: Effect of SDS surfactant concentration on low-shear-rate viscosity of 4-wt% $C_8H_{17}OIP(EtO)_{595}IPOC_8H_{17}$ (□) and 3-wt% $NPOIP(EtO)_{595}IPONP$ (■), with 1 cmc SDS = 8×10^{-3} M.
b: Effect of branched $C_{13}H_{27}O(EtO)_9H$ nonionic surfactant aqueous-solution concentration on low-shear-rate viscosity of 3-wt% $NPOIP(EtO)_{595}IPONP$; for reference, 1 cmc branched $C_{13}H_{27}O(EtO)_9H$ = 0.003 wt%.

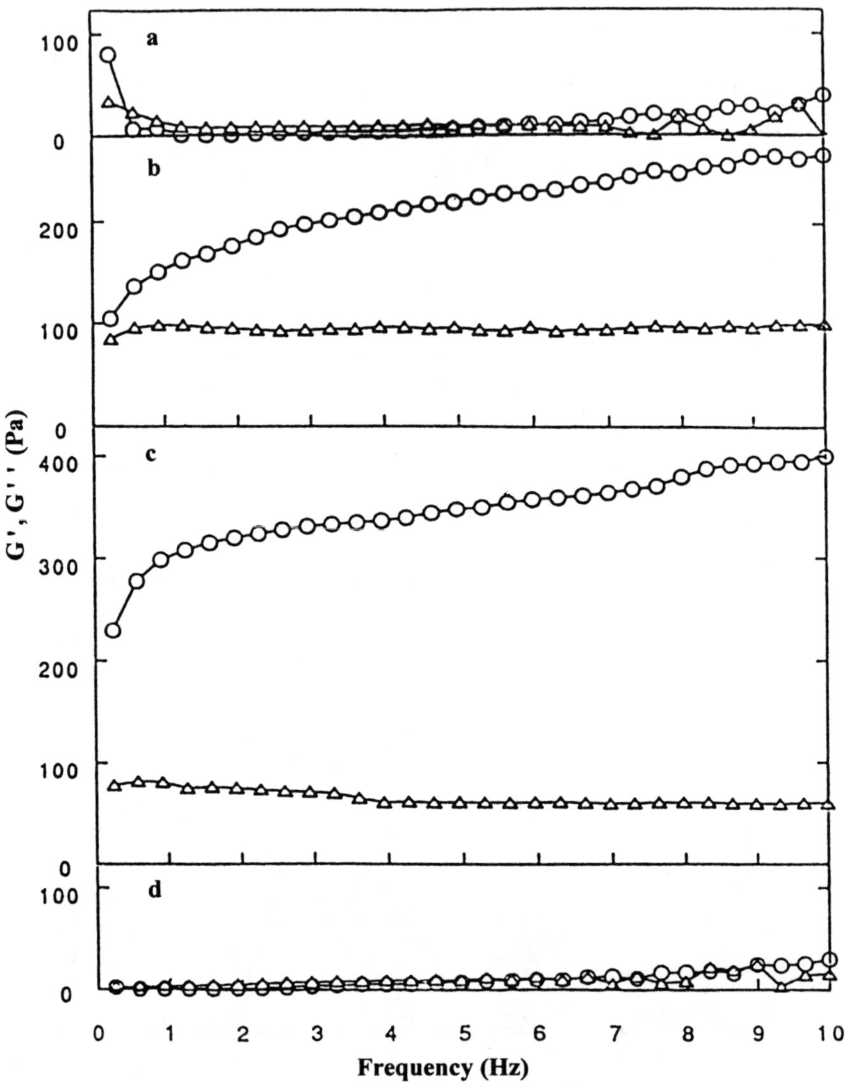

Figure 12. Storage modulus (G') (○) and loss modulus (G") (△) dependence on frequency of 3-wt% NPIP(EtO)$_{595}$IPNP solutions with various SDS concentrations: (a) 0 M; (b) 4 × 10^{-3} M; (c) 8 × 10^{-3} M; and (d) 2 × 10^{-2} M.

magnitude of the viscosity increase of a 3 wt% NPIP-adduct is greater than that of a 4 wt% $C_8H_{17}IP(EtO)_{595}IPC_8H_{17}$ adduct. Oscillatory data on the SDS solutions reveal an increasing elastic (storage modulus, Figures 12 and 13) response at the SDS concentrations that affect the viscosity increase. A viscosity increase is observed at high nonionic-surfactant (Figure 11b) concentrations (~300 cmcs) in NPIP-$(EtO)_{595}$IPNP solutions, and the presence of an elastic network in this nonionic combination is reflected in the high value for the storage

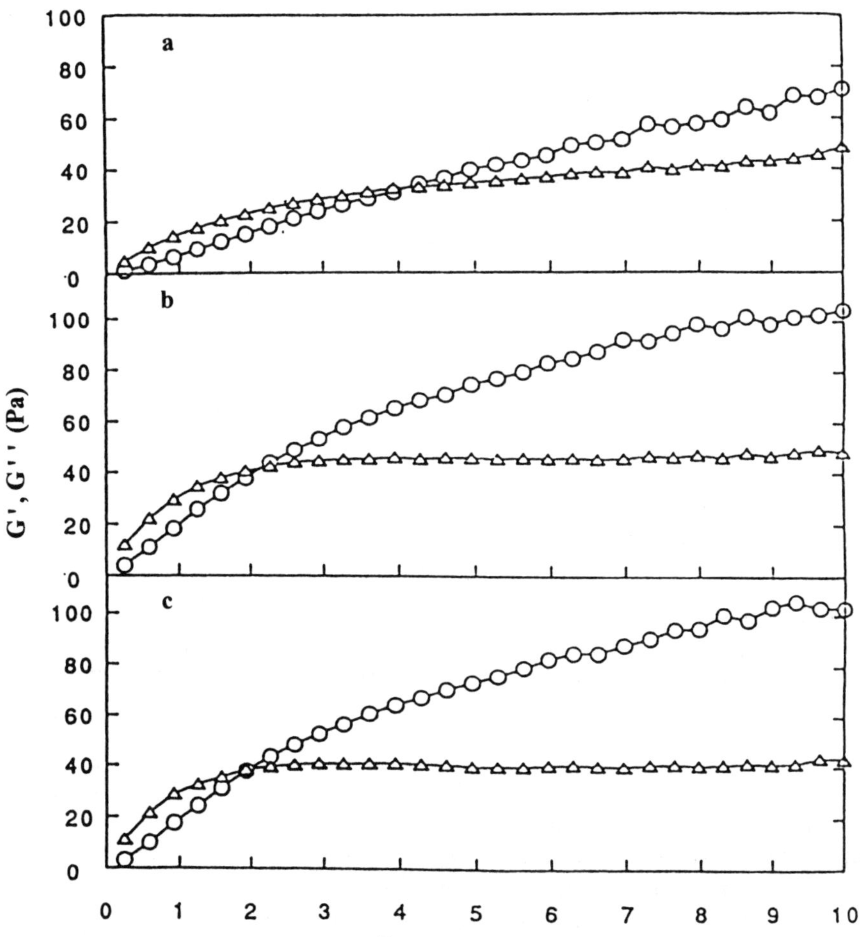

Figure 13. Storage modulus (G') (○) and loss modulus (G") (△) dependence on frequency of 4-wt% $C_8H_{17}IP(EtO)_{595}IPC_8H_{17}$ solutions with various SDS concentrations (a) 0 M; (b) 4 × 10^{-3} M; and (c) 8 × 10^{-3} M (0.02 M SDS solutions gave no oscillatory response).

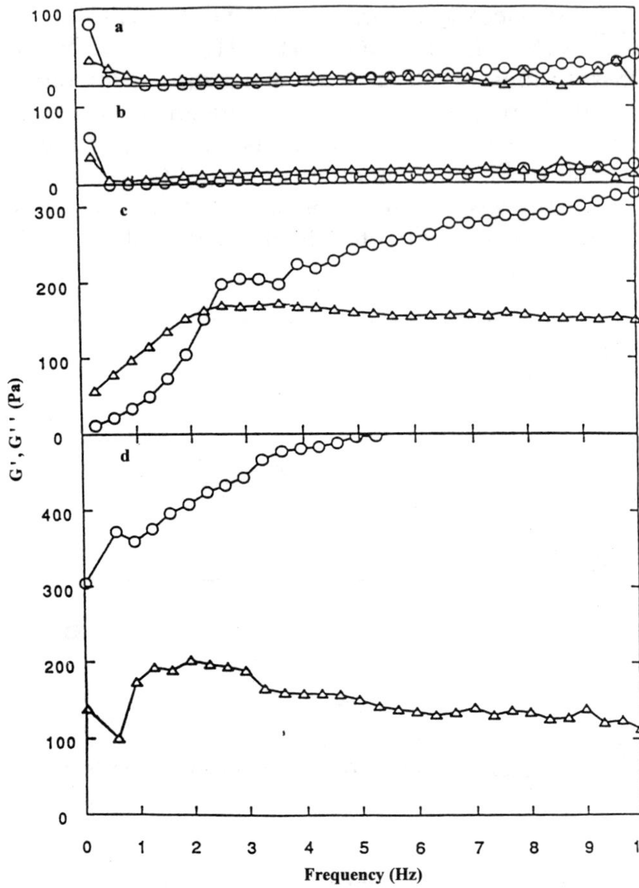

Figure 14. Storage modulus (G') (○) and loss modulus (G") (△) dependence on frequency of 3-wt% NPOIP(EtO)$_{595}$IPONP solutions with various branched $C_{13}H_{27}O(EtO)_9H$ concentrations: (a) 0 wt%; (b) 0.003 wt%; (c) 0.03 wt%; and (d) 0.30 wt%.

modulus (Figure 14) of the solutions in the area of the viscosity maximum. No viscosity maximum is observed over a broad range of nonionic-surfactant concentrations with the 4 wt% C_8H_{17}IP-(EtO)$_{595}$IPC_8H_{17}, and no elastic network is evident.

In the study of spacer lengths, the viscosity dependence of the fully modified $[C_{18}H_{37}NHC(O)O]_2(EtO)_x$ HEURs on the concentration of the anionic surfactant SDS and the nonionic surfactant $C_8H_{17}C_6H_4O(EtO)_{10}H$ is illustrated in Figures 15 and 16, respectively. With the nonionic surfactant, the lower surfactant concentration

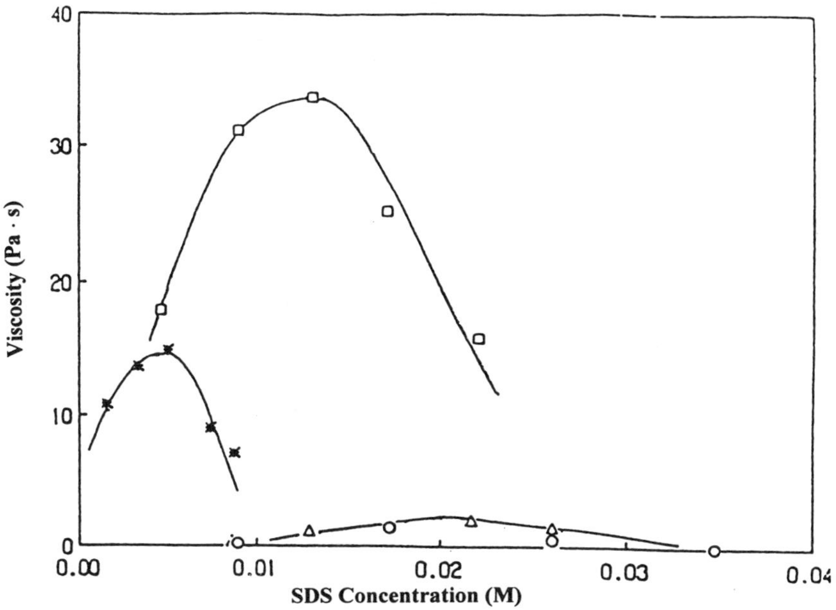

*Figure 15. Low-shear-rate viscosity (2 s^{-1}) of 2-wt% $C_{18}H_{37}$-modified POE as a function of SDS molar concentration. Key: ○, $(C_{18}H_{37}NHCOO)_2(EtO)_{182}$; △, $(C_{18}H_{37}NHCOO)_2(EtO)_{331}$; □, $(C_{18}H_{37}NHCOO)_2(EtO)_{531}$; and *, $(C_{18}H_{37}NHCOO)_2(EtO)_{663}$.*

represents that required to achieve solutions of soluble thickener. In this study, $C_8H_{17}C_6H_4O(EtO)_{10}H$ instead of branched $C_{13}H_{27}O(EtO)_9H$ is used as the nonionic surfactant. The aromatic-containing nonionic surfactant is used in our competitive adsorption studies on model latices (*31*) because of the analytical (UV) sensitivity of the phenyl unit. This procedure is discussed in Chapter 24. A molecular weight influence on the magnitudes of the maxima is evident with both surfactants. As the number of EtO units between the terminal hydrophobes is decreased, the concentration of terminal hydrophobes increases (in constant-weight-percent studies). An increase in viscosity would be expected on the basis of an increasing hydrophobe content. This increase is not observed because of prevalent intramolecular associations. The optimal spacer length for achieving the highest viscosity in the presence of surfactant appears to be approximately 500 EtO units in the $[C_{18}H_{37}NHC(O)O]_2(EtO)_x$ series. The highest viscosity in the NPIP series also occurs at this spacer length. Both nonionic surfactants effect viscosity increases similar in magnitude to that effected by SDS with the 500 EtO-unit HEURs. The viscosity in-

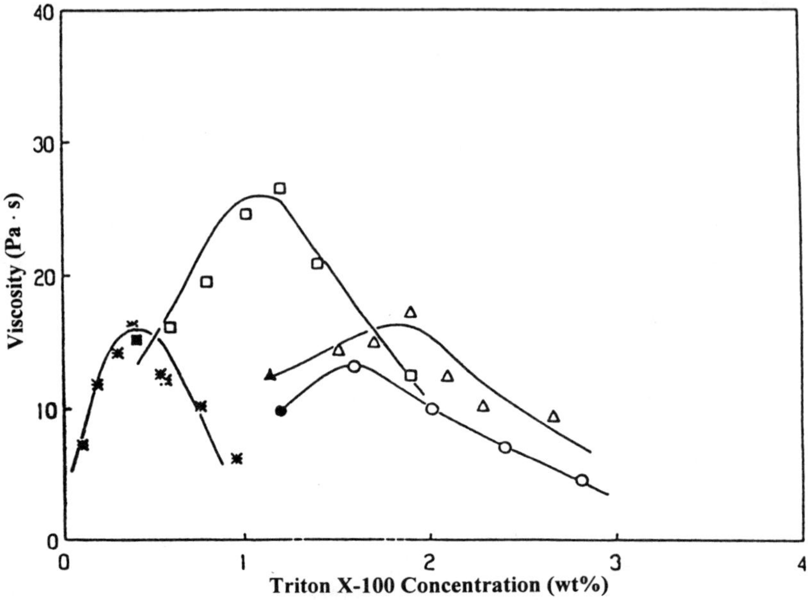

*Figure 16. Low-shear-rate viscosity (2 s^{-1}) of 2-wt% $C_{18}H_{37}$-modified POE as a function of Triton X-100 concentration. Key: ○ and ●, $(C_{18}H_{37}NHCOO)_2(EtO)_{182}$; △ and ▲, $(C_{18}H_{37}NHCOO)_2(EtO)_{331}$; □ and ■ $(C_{18}H_{37}NHCOO)_2(EtO)_{531}$; and *, $(C_{18}H_{37}NHCOO)_2(EtO)_{663}$. Closed symbols represent minimum amounts of nonionic surfactant needed to achieve solubility.*

creases with the nonionic surfactant are greater relative to SDS when the $(EtO)_x$ spacing is small.

Variations with HEUR Geometry. The HEURs synthesized in Scheme III differ from those in Schemes I and II; the EtO linkages are much shorter, and each contains a centralized nonhydrophilic unit. The positions of the surfactant maxima in HEUR thickeners are at much higher SDS surfactant levels (Figure 17). Larger separation distances are needed to elucidate the significance of multiple-arm geometries.

Variations of Viscosity with Shear Rate. The SDS concentrations discussed in this section correspond to the levels required to achieve a viscosity maximum under steady-state conditions with the HEUR studied. For brevity, only the largest hydrophobe is discussed, as a parallel to the coating formulations discussed in Chapter 24. Significant solution viscosity increases are observed when SDS is added to

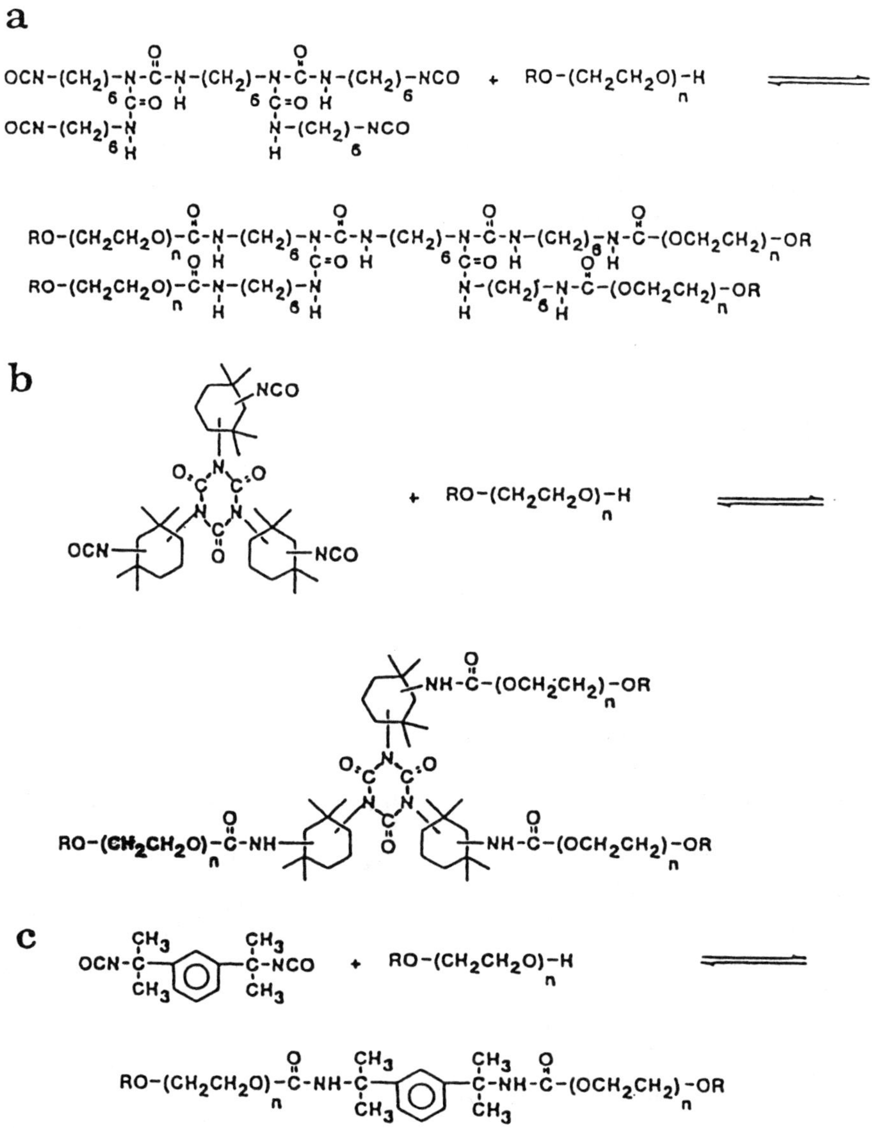

Scheme III. Synthesis of HEURs with internal hydrophobes by the addition of ethoxylated nonylphenol to biuret of HDI (a), isocyanurate of isophorone diisocyanate (b), and m-*tetramethylxylene diisocyanate (c).*

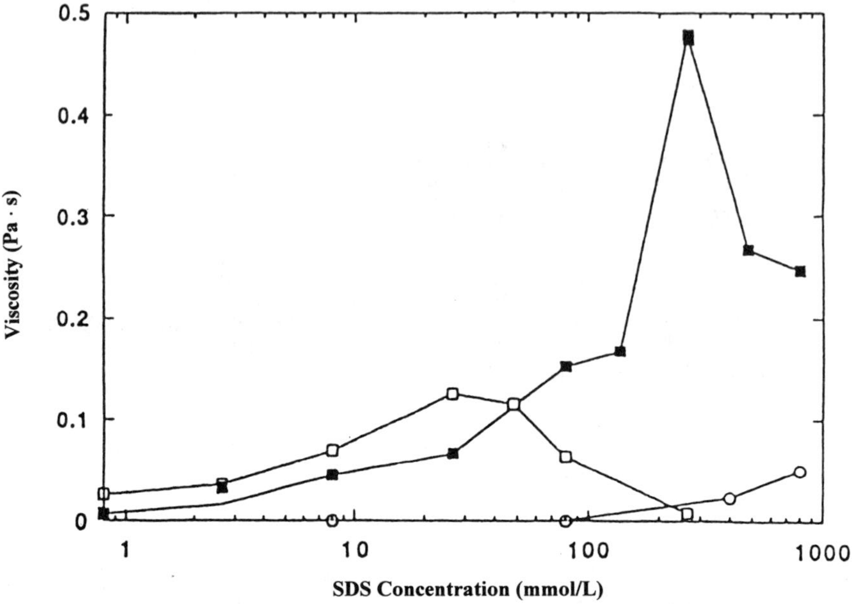

Figure 17. Effect of SDS concentration on viscosity (40 s^{-1}) of HEUR thickeners with 4-wt% $(C_9H_{19}O(EtO)_{100})_2$TMXDI solution (□), 4-wt% $(C_9H_{19}O(EtO)_{100})_4(HDI)_B$ solution (■), and SDS only (○).

partially substituted HEURs that contain mixtures of di- and monosubstituted HEURs and unreacted POE. For example, the viscosity variances of $[C_{18}H_{37}NHC(O)O]_{1.4}(EtO)_{331}$ with increasing shear rate (total scan for these deformation studies was 3 min) is given in Figure 18. Shear thickening is observed at moderate shear rates with this partially modified HEUR, particularly in the presence of SDS.

The addition of SDS to the fully substituted smaller-hydrophobe $(NPIP)_2(EtO)_{331}$ thickener effects a viscosity increase comparable to twice the concentration of this fully substituted telechelic HEUR without SDS (Figure 19). In the absence of SDS, this fully modified thickener has a swollen gel appearance, and at lower concentrations (e.g., 4 wt%), it exhibits a clear nonviscous upper layer not observed in the 8-wt% solution. The higher-concentration (i.e., 8-wt%) solution and the 4-wt% solution containing 0.024 M SDS appear to approach a yield stress. The surfactant solution network is unstable at higher deformation rates relative to the higher concentration of polymer without SDS (Figure 19). When surfactants are used at concentrations beyond that required for the viscosity maximum at low (2 s^{-1}) shear rates, the

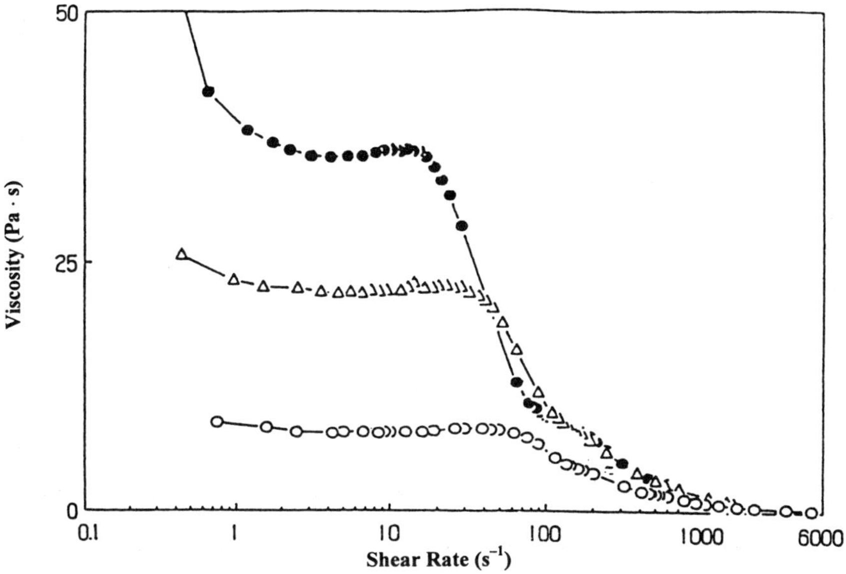

Figure 18. Viscosity as a function of shear rate (scan completed in 3 min) for $(C_{18}H_{37}NHCOO)_{1.4}(EtO)_{331}$ thickener. Key: ○, 4-wt% solution; △, 5-wt% solution; ●, 4-wt% solution with 0.012 M SDS.

HEUR thickener viscosity remains constant over a greater range of higher deformation rates (*32*).

Oscillatory Rheology Studies. The interpretations just given are supported by oscillatory measurements that allow the network structure to be estimated from the elastic component (storage modulus). With the larger hydrophobe [i.e., $(\text{octadecyl})_x(\text{POE})_y$], strong gels are observed with all spacer links, and soluble SDS solutions are very elastic. The partially substituted product $[C_{18}H_{37}NHC(O)O]_{1.4}$-$(EtO)_{331}$ promotes large viscosity increases but exhibits a dominant elastic response only at high oscillatory frequencies and high concentration (8 wt%; Figure 20). The addition of SDS promotes a significant network in the 4-wt% $[C_{18}H_{37}NHC(O)O]_{1.4}(EtO)_{331}$ solution (Figure 21), even though this solution is a mixture of components that include mono- and unsubstituted POE. The surfactant system exhibits a longer relaxation time (the reciprocal of the frequency at the storage/loss moduli crossover point [*33*]), and this longer time reflects a more complex aggregate than is formed by the HEUR at 8 wt% without SDS. This partially substituted "model" HEUR was selected to highlight the data

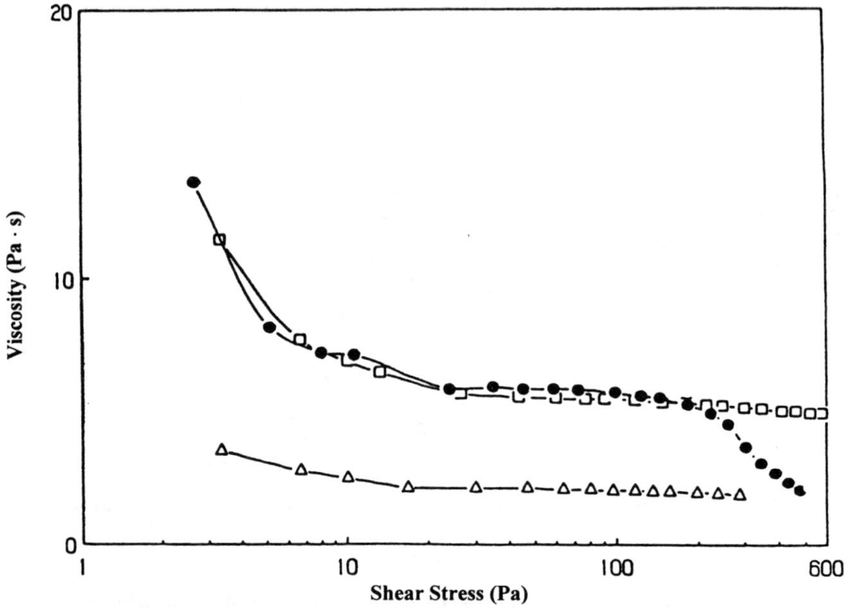

Figure 19. Viscosity as a function of shear stress for $(NPIP)_2(EtO)_{331}$ thickener (scan completed in 3 min). Key: △, *5-wt% solution;* □, *8-wt% solution; and* ●, *4-wt% solution with 0.024 M SDS.*

one can obtain from "inadequate models" that appear real enough to model associative thickener behavior.

These descriptions reflect only an overview of part of our studies of model HEUR associative thickeners, but they provide a point of comparison with some of our studies with other associative thickeners, HMHEC and HASE.

Other Associative Thickener Types. Hydroxyethyl cellulose (HEC) has been a commercial product on a major scale for three decades. It is a derivative of the world's most abundant polymer, cellulose. The synthesis and characteristics of HEC are described in Chapters 8 and 10. According to that characterization, HEC can easily be classified as a model thickener. The hydrophobe-modified version (HMHEC) contains three (*34*) hydrophobes, $C_{12}H_{25}$ to $C_{16}H_{33}$ in size, but their spacing among the repeating glucopyranosyl rings is unknown. With this understanding, we consider the response of HMHEC in aqueous surfactant solutions.

In the HEUR solutions just discussed, the maximum in viscosity with increasing SDS concentration is a function of the spacing between terminal hydrophobes and perhaps the geometry of the HEUR.

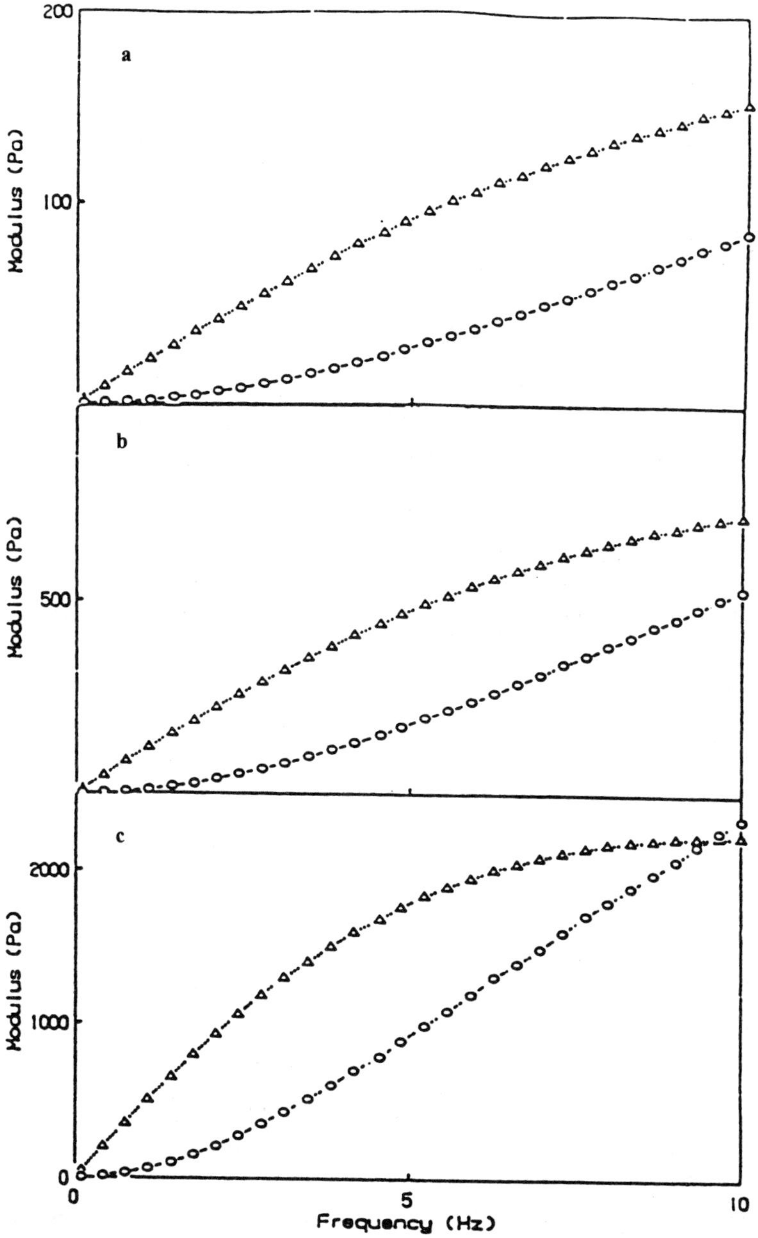

Figure 20. Storage and loss moduli as a function of frequency for $(C_{18}H_{37}NHCOO)_{1.4}(EtO)_{331}$. (a) 4-wt% solution; (b) 5-wt% solution; and (c) 8-wt% solution. Key: ○, storage modulus; and △, loss modulus.

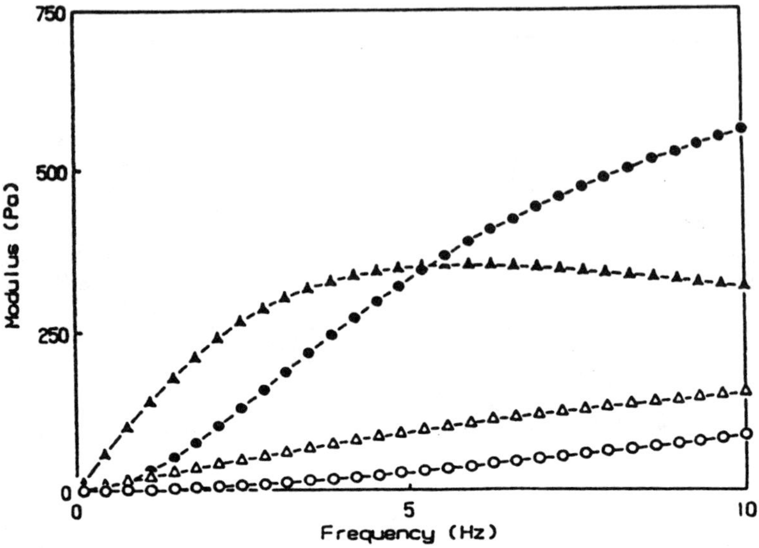

Figure 21. Storage and loss moduli as a function of frequency for 4-wt% $(C_{18}H_{37}NHCOO)_{1.4}(EtO)_{331}$. Key: ○ and ● storage modulus; and △ and ▲, loss modulus. Open symbols, no surfactant; closed symbols, 0.012 M SDS.

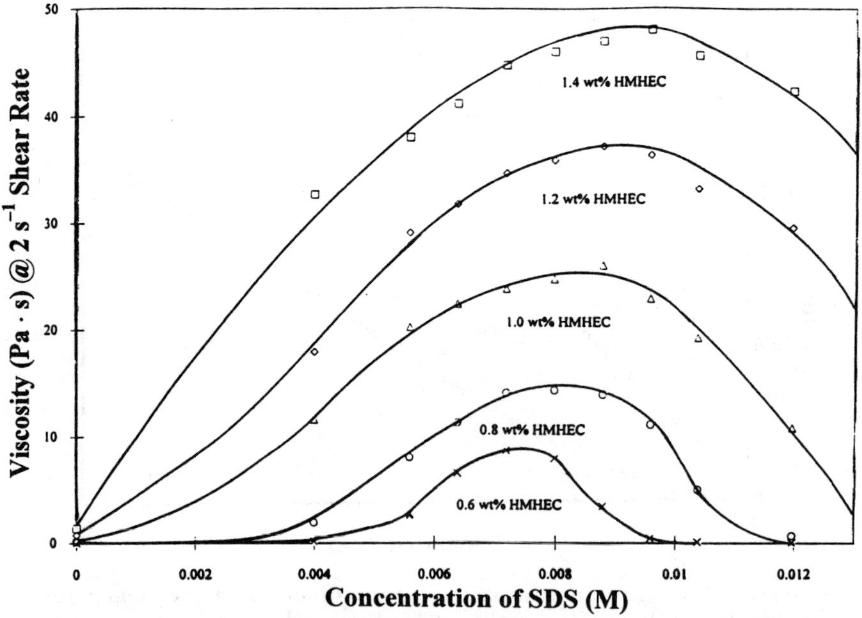

Figure 22. Effect of SDS surfactant concentration on low-shear-rate viscosity of different solids (given with each data set in figure) on HMHEC aqueous solution.

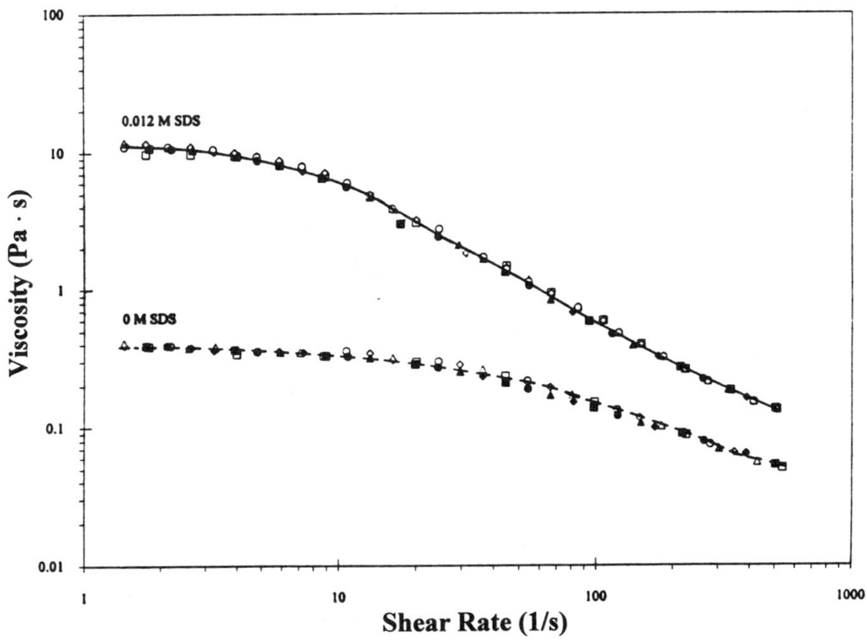

Figure 23. Rheology profile of 1-wt% HMHEC with 0 M SDS (0 cmc; broken line) and 0.012 M SDS (1.5 cmc; solid line) surfactant aqueous solution. Key: □ and ■, initial loop; ◇ and ◆, 5-min elapse; △ and ▲, 3-min elapse; and ○ and ●, 1-min elapse. Open symbols, ascending time was 3 min; closed symbols, descending time was 3 min.

The maximum in HMHEC (Natrosol Plus) occurs at approximately 1 cmc of an anionic surfactant (*21*). The maxima are also dependent on the thickener's concentration, as Figure 22 makes evident. The viscosity dependence on shear rate of a 1-wt% solution without and with 1.5 cmcs of SDS is illustrated in Figure 23 for the thickener in which the hydrophobe is, on average, only 4 EtO units from the segmentally rigid anhydroglucose repeating units. No thixotropy occurs in either solution. At a lower SDS concentration (0.5 cmc), a distinct thixotropic behavior develops (Figure 24). Shear thickening is not evident in the low shear deformation range, but an inflection at 100 s^{-1} is observed. The sample, as in the earlier studies of $C_{18}H_{37}$-HEUR samples, is allowed to rest after this initial thixotropic loop, for 5, 3, and then 1 min between cycles. With decreasing shear rate, the responses are similar for all samples, even with their different histories during the cycle of increasing shear rate. No such similar responses were observed in the lower-molecular-weight model

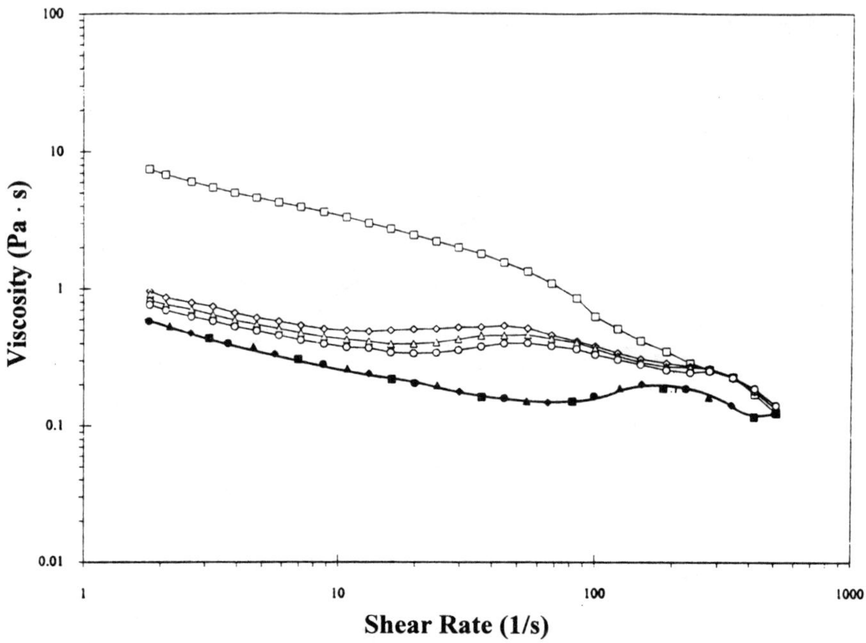

Figure 24. Hysteresis loop of 1-wt% HMHEC with 0.004 M (0.5 cmc) SDS surfactant aqueous solution. Key: □ and ■, initial loop; ◇ and ◆, 5-min elapse; △ and ▲, 3-min elapse; and ○ and ●, 1-min elapse. Open symbols, ascending time was 3 min; closed symbols, descending time was 3 min.

HEURs (Figures 4 and 5). With increasing shear rate, the HMHEC–0.5 cmc SDS solutions exhibit decreasing viscosity at 1 s^{-1} as the gestation time is decreased. As the SDS concentration is increased to 1 cmc, this distinctive pattern moves toward unity with narrower thixotropic loops. We are actively studying HMHEC with SDS in the low-cmc range.

The general characteristics of HASE associative thickeners are also given in Chapter 10. The model HASE examined in this chapter contains nonylphenol hydrophobes extended by an average of 50 EtO units from an aromatic monomer (35% by weight). Two other monomers are present in significant amounts: acrylic acid and ethyl acrylate. This combination of monomers ensures a balance of glass transition temperatures and monomer hydrophilicity that will promote the liberation of acid from the "particle's interior" upon neutralization (35) of the dispersion. A small amount of difunctional monomer is incorporated into the synthesis reaction to maintain a swollen network when

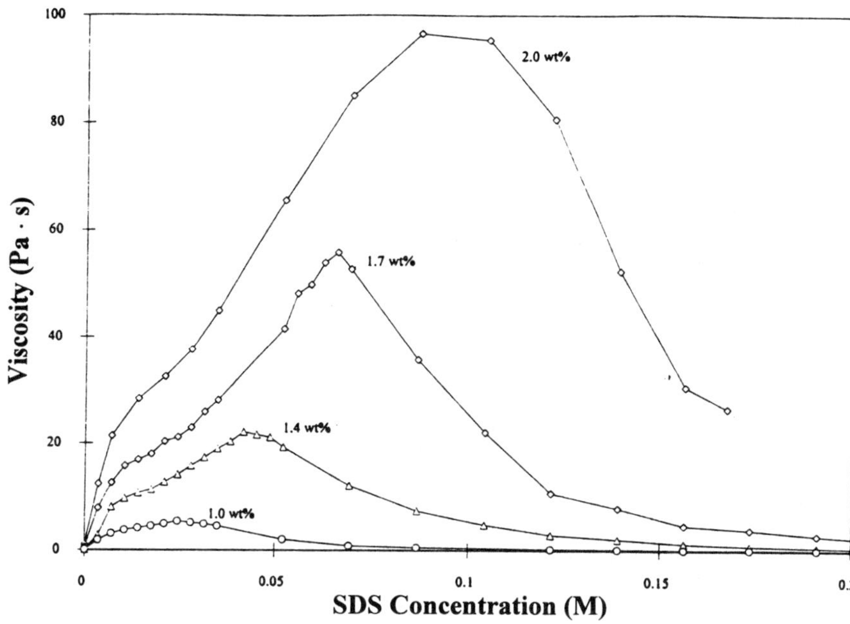

Figure 25. Viscosity of model HASE thickener (35) versus different SDS surfactant concentrations at pH = 9.2 and 2 s^{-1} shear rate and with different solids.

the acid groups are neutralized to obtain the thickened solution. This type of thickener, unlike HEURs and HMHEC, is very sensitive to electrolyte concentration (*20*).

The positions of the viscosity maxima (i.e., the amount of SDS needed) are dependent on the HASE thickener concentration (Figure 25) and its hydrophobe content (which is not demonstrated). The viscosity dependence on shear rate is illustrated in Figure 26. Little if any thixotropy is exhibited in any of the HASE–SDS combinations. The most notable influence is the shear thickening evident at SDS concentrations just above that needed for the viscosity maxima. At SDS concentrations well above that concentration, an extended Newtonian viscosity to ca. 100 s^{-1} in the 1.0-wt% HASE and to ca. 40 s^{-1} in the higher SDS levels in the 1.7-wt% HASE solutions is observed before shear thinning is observed.

In oscillatory studies (Figure 27), crossover occurs in G′ and G″ at ca. 3 Hz when the concentration of SDS is above that required to reach the viscosity maxima. This crossover would be expected if a network of associations is substantial. The decrease in the maximum viscosity observed is due, as has been suggested (*21*) for HMHEC, to

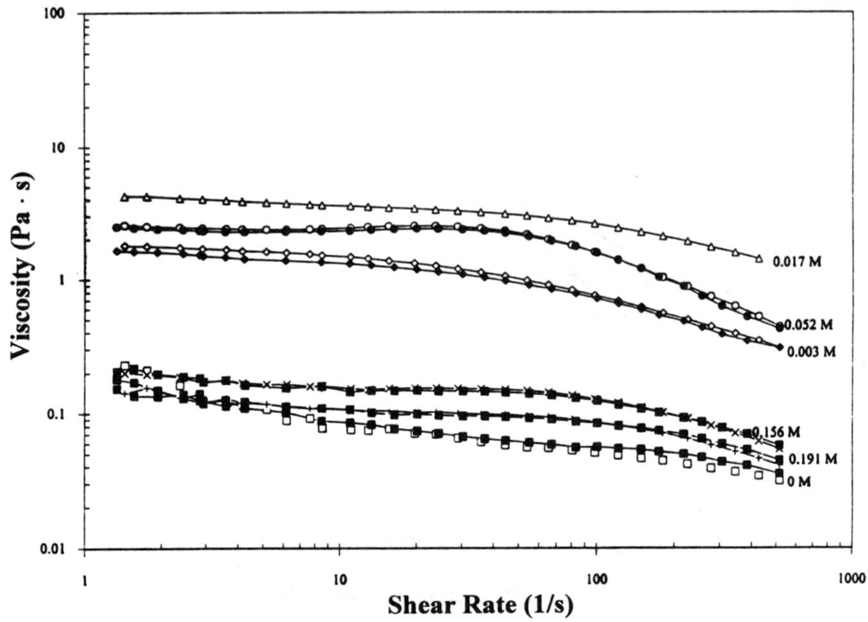

Figure 26. Viscosity versus shear rate of model HASE thickener (35) at 1.0-wt% solids with different concentrations of SDS surfactant. Key: □ and ■, 0 M; ◇ and ◆, 0.003 M; △ and ▲, 0.017 M; ○ and ●, 0.052 M; ×, 0.156 M; and +, 0.191 M. Open symbols, ascending time was 3 min; closed symbols, descending time was 3 min.

a decrease in hydrophobe participation from different HASE aggregates in an SDS micelle. Despite the high hydrophobe monomer content (35%) of this particular model HASE, electrostatic repulsions contribute significantly to viscosity build. The addition of ≥1-wt% NaCl results in a substantial viscosity drop. Another factor to consider is the influence of salinity on the shape of the SDS micelle. With increasing salinity, SDS micelles change from a spherical to a cylindrical form (25). An example of such a change is evident when the salinity is increased in nonionic HMHEC–SDS solutions. The viscosity increases until the SDS cmc level is reached, but above the cmc, the viscosity remains higher than at lower cmc levels. The magnitude of the viscosity in HMHEC above the cmc of SDS is proportional to the salinity level (36).

In Chapter 24, the behavior of some of these model associative thickeners in the presence of latices, titanium dioxide, and fully formulated latex coatings is discussed.

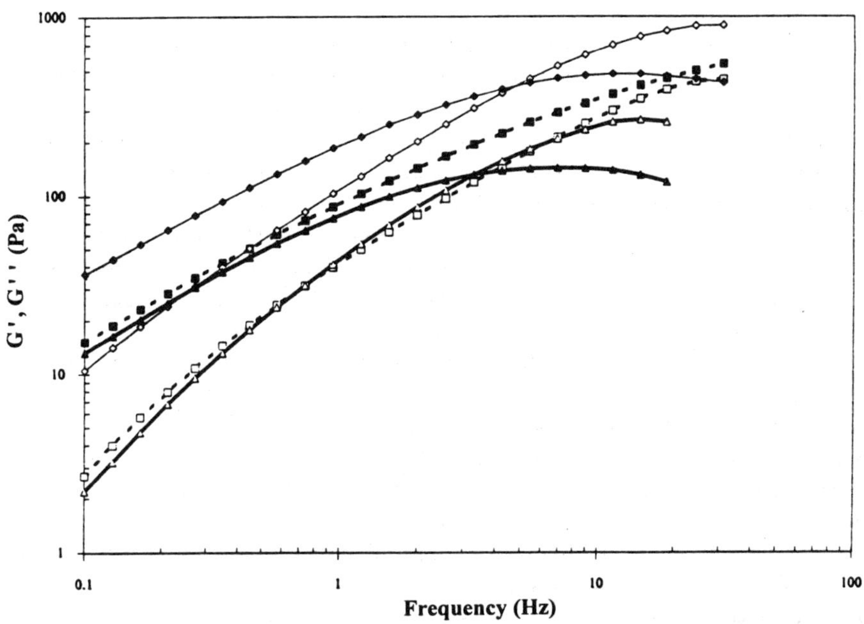

Figure 27. G′ and G″ versus frequency of model HASE thickener (35) at 1.7-wt% solids with different concentrations (M) of SDS surfactant. Key: □, 0.028 M; ◇, 0.055 M; △, 0.104 M. Open symbols, G′ (storage modulus); closed symbols, G″ (loss modulus).

Conclusions

In HEUR thickeners, the size of the terminal hydrophobe influences the magnitude of the aqueous-solution viscosity achieved. Increasing the EtO spacer length between the terminal hydrophobes also increases the solution viscosities by promoting less intramolecular and more intermolecular hydrophobic associations. This effect continues until the concentration of polymer hydrophobes with increasing EtO content beyond ca. 500 units falls below an effective level for maximum domain participation. The phenomenon also is evident with increasing anionic or nonionic surfactant concentration in studies near HEUR critical aggregation concentrations. Both surfactants form elastic networks with HEUR thickeners. When small spacer distances are used in the synthesis of HEURs with internal hydrophobes and narrow molecular weight distributions, phase separation in the aqueous solutions is observed. This phase separation is related to the dominance of intramolecular hydrophobic associations. The phenomenon can be eliminated with the addition of SDS but not of branched-chain nonionic hydrophobes at moderate concentrations. Phase separation of linear HEURs with narrow molecular weight distributions in the absence of dispersed phases is dependent on the size of the hydrophobe and the number of EtO units between the hydrophobes.

HEURs prepared by S-G polymerizations can be reproducibly synthesized if care and understanding of the causes affecting molecular weight in S-G process are understood. However, the products are a mixture of components that vary in molecular weight and hydrophobe content. Fluorescence (using an extrinsic probe) and viscosity measurements reveal that internal alkyl groups (the alkyl groups of the diisocyanates) do not build viscosity through hydrophobic aggregation. A dramatic increase in viscosity as a function of thickener concentration is observed with increasing effective terminal-hydrophobe size, defined as the combination of the long-chain alkyl amine and the alkyl group of the diisocyanate used to couple the amine to the POE. Increasing the molecular weights of the S-G HEURs while keeping the terminal-hydrophobe size constant effects lower viscosity efficiency in compositionally defined HEUR polymers. Phase separation is not observed in step-growth HEURs because of the broad molecular weight distributions and the presence of in situ generated nonionic surfactants.

Acknowledgments

The financial support of these studies by DuPont, James River, and DSM Companies is gratefully acknowledged.

References

1. Kaczmarski, J. P.; Glass, J. E. *Macromolecules* **1993,** *26,* 5149–5156.
2. Lundberg, D. J.; Brown, R. G.; Glass, J. E.; Eley, R. R. *Langmuir* **1994,** *10*(9), 19.
3. Fonnum, G.; Kaqkke, J.; Hansen, F. K. *Colloid Polym. Sci.* **1993,** *271,* 380.
4. Annable, T.; Buscall, R.; Ettelaie, R.; Whittlestone, D. *J. Rheol.* **1993,** *37*(4), 695.
5. Bock, J.; Siano, D. B.; Valint, P. L.; Pace, S. J.; Schulz, D. N. In *Polymers in Aqueous Media: Performance Through Association;* Glass, J. E., Ed.; Advances in Chemistry 223; American Chemical Society: Washington, DC, 1989; Chapters 21 and 22.
6. Sarrazin-Cartalas, A.; Lliopoulos, A. R.; Olsson, U. *Langmuir* **1994,** *10,* 1421.
7. Landoll, L. M. *J. Polym. Sci., Polym. Chem. Ed.* **1982,** *20,* 443.
8. Donbrow, M. In *Nonionic Surfactants Physical Chemistry;* Schick, M. I., Ed.; Surfactant Science Series 23; Marcel Dekker: New York, 1987; pp 1011–1072.
9. Abraham, R. J.; Fisher, J.; Loftus, P. *Introduction to NMR Spectroscopy;* John Wiley: New York, 1988.

10. Char, K.; Frank, C. W.; Gast, A. P.; Tang, W. T. *Macromolecules* **1987,** *20,* 1833.
11. Lundberg, D. J.; Glass, J. E.; Eley, R. R. *J. Rheol.* **1991,** *35*(6), 1255–1274.
12. Mutsuhisa, F.; Yokoyama, T. *J. Polym. Sci., Polym. Chem.* **1986,** *24,* 3291.
13. Thompson, C. M.; Taylor, S. G.; McGee, W. M. *J. Polym. Sci., Polym. Chem.* **1990,** *28,* 333.
14. Kaczmarski, J. P.; Glass, J. E. *Polym. Mater. Sci. Eng.* **1991,** *65,* 175.
15. Kaczmarski, J. P.; Glass, J. E. *Langmuir* **1994,** *10*(9), 3035.
16. Ikemi, M.; Noboyoki, O.; Shinohara, I.; Chiba, A. *Macromolecules* **1982,** *15,* 281.
17. Van Alstine, J. M.; Sharp, K. A.; Brooks, D. E. *Colloids Surf.* **1986,** *17,* 115.
18. Boussouira, B.; Ricard, A. *Polym. Bull.* **1988,** *19,* 193.
19. Billmeyer, F. W., Jr. *Textbook of Polymer Science;* John Wiley: New York, 1971; pp 84–90.
20. Tarng, M.-R. Ph.D. Thesis, North Dakota State University, 1995.
21. Sau, A. C.; Landoll, L. M. In *Polymers in Aqueous Media: Performance Through Association;* Glass, J. E., Ed.; Advances in Chemistry 223; American Chemical Society: Washington, DC, 1989; p 343.
22. Bergh, J. S.; Lundberg, D. J.; Glass, J. E. *Prog. Org. Coatings* **1989,** *17,* 155.
23. Kurzendorfer, C. P.; Schwuger, M. J.; Lange, H. *Ber. Bunsenges. Phys. Chem.* **1978,** *82,* 962.
24. Meguro, K.; Akasu, H.; Ueno, M. *J. Am. Oil Chem. Soc.* **1976,** *53,* 145.
25. Jones, M. J. *J. Colloid Interface Sci.* **1967,** *23,* 36.
26. Zana, R.; Lang, J.; Lianos, P. In *Microdomains in Polymer Solutions;* Dubin, P., Ed.; Plenum: New York, 1985; Chapter 20.
27. Cabane, B. *J. Phys. Chem.* **1977,** *81,* 1639.
28. Cabane, B.; Duplessix, R. *J. Phys. (Paris)* **1982,** *43,* 1529.
29. Cabane, B.; Duplessix, R. *Colloids Surf.* **1982,** *13*(1), 19–33, 66–72.
30. Kuo, C.; Provder, T.; Kah, A. F. *Paint Resins* **1983,** April.
31. Ma, Z.; Kaczmarski, J. P.; Glass, J. E. *Polym. Mater. Sci. Eng. Prep.* **1992,** *66,* 23.
32. Kaczmarski, J. P. Ph.D. Thesis, North Dakota State University, 1993.
33. Ferry, J. D. *Viscoelastic Properties of Polymers,* 3rd ed.; John Wiley: New York, 1980.
34. Goodwin, J. W. In *Polymers in Aqueous Media: Performance Through Association*; Glass, J. E., Ed.; Advances in Chemistry 223; American Chemical Society: Washington, DC, 1989; Chapter 19.
35. Hoy, K. L. *J. Coating Technol.* **1979,** *51,* 27.
36. Ma, Z. Ph.D. Thesis, North Dakota State University, 1993.

RECEIVED for review April 16, 1994. ACCEPTED revised manuscript May 10, 1995.

18

Model Hydrophobically End-Capped Poly(ethylene oxide) in Water

E. Alami, M. Rawiso, F. Isel, G. Beinert, W. Binana-Limbele, and J. François

Institut Charles Sadron Centre National de la Recherche Scientifique–Université Louis Pasteur, 6 rue Boussingault, 67083 Strasbourg Cedex, France

Model hydrophobically end-capped poly(ethylene oxides) were synthetized with protonated and deuterated end groups. Their associative properties in water as a function of their hydrophobicity were studied by viscosimetry, turbidimetry, fluorescence, and neutron scattering. Different concentration thresholds that allow the description of the different steps of association were identified. The neutron-scattering experiments can be interpreted by a model in which end groups are assumed to be gathered in spherical micelles. These micelles are organized in a liquid cubic lattice at higher concentrations.

Among the commercial polymers used today in aqueous paint formulations (ethylene oxide urethane copolymers (HEUR), polyacrylates, and cellulose derivatives), HEURs generally have a better efficiency despite their lower molecular weights (*1–4*). This higher efficiency seems to be the result of a subtle balance between polymer–polymer interactions and polymer–latex interactions (4–7) via hydrophobic tails. In the absence of other additives, the HEUR end groups are assumed to form micellelike clusters. Bridging of these clusters by polymer chains induces a large increase in viscosity. Despite the large number of studies of these systems, information about the sizes and shapes of the polymer aggregates and about the aggregation numbers of the hydrophobic microdomains that play the role of

0065–2393/96/0248–0343$12.00/0

cross-links is lacking. In fact, most of the studies are based on rheological measurements that cannot provide such information directly. Besides, most of the polymer samples studied have been obtained by condensation (5). Their polydispersity is consequently rather large (*6*, *8*), and the polymer chains contain urethane groups, which may have some influence on the association phenomena. These observations led some authors to undertake a series of studies restricted to the binary (polymer–water) and ternary (polymer–water–surfactant) systems but using a wide range of techniques, including fluorescence (*9–12*), light scattering (*8*, *13*), rheology (*6*), solubilization experiments (*8–10*, *13*, *14*), and NMR self-diffusion measurements (*15*, *16*).

This chapter presents results obtained with model associative polymers of low polydispersity and with no urethane groups. In some cases, the paraffinic chain ends were deuterated to determine the shapes and sizes of the hydrophobic microdomains by neutron scattering. Other, more classical properties of these samples are also discussed.

Experimental Details

Materials. The model polymers exhibit a simple chemical structure (**1**).

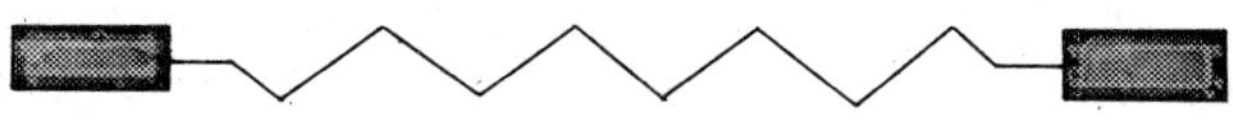

1

In this structure, the poly(ethylene oxide) (PEO) chain () has a molecular weight M_{PEO} ranging from 2,000 to 35,000, and the number of carbon atoms in the paraffinic chain (N_p; ) ranges from 8 to 18. Chains are denoted $C_{H_{N_p}}$–M_{PEO}–$C_{H_{N_p}}$ or $C_{D_{N_p}}$–M_{PEO}–$C_{D_{N_p}}$ depending on whether the extremities are protonated or deuterated, respectively.

The samples with protonated chain ends were prepared in two steps: (1) preparation of *n*-dodecyl *p*-toluene sulfonate (DPS) by the action of *p*-toluenesulfonyl chloride on a 1-alcohol in the presence of pyridine, according to a classical procedure (*17*, *18*); (2) reaction between DPS and an α,ω-dihydroxylated PEO previously metalled by potassium diphenylmethane:

$$K^+\text{–(–O–}CH_2\text{–}CH_2\text{–)}n\text{O–}K^+ + 2\ CH_3\text{–}\varphi\text{–}SO_3\text{–}(CH_2)_{N_p-1}\text{–}CH_3$$

$$\downarrow$$

$$CH_3\text{–}(CH_2)_{N_p-1}\text{–(O–}CH_2\text{–}CH_2\text{–)}n\text{–O–}CH_2)_{N_p-1}\text{–}CH_3 + 2CH_3\text{–}\varphi\text{–}SO_3K$$

Samples with deuterated chain ends were prepared by the same procedure except that DPS was replaced with an *n*-deuterated alkyl bromide (Williamson reaction).

The samples were characterized by size exclusion chromatography in tetrahydrofuran. The chromatograms exhibit only one peak, which corresponds to that of the unmodified PEO as opposed to those of the commercial samples prepared by polycondensation.

The degree of substitution, τ, of the hydroxyl groups were determined by ^{13}C, ^{1}H, and ^{2}D NMR spectroscopy and elemental analysis. The results were confirmed by a determination of the residual OH groups by UV spectroscopy after reaction of the polymer with naphthyl isocyanate. The value of τ ranged from 80 to 95%.

Physicochemical Methods. *Turbidimetry.* The polymer solutions were prepared in sealed glass tubes and heated in a thermostatic bath. The cloud point was determined visually.

Static Fluorescence. Fluorescence spectra were recorded on a Hitachi F-4010 spectrofluorometer between 350 and 500 nm. Pyrene with an excitation wavelength of 335 nm was used as a fluorescent probe.

This method was used to determine the onset of polymer association, called critical micellar concentration (cmc), by analogy with low-molecular-weight surfactants and the aggregation number of the micelles or hydrophobic microdomains. The vibronic fluorescence spectra of pyrene exhibit five peaks (numbered 1 to 5). It is well known that the ratio I_1/I_3 (peak 1 intensity/peak 3 intensity) correlates well with the polarity of the immediate environment of pyrene. In a first series of experiments, I_1/I_3 was measured as a function of polymer concentration to follow the association process and to determine an order of magnitude of the cmc. In a second series, the aggregation number of the hydrophobic microdomains could be obtained from the decrease in fluorescence of the probe as a function of concentration $[Q]$ of a fluorescence quencher. The fluorescence intensity I_Q in the presence of the quencher (*19–21*) is given by

$$I_Q = I_0 \exp(-[Q]/[M]) \quad (1)$$

where I_0 is the fluorescence intensity in the absence of quencher, and $[M]$ is the concentration of micelles. $[M]$ is directly obtained from the slope of a $\ln(I_Q/I_0) = f([Q])$ plot, and the aggregation number N may be theoretically calculated from

$$N = (C_p - \text{cmc})/[M] \quad (2)$$

where C_p is the molar concentration of end groups.

Dodecylpyridinium chloride was used as a quencher, and the experiments were carried out by keeping the pyrene concentration constant (saturated solutions) and increasing the $[Q]$.

Viscosimetry. A low-shear Couette flow-type rheometer, Contraves LS30, was used to determine the solution viscosity of the samples in a shear rate range from 0.017 to 128 s^{-1}.

Small-Angle Neutron Scattering (SANS). The SANS experiments were performed on the SANS instrument (PACE) of the L. Brillouin laboratory (LLB Saclay, France). In a first configuration, the wavelength λ and the sample–detector distance D were set at 6 Å and 1 m, respectively. In a second configuration, λ = 10.5 Å and D = 3 m. The whole range of scattering vector Q covered was $0.006\ \text{Å}^{-1} < Q < 0.32\ \text{Å}^{-1}$ ($Q = 4\pi \cdot \sin(\theta/2)\lambda$), where θ is the scattering angle.

We investigated the concentration regime $2\% < c < 55\%$ w/w at 25 °C. The solutions were prepared in an H_2O/D_2O mixture (17.6% D_2O), which allows us to match (22–25) the PEO chain contrast. Information about PEO chain conformation can also be obtained with other solvents (that study will be published elsewhere).

Absolute values of the scattered intensity $I(Q)$ were obtained from the following expression:

$$I(Q) = \frac{\dfrac{1}{V}\left(\dfrac{I^b(Q)}{T} - \dfrac{I_{cv}{}^b(Q)}{T_{cv}}\right)}{\dfrac{1}{V_{water}}\left(\dfrac{I_{water}{}^b(Q)}{T_{water}} - \dfrac{I_{cv}{}^b(Q)}{T_{cv}}\right)}\left(\frac{d\Sigma}{d\Omega}\right)_{water} \tag{3}$$

$I_{cv}{}^b(Q)$, $I^b(Q)$, and $I_{water}{}^b(O)$ are the scattered intensities of the empty cell, the cell filled with sample, and the cell filled with water (used as a reference), respectively. T_{cv}, T, and T_{water} are the transmission values for the same cells. V and V_{water} are the scattering volumes of the sample and of water, respectively. $(d\Sigma/d\Omega)_{water}$ is the effective cross-section of the water sample at the incident wavelength.

If ΔI is the excess of coherent scattering intensity of the solution with respect to the solvent, then ΔI is calculated by

$$\Delta I(Q) = I(Q) - (1 - \Phi_p)I(Q)_{D_2O} - \Phi_p I_p{}^{inc}(Q) \tag{4}$$

where Φ_p is the volume fraction of the polymer, and $I_p{}^{inc}(Q)$ is the incoherent scattering intensity of the pure PEO. $I_p{}^{inc}(Q)$ is measured from the blanks of increasing concentration of unmodified PEO in a 17.6% D_2O/H_2O mixture.

The specific scattering-length densities of the solvent ($\rho_s = 0.66 \times 10^{10}\ cm^{-2}$) and of the paraffinic deuterated end group ($\rho = 7.33 \times 10^{10}\ cm^{-2}$) are obtained from the tabulated atomic scattering lengths (*26*) and atomic numbers (*27*).

Results

Solubility. The solubilities of the model polymers were measured by variations of the cloud point plotted against concentration. Some examples of the results obtained for the series C_{H12}–PEO–C_{H12}

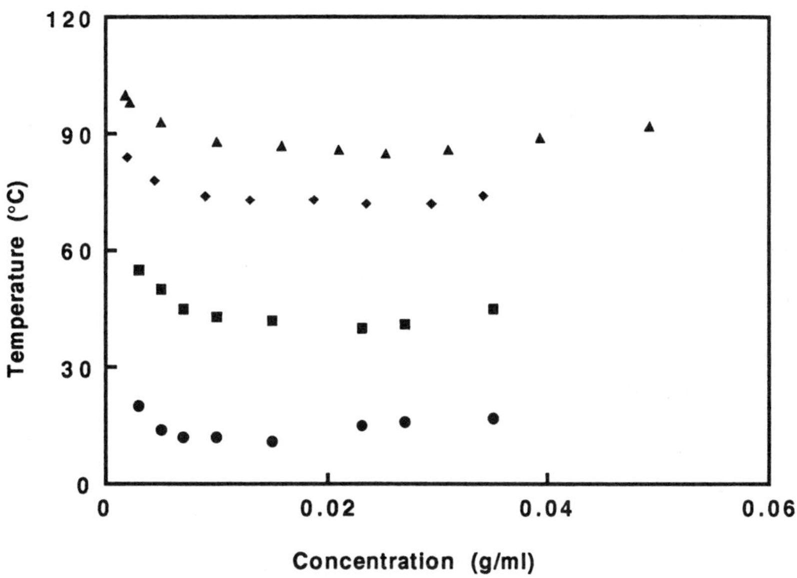

Figure 1. Cloud points of model polymer solutions. From bottom to top: C_{H12}–6000–C_{H12}(●), C_{H12}–10000–C_{H12}(■), C_{H12}–20000–C_{H12}(◆), and C_{H12}–35000–C_{H12}(▲).

are given in Figure 1. The demixing curves exhibit the same features as those of homopolymers and low-molecular-weight nonionic surfactants (*28–30*). When M_{PEO} increases at a given N_p, the lower critical solution temperature (LCST) tends toward 100 °C, which is the value for a PEO corresponding to an infinite-molecular-weight sample. In Figure 2 the demixing curves obtained with the unmodified PEO (M = 6000) (*30*) and the modified C_{D12}–6000–C_{D12} can be compared. The deuteration of the chain ends favors solubilization: the LCST of the deuterated C_{D12}–6000–C_{D12} is 30 °C, which can be compared to the 13 °C obtained for the hydrogenated homologous sample. Nevertheless, such a comparison should be made for the same degree of substitution τ, which is not exactly the case. Besides, the solubilities of these polymers are not simply related to the average hydrophobic–hydrophilic balance. In Figure 3 the LCSTs are plotted as a function of the hydrophobicity expressed by M_o/M (M_o is the molecular weight of the two paraffinic chains, and M is the molecular weight of the polymer) for our model polymers together with diblock nonionic surfactants of the $OH(CH_2–CH_2–O)_x–C_{H12}$ type. Triblock copolymers are insoluble in water in a range of hydrophobicity at which diblock polymers are expected to exhibit an LCST higher than 100 °C. This result illustrates the effect of the triblock structure quite well.

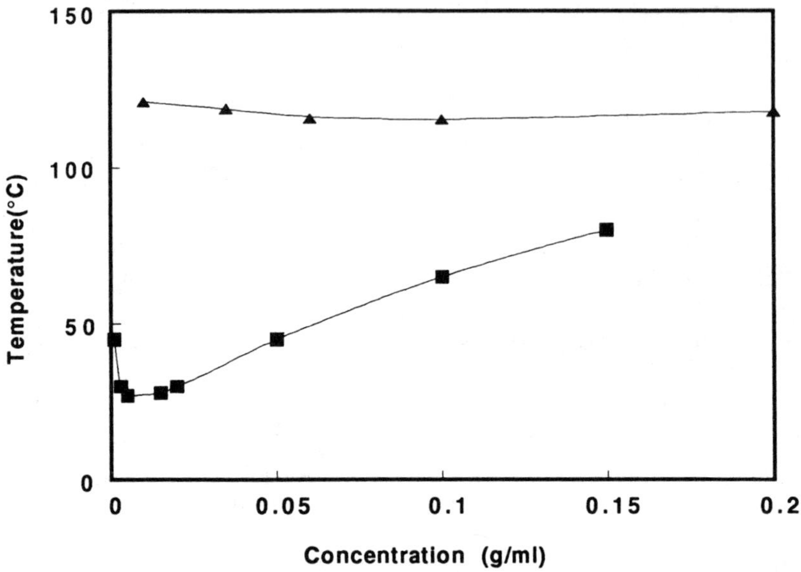

Figure 2. Cloud points of the unmodified PEO (M = *6000) (▲) and the model polymer C_{D12}–6000–C_{D12}(■).*

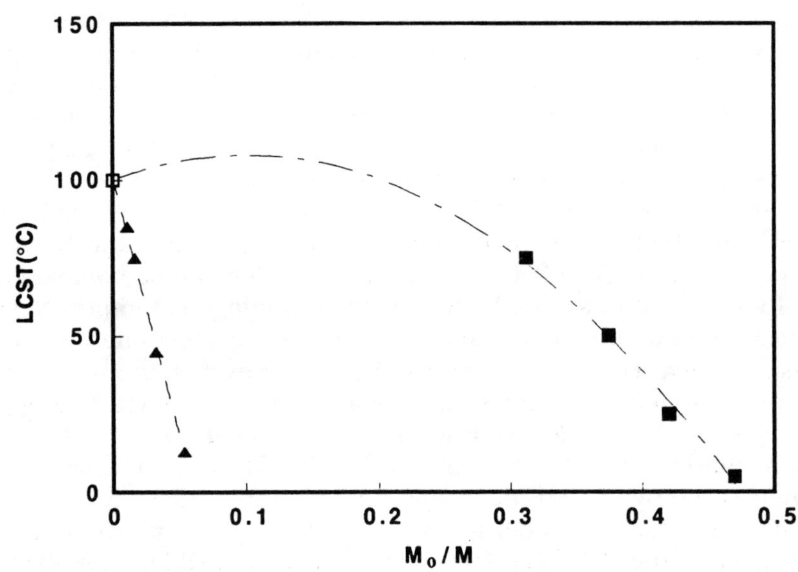

Figure 3. Variation in LCST versus M_0/M *for diblock nonionic surfactants* $OH(CH_2–CH_2–O)_x–C_{H12}$(▲) *and triblock model polymers* $C_{N_p}–M_{PEO}–C_{N_p}$(■).

Table I. Aggregation Numbers of Hydrophobic Microdomains as a Function of Nature and Concentration of Polymers

C_{H12}–M_{PEO}–C_{H12}	C_p (g/mL)	cmc domains (g/mL)	N
6000 (D)[a]	1.8×10^{-3}	3×10^{-4}–4×10^{-3}	6
6000	2.1×10^{-2}	3×10^{-4}–4×10^{-3}	12
10000	2.3×10^{-2}	4.6×10^{-4}, 6.7×10^{-3}	22
20000	5×10^{-2}	2×10^{-3}, 3×10^{-2}	20
35000	5×10^{-2}	3×10^{-3}, 6×10^{-2}	14

[a] The symbol (D) indicates that the polymers have deuterated end groups.

cmc and Aggregation Number *N*. The micellization of the model polymers was investigated by static fluorescence; the ratio I_1/I_3 is a good indicator of the polarity of the probe microenvironment, particularly in hydrophobic microdomains. It is well known that the micellization of low-molecular-weight surfactants is accompanied by an abrupt decrease in I_1/I_3 (over half a magnitude of concentration). This decrease indicates a cooperative process. Moreover, this ratio reaches a plateau of around 1.2 at higher concentrations. By comparison, the change in I_1/I_3 for our model polymers occurs over a wider range of concentration (see Table I and Figure 4), and the final value

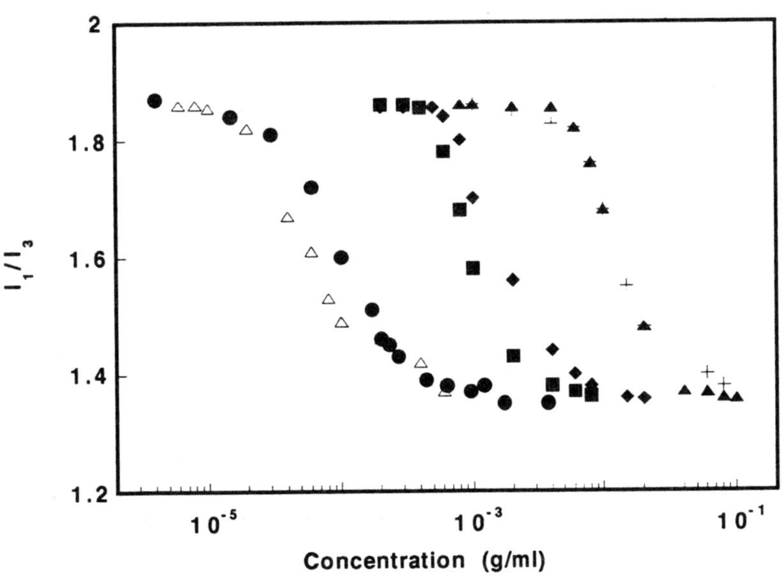

Figure 4. Variation in the intensity ratio I_1/I_3 for pyrene in mixtures of water with C_{D12}–6000–C_{D12}(■), C_{H12}–6000–C_{H12}(●), C_{H12}–10000–C_{H12}(◆), C_{H12}–20000–C_{H12}(▲), C_{H12}–35000–C_{H12}(+), and a commercial sample (△).

is 1.4. The association of the hydrophobic parts is a less cooperative process than that of low-molecular-weight surfactants, and the structure of the hydrophobic microdomains is open enough to allow the probe to have some contact with water. On the other hand, the micellization covers 1 order of magnitude of concentration, while the change in I_1/I_3 is observed over 3 orders of magnitude for a commercial polymer previously characterized (*8*). This result shows the large influence of sample polydispersity on the cooperativity of the association. Moreover, the CMC of the deuterated sample (C_{D12}–6000–C_{D12}) is 10 times greater than that of the hydrogenated one, in agreement with the higher LCST. Figure 5 shows that the CMC of the triblock polymers decreases more abruptly with M_0/M than does that of diblock surfactants. This result is consistent with the lower solubility demonstrated by Figure 3.

A few examples of $\ln(I_Q/I_0) = f([Q])$ plots are given in Figure 6. The aggregation numbers obtained from the slopes are given in Table I. The cmcs in equation 2 were taken as the onset of the change in I_1/I_3.

These results demonstrate well the formation of hydrophobic microdomains whose aggregation is a decreasing function of M_{PEO} at a given concentration and Np.

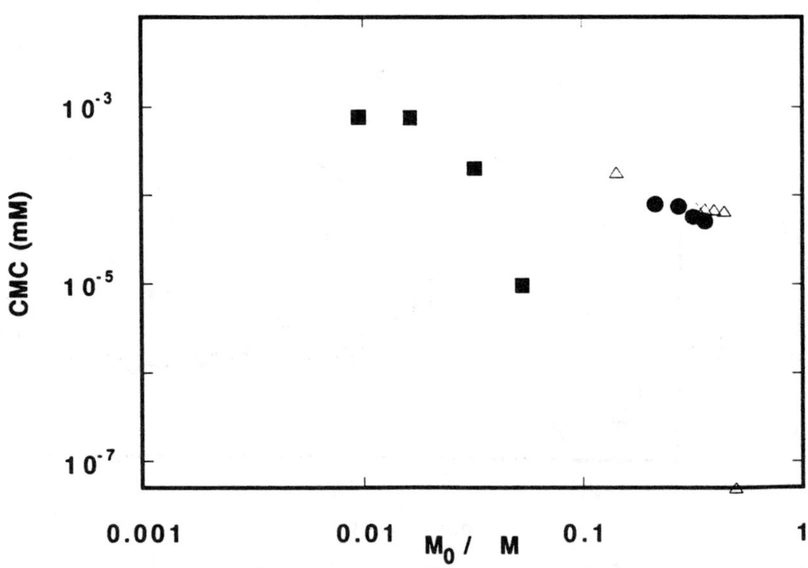

Figure 5. Variation in cmc versus M_0/M *for diblock nonionic surfactants* $OH(CH_2–CH_2–O)_x–C_{H12}$ *(△), ethoxylated nonylphenol (●), and for triblock model polymers* $C_{N_p}–M_{PEO}–C_{N_p}$*(■).*

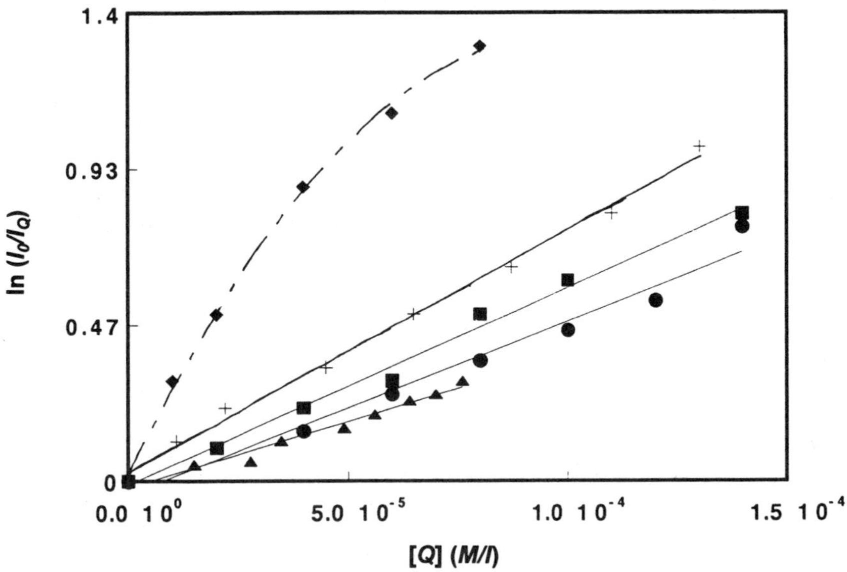

Figure 6. Variation in fluorescence intensity of pyrene versus quencher concentrations [Q] for different polymer solutions at 25 °C: C_{D12}–6000–C_{D12} (C_p = 1.8 × 10^{-3} g/mL [◆] and 2 × 10^{-2} g/mL [▲]; C_{H12}–10000–C_{H12} (C_p = 2.3 × 10^{-2} g/mL [■]), C_{H12}–20000–C_{H12} (C_p = 5 × 10^{-2} g/mL [●]), and C_{H12}–35000–C_{H12} (C_p = 5 × 10^{-2} g/mL [+]).

Viscosity. Figure 7 shows the concentration dependence of the low-shear reduced viscosity, η_{red}, of the polymer solutions of C_{H12}–10000–C_{H12}, C_{H12}–20000–C_{H12}, C_{H12}–35000–C_{H12}, at 25 °C. This temperature corresponds to an increasing difference with respect to the LCST of each compound. As described in all previous work, a divergence from the quasi-linear law is obtained with unmodified samples for a concentration $C\eta$, which is an increasing function with M_{PEO}. At a given C_p, the η_{red} of model polymers in this range of molecular weight decreases when M_{PEO} increases, as described elsewhere (*6*). Figure 8 allows us to compare the concentration dependence of η_{red} for C_{D12}–6000–C_{D12} and its unmodified homolog in a larger range of C_p. For $C_p < 6\%$, η_{red} of this polymer is lower than that of the three samples considered in Figure 7. A maximum η_{red} is found when M_{PEO} increases at constant N_p (Figure 9). This phenomenon was reported by Glass (*31*) for the model HEUR. A maximum was also found as N_p varied at constant M_{PEO} (*32*). This maximum may be explained by the two opposing effects of decreased hydrophobicity: a loss of associativity and an increase in solubility. For the lower molecular weight, the

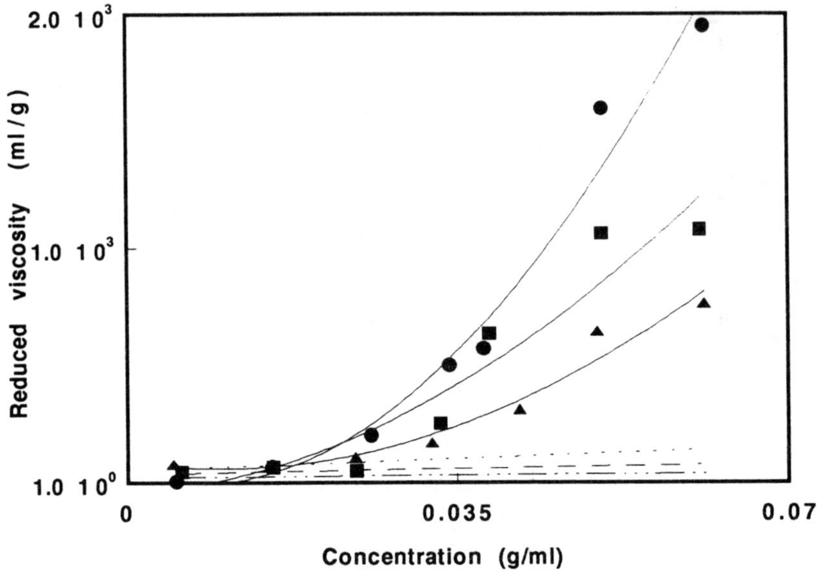

Figure 7. Reduced viscosities of model polymers in water C_{H12}*–10000–*C_{H12} *(●),* C_{H12}*–20000–*C_{H12} *(■), and* C_{H12}*–35000–*C_{H12} *(▲) and of their unmodified homologs PEO10000 (- - - - - - -), PEO20000 (–––), and PEO35000 (----) at 25 °C.*

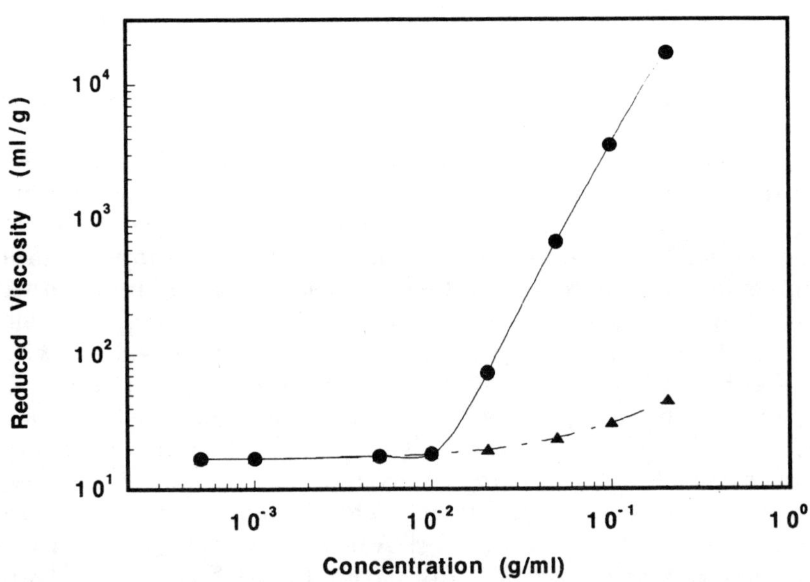

Figure 8. Reduced viscosity in water of the model polymer C_{D12}*–6000–*C_{D12}*(●) and its unmodified homolog PEO6000 (▲).*

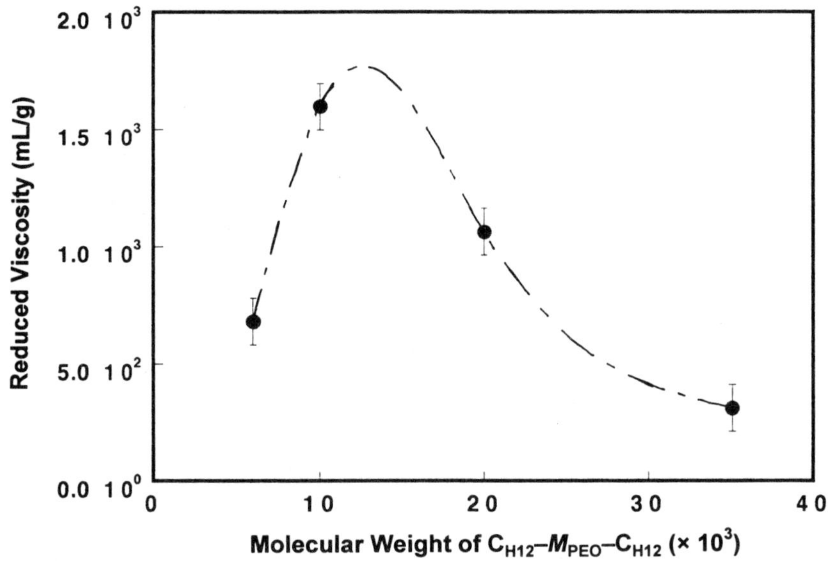

Figure 9. Variation in reduced viscosity with various molecular weights of model polymer C_{H12}–M_{PEO}–C_{H12} for C_p = 5% at 25 °C.

hydrophobicity is high enough to depress the LCST and to induce, at room temperature, the formation of very compact and quasi-insoluble aggregates whose contribution to the viscosity is negligible. When M_0/M increases, LCST increases. If the molecular weight is very large, no association is expected and the difference arising from the presence of chain ends is very small.

In Table II, the two critical concentrations cmc and $C\eta$ obtained with the C_{H12}–M_{PEO}–C_{H12} (by fluorescence and viscosity, respectively) are compared with the critical overlap concentration C^* of the unmodified PEO. C^* is calculated from the simple relation $C^* = 1/[\eta]$, where $[\eta] = 0.054 \times M^{0.66}$. The cmc and $c\eta$ are close to C^* for the

Table II. cmc Versus C_η

C_{H12}–M_{PEO}–C_{H12}	*cmc range (g/mL)*	$C\eta$ *(g/mL)*	C^* *(g/mL)*
6000	3×10^{-4}–4×10^{-3}	10^{-2}	5.9×10^{-2}
10000	4.6×10^{-4}–6.7×10^{-3}	10^{-2}	4.2×10^{-2}
20000	2×10^{-2}–3×10^{-2}	1.5×10^{-2}	2.6×10^{-2}
35000	3×10^{-3}–6×10^{-2}	2×10^{-2}	1.8×10^{-2}

NOTE: Measured values of cmc and C_η were obtained with model polymers C_{H12}–M_{PEO}–C_{H12} (by fluorescence and viscosity, respectively) at 25 °C along with the critical overlap concentration C^* of their unmodified homolog.

two higher molecular weights (M = 20,000 and 35,000). Association thus requires overlap of the chains. For the two other samples of lower molecular weight (M = 10,000 and 6,000), the association occurs for C_p = cmc, much lower than C^*, and a divergence of viscosity is observed for an intermediate C_p. In fact, $C\eta$ can be considered the critical overlap concentration of the aggregates that begin to develop at $C_p <$ cmc. If all the polymer molecules were involved in long linear aggregates at infinite dilution, the $C\eta$ would be expected to be lower than the cmc. That difference would correspond to the formation of hydrophobic microdomains in which many more than two end groups are included. The fact that $C\eta >$ cmc indicates a rather compact structure of the aggregates even if $C_p <$ cmc. This result is in agreement with a step-by-step aggregation, as deduced from fluorescence experiments.

SANS Experiments. The structures of aqueous solutions of model polymers C_{D12}–6000–C_{D12}, C_{D12}–10000–C_{D12}, and C_{D12}–20000–C_{D12}, have been investigated by SANS as a function of concentration. This chapter focuses mainly on the aggregation of deuterated paraffinic chain ends.

The SANS results were corrected according to standard procedures and represented on an absolute scale. Some typical scattering curves are shown in Figures 10 and 11. At small Q values, they exhibit one or two peaks that are related to correlations between the paraffinic microdomains.

From the plots obtained with C_{D12}–6000–C_{D12}, we observe that as the concentration increases, a disorder–order transition takes place. This transition is illustrated by the presence of only one peak at lower concentrations and the appearance of a second peak at higher concentrations. The ratio of Q values at the maxima of each peak (Q_{max2}/Q_{max1}) is close to $\sqrt{3}$ and suggests a local liquid cubic lattice. This hypothesis is supported by recent X-ray scattering studies of C_{H12}–2000–C_{H12}, and C_{H12}–4000–C_{H12} that reveal several peaks corresponding unambiguously to a crystalline cubic structure (*33*). Figure 11 also shows that the ordering in the solutions increases when the molecular weight decreases at constant N_p.

In a first approximation, the mean interparticle spacing d is given by $d = 2\pi/Q_{max1}$. The variations of d with respect to the concentration C_{Hyd} of the hydrophobic parts is represented in Figure 12 for the three polymers. These variations can be roughly expressed in terms of power laws with an absolute value for the exponent of about 1/5, which is lower than the value of 1/3 expected for a homogeneous array of particles of constant size.

In the higher Q range, oscillations of the scattered intensity are superimposed on Q^{-4} behavior, shown by plotting $Q^4 \Delta I(Q) = f(Q^4)$

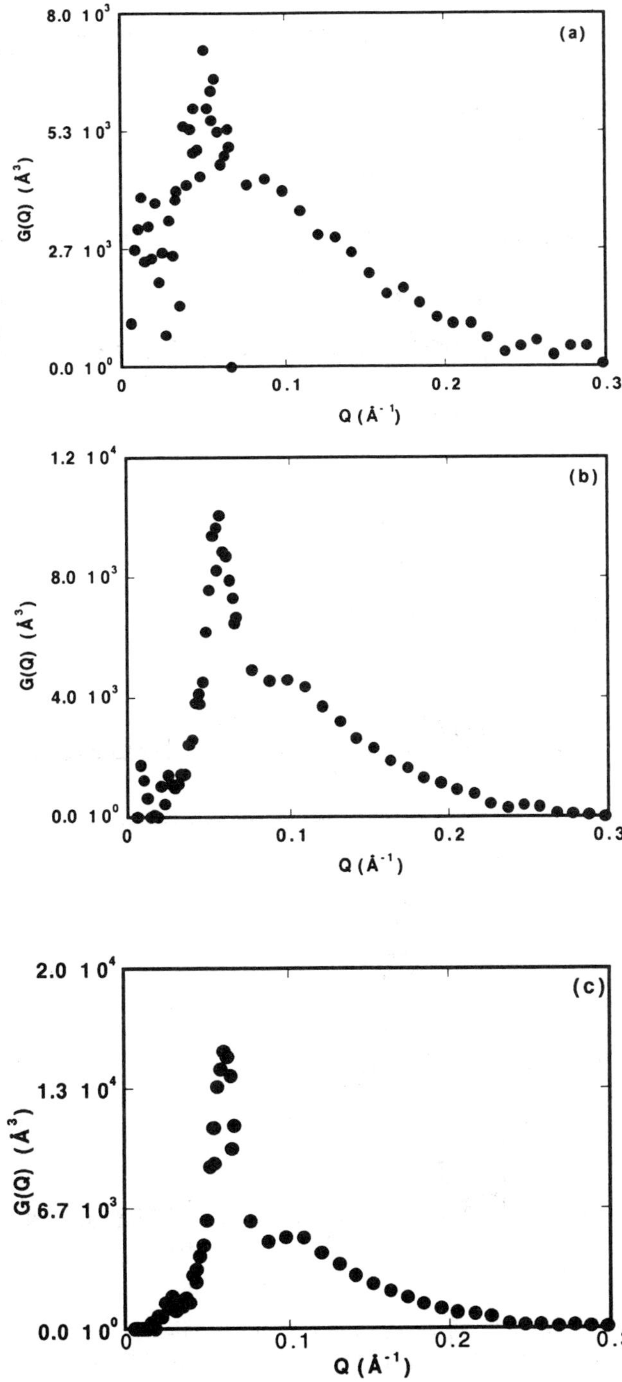

Figure 10. Neutron-scattering curves obtained with C_{D12}–6000–C_{D12} at 5% (a), 10% (b), and 15% (c).

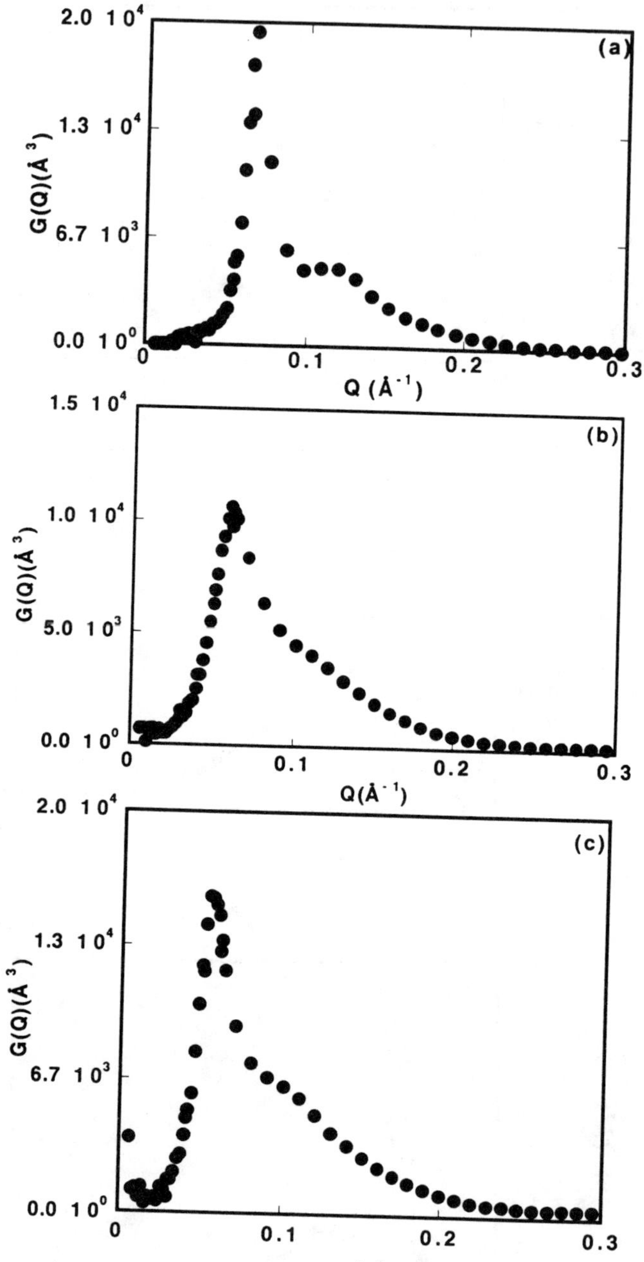

Figure 11. Neutron-scattering curves obtained with C_{D12}–6000–C_{D12} (30%) (a), C_{D12}–10000–C_{D12} (54%) (b), and C_{D12}–20000–C_{D12} (50%) (c).

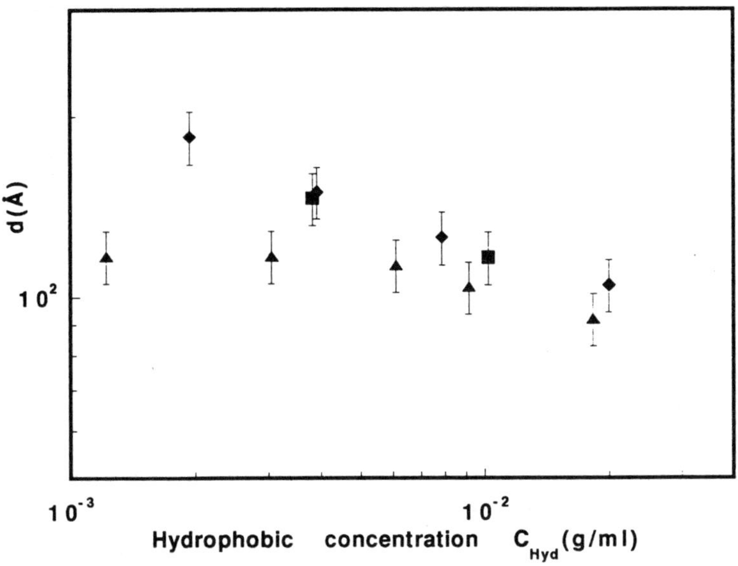

Figure 12. Variation in interparticle spacing versus concentrations of hydrophobic groups: C_{D12}–6000–C_{D12}(▲), C_{D12}–10000–C_{D12}(◆), and C_{D12}–20000–C_{D12}(■).

(Figure 13). The Q^{-4} dependence is predicted from Porod's law for a sample that exhibits a sharp boundary between two levels of scattering length density. We can deduce that the paraffinic deuterated chain ends are clustered in particles that have uniform density and sharp boundaries.

The SANS results can be interpreted on the basis of the theory of hard spherical particles. For a system of homogeneous, monodisperse spherical particles,

$$\Delta I(Q)/(\rho - r_s)^2\phi v = G(Q) \tag{5}$$

where $G(Q) = V_p P(Q) S(Q)$, and V_p is the volume of the particle.

The form factor for solid spheres of radius R is (24)

$$P(Q) = \left(\frac{3[\sin(QR) - QR\cos(QR)]}{(QR)^3}\right)^2 \tag{6}$$

The structure factor for particles at a distance r apart is given by

$$S(Q) = 1 + 4\pi n \int_0^\infty [g(r) - 1] \frac{\sin(Qr)}{Qr} r^2 \, dr \tag{7}$$

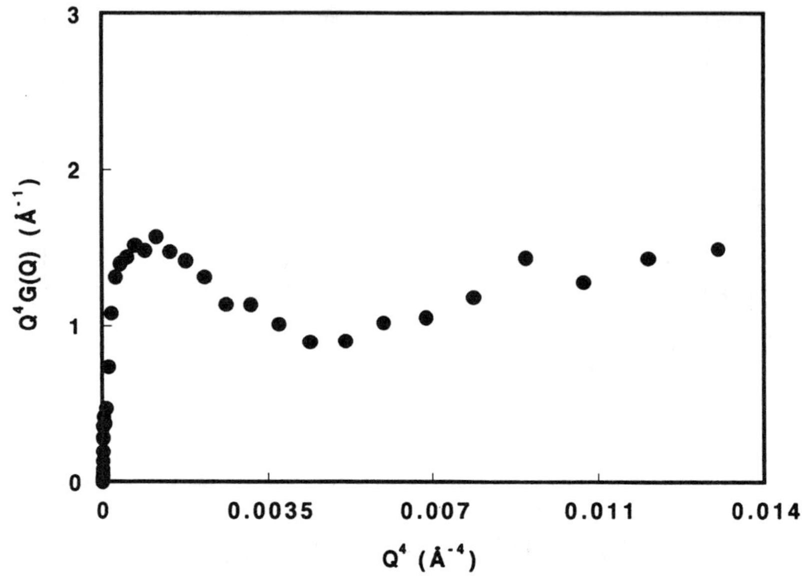

Figure 13. $Q^4G(Q) = f(Q^4)$ *plot for* C_{D12}*–6000–*C_{D12} *(30%).*

where *n* is the number of particles per unit volume, and *g(r)* is the radial distribution function, which is directly related to the interparticle potential (*34*).

For higher *Q*, *S(Q)* tends toward 1. The intermolecular correlations are negligible, and *G(Q)* becomes $G(Q) = V_pP(Q)$. By fitting the *G(Q)* values at higher *Q* using expressions (6) and (7), the radii of the particles can be deduced. Figure 14 shows an example of such a fit in which the full curve is calculated with a Schultz–Zimm distribution of the radii. Additionally, the number-averaged value of *R* can be obtained from the plateau value in the high *Q* range, in the $Q^4G(Q) = f(Q^4)$ representation. In Table III, the values of *R* calculated from the fit (R_{fit}) and the values obtained from this plateau (R_{plat}) are compared. They are in good agreement. The sizes of the hydrophobic microdomains increase when concentration is increased, although the investigated concentrations are much higher than the cmc. This result confirms the poor cooperativity of the association. *R* reaches a limit value of 17 ± 1 A at higher concentration. This *R* value corresponds to the length of the extended paraffinic C_{12} chain (*28*) and is compatible with a spherical model. The *R* values obtained at the lower concentrations correspond to an aggregation number *N* of about 8 and are consistent with fluorescence results (Table I). The polydispersity index (R_w/R_n) of the spheres is close to 1.2 for the higher concentrations and increases when the polymer concentration decreases.

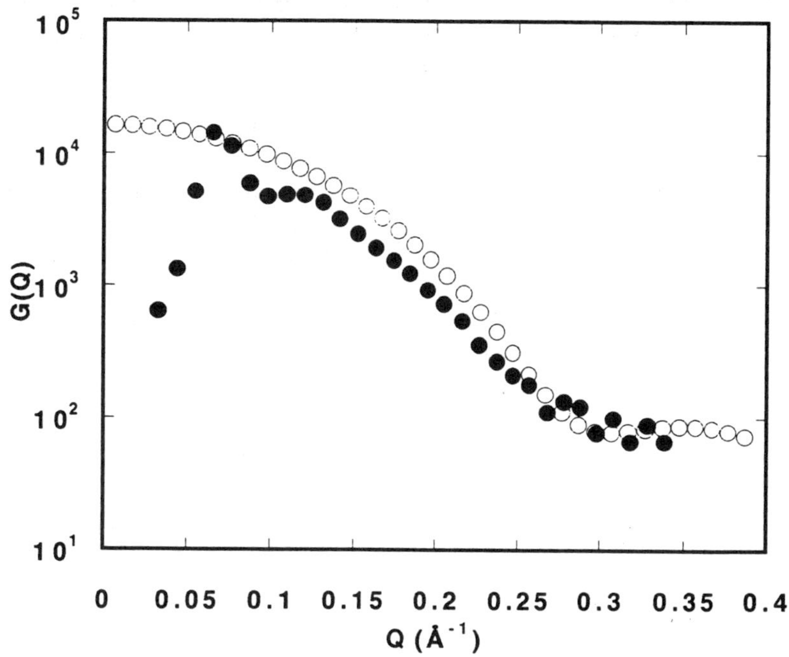

Figure 14. Example of the high-Q scattered-intensity curve fit for C_{D12}–6000–C_{D12} (30%). The full curve was calculated from expressions 6 and 7 with a polydispersity index of 1.2.

Table III. Values of Structure Parameters Obtained from Neutron-Scattering Experiments

C_{D12}–M_{PEO}–C_{D12} *(concentration)*	d (Å)	R_{plat} (Å)	R_{fit} (Å)	R_{calc} (Å)
6000 (5%)	117	10.5	12	10.7
6000 (10%)	113	12.5	14	13
6000 (15%)	104	13	15	14
6000 (30%)	92	16	17	15
10000 (5%)	184	15	14	14.5
10000 (10%)	149	15	17	14.8
10000 (20%)	125	17	17	15.6
10000 (54%)	104	17	17	17.7
20000 (20%)	146	9		
20000 (54%)	116	12		

The increase in R with increasing polymer concentration may explain the low exponent value in the power law $d = f(C_{Hyd})$. Nevertheless, in order to check the homogeneity of the solution, we have calculated, within the hypothesis of a cubic lattice, the R value (R_{calc}) expected from the measured interparticle spacing d, shown in Table III. The calculated values are of the same order of magnitude as but are systematically lower than the measured values. If such discrepancies can be considered significant, one can propse an inhomogeneous structure with large aggregates made of several hydrophobic microdomains organized in a cubic lattice for these solutions, as in Figure 15. Such a model imples that the micelles are connected by PEO chains and that d increases when the PEO chain length increases. These possibilities seem to be confirmed by Figure 12.

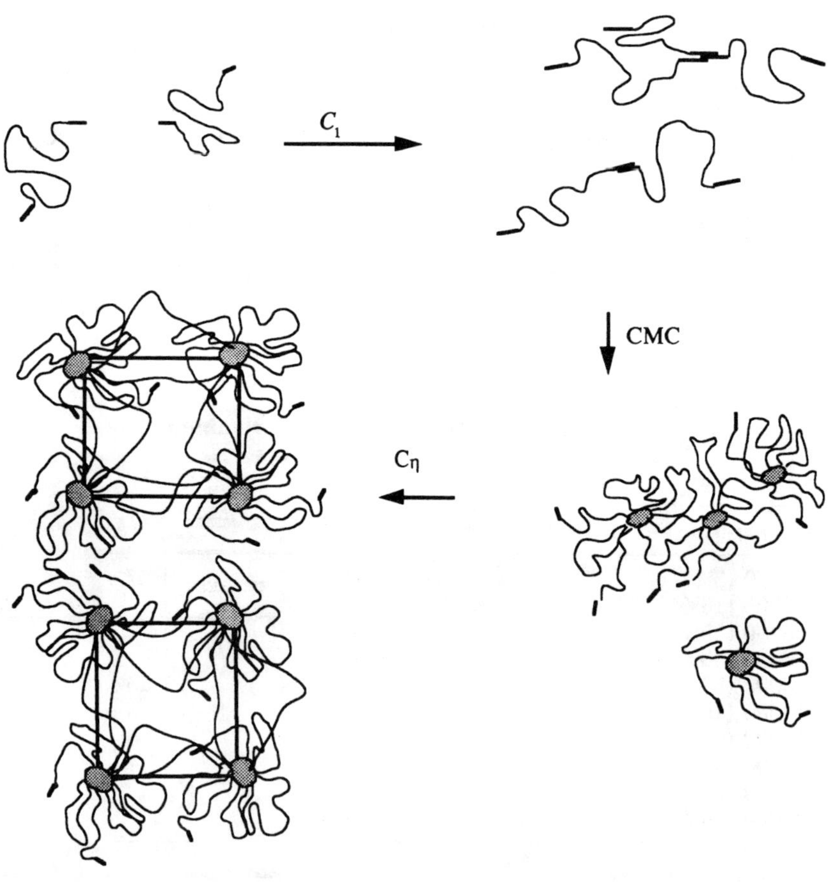

Figure 15. The steps in association.

Conclusion

Figure 15 summarizes the conclusions of this chapter, in which we attempted to describe the different steps of association of the model polymers according to hydrophobicity. Preliminary light-scattering results reveal high molecular weights for C_p < cmc. This result suggests an initial critical concentration C_1 at which isolated molecules begin to associate with a very small aggregation number of hydrophobic end groups. The increase in concentration induces the formation of hydrophobic microdomains, or micelles, at C_p = cmc as measured by fluorescence. At this stage, no significant enhancement of viscosity is observed. At higher concentration, that is, $C_p > C\eta$ (probably corresponding to the critical overlap concentration of the aggregates), the viscosity strongly diverges from that of unmodified polymer solutions. A further increase in concentration results in a disorder–order transition. Finally, around a concentration of 50%, the structure of the polymer solution can be described as consisting of large aggregates in which the micelles are spherical and are organized in a cubic lattice. The concentration thresholds between these different stages increase when the length of the PEO chain increases for the same hydrophobic group. While the association of these triblock polymers is much stronger than that of the diblock ones with the same hydrophobicity (lower cmc and LCST values), the triblock polymers appear to associate by a poorly cooperative process.

Acknowledgment

This work is dedicated to the memory of F. Clouet, who was an active participant in this project.

References

1. *Polymers in Aqueous Media: Performance through Association;* Glass, J. E., Ed.; Advances in Chemistry 223; American Chemical Society: Washington, DC, 1989.
2. *Polymers as Rheology Modifiers;* Schulz, D. N.; Glass, J. E., Eds.; Advances in Chemistry 462; American Chemical Society: Washington, DC, 1991.
3. Back, J.; Valint, P. L., Jr.; Pace, S. J.; Siano, D. B.; Schultz, D. N.; Turner, S. R. In *Water-Soluble Polymers for Petroleum Recovery;* Schultz, D. N., Ed.; Plenum: New York, 1986; Vol. 147.
4. Nae, H. N.; Reichert, W. W. *Polym. Mater. Sci. Eng.* **1989,** *61*, 624.
5. Karunasena, A.; Glass, J. E. *Prog. Org. Coat.* **1989,** *17*, 301.
6. Jenkins, R. D. Ph.D. Dissertation, Lehigh University, 1990.
7. Santore, M. M., Ph.D. Dissertation, Princeton University, 1990.

8. Maechling-Strasser, C.; François, J.: Clouet, F.; Tripette, C. *Polymer* **1992**, *33*, 627.
9. Binana-Limbele, W.; Clouet, F.; François, J. *Colloid Polym. Sci.* **1993**, *271*, 748.
10. Binana-Limbele, W.; Clouet, F.; François, J. *Colloid Polym. Sci.*, in press.
11. Yekta, A.; Duhamel, J.; Adiwidjaja, H.; Brochard, P.; Winnik, M. A. *Langmuir* **1993**, *9*, 881.
12. Yekta, A.; Duhamel, J.; Brochard, P.; Adiwidjaja, H.; Winnik, M. A. *Macromolecules* **1993**, *26*, 1829.
13. Maechling-Strasser, C.; Clouet, F.; François J. *Polymer* **1992**, *33*, 1021.
14. Nyströn, B. R.; Walderhaug, H.; Hansen, F. K. *J. Phys. Chem.* **1993**, *97*, 7743.
15. Person, K.; Abrahmsen, S.; Stilbs, P.; Hansen, F. K.; Walderhaug, H. *Colloid Polym. Sci.* **1992**, *270*, 465.
16. Walderhaug, H.; Hansen, F. K.; Abrahmsén, S.; Person, K.; Stilbs, P. *J. Phys. Chem.* **1993**, *97*, 8336.
17. Shirley, D. A.; Reedy, W. H. *J. Am. Chem. Soc.* **1951**, *73*, 458.
18. Drahowzal, F.; Klamann, D. *Monatsh. Chem.* **1951**, *82*, 460.
19. Kalyianasundaran, K.; Thomas, J. K. *J. Am. Chem. Soc.* **1977**, *99*, 2039.
20. Turro, N. J.; Yekta, A. *J. Am. Chem. Soc.* **1978**, *100*, 5951.
21. McNeil, R.; Thomas, J. K. *J. Colloid Interface Sci.* **1981**, *83*, 57.
22. Cotton, J. P.; Benoit, H. *Phys.* **1975**, *36*, 905.
23. Ionescu, L.; Picot, C.: Duval, M.; Duplessix, R.; Benoit, H.; Cotton, J. P. *J. Polym. Sci.* **1981**, *19*, 1019.
24. Jacrot, B. *Rep. Prog. Phys.* **1976**, *39*, 911.
25. Stuhrmann, H. B. *J. Appl. Crystallogr.* **1974**, *7*, 173.
26. Koester, L.; Steyerl, A. *Neutron Physics;* Springer: Berlin, Germany, 1977.
27. *Handbook of Chemistry and Physics,* 57th ed.; Weast, R. C., Ed.; CRC: Cleveland, OH, 1983–1984.
28. Corti, M.; Minero, C.; Degiorgio, V. *J. Phys. Chem.* **1984**, 88, 309.
29. Gantz, G. M. In *Nonionic Surfactants;* Schick, J. J., Ed.; Dekker: New York, 1967; p 736.
30. Saeki, S.; Kuwahara, N.; Hakata, Kaneko, M. *Polymer* **1976**, *17*, 685.
31. Glass, J. E. Proceedings of the American Chemical Society Symposium, Denver, 1993.
32. Binana Limbele, W., unpublished results.
33. Abrahmsén, S. accepted in *Colloid Interface Sci.*
34. Croxton, C. A. *Introduction to Liquid State Physics;* Wiley: New York, 1979.

RECEIVED for review November 18, 1993. ACCEPTED revised manuscript March 28, 1995.

19

Urethane-Coupled Poly(ethylene glycol) Polymers Containing Hydrophobic End Groups

NMR Characterization as a Step Toward Determining Aggregation Numbers in Aqueous Solutions

Ahmad Yekta, Thierry Nivaggioli, Sabeshan Kanagalingam, Bai Xu, Zahra Masoumi, and Mitchell A. Winnik*

Department of Chemistry and Erindale College, University of Toronto, Toronto, Ontario, Canada M5S 1A1

Urethane-coupled poly(ethylene glycol) polymers containing hydrophobic end groups are aqueous systems with remarkable rheological properties. Many of these polymers have been synthesized by using isophorone diisocyanate (IPDI) to couple segments of poly(ethylene glycol) with hydrophobic aliphatic alkyl chains (R) through urethane linkages (IPDU). This chapter discusses the importance of chemical characterization of these polymers for mechanistic experiments and develops a ^1H NMR method for this purpose. Two model compounds, reaction products of IPDI with methanol and 1-hexadecanol, are synthesized and used to understand the ^1H NMR spectroscopy of these urethanes. Three independent linear equations make it possible to relate integrated intensities of different regions of the ^1H NMR spectrum to the number of moles of R and IPDU per

*Corresponding author

0065–2393/96/0248–0363$12.00/0

gram of sample independently of knowledge of the polymer molecular weight. Six samples of HEUR are characterized by this method. We show how such information can affect the planning of synthetic strategies. For two samples, we also show how this information can revise our knowledge of the physical characterization of the micellar aggregation numbers inferred from fluorescent-probe experiments.

URETHANE-COUPLED POLY(ETHYLENE GLYCOL) POLYMERS containing hydrophobic end groups (HEUR polymers) constitute a class of associating polymers (AT) widely used as rheology modifiers in paints and other coatings to improve the performance of the paints. The AT polymers are believed to form micellelike clusters in water through self-association and are able to adhere to the surfaces of latex particles. Results from recent work (*1–5*) have added much to the qualitative understanding of the association structure of HEUR telechelic associative thickeners. However, lack of definitive structural characterization has hampered progress in understanding the more quantitative aspects of the association phenomena involved (*5*). In this respect, a key quantity that must be determined is the number of hydrophobic groups per micellelike cluster, that is, the mean aggregation number N_R. Only through a correlation of N_R (together with some of the related dynamic time constants) with well-defined polymer structures can we hope to shed enough light to obtain a microscopic-level rationale for some of the very interesting rheological properties that aqueous solutions of ATs display (*6–11*).

Two research groups recently reported N_R values for AT samples that were similar in structure (*1, 4*). The two polymers had different hydrophobic end groups, and researchers used different types of experiments to evaluate the N_R. Nevertheless, much of the discrepancy between the two sets of results might be explained by lack of adequate characterization of the samples. We used model AT compounds from Union Carbide that were carefully purified and characterized. In order to illustrate the importance of structural characterization of AT samples in determining their association mechanism, we compare results here for a set of AT samples in which no sample contains 100% targeted double-end-capped chains. A ^{1}H NMR method is used for determining the end-group content of HEUR telechelic ATs. That information is used to correct the apparent N_R values determined by fluorescence quenching experiments in water.

The generally assumed structure of hydrophobically end-capped (HEUR) ATs can be represented by structure **1**.

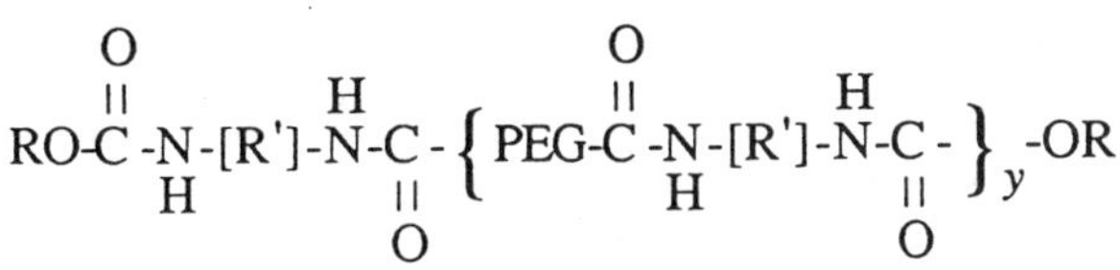

1

The overall synthetic reaction can be written as

$$2\text{ROH} + y\ \text{HO–PEG–OH} + (y + 1)\ \text{OCN–[R']–NCO} \rightarrow \mathbf{1}$$

Here R represents the hydrophobic end group typically derived from a fatty alcohol ROH. In the systems of interest to us, R is a linear alkyl chain and can vary from $C_{12}H_{25}$ to $C_{18}H_{37}$. The ATs we examine in this work have R = $C_{16}H_{33}$. The poly(ethylene glycol) (PEG) unit used in the synthesis has mostly intermediate molecular weights of 8,200 or 35,000 and is nearly monodisperse in chain length. However, the resulting AT can be highly polydisperse. The step growth size y can range from 1 to 10, and the magnitude of y depends on the mole equivalent ratios of the reacting constituents used to prepare the polymer. In addition, y also depends on the strategy employed for the synthesis of the AT. When controlled measures are employed to obtain $y = 1$, polymers with a narrow size distribution can be obtained. Values of y greater than unity imply significant polydispersity in the step growth process.

Here [R′] represents the structural unit of the parent diisocyanate (i.e., ONC–[R′]–NCO) used in the synthetic reaction. The diisocyanate often used to synthesize HEUR polymers is isophorone diisocyanate (IPDI, structure **2**).

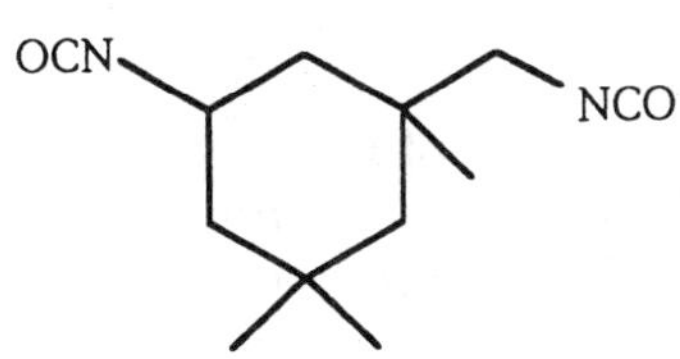

2

Among various diisocyanates, IPDI is sometimes considered the coupling agent of choice because the resulting urethane linkage affords good hydrolytic stability. For this reason this chapter focuses on the characterization of IPDI-based ATs. These materials are identical to those originally reported by Jenkins (*11*). These same samples have been investigated by light scattering (*12*), neutron scattering (Ottewill,

R., personal communication), fluorescence probes (*1–3*), and field gradient NMR techniques (*13*). We also report on two related polymers prepared by S. Menchen at Applied Biosystems.

The information we would like to obtain by ^{1}H NMR analysis of **1** is the number of moles of both R and IPDU (the corresponding diurethane of IPDI) per gram of polymer. Such information is particularly attractive because it can be obtained independently of our knowledge of the polymer number-average molecular weight (M_n), a quantity that is also difficult to determine. This simplifies the calculation of "true" N_R values. In addition, the method developed in this analysis can guide the chemist at each step of synthesis and purification so that a product faithful to **1** can be obtained.

End-group analysis of polymers by ^{1}H NMR spectroscopy is beset by several difficulties. First, the expected ^{1}H NMR signals from the R and IPDU groups are small (2–3% of the total signal). Therefore any contamination from residual solvents or unreacted small molecules contributes significant errors to an analysis of the spectra. Second, resonances from IPDU overlap those of the end-group R and are complicated because of the presence of *cis* and *trans* isomers of IPDU. Commercially available IPDI has about a 75:25 *cis*:*trans* ratio, where *cis* refers to the relative position of the two isocyanate groups. The complexity of the ^{1}H NMR spectrum of IPDI was the principal reason behind an initially erroneous determination of the *trans* isomer as the major constituent of commercial IPDI samples (*14*). More recent ^{1}H and ^{13}C NMR experiments along with X-ray structure analysis established the *cis* configuration as the major isomer (*14*). Over the past several years, further investigations have provided additional support for this assignment (*14–18*).

In this chapter we show how contamination problems can be overcome by appropriate and simple "purification" procedures, while the extent of peak overlap can be recognized and accommodated by proper reference to model compounds. The two model compounds we prepare are the reaction products of IPDI with excess methanol and with excess 1-hexadecanol. We also show how the results of ^{1}H NMR characterization of our HEUR samples can be used to correct micellar aggregation numbers obtained from fluorescence quenching measurements.

Experimental Methods

Materials. The polymeric associative thickeners we examined were gifts from Union Carbide and Applied Biosystems. All polymers have an M_n in the range of 25,000–51,000. The synthetic procedures and purification methods are briefly described later in this chapter. Union Carbide's

Table I. Structural Parameters of AT Samples Studied: R–IPDU–[PEG–IPDU]$_y$–R

Sample	M_n*(PEG)*	y
46RCHX22-2	8,200	4
46RCHX22-3	8,200	6
6RDJY107 and 6RDJY110	8,200	3
SMM9267-3-2	35,000	1
SMM9440-23-21	35,000	1

samples (46RCHX22-2, 46RCHX22-3, 6RDJY107, and 6RDJY110) are polydisperse in chain length and were synthesized from Carbowax 8000 (a PEG of M_n 8,200) (*11*). 6RDJY110 is the deuterated analog of 6RDJY107 with $-C_{16}D_{33}$ as the end group. SMM9267-3-2 and SMM9440-23-21 were prepared by S. Menchen of Applied Biosystems. These two samples have very narrow molecular-weight distributions and were synthesized from a PEG of M_n = 35,000. The nominal structural parameters for each sample (i.e., M_n(PEG) and y) are given in Table I.

For the synthesis of the model compounds, 98% IPDI, 99+% *p*-difluorobenzene, 99% 1-hexadecanol, and calcium hydride granules were purchased from Aldrich and used without further purification. $CDCl_3$, 99.8% D, was purchased from Cambridge Isotope Laboratories and stored over molecular sieves. All solvents, glass distilled or spectrograde, were purchased from Caledon. Methanol was made anhydrous by distillation from calcium hydride. Toluene was dried by simple distillation, and the initial distillate was discarded. All reflux–distillations were carried out under a N_2 atmosphere.

Synthesis of the Model Compounds. ***Methyl Diurethane of IPDI.*** This is $(MeO)_2$IPDU, carbamic acid, [3-[(methoxycarbonyl)amino]methyl]-3,5,5-trimethylcyclohexyl] methyl ester; CAS Reg. No. 112343–34–7. A 0.54-g volume of liquid IPDI was added to 20 mL of anhydrous MeOH. The mixture was refluxed under an N_2 atmosphere for 48 h. Excess MeOH was rotorevaporated at room temperature, and the resulting crystalline powder was dried under high vacuum overnight. The product had an mp of 106–113 °C and was not purified further so as not to disturb the *cis–trans* isomeric ratio. Gas chromatography–mass spectrometry (GC–MS) showed that the product was 96% pure. GC–MS was run on a VG 70-250S double-focusing spectrometer coupled to an HP5890 gas chromatograph with the following conditions: capillary column, J&W Scientific DB-5, 30 m; He flow, 1 mL/min; temperature 170–290 °C; and temperature gradient, 2.5 °C/min. Results are expressed in three parts: number of seconds of retention time over 11.5 min, percent of total GC signal, and M^+ parent mass. The observed values are 13 s, 0.6%, M^+ 211; 20 s, 1.4%, M^+ 226; 87 s, 27.6%, M^+ 286; and 107 s, 68.8%, M^+ 286. In the order of increasing GC signal, we assign the two major peaks to *trans* and *cis* isomers of $(MeO)_2$IPDU, respectively. The *cis–trans* ratio given previously agrees with that in the parent IPDI and is further corroborated by their NMR spectra discussed later in this chapter.

Hexadecyl Diurethane of IPDI. This is $(C_{16}H_{33}O)_2$IPDU; carbamic acid, [3-[(hexadecyloxycarbonyl) amino] methyl]-3,5,5-trimethylcyclohexyl] hexadecyl ester. Liquid IPDI (1.0 g, 4.6 mmol) and cetyl alcohol (5.2 g, 22 mmol, 140% in excess) were added to freshly distilled toluene (20 mL). The mixture was refluxed under N_2 for 24 h and then cooled to room temperature. CH_3CN (40 mL) was slowly added. The resulting precipitate, filtered with suction to dryness, weighed 4.0 g. In order to guide purification of the crude product, we noted that ^{1}H NMR spectroscopy of the α-methylene protons of cetyl alcohol in $CDCl_3$ shows a distinct triplet peak centered at 3.62 ppm. Consequently, the precipitated product has a cetyl alcohol contamination level of ca. 40%.

Thin-layer chromatography (TLC) (silica gel, Merck 60 F_{254}) of the precipitated powder, dissolved in ethyl acetate, showed two spots with R_f values of 0.74 and 0.59. The corresponding R_f for cetyl alcohol was 0.44. This powder (2.0 g in 100 mL of ethyl acetate) was chromatographed in a column packed with 800 mL (400 g) of silica gel (35–70 m, Toronto Research Chemicals). The eluent was divided into 12 fractions. Fractions 2–5 showed one spot on the TLC (R_f 0.78); fractions 6–12 showed two spots (R_f 0.79 and 0.60). Fractions 2–5 were rotor-evaporated and dried in a vacuum oven at 50 °C for 48 h. ^{1}H NMR spectroscopy of the product showed a residual 5% cetyl alcohol contamination level. This product was rechromatographed twice on a silica gel column using 60:40 vol/vol cyclohexane–EtOAc as the eluent to obtain a product that showed only one spot by TLC and no cetyl alcohol contamination by NMR spectroscopy.

Purification of the Associative Thickener Polymers. The purification strategy takes advantage of the temperature-dependent solubility of PEG in methanol (−15 °C, fast) or ethyl acetate (10 °C, slow overnight storage). A typical procedure is to dissolve the sample in the warm solvent (45 °C, 5 wt%) and then cool the solution to the appropriate temperature to precipitate (recrystallize) the PEG-based polymer but not the small-molecule contaminants. With methanol, separation of the precipitate requires low-temperature centrifugation followed by decanting of the solvent. With ethyl acetate, the procedure is simpler, and filtration at room temperature is adequate. The procedure just described was repeated three times. Finally, the last traces of solvent were removed from the sample by dissolving the sample (10 wt%) in benzene and then freeze-drying it. NMR spectroscopy showed no trace of benzene protons in the sample. The last traces of volatile solvents could also be removed by overnight partial melting (55 °C) of the sample under high vacuum.

NMR Spectroscopy. ^{1}H, ^{2}H, and ^{13}C spectra were run on either a Varian VXR 400S or a Varian Unity 500-MHz spectrometer with solutions containing 1–3 wt% of polymer in $CDCl_3$. Deuterated end-group sample 6RDJY110 was also examined in $CHCl_3$. To obtain quantitative integration of the signals, the spectra were run at different pulse delay times until constant results were obtained. In this way, nonlinear integration effects due to different proton relaxation times were avoided. A pulse delay time of 20 s was satisfactory for all of the systems under study. In most cases, 30–100 pulses were averaged. An internal standard allows determination of the moles of proton per gram of polymer: Samples contained a known weight of polymer plus a precisely known quantity of *p*-difluorobenzene (ca. 1.6 wt%). *p*-Difluorobenzene has ^1H NMR resonances (triplet centered at 6.99 ppm) far away from those of the polymeric samples, so relative integrations are easily evaluated. 6RDJY110 was analyzed by ^2H NMR spectroscopy on a Varian VXR 400S at 61.3 MHz. Here a known amount of $CDCl_3$ was added as an internal standard to calculate the amount of end-group $-C_{16}D_{33}$.

Results and Discussion

Separation of IPDU and Alkyl R Resonances. Figures 1–3 show that except for the intense singlet due to PEG protons (ca. 3.65 ppm), the main spectral region of interest lies between 0.7 and 2.0 ppm, where resonances due to IPDU and the alkyl R protons overlap. Figure 1 shows the ^1H NMR spectrum of the model compound $(CH_3O)_2$IPDU. Peaks further downfield, spanning 2 to 5 ppm, are not as useful for our purpose. Figure 1 shows that we have distinct resonances due to IPDU in the range of 2 to 5 ppm: 2.8–3.4 ppm, two H of position h that are well separated for the *cis* and *trans* isomers; 3.6–3.9 ppm, one H of position b; 4.6–5.0 ppm, two H of positions i and j. The resonances at 4.6–5.0 ppm are weak in intensity and somewhat variable in position (Table II).

It follows that quantitative information is best obtained by separating contributions from IPDU and alkyl R resonances in the region 0.7–2.0 ppm. In Figure 1 the CH_3O peak appears at 3.6 ppm with no

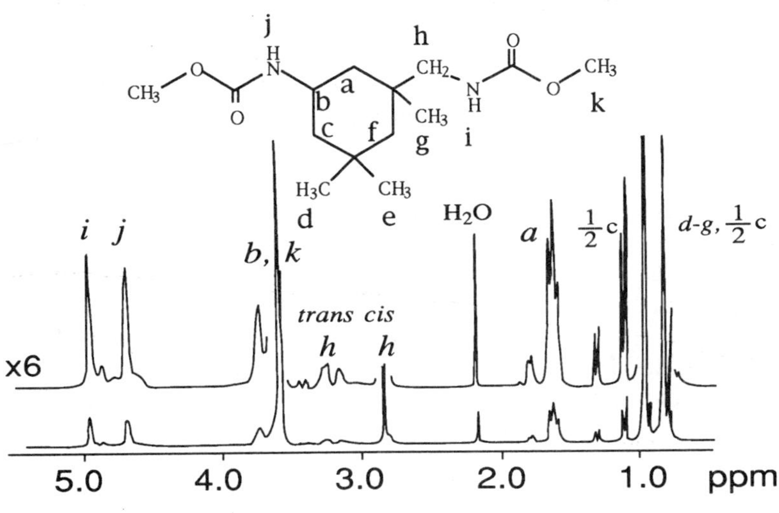

Figure 1. The 500-MHz Fourier transform ^1H NMR spectra of $(CH_3O)_2$IPDU in $CDCl_3$. Italic letters correspond to assigned proton positions. Table I presents relevant integration values.

Table II. Integration Results with Model Compound $(CH_3O)_2$IPDU for Data in Figure 1

Integration Range (ppm)	*Integrated No. of H*[a]	*Assigned Positions*[b]
0.70–1.08	12.08	d–g,
1.08–1.35	1.03	(1/2)c
1.55–1.85	2.04	(1/2)c
2.75–2.90	1.40	a
3.10–3.30	0.45	*cis* h
3.50–3.80	7.30	*trans*
4.60–4.75	0.88	h
4.90–5.00	0.96	b, k
		j
		i

[a] Number is relative to that for the internal standard *p*-difluorobenzene.
[b] The appropriate structure is shown with Figure 1.

interference for IPDU resonances in the 0.70–2.0-ppm region. The main contribution from IPDU lies in the region of 0.70–1.10 ppm, where 12 H appear (cf. Table II). These must be the three CH_3 groups and three H from the ring. In the range of 1.10–1.35 ppm, two doublets sum to a single proton. We conclude that these doublets are half of an ABX pattern, with the corresponding partner appearing below the methyl signals at 0.70–1.10 ppm. We tentatively assign the two doublets to one of the protons at position c. Although we do make positional assignments in Table II, our analytical conclusions do not depend on the correctness of such assignments.

Figure 2 shows the ^1H NMR spectrum of the model compound $(C_{16}H_{33}O)_2$IPDU. As expected, the CH_3 resonances from the hexadecyl R groups appear at ca. 0.8 ppm. Table III shows that the portion of the spectrum that appears at 0.70–1.10 ppm integrates ($I_{(0.70-1.10)}$) to 12 protons per IPDU and three protons per alkyl R. Equation 1 gives an algebraic expression of this observation. In the region 1.10–1.40 ppm, where three protons are due to IPDU, we also observe resonances due to 13 CH_2 groups from each hexadecyl group (broad singlet centered at about 1.24 ppm). Furthermore, in the region 1.5–1.9 ppm, where two protons of IPDU appear, we also find the peak for the alkyl CH_2 positioned β to the oxygen atom.

These observations can be expressed by three independent linear equations:

$$I_{(0.70-1.10)} = 12\ \mathrm{IPDU} + 3\ \mathrm{R} \quad (1)$$

$$I_{(1.10-1.90)} = 3\ \mathrm{IPDU} + 28\ \mathrm{R} \quad (2)$$

$$I_{(1.10-1.40)} = \mathrm{IPDU} + 26\ \mathrm{R} \quad (3)$$

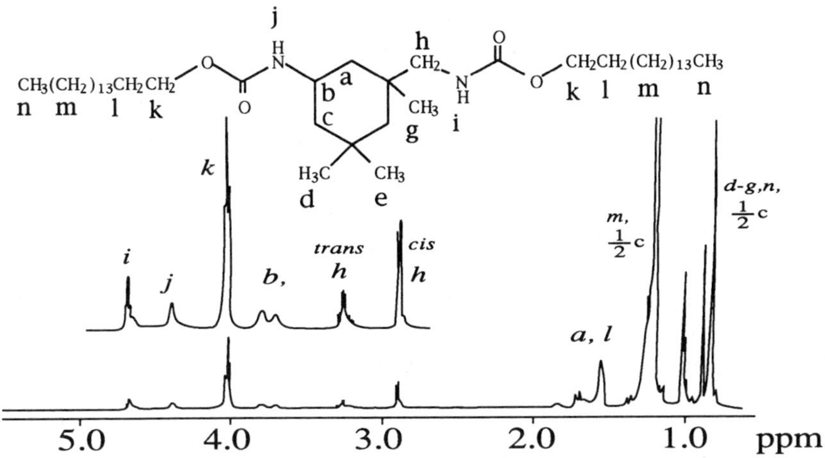

Figure 2. The 500-MHz Fourier transform ^1H NMR spectra of $(C_{16}H_{33}O)_2IPDU$ in $CDCl_3$. Italic letters correspond to assigned proton positions. Table II presents relevant integration values.

Two of these equations are sufficient for solving the unknown contents of R and IPDU. We prefer equations 1 and 2 because they cover a greater portion of the spectrum. However, there is a potential disadvantage in the use of equation 2. In some samples, a variable-resonance peak due to a trace of water appears in the range of 1.5–2.5 ppm, and this peak can interfere with proper integration of the relevant peaks. Furthermore, absolute quantities of R and IPDU can be calculated from equations 1–3 only if the integrals are evaluated relative

Table III. Integration Results with Model Compound $(C_{16}H_{33}O)_2$IPDU for Data in Figure 2

Integration Range (ppm)	*Integrated No. of H*[a]	*Assigned Positions*[b]
0.70–1.10	17.30	d–g, n, (1/2)c
1.15–1.42	53.6	m, (1/2)c
1.55–1.62	3.57	l
1.65–1.90	1.94	a
2.80–3.00	1.31	*cis* h
3.15–3.40	0.77	*trans* h
3.60–3.85	0.99	b
3.90–4.12	4.17	k
4.35–4.44	0.89	j
4.62–4.72	0.88	i

[a] Numbers are relative to that for the internal standard *p*-difluorobenzene.
[b] The appropriate structure is shown with Figure 2.

to that of an internal standard such as *p*-difluorobenzene. Otherwise, only relative quantities of R/IPDU can be obtained. Integration of the PEG protons (3.65 ppm) allow one to quantify the number of ethylene oxide units (CH_2CH_2O) per IPDU linkage. Evaluation of the quantity of spacer PEG units per IPDU linkage would require additional knowledge of M_n for the PEG that was utilized in the synthesis. In any case, one cannot infer the extent of unreacted PEG from the corresponding ^{1}H NMR signal.

Application to HEUR Polymers. The samples analyzed in this work were synthesized with different strategies in mind. Union Carbide samples from Jenkins (*11*) were prepared using monodisperse PEG of M_n 8,200 (Carbowax 8000), with the intention of making structure **1** with y = 3, 4, or 6 (cf. Table I). As an example, for synthesis of sample 46RCHX22-2, a reaction stoichiometry of 100% efficiency was presumed to obtain the structure **1**: That is, 4 mol of PEG (in the melt) was reacted with 5 mol of IPDI and 2 mol of hexadecanol. The calculated value of M_n is 34,000. Experimentally, gel permeation chromatography analysis of the sample in tetrahydrofuran shows a polydispersity of ca. 2. We had difficulty with accurate determinations of M_n, but the results are not very far from the calculated M_n (+25%). ^{1}H NMR analysis of the *crude* reaction product according to the methods described shows that 1.00 g of the crude sample (vacuum dried overnight, 55 °C) contains 136 mmol of IPDU and 56.0 mmol of $C_{16}H_{33}$ alkyl groups. This finding translates into an IPDU/R ratio of 2.43 and an IPDU/PEG ratio of 1.24. For the unpurified polymer, the closeness of our results to the stoichiometrically expected ratios is not surprising, because our NMR results only confirm the initial molar quantities of R, PEG, and IPDI that were added to the mixture. We recovered in a different arrangement what was put into the reaction mixture. This result is interesting only in that it gives us confidence in the fidelity of our analytical characterization method.

When we recrystallize 46RCHX22-2 according to procedures described, every 1.00 g of the *purified* polymer contains 139 mmol of IPDU and 42.6 mmol of $C_{16}H_{33}$ groups (Figure 3). These numbers translate into an IPDU/R ratio of 3.26 and an IPDU/PEG ratio of 1.25. If the differences are to be considered significant, during the synthesis, some 20% of cetyl alcohol remained unreacted or formed low-molecular-weight byproducts that were removed by the purification procedure. Formation of macrocycles of the type (PEG-IPDU)$_n$ can be by a pathway that results in excess unreacted cetyl alcohol. Some dihexadecyl IPDU will also form during polymer synthesis.

A second series of HEUR ATs was synthesized with the intent of preparing structure **1** as a monodisperse polymer with $C_{16}H_{33}O$– end

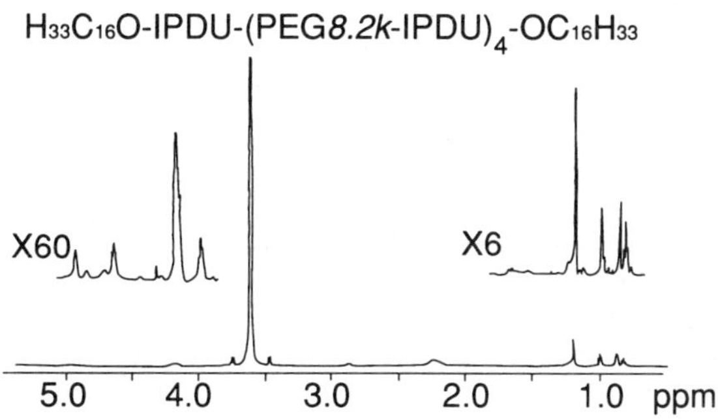

Figure 3. The 500-MHz Fourier transform ^1H NMR spectra of purified associative thickener 46RCHX22-2 in $CDCl_3$.

groups, i.e., $y = 1$. For pedagogic value, we choose here one of the earlier samples, which was prepared before the reaction conditions were optimized. For the synthesis of sample SM9267-3-2, 1-hexadecanol was reacted with a large excess of IPDI in the absence of a catalyst, and the excess IPDI was removed under vacuum. The product, presumably monofunctionalized $C_{16}H_{33}O$–IPDI, was added in excess to a known amount of PEG (Fluka, M_n 35,000) in the presence of a trace of a catalyst (dibutyltin laurate) in CH_2Cl_2 solution. The final product was purified as already described. If the presumed adduct of the first stage were pure, one would expect an IPDU/R ratio of 1.0 in the polymer. We find IPDU/R = 0.94 within an experimental error of unity. On the other hand, we find 26.5 mmol of $C_{16}H_{33}$ per g of polymer, which corresponds to 0.96 groups of $C_{16}H_{33}$ per PEG segment of the polymer. In other words, the sample was, on average, monofunctionalized only. The second step of the synthesis had not gone to completion.

Table IV summarizes the NMR characterization of the AT samples studied. The extent of alkyl group labeling is consistently lower than that expected of the target structure **1**.

Obtaining Aggregation Numbers

Fluorescence quenching experiments do not in fact yield values of N_R directly. From analysis of the quenching data, one obtains the total number of micellelike clusters in the sample under study. The number of clusters must be divided by the total number of associating hydrophobic groups in the sample so that N_R can be evaluated. Any uncertainty in the hydrophobic group content of the sample will be reflected as an uncertainty in the calculated value of N_R. We now consider how

Table IV. NMR Characterization of AT Samples

Sample	Content (mmol/g of AT)			Alkyl Groups/AT Chain
	IPDU	R	PEG	
46RCHX22-2 purified	139	42.6	111	1.4
46RCHX22-2 as received	136	56.0	110	
46RCHX22-3 purified	147	33.3	132	1.7
46RCHX22-3 as received	158	45.6	132	
6RDJY107 purified	138	50	120	1.3
6RDJY107 as received	160	77	120	
6RDJY110 purified	148	56	108	1.5
6RDJY110 as received	174		118	
SMM9267-3-2 purified	24.9	26.5		0.96
SMM9440-23-21 purified	68	41	28	1.4

structural characterization affects evaluation of the aggregation number N_R. The details of this work will be published elsewhere (*20*). Here we give only the salient features of the evaluation procedure. In fluorescence measurements of N_R by pyrene (Py) self-quenching, one dissolves known quantities of pyrene in micellar solutions of the polymer while keeping a fixed concentration of the polymer ($C_{polymer}$ in grams per liter). From the fit of pyrene fluorescence decays to an appropriate expression, one extracts the mean number of pyrene molecules per micelle, $\langle n \rangle$ = [Py]/[micelle]. A linear plot of the experimentally available values of $\langle n \rangle$ versus [Py] yields the micelle concentration as the reciprocal of the slope. If the polymer structure conforms to that of structure **1** and we know the molecular weight of the polymer, then

$$[\text{micelle}] = [C_{16}H_{33}O]/N_R = 2[\text{Polymer}]/(N_R)$$

$$[\text{micelle}] = 2\,(C_{\text{polymer}})/(N_R)(M_n)$$

and we are able to evaluate N_R. The multiplying factor 2 relates to the fact that structure **1** contains two hydrophobic groups for each polymeric chain. If the polymer does not fully conform to structure **1** and we do not know the exact value of M_n, then an alternative approach is needed. The information necessary to calculate N_R is the number of moles of hydrophobic groups per gram of polymer.

The following expression applies:

$$[\text{micelle}] = [C_{16}H_{33}O]/N_R$$

$$[\text{micelle}] = (C_{\text{polymer}})(\text{moles of } C_{16}H_{33}O/\text{g of polymer})/N_R$$

As an example, for the 10.0-g/L aqueous polymeric sample 46RCHX22-2 (purified), our fluorescence experiments yield a slope of $\langle n \rangle$ versus [Py] that equals $[\text{micelle}]^{-1} = 3.77 \times 10^4$ L/mol. Assuming structures of type **1**, one obtains an N_R of 22. Instead, if we modify the results with the data from the ^{1}H NMR characterization described earlier, we obtain an N_R of 18 $\pm$ 2 (error limits relate to reproducibility of the ^{1}H NMR results). Similarly, for the purified sample of SMM9267-3-2, we obtain $[\text{micelle}]^{-1} = 1.03 \times 10^4$ L/mol, an equation that leads to $N_R = 59$ based on an assumption of structure **1**. In this case, modifying the result with data from the ^{1}H NMR characterization gives an N_R of 28 $\pm$ 2, which is an appreciable correction. Clearly, assumption of a structure based on reaction stoichiometry can lead to serious errors in our attempts to achieve a better understanding of association phenomena in aqueous polymeric HEUR solutions.

Conclusions

We have developed an understanding of the ^{1}H NMR spectroscopy of HEUR polymers comprised of PEG diols coupled with IPDI and end capped with 1-hexadecanol. Three equations that relate integrated areas of portions of the spectra to the chemical composition of the polymers are presented. These equations should have a broad applicability in that modification can be made in a straightforward way for similar polymers having different *n*-alkyl end groups.

We use the analysis of two polymer samples to show how one can obtain proper N_R values from pyrene fluorescence self-quenching data by correcting for the actual hydrophobic group content of the polymer.

Acknowledgments

We thank Union Carbide, Aquelon, and NSERC Canada for their support of this research and S. Menchen of Applied Biosystems Inc. for polymer samples. In addition, we thank Nick Plavac for all his assistance in obtaining the NMR spectra.

References

1. Yekta, A.; Duhamel, J.; Adiwidjaja, H.; Brochard, P.; Winnik, M. A. *Langmuir* **1993,** *9,* 881.
2. Yekta, A.; Duhamel, J.; Brochard, P; Adiwidjaja, H.; Winnik, M. A. *Macromolecules* **1993,** *18,* 29.
3. Wang, Y.; Winnik, M. A. *Langmuir* **1990,** *6,* 1437.
4. Richey, B.; Kirk, A. B.; Eisenhart, E. K.; Fitzwater, S.; Hook, J. *J. Coatings Technol.* **1991,** *63,* 31.
5. Kaczmarski, J. P.; Glass, J. E. *Macromolecules* **1993,** *26,* 5149.

6. Annable, T.; Buscall, R.; Ettelaie, R.; Whittlestone, D. *J. Rheol.* **1993,** *37,* 695.
7. Annable, T.; Buscall, R.; Ettelaie, R.; Shepherd, P. *Langmuir* **1994,** *10,* 1060.
8. *Water-Soluble Polymers: Beauty with Performance;* Glass, J. E., Ed.; Advances in Chemistry 213; American Chemical Society: Washington, DC, 1986.
9. *Polymers in Aqueous Media: Performance Through Association;* Glass, J. E., Ed.; Advances in Chemistry Series 223; American Chemical Society: Washington, DC, 1989.
10. *Polymers as Rheology Modifiers;* Schulz, D. N.; Glass, J. E., Eds.; ACS Symposium Series 462; American Chemical Society: Washington, DC, 1991.
11. Jenkins, R. D., Ph.D. Thesis, Lehigh University, 1990.
12. Ou-Yang, H. D.; Gao, Z. *J. Phys. II* **1991,** *1,* 1375.
13. Chapter 19 in this book.
14. Wendlich, D.; Reiff, H.; Dietrich, D. *Angew. Makromol. Chem.* **1986,** *141,* 173.
15. Auf der Heyde, W.; Hubel, W.; Boese, R. *Angew. Makromol. Chem.* **1987,** *153,* 1.
16. Hatada, K.; Ute, K. *J. Polym. Sci., Polym. Lett.* **1987,** *25,* 477.
17. Born, L.; Wendlich, D.; Reiff, H.; Dietrich, D. *Angew. Makromol. Chem.* **1989,** *171,* 213.
18. Hatada, K.; Ute, K.; Oka, K.-I. *J. Polym. Sci., Polym. Chem.* **1990,** *28,* 3019.
19. Bialas, N.; Hocker, H. *Makromol. Chem.* **1990,** *191,* 1843.
20. Yekta, A.; Xu, B.; Duhamel, J.; Adiwidjaja, H.; Winnik, M. A. *Macromolecules* **1994,** *25,* 956.

RECEIVED for review November 18, 1993. ACCEPTED revised manuscript November 29, 1994.

20

Self-Diffusion Coefficients of Associating Polymers from Pulsed-Gradient Spin-Echo Nuclear Magnetic Resonance Spectroscopy

Peter M. Macdonald, Yoshimitsu Uemura, Lara Dyke, and Xiaoxia Zhu

Department of Chemistry and Erindale College, University of Toronto, Toronto, Ontario M5S 1A2, Canada

The self-diffusion coefficients of hydrophobic ethoxylated urethane associating polymers of different molecular weights but a constant length of the hydrophobic end cap were measured in aqueous solution by a pulsed-gradient spin-echo NMR technique. With increasing concentration, all polymers registered a decrease in their mean diffusion coefficient accompanied by an increase in the dispersion of the diffusion coefficient about the mean. Both effects correlated with the pronounced concentration-dependent viscosity increase characteristic of such polymers. The presence of hydrophobic end caps decreased the diffusion coefficient by more than an order of magnitude relative to that of control polymers of identical size but lacking such modifications. The results are consistent with the predictions of the transient micellar network model of network formation by such polymers and yield an estimate of the size (20-nm radius) and aggregation number (one micelle contains 20 hydrophobic chains) for the associated clusters.

THE RHEOLOGICAL PROPERTIES OF ASSOCIATING POLYMERS are being intently scrutinized both experimentally, in aqueous solution and in

0065–2393/96/0248–0377$12.00/0

complexes with latex and surfactant, and theoretically with a view to developing a detailed understanding of the molecular basis of the rheological behavior (*1–3*). The hydrophobic ethoxylated urethane (HEUR) associating polymers, which consist of poly(ethylene glycol) (PEG) chain extended by diisocyanates and end capped by long-chain alkanols, have received particular attention (*4*). They exhibit the desired combination of Newtonian behavior at low shear rates and shear thinning only at relatively high shear rates. Furthermore, because the number and locations of the associating groups are fixed, the possible states of the associated chains and the topologies of association are relatively limited, which simplifies modeling of the associated state. Recent spectroscopic and light-scattering studies (*5*) suggest that the associated state consists of a three-dimensional network of small micelles (aggregation number, 10). The micelles are formed by the associated hydrophobic groups at the ends of looped chains. The network is formed when micelles are linked to one another by bridging chains. Most of the rheological properties of HEUR associating polymers can be reproduced by a model in which the network becomes transient as a result of the finite rate of disengagement of the chain ends from the micelles while the proportion of bridging chains to looping chains gradually increases as the polymer concentration increases (*6*). The transient micellar network model leads to certain predictions regarding the topology of the associated state and, by extension, the diffusion of the associating polymers making up the network. In particular, at low polymer concentrations the diffusing species should consist of a well-defined micelle of a characteristic size. Pulsed-gradient spin-echo (PGSE) NMR spectroscopy is a particularly useful technique for measuring self-diffusion coefficients and has been applied to the study of diffusion in a broad spectrum of physical situations (*7, 8*), including HEUR-type associating polymers (*9–11*). Walderhaug and co-workers (*9–11*) investigated the effects of varying the length of the hydrophobic end cap and demonstrated that the self-diffusion coefficients of the associating polymers are best described in terms of a distribution that broadens continuously with increasing concentration. Likewise, the mean self-diffusion coefficient simultaneously undergoes a dramatic decrease at a concentration at which rheology data indicate that the solution viscosity rises abruptly. Furthermore, these investigators explained how to extract information regarding the strength of the intermolecular couplings and the aggregation numbers of the micellar aggregates from the diffusion data.

In this chapter we describe PGSE NMR measurements of the self-diffusion coefficients of HEUR associating polymers with different molecular weights but a constant chain length of the hydrophobic end cap. These polymers were extensively characterized in other laboratories with regard to both rheological behavior (*12*) and molecular properties (*5*). We compare the concentrations and molecular weight dependences of their self-diffusion coefficients with those of control polymers lacking hydrophobic end caps. We demonstrate that the measured self-diffusion coefficients conform to the predictions of the transient micellar network model of the associated state.

Experimental Details

The HEUR associating polymers and the control polymers lacking hydrophobic end caps were obtained from Richard D. Jenkins, Union Carbide Chemicals and Plastics Co. Inc., UCAR Emulsion Systems, Cary, NC. Their number-average molecular weights (M_n) are listed in Table I. In each case a stock solution consisting of 1 wt% polymer in 99.9% deuterium oxide was prepared by hydrating the dried polymer for 24 to 48 h with the appropriate volume of deuterium oxide. Care was taken to avoid exposing the solutions to light. Solutions with lower polymer concentrations were prepared by serial dilution of the stock solution directly into an NMR tube. The NMR tubes were then sealed and stored in the dark at 4 °C until use.

PGSE NMR measurements were made with a magnetic resonance imaging probe with actively shielded gradient coils (Doty Scientific, Columbia, SC) installed in a Chemagnetics CMX 300 NMR spectrometer operating at 300 MHz for protons. A standard Stejskal–Tanner PGSE sequence (90°_x–τ–180°_y–τ, with gradient pulses during τ) (*13*) was used as illustrated in Figure 1. The 90 ° pulses were 18 μs long, the interpulse delay was 250 ms, the recycle delay was 30 s, the spectral widths were 10 kHz, the data size was 4 K, line broadening was 10 Hz, and the number of acquisitions was 16–32 scans. In this study, PGSE experiments were performed at 23 °C, and the gradient pulse was applied to the *z* direction only. Several levels of gradient strength between ca. 20 and 200 G/cm were necessary, depending on the polymer concentration. The lower range of gradient strengths was calibrated by using the diffusion coefficient of 2 vol% H_2O in D_2O ($D = 1.9 \times 10^{-9}$ m²/s) (*14*) (see equation 1 in the following section). For higher gradients, a sample of 10 wt% PEG in deuterium oxide, which could be measured with both low and high gradient strengths, was used as the calibration standard. The precision of the mea-

Table I. Distinguishing Features of HEUR Polymers

Polymer Designation	*Molecular Weight* (M_n)	*End Group*
6RDJY-107	26,800	$C_{16}H_{33}$
46RCHX22-2	34,200	$C_{16}H_{33}$
3RDJY-14	48,000	$C_{16}H_{33}$
46RCHX23-1	16,600	H
46RCHX23-2	34,000	H
46RCHX23-4	67,000	H

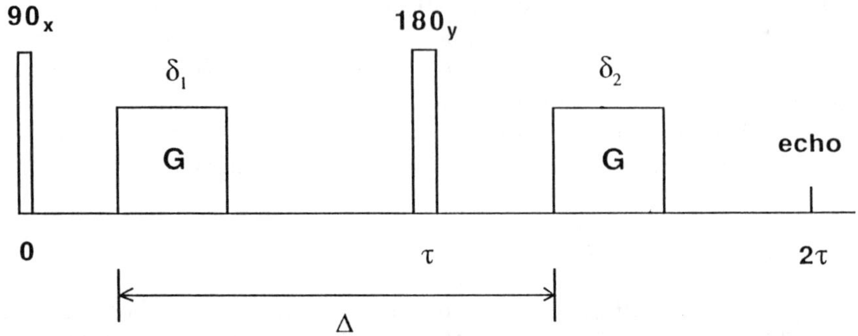

Figure 1. PGSE NMR pulse sequence. G is the field gradient amplitude, δ is the gradient pulse duration, Δ is the gradient pulse separation, and τ is the rf pulse separation. Normally, Δ is set equal to τ. At low gradient amplitudes the first and second gradient pulse durations are set equal to one another. At high gradient amplitudes it is sometimes necessary to adjust the second gradient pulse duration to ensure that the signal echo appears at the proper time (2τ) and with the proper phase. Fourier transformation of the echo signal starting from its peak amplitude at 2τ yields the frequency spectrum.

sured self-diffusion coefficients of solvent was very good at low gradient strength, the error being estimated at less than 5%; at higher gradient strength, the error was larger because of eddy currents but in all cases was less than 15%. The presence of eddy currents can both shift the position of the echo maximum and distort its phase (*15*). To overcome this problem, we implemented a method proposed by Hrovat and Wade (*16*) in which the duration of the second gradient pulse is increased to compensate for imbalances arising from induced eddy currents. When proper eddy current compensation is applied, the echo position and phase are correct, and Fourier transformation of the echo beginning at its maximum yields the corresponding spectrum.

Results and Discussion

The PGSE NMR pulse sequence is shown in Figure 1. The sequence consists of a pair of magnetic-field gradient pulses (duration δ and amplitude G); the first is applied between the two radio frequency (rf) pulses of the usual spin-echo NMR sequence, and the second is applied between the last rf pulse and the appearance of the spin-echo signal. For isotropic diffusion characterized by a single diffusion coefficient in a homogeneous magnetic field in which the residual gradient G_0 is negligible, the intensity $I_{2\tau}$ of the NMR signal at time 2τ following the start of the PGSE pulse sequence is related to the diffusion coefficient D according to equation 1:

$$\frac{I_{2\tau}}{I_0} = \exp[-(\gamma G\delta)^2(\Delta - \delta/3)D] \quad (1)$$

where G is the pulsed-gradient strength, δ is the duration of the gradient, Δ is the interval between the gradients, τ is the rf pulse interval, and γ is the magnetogyric ratio (*13*). The effect of T_2 (spin–spin relaxation time) is constant when τ is kept constant and is contained within the term I_0. The diffusion coefficient can be derived from the slope of a plot of the logarithm of the signal intensity as a function of $\delta^2(\Delta - \delta/3)$ once the gradient strength G is known. The gradient strength is obtained by calibration with a sample of known diffusion coefficient, as described in ***Experimental Details***.

Figure 2 shows a series of ^{1}H NMR spectra of the M_n 48,000 C_{16} HEUR associating polymer (3RDJY-14) at 0.5 wt% in deuterium oxide for various durations of the field gradient pulse. The left-most spectrum was obtained with a very short gradient pulse and shows two

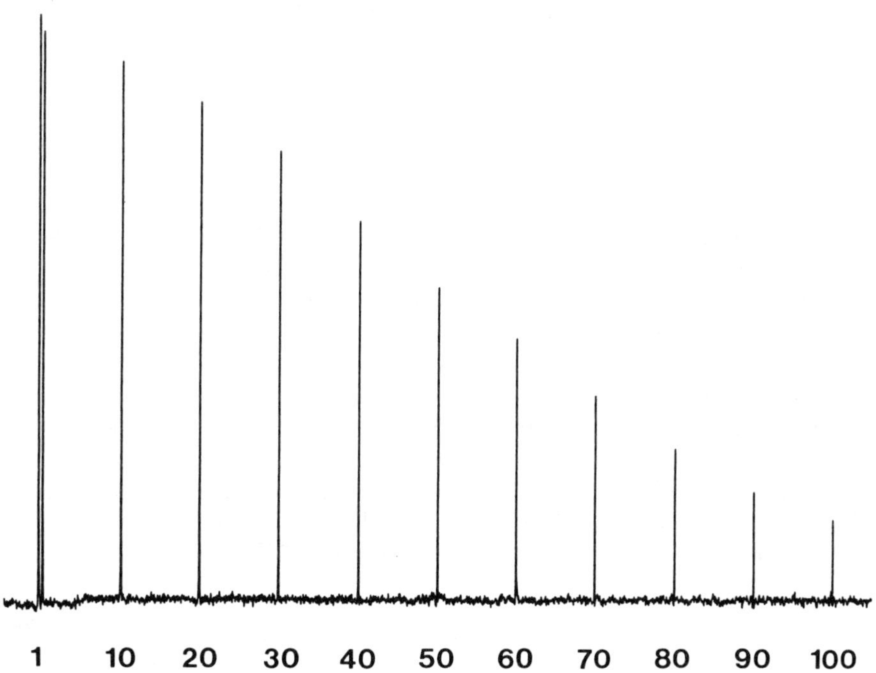

Figure 2. Proton NMR spectra of the M_n 48,000 C_{16} HEUR associating polymer at 0.5 wt% in deuterium oxide at room temperature, showing attenuation of the proton signals with increasing duration of the gradient pulse δ, indicated in milliseconds. In this instance the gradient pulse amplitude was 58.0 G/cm and Δ = τ = 250 ms.

resonances: One arises from the residual protons in deuterium oxide, and the other arises from the ethylenic protons of the PEG chains of the HEUR associating polymer. Other expected resonances, such as those of the diisocyanate chain extenders or the methylenes of the hexadecanol hydrophobic end caps, do not appear, either because their numbers are small relative to the number of PEG protons or because of their shorter spin–spin (T_2) relaxation times relative to those of PEG protons. The methylenes of the hexadecanol end caps are readily apparent, however, in the ^1H NMR spectra of C_{16} HEUR associating polymers dissolved in chloroform (*17*). If the associated state that forms in aqueous solution involves hindered molecular motion, such as might be expected within the micellar cluster, this hindrance shortens the T_2 and increases the NMR line width. Thus for the τ value used in these measurements (250 ms), the resonance signal of the hydrophobic end caps could well be largely suppressed. With increasing gradient pulse duration, the DO resonance intensity quickly decreases to zero, which is expected given the rapid diffusion of water and the rather high gradient strengths used in this particular experiment. The remaining PEG resonance intensity decays exponentially with increasing duration of the gradient pulse.

Figure 3 illustrates typical plots of the logarithm of the signal intensity of the PEG protons as a function of the gradient pulse duration for two concentrations of the M_n 34,200 C_{16} HEUR associating polymer (46RCHX22-2) (0.1 and 1.0 wt%) in deuterium oxide. At 0.1 wt% the decrease in intensity is apparently monoexponential, indicating a common, well-defined, effective rate of diffusion for all polymer molecules in the sample. At 1.0 wt% the intensity decrease shown in Figure 3 is quite evidently curved. Some possible origins of this nonlinearity include restricted diffusion and polydispersity of molecular weight. Similar nonlinearities are obtained at high concentrations for all the C_{16} HEUR associating polymers. In contrast, the control polymers lacking hydrophobic end caps all displayed apparently monoexponential decay of the signal intensity in the PGSE NMR experiment.

The least ambiguous method for analyzing such multiexponential decays is to use a "stretched exponential" of the form shown in equation 2 (*10*).

$$\frac{I_{2\tau}}{I_0} = \exp-[(\gamma G\delta)^2(\Delta - \delta/3)D_e]^\beta \qquad (2)$$

The parameter β is a measure of the width of the distribution of the diffusion coefficients about the mean value, where $0 \leq \beta \leq 1$, such that when $\beta = 1$, the diffusion coefficient is monodisperse. The parameter D_e is an effective self-diffusion coefficient and is related to the

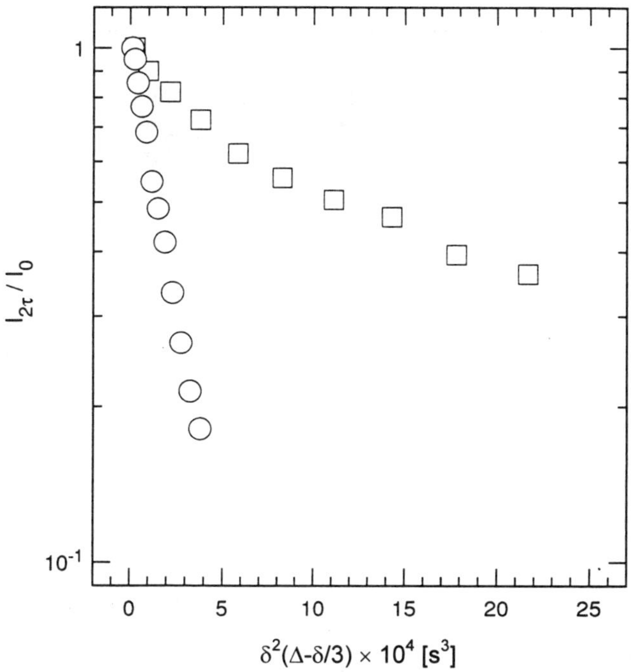

Figure 3. Logarithm of the proton NMR signal intensity (ln I*) versus* $\delta^2(\Delta - \delta/3)$ *for* M_n *34,200* C_{16} *HEUR associating polymer at 0.1 wt% in deuterium oxide with a gradient amplitude of 58 G/cm and* $\Delta = 250$ *ms (○) and at 1.0 wt% in deuterium oxide with a gradient amplitude of 181 G/cm and* $\Delta = 250$ *ms (□).*

mean diffusion coefficient D_m via the gamma function Γ according to equation 3:

$$\frac{1}{D_m} = \int_0^{\infty} \exp-(XD_e)^{\beta}\, dX = (1/\beta)(1/D_e)\Gamma(1/\beta) \tag{3}$$

where $X = (\gamma G\delta)^2(\Delta - \delta/3)$. The values of β and D_e are obtained from a two-parameter nonlinear least-squares fit of equation 2 to the diffusion data or from plotting $\ln[\ln(I_0/I_{2\tau})]$ versus $\ln(X)$, which yields a line with a slope equal to β and an intercept equal to $\beta\ln(D_e)$. For the polymer concentrations in Figure 3, the values of β are 0.987 and 0.573 for 0.1 and 1.0 wt% solutions, respectively, of the M_n 34,200 C_{16} HEUR associating polymer, and the corresponding values of D_e are 8.05×10^{-12} and 9.57×10^{-14} $m^2\ s^{-1}$. Therefore if equation 3 is used, the mean diffusion coefficients D_m for these two cases are 8.00×10^{-12} and 5.98×10^{-14} $m^2\ s^{-1}$, respectively.

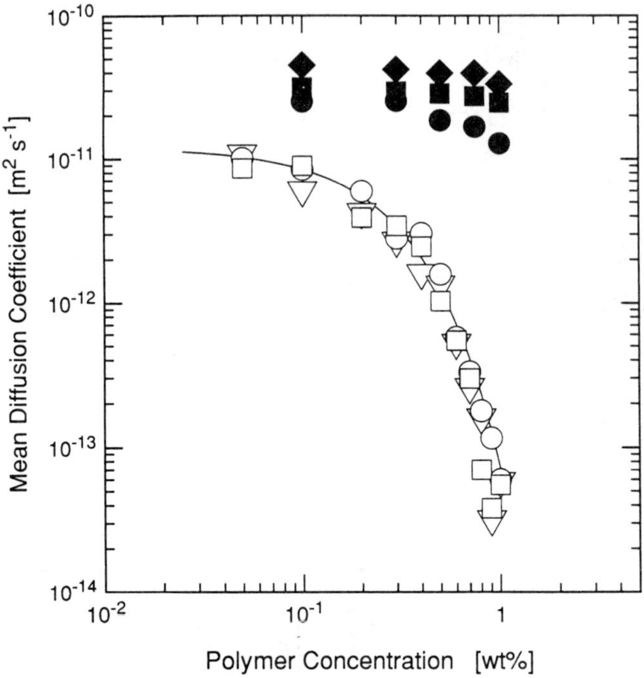

Figure 4. Mean diffusion coefficients as a function of concentration for both the control and the C_{16} HEUR associating polymers. Key: □, M_n *26,800 C_{16} HEUR associating polymer;* ○, M_n *34,200 C_{16} HEUR associating polymer;* ▽, M_n *48,000 C_{16} HEUR associating polymer;* ◆, M_n *16,600 control polymer;* ■, M_n *34,000 control polymer;* ●, M_n *67,000 control polymer.*

Figure 4 illustrates the dependence of the mean diffusion coefficients D_m on polymer concentration for both the C_{16} HEUR associating polymers and the control polymers. The control polymers exhibit a progressive decrease in diffusion coefficient with increasing concentration. The higher-molecular-weight species diffuse at a slower rate. The diffusion coefficients of the C_{16} HEUR associating polymers are all markedly lower than those of the control polymers, exhibit little dependence on the molecular weight of water-soluble backbone, and show evidence that two concentration regimes exist. This behavior is essentially identical to that reported by Persson et al. (9) and Walderhaug et al (*10*) for their HEUR-type associating polymers as determined by using PGSE NMR.

Figure 5 compares the values of β for the various C_{16} HEUR associating polymers and their controls over the same range of concentrations shown in Figure 4. In general, with increasing polymer concen-

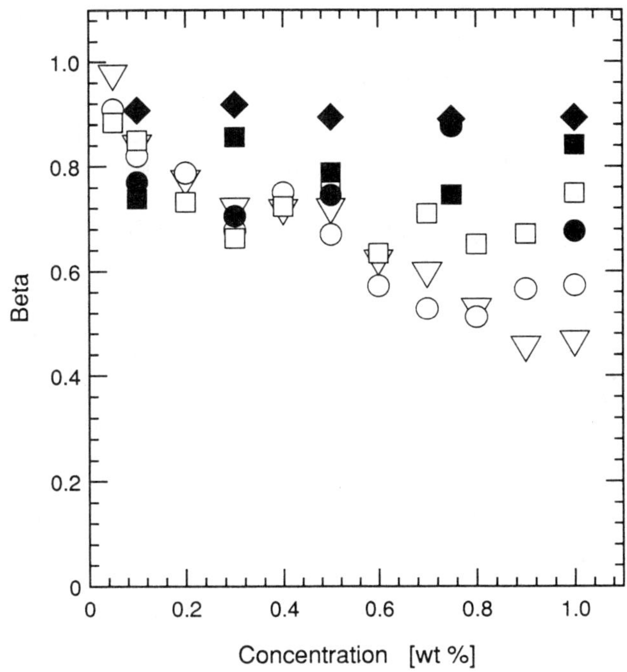

Figure 5. Distributions of diffusion coefficients about mean values as a function of concentration for the control and the associating polymers. Symbol designations are the same as in Figure 4.

tration, the dispersion of the diffusion coefficients about the mean value increases. However, any differences between C_{16} HEUR associating polymers of different molecular weights are small relative to the concentration dependence of the quantity β for any one molecular weight. For the control polymers, in contrast, the polymer molecular weight appears to be the important factor determining β, but the concentration has little impact, at least in the range of concentrations investigated here. Remarkably, the dispersion of the diffusion coefficients for the HEUR associating polymers is narrower than that of the control polymers at lower concentrations but wider at higher concentrations.

For the control polymers, the concentration dependence of the mean diffusion coefficient and the distribution about the mean are readily accounted for in terms of classic hydrodynamic theory and the polydispersity of molecular weight of these samples. Specifically, standard dilute hydrodynamic theory states that the observed diffusion coefficient extrapolated to zero concentration should scale according to M^a, with $0.5 \leq a \leq 0.6$, depending on the quality of the solvent (*18*).

For the control polymers, we find that $a = 0.5 \pm 0.05$. Considering that only three different molecular weights were tested, this value is in reasonable agreement with expectations for a good solvent. With regard to the rather broad distribution of the diffusion coefficients about the mean value, the polydispersity of molecular weight for these control polymers is probably sufficient to account for the observed dispersion. Polydispersity does indeed manifest itself in the intensity decays in the PGSE NMR experiment (*19*). Unfortunately, the polydispersity of molecular weight for the control polymers has not been measured. However, the molecular weight distribution of the M_n 34,200 C_{16} HEUR associating polymer is rather broad ($M_w/M_n = 1.7$) (*5*). Given the similarities in the method of synthesis of the control and C_{16} HEUR polymers, polydispersity of the control polymer molecular weights should be comparable to that of the C_{16} HEUR associating polymers. Furthermore, for the case of a step–step polymerization process such as that used to synthesize the C_{16} HEUR associating polymers (*4*), Carothers–Flory statistical theory indicates that M_w/M_n is expected to increase with increasing polymer molecular weight (*18*). For the control polymers at the concentrations investigated here, $c << c^*$, so no associations or entanglements are anticipated. Hence the expected polydispersity of molecular weight for the control polymers can account for the observed distribution of control polymer diffusion coefficients and the facts that this distribution is virtually independent of concentration over the range investigated and that β is molecular weight dependent.

Before we discuss the interpretation of the PGSE NMR results for the C_{16} HEUR associating polymers, some comments regarding the details of the PGSE NMR experiment are warranted. The interpretation of PGSE NMR-derived diffusion coefficients is rather straightforward when the inequality in equation 4 is satisfied:

$$2Dt >> \langle r^2 \rangle \quad (4)$$

where D is the self-diffusion coefficient, $t = (\Delta - \delta/3)$ is the diffusion time for the PGSE NMR experiment, and $\langle r^2 \rangle$ is the square of the dimension of the object that is diffusing (*8*). When $2Dt$ is substantially larger than the molecular dimensions, the experiment measures true center-of-mass diffusion. When $2Dt$ is smaller than the molecular dimensions, the observed diffusion contains contributions from other, more localized mechanisms such as segmental motions. One consequence of any failure to satisfy this inequality is that the observed decay of the NMR signal intensity will appear to be multiexponential, seemingly correlating with a "fast" and a "slow" contribution to the observed diffusion. However, multiexponential decay of the NMR signal in the PGSE NMR experiment can originate from other sources, such as polydispersity of molecular weight (*19*). One can test whether the inequality is satisfied in a particular situation by probing the dependence of the intensity decay on the universal time parameter $(\gamma G\delta)^2(\Delta - \delta/3)$ (*11*). Thus for purely Brownian center-of-mass diffusion, the intensity decays with increasing values of δ for different values of G or Δ should be identical. When we perform the PGSE NMR experiment at different values of Δ, the inequality in equation 4 is satisfied at all measured concentrations for the control polymers but only at concentrations of less than about 0.5 wt% for the C_{16} HEUR associating polymers. That the diffusion coefficients of the C_{16} HEUR associating polymers and the viscosities of such solutions undergo a dramatic change at this concentration is no coincidence.

For purely center-of-mass diffusion, a situation pertaining to the low-concentration regime for the C_{16} HEUR associating polymers, the diffusion coefficients obtained from PGSE NMR provide information regarding the sizes of the diffusing species. From the Stokes–Einstein equation, the hydrodynamic radius of the diffusing species, R_h, depends on the self-diffusion coefficient D, the Boltzmann constant k, the absolute temperature T, and the solution viscosity η according to equation 5:

$$R_h = \frac{kT}{6\pi\eta D} \tag{5}$$

For the M_n 34,000 control polymer, for instance, when we extrapolate to infinite dilution, at which the solution viscosity should approach that of water, the calculation yields an R_h of 7.9 nm. Values of R_h for the other control polymers are listed in Table II. Devenand and Selser

Table II. R_h of HEUR Polymers at Infinite Dilution

Polymer–End Group	D_0[a] $(\times 10^{-11}\ m^2\ s^{-1})$	R_h[b] *(nm)*
26,800–$C_{16}H_{33}$	1.4	17.5
34,200–$C_{16}H_{33}$	1.2	20.4
48,000–$C_{16}H_{33}$	1.1	22.7
16,600–H	4.7	5.2
34,000–H	3.1	7.9
67,000–H	2.3	10.5

[a] Determined by extrapolating the mean diffusion coefficient to zero concentration.
[b] Determined from equation 6 by assuming that the solution viscosity equals that of pure water.

(*20*) reported that the molecular weight dependence of the radius of gyration of PEG in aqueous solution obeyed the relationship shown in equation 6.

$$R_h(\text{nm}) = 0.0145 M_w^{0.571 \pm 0.009} \quad (6)$$

When $M_n = 34{,}000$ and $M_w/M_n = 1.7$ (*5*), equation 6 predicts an R_h of 7.6 nm, which is in good agreement with our results from PGSE NMR. In fact the agreement is reasonable for all three control polymers.

For the M_n 34,200 C_{16} HEUR associating polymer, equation 5 yields an R_h of 20.4 nm, a value that is in excellent accord with the results of dynamic light-scattering experiments on the identical compounds (*5*) as well as with PGSE NMR measurements on closely related polymers (*10*). Values of R_h for the other C_{16} HEUR associating polymers are listed in Table II. The larger sizes of the diffusing species of HEUR associating polymers relative to those of the controls indicate that these polymers assume an aggregated state even at very low concentrations. In fact, the onset of aggregation occurs at concentrations well below the lowest concentration measured in the present study (*4*).

From the sizes of the aggregated species one can extract the aggregation number for the micelle, which is the number of hydrophobic end groups that associate to form a micelle. Distinguishing between the cluster size and the aggregation number is important. As we use it, the cluster size refers to the number of polymer chains associating with a micelle in the concentration regime in which virtually no bridging chains exist. The aggregation number always refers to the number of hydrophobic groups associated with a hydrophobic site. For a hard sphere, the mass of the aggregated species M is related to its radius r through the intrinsic viscosity $[\eta]$ via equation 7.

$$[\eta]M = (10\pi/3)N_A r^3 \quad (7)$$

The intrinsic viscosity of solutions of the M_n 34,200 C_{16} HEUR associating polymer is 1.1 dL/g (*5*). Applying equation 7 yields an M of approximately 460,000. With $M_n = 34{,}200$ and $M_w/M_n = 1.7$ (*5*), the cluster size is between 8 and 13 polymer chains, depending on whether one employs M_n or M_w in the calculation. So the aggregation number falls between 16 and 26 hydrophobic chain ends per micelle.

The question of the aggregation number in micelles of associating polymers is central to understanding the nature of the association process. Two different views have emerged. The "open" association model proposes that aggregates are formed through successive accretion of unimers into loosely defined hydrophobic sites. This model is based on light-scattering studies that indicate that the apparent weight-average molecular weight of the associating polymers in aqueous solution increases with increasing polymer concentration (*21, 22*), and predicts rather low aggregation numbers for the micelles of associating polymers. In contrast, the "closed" association models (also referred to as the microgel [5] and transient micellar network models [6]) propose that well-defined micelles with higher aggregation numbers are first formed at low concentrations by the aggregation of hydrophobic groups at the ends of looped chains. With increasing concentration, a network of micelles linked to one another by bridging chains forms. Therefore a major difference between the two models concerns their predictions regarding the aggregation number. The aggregation numbers we extract from our PGSE NMR data support the closed association model.

Aggregation numbers lower by a factor of 2 than those reported here have been estimated for HEUR associating polymers from PGSE NMR results (*9, 10*). In extracting aggregation numbers from their diffusion data, those researchers assumed that the R_h of both the associated and the nonassociated species scale according to $M^{0.53}$. This assumption is reasonable for the control polymers but not immediately evident for strongly chain-end-associated polymers. Let us assume that in the low concentration regime the size of the micelles follows a scaling law similar to that expected for free polymers in dilute solution, that is,

$$R = K(nM)^a \tag{8}$$

where M is the molecular weight of a single polymer chain, and n is the number of chains contributing to a cluster. If K and n are independent of molecular weight, a graph of $\ln(R_h)$ versus $\ln(M)$ will be linear with slope a and an intercept of $\ln(Kn^a)$.

Figure 6 compares the results of such an analysis for the controls versus the C_{16} HEUR associating polymers. The linearity is reasonable for both cases, with $a = 0.50$ (± 0.05) and 0.44 (± 0.04) for the control and the HEUR polymers, respectively. Exponent a for the control polymers is exactly that anticipated from the molecular weight dependence of D_o discussed previously. On the other hand, the scaling exponent for the C_{16} HEUR associating polymers is smaller than expected even if water were a theta solvent for PEG. However, such a low value for the molecular weight scaling constant for HEUR associating polymers is not without precedent. Specifically, Ou-Yang and Gao (*23*) in their study of the binding of C_{12} and C_{16} HEUR associating polymers to polystyrene latex spheres observed that the thickness of

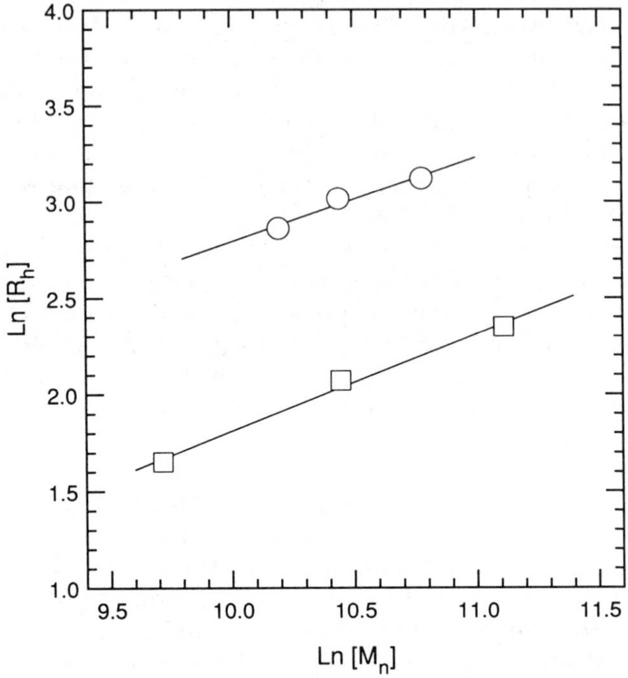

Figure 6. Scaling law describing the dependence of the size of the diffusing species on the molecular weight of the constituent polymers for the C_{16} HEUR associating polymers, ○, and the control polymers, □. The solid lines represent fits to the data using equation 8 and the parameters described in the text.

the adsorbed layer at saturation scaled according to $a = 0.46$. Although both the study by Ou-Yang and Gao (*23*) and the present report investigated only a limited range of molecular weights, so that only qualitative conclusions are warranted, these conclusions are nevertheless compelling. First, any calculation of the aggregation number based on the assumption that $a = 0.53$ for both the unimer and the micelle tends to yield an underestimate relative to a calculation based on the observed values of a. This underestimate might explain the differences in the aggregation numbers reported by Stilbs and co-workers (*9, 10*) and by the present report. Second, the lower value of a for the HEUR associating polymers relative to that for the controls suggests a reduction in chain configurational freedom that is not unexpected given the conformational restraints imposed by the associated state. Substituting equation 8 into equation 7 leads to the prediction that the intrinsic viscosity of the HEUR associating polymers scales according to $M^{1/3}$, and this prediction can be verified experimentally. The inference is that because of associations, the size of the diffusing species for the HEUR associating polymers grows with increasing molecular weight at a slower rate than one expects for a random-flight model of chain configurations. Third, the apparent success of equation 8 in describing the dependence of the sizes of the micelles on the molecular weights of the component polymer chains suggests that the cluster size n is relatively constant over the range of molecular weights available. This prediction, too, is experimentally verifiable.

To apply the Stokes–Einstein equation at concentrations other than infinite dilution, one cannot assume that the solution viscosity simply equals that of water. Several groups have reported viscosity data for the HEUR associating polymers (*12*), including the dilute concentration regime for the M_n 34,200 C_{16} HEUR associating polymer (*5*). If the inequality in equation 4 is not violated, the Stokes–Einstein equation provides the corresponding R_h. When the viscosity data of Yekta et al. (*5*) and the mean diffusion coefficients reported here are used, the calculated R_h increases from 20.4 nm at infinite dilution to a maximum of 42.7 nm at 0.3 wt% polymer concentration. Above this concentration, for the conditions used in our PGSE NMR measurements, the R_h appears to decrease. However, the contribution of local motions to the observed decay of the signal intensity and the derived diffusion coefficient begin to make themselves felt in this concentration range, so the Stokes–Einstein equation cannot be applied reliably. This increase in the size of the diffusing species with increasing concentration of associating polymer confirms the observations made by others who used dynamic light scattering (*21, 22*).

The distribution of diffusion coefficients about the mean value and the dependence of that distribution on the polymer concentration differ between the associating polymers and the controls. As already discussed, polydispersity of molecular weight appears to largely account for the breadth of the distribution of diffusion coefficients observed for the control polymers. For the HEUR associating polymers we have the additional complication of chemical heterogeneity, since ^1H NMR characterization reveals that approximately 20% of the M_n 34,200 C_{16} HEUR associating polymer chains have only a single, rather than a double, hydrophobic end cap (*17*). Nevertheless, PGSE NMR measurements indicate that the aggregates formed by the HEUR associating polymers at low concentrations possess a narrower distribution of diffusion coefficients than the corresponding control polymers. This situation is the one that would be predicted if we assume that (1) all polymer chains are associated with a micellar cluster and (2) the molecular weight distribution within any one cluster mirrors that of the polymer chain population as a whole. In other words, we assume that the tendency to associate is dictated by the hydrophobic end cap rather than by the length of the water-soluble backbone. Random aggregation statistics then predict a narrowing of the size distribution of the aggregated state relative to that of the nonaggregated state.

When considered alongside the increase in the mean size of the diffusing species, the increasing breadth of the size distribution with increasing concentration suggests the presence of "supra-aggregates" consisting of two or more micellar clusters joined by bridging chains. If the increasing sizes of the diffusing species with concentration are considered to arise by accretion of polymer chains into existing clusters, that is, the aggregation number increases with increasing concentration, then we have no reason to expect a priori that the cluster size dispersion would change markedly. Moreover, the data of Yekta et al. (5) suggest that the aggregation number is constant in this concentration range.

Concluding Remarks

These PGSE NMR diffusion studies of C_{16} HEUR associating polymers indicate that the length of the polymer backbone is secondary to the length of the hydrophobic end cap as an effector of associating behavior. The PGSE NMR data are characteristic of a distribution-of-diffusion coefficient and are best analyzed by using a stretched exponential as described by Walderhaug et al. (*10*) and Nystrom et al. (*11*). At low concentrations the associating polymers aggregate to form micelles composed of approximately 10 polymer chains each. The dependence of the diffusion coefficients and the dispersion of the diffusion coefficients on polymer concentration supports the transient micellar network model of association as propounded independently by Yekta et al. (5) and Annable et al. (6).

Acknowledgments

We thank Richard Jenkins of Union Carbide Co., UCAR Emulsions Systems, Cary, NC, for providing us with the C_{16} HEUR associating polymers and control polymers. We are grateful to the Natural Sciences and Engineering Research Council of Canada for their support of this research through the provision of Operating, Equipment, and Strategic research grants. We are indebted to A. Yekta and M. A. Winnik for their critical comments on the manuscript.

References

1. *Water-Soluble Polymers;* Glass, J. E., Ed.; Advances in Chemistry 213; American Chemical Society: Washington, DC, 1986.

2. *Polymers in Aqueous Media;* Glass, J. E., Ed.; Advances in Chemistry 223; American Chemical Society: Washington, DC, 1989.
3. *Polymers as Rheology Modifiers;* Schulz, D. N., Glass, J. E., Eds.; ACS Symposium Series 462; American Chemical Society: Washington, DC, 1991.
4. Jenkins, R. D. Ph.D. Thesis, Lehigh University, Bethlehem, PA, 1990.
5. Yekta, A.; Duhamel, J.; Adiwidjaja, H.; Brochard, P.; Winnik, M. A. *Langmuir* **1993,** *9,* 881.
6. Annable, T.; Buscall, R.; Ettelaie, R.; Whittlestone, D. *J. Rheol.* **1993,** *37,* 695.
7. Stilbs, P. *Prog. Nucl. Magn. Reson. Spectrosc.* **1987,** *19,* 1.
8. Von Meerwall, E. D. *Adv. Polym. Sci.* **1983,** *54,* 1.
9. Persson, K.; Abrahmsen, S.; Stilbs, P.; Hansen, F. K.; Walderhaug, H. *Colloid Polym. Sci.* **1992,** *270,* 465.
10. Walderhaug, H.; Hansen, F. K.; Abrahmsen, S.; Persson, K.; Stilbs, P. *J. Phys. Chem.* **1993,** *97,* 8336.
11. Nystrom, B.; Walderhaug, H.; Hansen, F. K. *J. Phys. Chem.* **1993,** *97,* 7743.
12. Jenkins, R. D.; Silebi. C. A.; El-Asser, M. S. In reference 3, p 222.
13. Stejskal, E. O.; Tanner, J. E. *J. Chem. Phys.* **1965,** *42,* 288.
14. Mills, R. *J. Phys. Chem.* **1973,** *77,* 685.
15. van Halls, J. J.; Bergman, A. H. *J. Magn. Reson.* **1990,** *90,* 52.
16. Hrovat, M. I.; Wade, C. G. *J. Magn. Reson.* **1981,** *44,* 62.
17. Yekta, A.; Kanagalingam, S.; Winnik, M. A. *In Techniques in Polymer Characterization;* Glass, J. E., Ed.; Advances in Chemistry 248; American Chemical Society: Washington, DC, 1994.
18. Flory, P. J. *Principles of Polymer Chemistry;* Cornell University: London, 1953.
19. Von Meerwall, E. D. *J. Magn. Reson.* **1982,** *50,* 409.
20. Devanand, K.; Selser, J. C. *Macromolecules* **1991,** *24,* 5943.
21. Maechling-Strasser, C.; Francois, J.; Clouet, F.; Tripette, C. *Polymer* **1992,** *33,* 627.
22. Maechling-Strasser, C.; Clouet, F.; Francois, J. *Polymer* **1992,** *33,* 1021.
23. Ou-Yang, H. D.; Gao, Z. *J. Phys. II France* **1991,** *1,* 1375.

RECEIVED for review November 18, 1993. ACCEPTED revised manuscript August 8, 1994.

21

Transient Polymeric-Bridging Dynamics of Colloids

L. E. Dewalt, Z. Gao, and H. D. Ou-Yang

Department of Physics and Polymer Interfaces Center, Lehigh University, Bethlehem, PA 18015

Telechelic poly(ethylene oxide) (PEO) polymers in aqueous solution can form bridges between colloidal surfaces by attaching one hydrophobe onto each surface. Colloidal clusters can form because of this polymeric-bridging-induced attraction between the particles. Using a suspension of 28-nm polystyrene colloidal particles and linear PEO polymers terminated on both ends by $C_{20}H_{41}$ hydrophobes, we investigated the relaxation of transient aggregated colloidal clusters into smaller clusters or singlets. Polarized and depolarized static light scattering and dynamic light scattering were used to determine the sizes and molecular weights (in terms of the singlet mass) of the clusters as a function of time. After the transient clusters were made, a short, initial period of about 300 was dominated by multiple scattering, which indicates the presence of very large clusters. After that, two distinct relaxation regimes could be identified: one, the intermediate time regime, had a characteristic time of about 800 s, and the second, the long time regime, had a characteristic time of 10,000 s. The long time regime is believed to be the time when doublets relax to singlet particles. The temperature dependence of relaxation dynamics indicates that the breaking up of the low-order multiplets follows Arrhenius behavior with an activation energy of about 30 k_BT.

WATER-BASED ASSOCIATIVE POLYMERS are used in coating, adhesion, paper making, mining, water treatment, and various other material-

0065–2393/96/0248–0395$12.00/0

processing applications that involve colloidal suspensions as raw materials. In many of these applications, controlled polymeric-bridging flocculation from a marginally stabilized colloidal state is one of the primary concerns. Associative polymers used in commercial flocculation processes have very high molecular weights and functional groups that can strongly adsorb onto the colloidal surface (*1–4*). The phenomenon of incipient flocculation and a review of the related research were discussed extensively by Napper (*5*). Computer simulations of bridging flocculation by homopolymers were investigated recently (*6*). An increasingly interesting area of flocculation for commercial applications and research is the use of polyelectrolytes (*7–11*). However, the regime near the onset of flocculation and, in particular, the kinetics of polymer-induced flocculation are still open fields for fundamental understanding, and they prompted this work (*12–14*).

In this chapter we discuss studies of transient, finite clusters of polystyrene (PS) latex particles aggregated by water-soluble associative polymers. The model polymer we use is a telechelic polymer composed of a water-soluble, linear poly(ethylene oxide) (PEO) backbone with both ends terminated by a $C_{20}H_{41}$ hydrophobe. Because of its moderate association strength per chain and relatively low molecular weight, this model telechelic polymer is probably not the most favorable one for commercial flocculation applications. However, the relatively low association strength per chain allows the flocculation process to be reversible, and the simplicity of the polymer's linear, symmetric structure permits comparison of experimental observations with a simple model.

For the model telechelic polymers we are studying, flocculation can be induced at a fixed particle concentration and polymer molecular weight by varying the polymer concentration. At low polymer and colloid concentrations, repulsion between individual polymer-coated particles renders the suspension stable. The suspension is also stable at very high polymer and colloid concentrations, at which an extended network of polymer forms throughout the system. In the intermediate concentration regime, when particle surfaces are partially covered by adsorbed polymers, finite clusters of particles can form because of polymer bridging.

It is also relevant to introduce results from previous experiments on interactions of this type of telechelic PEO polymer with PS latex particles. At low particle concentration, such that the interparticle distance is much larger than the stretched polymer chain length, telechelic PEO chains in the adsorbed layer can form a dense brush with both $C_{20}H_{41}$ hydrophobic end groups attached on the PS surface and loops formed by the water-soluble PEO backbone extending into the solvent (*15*). These extended polymeric brushes give rise to an effective repulsive interaction between the particles that renders PS stable. However, at lower polymer surface coverage, exchange of polymer chains between colloidal surfaces is substantial. Strong evidence suggests that the exchange of polymer chains between colloidal surfaces occurs by direct transfer from one surface to another rather than by desorption of chains into the solution first and then reabsorption onto another surface (*16, 17*). This evidence implies that a transient bridged state of two (or more) colloidal particles exists during the chain exchange process. The bridged pairs or clusters must be transient, because the overall particle concentration was kept low enough that thermodynamics dictate that the suspension remain dominated by singlet particles. However, if the adsorption interaction is stronger or the particle concentration is higher, the system will remain flocculated. A general strategy for investigating the flocculation phenomenon is to approach the critical flocculation point (CFPT) from the stable phase. Our intention here is to investigate the cluster and the transient lifetimes of the aggregated doublets.

The present study focuses on the regime of low particle concentration and low polymer concentration so that the system is below the CFPT. The challenge is to create transient flocculated colloidal clusters whose lifetimes are long enough for observation. By injecting a drop of high-concentration polymer solution into a dilute, uniform PS latex particle solution, we are able to locally quench the system above the CFPT, thus creating flocculated clusters. This process of making transient flocculated clusters also works when we reverse the mixing procedure, that is, when we inject a small amount of concentrated PS latex particle solution into a dilute, uniform polymer solution. Since the overall colloidal suspension is thermodynamically stable, flocculated colloidal clusters are only transient. Using this approach, we were able to observe the transient behavior of low-order colloidal clusters (multiplets) during relaxation of the system to the stable suspension state (singlet-dominated state).

Light-scattering measurements are used in this study to monitor the forming and breaking of the multiplets. Since the flocculated clusters are typically anisotropic, they give rise to depolarized (VH) scattering. We measure both the polarized (VV) and the VH scattering intensities as a function of time. We also obtain VV dynamic-light-scattering data to provide additional information on the size of the multiplets. From the decay of low-order multiplets, we are able to obtain the lifetimes of transient doublets. Modeling the breaking of the bridged pairs as an activation process, we can obtain a measure of the activation energy from temperature-dependent data.

Experimental Details

The model associative polymers used in this study are telechelic copolymers with a molecular structure of R-O-(DI-PEO)$_m$-DI-O-R, as obtained from Union Carbide. Here PEO is a PEO segment with a nominal molecular weight of 8,200 g/mol, DI is an isophorone diisocyanate group that links the PEO segments with a polymerization index *m* of 12, and R is a terminating hydrophobic end group ($C_{20}H_{41}$). For simplicity, the sample is labeled according to the number of carbons in the end group as well as the overall molecular weight of the chain (in thousands of grams per mole). Therefore, C20-116 is a sample with $C_{20}H_{41}$ end groups and a total molecular weight of 116,000 g/mol. From previous studies (*18*) we learned that both the PEO backbone and the DI linkers adsorb to PS particles in water. However, the adsorption strength of the end groups dominates, so we can assume that the model polymer effectively has the A–B–A telechelic architecture.

The adsorption substrate for these experiments is 28-nm-diameter PS spheres (Duke Scientific) in aqueous suspension. In all of the experiments, we compared the sizes of the particles with adsorbed polymer to the sizes of bare particles as obtained by dynamic light scattering; the sizes of the bare particles agreed with the manufacturer's stated size of 28 nm. Our purposes here were to create transient polymer-bridged colloidal clusters and to study the relaxation of the cluster decay through light-scattering observations. A transient cluster can by produced by one of two means: addition of a small amount of high-concentration polymer solution to a uniform, dilute suspension of colloidal particles or addition of a small amount of high-concentration particle suspension to a uniform, dilute polymer solution. These two processes yield different initial clusters sizes, but the long-time relaxation is very similar. Throughout this study we followed the first procedure. When 40 μl of 0.5% (by weight stock) solution of C20-116 polymer chains was mixed with a 28-nm PS particle suspension at a PS volume fraction of 6×10^{-3} (this mixture is a solution containing about 10 chains per sphere, on average), a slightly opaque sample was initially produced. The opacity of the sample decayed within a few minutes.

We simultaneously monitored, at symmetric scattering angles of 45°, the VV and VH scattering intensities. The VV and VH intensity data were accumulated by two Brookhaven Instruments BI-2030 autocorrelators and averaged over periods of 30 to 60 s. This time was chosen to provide adequate time resolution for the kinetics of relaxation. For a fixed particle concentration, the VV scattering intensity in the low-concentration limit is linearly proportional to the square of the mass of clusters in solution and is therefore very sensitive to clustering of the particles. VH scattering comes from two effects: multiple scattering and rotations of anisotropic particles (*19*).

Figures 1 and 2 show the normalized VV scattered intensity versus time for two samples prepared as described above. Time $t = 0$ is the point at which the collection of data began; this time was, at most, several seconds after addition and subsequent mixing of the polymer with the uniform colloidal suspension. The normalized intensity data from Figure 2 are replotted in Figure 3 on a semilog scale to reveal the relaxation times. The VH scattering of the event shown in Figure 2 is recorded by

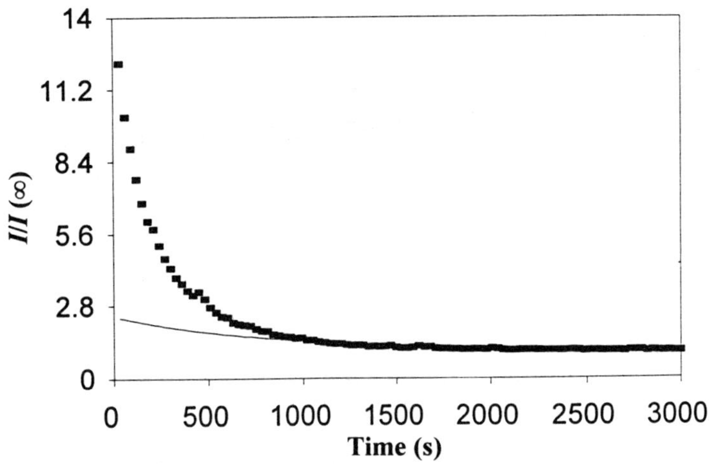

Figure 1. Normalized VV scattering intensity I(t)/I(∞) = n(t) *for a mixture of 10 C20-116 chains per 28-nm PS particle at 25 °C. The solid line represents a single exponential in the range of* n = 2 → *1.*

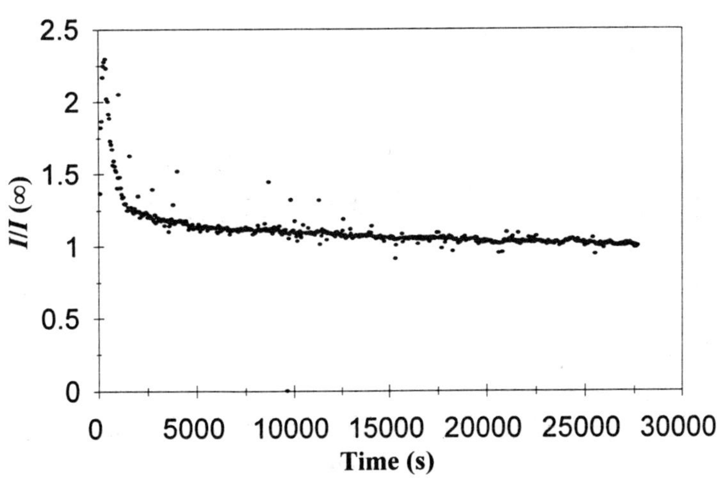

Figure 2. Normalized VV scattering intensity for a mixture of 10 C20-116 chains per 28-nm PS particle at 25 °C.

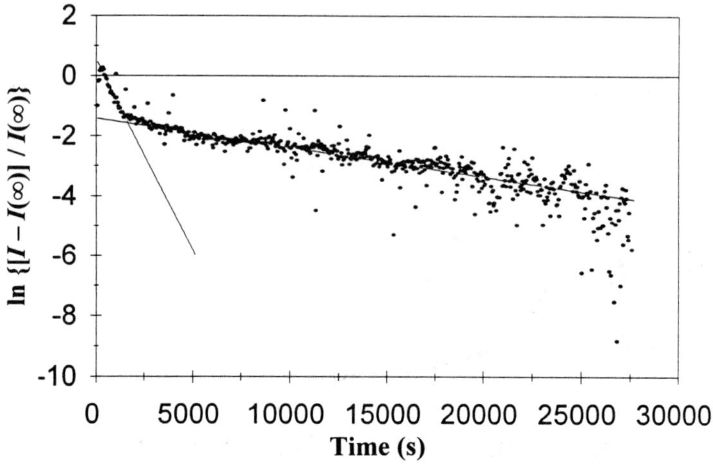

Figure 3. Semilog plot of the data in Figure 2, showing two characteristic decay times. The fast decay time is 800 s, and the slow decay time is 10,000 s.

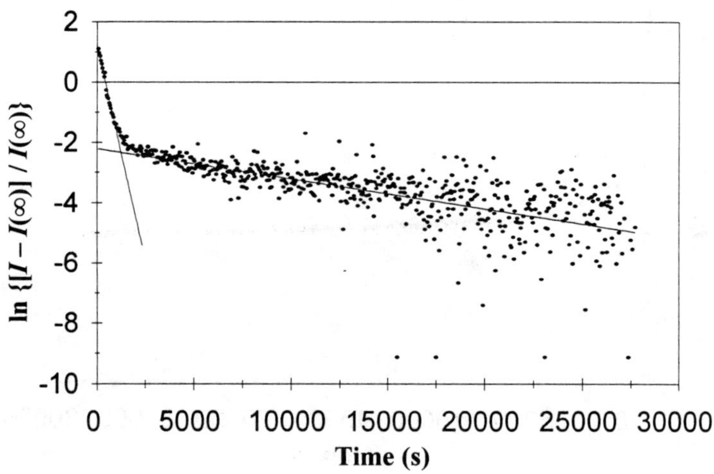

Figure 4. Semilog plot of VH scattering for the same sample as in Figure 2, also showing two characteristic decay times. The fast decay time is 350 s, and the slow decay time is 10,000 s.

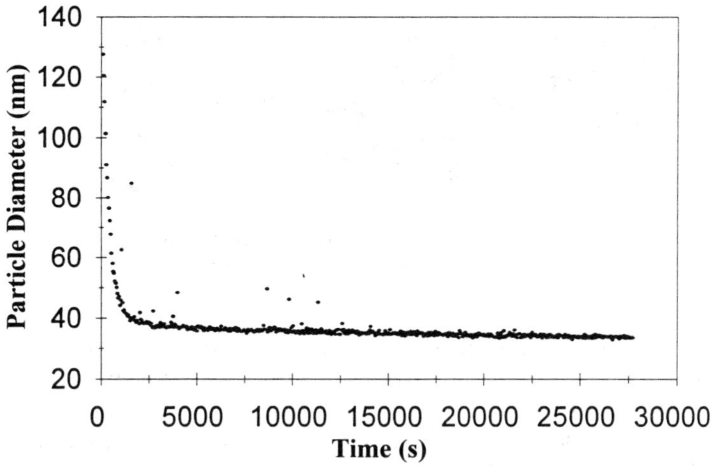

Figure 5. Particle size versus time plotted for the same sample as in Figures 2–4.

a second photomultiplier tube and shown in Figure 4 on a semilog plot. The VV and VH data are synchronized within an error of about 5 s, a short time compared to the relaxation time scale. The VV scattering data were also used to calculate the intensity–intensity autocorrelation function averaged over the collection period of about 60 s. The autocorrelation functions were then used to determine the average hydrodynamic radius of the scattering clusters by cumulative expansion of the correlation function. Figure 5 shows the cluster size obtained from dynamic-light-scattering analysis of the VV scattering data. Each point in Figure 5 was calculated from an individual correlation function.

Data Analysis and Discussion

The initial observation that the sample increases in opacity when the polymer is introduced suggests that significant flocculation is occurring. However, this flocculation is transient, since the sample becomes transparent again within several minutes. From the facts that at long times the suspension remained transparent and that light-scattering intensity carefully measured at long times confirmed that individual singlet particles dominate the suspension, we conclude that the free energy favors colloidal particles in the singlet form under the mixing conditions just described. Thus, by mixing polymer solution with the 28-nm-diameter PS particles in solution, we create a transient nonequilibrium state in which large flocculated clusters first form and then relax to an equilibrium state composed of single particles with polymer adsorbed to the surface.

By analyzing the decay in the light scattering, we can follow the

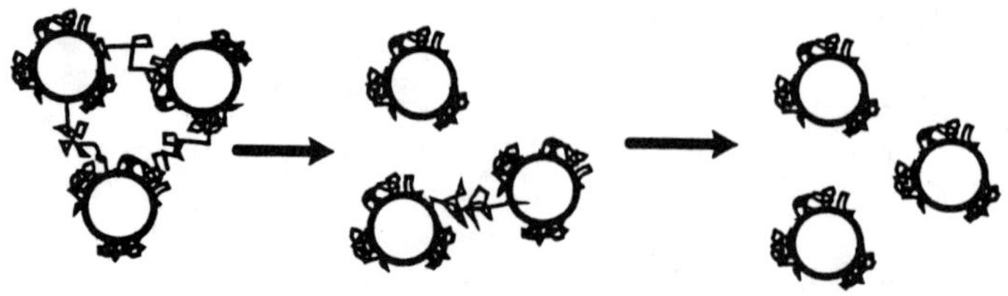

Figure 6. Relaxation model of the transient bridged state from the intermediate to the long time regime.

relaxation process of the cluster breakup. From relaxation data shown in Figures 3 and 4, we define three time regimes: the short time covers the first 300 s, intermediate times are from 300 to about 3,000 s, and long times are 3,000 to 30,000 s. The data presented thus far suggest the model indicated in Figure 6. Just after the concentrated polymer solution is mixed into the particle suspension, a population of transient polymer-bridged colloidal clusters is formed. In the very short time regime, the very large clusters start to break up quickly, in about 300 s. The finite, smaller clusters begin to break up during the intermediate time regime. Finally, in the long time regime, the relaxation is dominated by the breaking of doublets into single particles. We will analyze the data according to this model.

Quantitatively interpreting the data from the short time regime is difficult because reproducing the initial conditions from run to run is difficult (discussed later). Also, the sample appears milky because of multiple scattering and is difficult to analyze by conventional light-scattering techniques. However, the sample is transparent in the intermediate and long time regimes. Using conventional light-scattering theory, we can analyze the relaxation process by the following way. The small-angle VV light scattering at a fixed colloid concentration can be expressed as

$$I_{\mathrm{cluster}} = n\, I_{\mathrm{singlet}} \tag{1}$$

where n is the number of singlet particles per aggregated cluster (*19*). Normalizing the VV scattering intensity data to the scattering intensity at long times, the vertical axes of Figures 1 and 2 can be considered to be the number of particles in the scattering clusters. As stated earlier, in the short time regime of Figure 1, we see very large clusters, but in the same regime of Figure 2, we do not. Thus the nonreproduceability of the mixing process is shown.

Looking at the intensity relaxation data in Figure 3, we clearly see two relaxation times, one for the intermediate and one for the long time regimes; hence the definition of the time regimes. The intermediate characteristic decay time is 800 s, a time that is similar to the characteristic time found in Figure 1 (solid line). The long decay time has a characteristic time of 10,000 s. In the intermediate regime, in which the average number of particles in a cluster is on the order of 2 to 5, several relaxation channels in which the clusters can break up are present. For example, a cluster of 5 particles can relax by breaking up into clusters of 1, 2, 3, or 4 particles. In contrast, the long time relaxation occurs in the time domain in which the average cluster size is relaxing from about 1.2 to 1 particles. If we assume that the population of clusters in this time range is dominated by doublets and singlets, then only one decay channel exists, that is, from doublets to singlets. However, at this point we cannot exclude the possibility of a recombination channel yielding a finite distribution of doublets and singlets because the colloidal singlets are not completely sterically stabilized at 10 chains per particle. According to the size and intensity analyses for long times, the doublet population, if it exists, is very small.

The dynamic-light-scattering data in Figure 5 confirm the existence of clusters in the intermediate time regime. After about 2000 s, singlets begin to dominate the scattering. At very long times, the average particle size approaches 32 nm. This size implies that the 10 polymer chains form an adsorbed layer 2 nm thick, a value that is not an unreasonable according to our experience of polymer adsorption in a similar system. The polydispersity of the diffusion constant analyzed at long times shows a value of about 13%, which is comparable to that obtained from bare particles. This value also supports the point that at long times, singlet particles dominate. In addition, the particle diameter at each point is taken from a time-averaged correlation function over 60 s, which is somewhat shorter than the ideal length of 200 s or longer. However, the lack of scatter in the data at long times suggests that a clear trend was obtained in this measurement.

VH scattering, shown in Figure 4, supports the relaxation picture of our model. At very short times, we anticipate very large clusters, which are indicated by the milky appearance and multiple scattering. The first decay at a very short time in the multiple scattering regime has a relaxation time of 350 s. At the very long time the relaxation is due to the decay of doublets into singlets. As a matter of fact, the VH scattering from singlets should be very small, because the singlets are isotropic. This relaxation time is the same as that determined from VV scattering. The intermediate time scale has a relaxation time between 350 and 10,000 s but has not been analyzed in detail.

For the intermediate regime we estimated the lifetimes of the bound particles by assuming that particles leave multiplets by an activation process. The temperature dependence of the characteristic

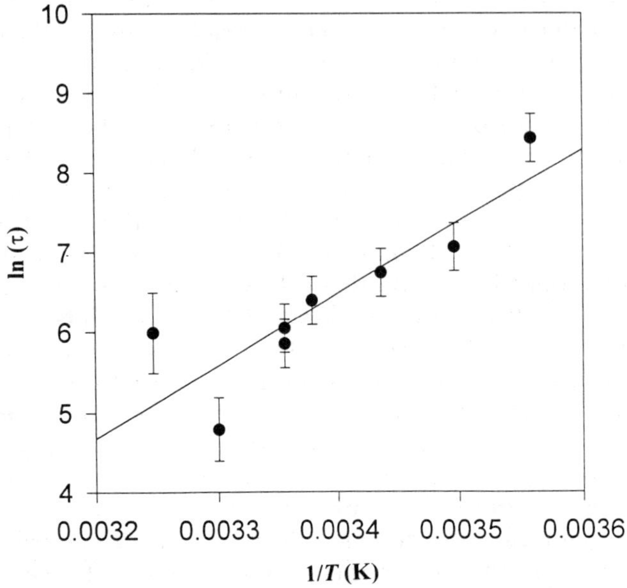

*Figure 7. Semilog plot of relaxation time versus 1/*T *for results from conventional VV static scattering. The solid line is a fit to the Arrhenius function with an activation energy of 30* k_BT.

decay time obtained from VV scattering is plotted versus $1/T$ in Figure 7. The Arrhenius plot indicates an activation energy of 30 k_BT. The error bars shown in Figure 7 are an indication of the reproduceability of the data.

We are in the process of determining the temperature dependence of the relaxation in the long time regime. This calculation should provide the effective binding energy for a pair of bridged particles, and this energy can be compared to the binding energies of the individual hydrophobes. Work is also being done on the effects of particle and polymer concentrations. These results will be reported elsewhere.

Conclusions

In this chapter we show the relaxation behavior of low-order colloidal multiplets formed by transient bridging flocculation. By using VV and VH light-scattering techniques, we found three relaxation regimes. The fast regime is a fast relaxation from large clusters in the multiple scattering regime. In the intermediate regime the large clusters have broken up into small clusters, probably with 1 to 5 particles per cluster. Here, the clusters still have several decay channels, and these channels allow a faster relaxation to equilibrium. In the slow regime, only one decay channel is left (doublet to singlet), and the relaxation time becomes significantly longer. The Arrhenius behavior in the intermediate time regime indicates an activation energy of 30 k_BT, a value 3 to 4 times that for the interaction between a $C_{20}H_{41}$ hydrophobe and a PS particle.

Acknowledgments

We acknowledge the Donors of the Petroleum Research Fund, administered by the American Chemical Society, for partial support of this research. We are also thankful for support from the NSF-IUCRC for Polymer Interfaces at Lehigh University. L. E. Dewalt is supported by a graduate fellowship from the U.S. Department of Education.

References

1. Russel, W. B. In *Colloid–Polymer Interactions;* Dubin, P.; Tong, P., Eds., ACS Symposium Series 532; American Chemical Society: Washington DC, 1992.
2. Scheutjens, J. M. H. M.; Fleer, G. J.; Cohen Stuart, M. A. *Colloids Surf.* **1986,** *21,* 285.
3. Gregory, J. *Colloids Surf.* **1987,** *31,* 231.
4. Dickinson, E.; Eriksson, L. *Adv. Colloid Interface Sci.* **1991,** *34,* 1.
5. Napper, D. H. *Polymeric Stabilization of Colloidal Dispersions;* Academic: London, 1983.
6. Dickinson, E.; Euston, S. R. *J. Chem. Soc. Faraday Trans.* **1991,** *87,* 2193.
7. Mabrire, F.; Audebert, R.; Quivoron, C. *J. Colloid Interface Sci.* **1984,** *97,* 120.
8. Wang, T. K.; Audebert, R. *J. Colloid Interface Sci.* **1987,** *119,* 459.
9. Durand-Piana, G.; Lafma, F.; Audebert, R. *J. Colloid Interface Sci.* **1987,** *119,* 474.
10. Muthukumar, M. *J. Chem. Phys.* **1987,** *86,* 7230.
11. Wang, T. K.; Audebert, R. *J. Colloid Interface Sci.* **1988,** *121,* 32.
12. Pelssers, E. G. M.; Cohen Stuart, M. A.; Fleer, G. J. *J. Chem. Soc. Faraday Trans.* **1990,** *86,* 1355.
13. Pelssers, E.; Cohen Stuart, M. A.; Fleer, G. J. *Colloids Surf.* **1989,** *38,* 15.
14. Virden, J. W.; Berg, J. C. *J. Colloid Interface Sci.* **1992,** *149,* 528.
15. Ou-Yang, H. D.; Gao, Z. *J. Phys. II* **1991,** *1,* 1375.
16. Gao, Z. Ou-Yang, H. D. In *Colloid–Polymer Interactions;* Dubin, P.; Tong, P., Eds.; ACS Symposium Series 532; American Chemical Society: Washington, DC, 1992.
17. Gao, Z.; Dewalt, L.; Ou-Yang, H. D. *Colloids Surf. A* **1994,** *86,* 255.
18. Gao, Z. Ph.D. Dissertation, Lehigh University, August 1993.
19. Berne, B.; Pecora, R. *Dynamic Light Scattering;* John Wiley: New York, 1976.

RECEIVED for review April 16, 1994. ACCEPTED revised manuscript May 16, 1995.

Associating Polymers

B. Performance in Solution

3. Hydrophobe-Modified, Alkali-Swellable and Hydrophobe-Modified Hydroxyethyl Cellulose Thickeners

22

Fluorescence Studies of Cellulose Ethers

Synthesis, Characterization, and Spectroscopic Properties of Labeled Polymers

Françoise M. Winnik

Department of Chemistry, McMaster University, 1280 Main Street West, Hamilton, Ontario L8S 4M1, Canada

Examples are given of the use of pyrene-labeled polymers as models for the study of solution properties of hydrophobically modified polymers acting as gelling agents or associative thickeners in waterborne fluids. Labeled polymers were obtained from three commercial cellulose ethers: Tylose M-300 (hydroxyethylmethylcellulose), Methocel-50-F (hydroxypropylmethylcellulose), and Klucel-L (hydroxypropylcellulose). The polymers contain on average one to five pyrene groups per chain. Procedures for labeling, purification, and characterization of the polymers are described. Spectroscopic parameters such as pyrene excimer-to-monomer intensity ratios, absorption spectra, excitation spectra, and excited-state lifetimes are reported for solutions of the polymers in water and in an organic medium (dichloromethane–methanol). These parameters present evidence that in water pyrene groups form hydrophobic clusters that are destroyed at the temperature corresponding to the cloud point. These clusters form in samples that undergo heat-induced phase separation, such as pyrene-labeled hydroxypropylcellulose in water.

0065–2393/96/0248–0409$12.00/0

CELLULOSE ETHERS HAVE A WIDE RANGE of industrial applications. They are used as additives in materials such as paints, inks, cosmetics, pharmaceuticals, foods, and ceramics (*1*). They may function as thickeners, flow control agents, film-forming agents, or packaging materials for controlled drug release. The solution properties of cellulose ethers are dictated primarily by the chemical structure of the ether substituent and the degree of substitution of the cellulose and to a lesser extent by the molecular weight of the polymer. Methylcellulose, hydroxypropylcellulose, and hydroxyalkylmethylcellulose at modest concentrations are soluble in cold water. Upon being heated, their aqueous solutions undergo characteristic phase changes. Methylcellulose solutions form gels (*2*). Solutions of hydroxypropylcellulose that are clear at room temperature become opaque above a well-defined temperature, the cloud point (*3*). Both gelation and clouding are reversible. With methylcellulose, a negative hysteresis was observed between rising and decreasing transition temperatures (*4*). Hydroxyalkylmethylcelluloses in water exhibit various reversible thermally induced phase changes that depend on the chemical structures of the hydroxyalkyl groups and their level of incorporation along the polymer backbone.

The properties of cellulose ethers are also influenced by the fabrication process. A typical procedure for the production of methylcellulose involves reacting methyl chloride with wood pulp pretreated with a caustic solution (*1*). In the manufacture of hydroxyalkyl methylcelluloses, an alkyl oxide such as ethylene oxide or propylene oxide is used in addition to methyl chloride to attach the hydroxyalkyl groups to the anhydroglucose units. The degree of substitution (i.e., the average number of substituted hydroxyl positions per glucopyranose unit) can be varied over a wide range of values, giving a large number of commercial grades. Moreover, hydroxyalkylation can take place at the terminal substituent hydroxyl position and lead to materials with molar substitution values (i.e., the average number of hydroxyalkyl substituents per glucose unit) larger than 3, the initial number of free hydroxyl groups on each anhydroglucose monomer. If methylation occurs after hydroxyalkylation, end capping of hydroxyalkyl substituents may occur as well. This heterogeneous chemical structure may be responsible for the unique properties of cellulose ethers.

Industrial users require materials that meet strict specifications. These specifications are controlled by the molecular characteristics of the polymers, so it is important to analyze not only macroscopic properties but also interactions taking place on the molecular level. Analytical techniques frequently used for the characterization of cellulose ethers include ^{13}C NMR (*5*) and EPR (*6*) spectroscopy, calorimetry (*7*), viscometry (*8, 9*), and interfacial measurements (*10*). Fluorescence spectroscopy has been applied to a limited extent to the study of solution properties of cellulose ethers either by means of dyes added to the polymer solutions (*11, 12*) or by fluorescent tags attached to the polymer itself (*13, 14*). The fluorescent-tag method yields direct information on the molecular properties of the polymers, but it is more difficult to carry out in practice. Two major problems are associated with the use of fluorescence-labeled water-soluble polymers: (1) Their synthesis and purification are often difficult to carry out, and (2) the fluorescent labels may act as hydrophobic modifiers and alter the properties of the starting material. This second drawback can be turned into an advantage if one sets as a goal the study of hydrophobically modified cellulose ethers, a class of polysaccharides gaining industrial importance (*15*). These polymers incorporate a small number of hydrophobic groups that are grafted along their backbones. Solutions of these polymers in water exhibit significantly enhanced viscosity compared to that of the original polymer as a result of the clustering of the hydrophobic pendant groups. Most intensely studied to date are the hydrophobic derivatives of hydroxyethylcellulose (*11, 16–18*), but other hydrophobically modified cellulose ethers will probably become available, given their low cost, lack of toxicity, and biodegradability.

The synthesis and characterization of pyrene-labeled cellulose ethers prepared from hydroxyethylmethylcellulose, hydroxypropylmethylcellulose, and hydroxypropylcellulose are discussed in this chapter. Pyrene was selected as the fluorescent tag because its spectroscopy is well documented (*19*). Particular attention is given to the preparation of materials and to purification procedures. Then the spectroscopic characterizations of the polymers in aqueous solution and in organic solvents are compared by using the absorption and excitation spectra, the pyrene excimer-to-monomer emission intensity ratios, and the lifetimes of the pyrene monomer and excimer as observable experimental parameters. Previously reported applications of fluorescence spectroscopy to the study of cellulose ethers are reviewed briefly to illustrate how the technique provides information on the molecular events associated with macroscopic phase changes.

Experimental Methods

Materials. Hydroxyethylmethylcellulose (Tylose, grade MH 300, Werk Kalle-Albert) was purchased from Fluka; hydroxypropylcellulose (Klucel-L, Hercules) was purchased from Aldrich Chemical Corp. Hydroxypropylmethylcellulose (Methocel-50-F) was a gift from Dow Chemicals. Pyrene-labeled hydroxypropylcellulose (HPC-Py/216) was prepared as described previously (*13*). Chemicals for the syntheses were purchased from Aldrich Chemical Corp. Water was deionized with a Millipore Milli-

Q water purification system. Dimethylformamide (DMF) was dried by reflux over calcium hydride followed by distillation at reduced pressure. Spectroscopy-grade solvents were used for all spectroscopic measurements.

Instrumentation. UV spectra were recorded with a Varian UV-Vis-NIR Cary-5 spectrophotometer. Steady-state fluorescence spectra were measured with a SPEX Fluorolog 212 fluorescence spectrometer equipped with a DM3000F data analysis system. Fluorescence lifetimes were determined with an LS-1 instrument from Photon Technology International equipped with a thyratron-gated N_2 lamp and a proprietary analog stroboscopic boxcar detection system. Gel permeation chromatography (GPC) measurements were performed with a Shimadzu size exclusion chromatography system equipped with an SPD-6A UV–visible light detector and an RID-6A differential refractive index detector. Data analysis was performed with Shimadzu Chromatopac software provided by the manufacturer. Two Progel brand columns (G3000 and G5000, Toyo-Soda) were used. Separations were carried out at 34 °C. The eluent (flow rate of 0.6 mL min^{-1}) was 0.1 M $NaNO_3$.

Fluorescence Measurements. Emission spectra (not corrected) were obtained with an excitation wavelength of 343 nm. Excitation spectra were monitored in the ratio mode at 480 nm for the excimer and 390 nm for the monomer. The excimer-to-monomer emission ratio (I_E/I_M) was calculated as the ratio of the emission intensity at 480 nm to the half-sum of the emission intensities at 376 and 396 nm. The concentration of the solutions for fluorescence analysis was 0.1 g L^{-1}. Under these conditions, the absorbance of the solutions at 343 nm was kept below 0.08 in all experiments.

Syntheses. ***4-(1-Pyrenyl)butyl Tosylate.*** To a solution of 4-(1-pyrenyl)butanol (*20*) (1.0 g, 3.65 mmol) in chloroform (10 mL, flushed through alumina) were added first pyridine (0.58 g, 7.3 mmol) and then *p*-toluenesulfonyl chloride (1.04 g, 5.48 mmol) in small portions over a period of 5 min. The mixture was stirred at 22 °C in the dark under nitrogen for 1.5 h. The resulting mixture, diluted to 20 mL with diethyl ether, was extracted successively with 10% aqueous HCl (twice), water (twice), and saturated brine (twice). The organic layer was dried over $MgSO_4$ and evaporated to yield an oil. Crystallization from ethyl acetate–hexane (25 mL, 1/2 vol/vol) yielded 4-(1-pyrenyl)butyl tosylate (1.34 g, 86%) (mp, 91–92 °C; λ_{max}(ϵ, tetrahydrofuran), 343 nm (38,500), 327 nm (25,000)).

Pyrene-Labeled Tylose (Ty-Py/190). To a gel of Tylose MH-300 (2.0 g, dried by azeotropic distillation of toluene) in DMF (30 mL) kept under nitrogen was added a solution of 4-(1-pyrenyl)butyl to sylate (0.25 g, 0.58 mmol) in DMF. The mixture was stirred at room temperature for 2 h. A suspension of sodium hydride (250 mg, 60% dispersion in oil, washed twice with dry hexane) in DMF (1 mL) was added. The reaction mixture was kept in the dark at room temperature under nitrogen for 3 days. Excess base was destroyed by the addition of dilute acetic acid (5 mL, $CH_3CO_2H–H_2O$, 1/2 vol/vol). A fluid solution formed immediately upon acidification. The fluid was stirred for 1 h at room temperature. The solvent was removed by high-vacuum distillation (bath temperature, 50 °C). The resulting amber foamy material was dissolved in CH_2Cl_2–MeOH (50 mL, 3/1 vol/vol). The polymer was isolated by precipitation of this solution into hexane (500 mL) and was purified by three successive precipitations from CH_2Cl_2–MeOH into hexane. It was dried in vacuo at 30 °C. Redissolution of the polymer in water (100 mL) and freeze-drying yielded Ty-Py/190 (1.78 g) as a slightly tan solid: [Py] = 3×10^{-5} mol g^{-1} of polymer or ca. 1 pyrene per 190 anhydroglucose units as determined by UV absorption in CH_2Cl_2–MeOH (3/1 vol/vol) using 4-(1-pyrenyl)butanol in the same solvent as was used with the reference material.

Pyrene-Labeled Methocel (Me-Py/250). A solution of Methocel-50-F (8.0 g, previously dried by azeotropic distillation of toluene) in DMF (50 mL) was prepared by stirring at room temperature under nitrogen for 24 h. This solution was then reacted with 4-(1-pyrenyl)butyl tosylate (1.0 g, 2.3 mmol) under the same conditions used with pyrene-labeled Tylose (described previously). An amber foamy material was recovered after evaporation of DMF. This material was dissolved in CH_2Cl_2–MeOH (50 mL, 3/1 vol/vol). The polymer was isolated by precipitation of this solution into hexane (500 mL) and was purified by three successive precipitations from CH_2Cl_2–MeOH into hexane. It was dried in vacuo at 30 °C. Redissolution of the polymer in water (100 mL) and freeze-drying yielded Me-Py/250 (7.24 g) as a white solid: [Py] = 6.4×10^{-6} mol g^{-1} or ca. 1 pyrene per 250 anhydroglucose units.

Pyrene-Labeled Methocel (Me-Py/100). The same procedure was employed starting with Methocel-50-F (5.0 g), 4-(1-pyrenyl)butyl tosylate (0.25 g, 0.58 mmol), and NaH (250 mg) in dry DMF (25 mL) to yield a labeled sample with [Py] = 4.16 mol g^{-1} or ca. 1 pyrene per 100 anhydroglucose units.

Results and Discussion

Synthesis and Characterization of Pyrene-Labeled Polymers. The fluorescent labels were attached via ether linkages to three cellulose ethers: hydroxyethylmethylcellulose (Tylose MH-300), hydroxypropylmethylcellulose (Methocel-50-F), and hydroxypropylcellulose (Klucel-L). The molecular characteristics of the starting polymers are listed in Table I, and their structures, together with those of the corresponding labeled derivatives, are represented in Figure 1. The pyrene groups were attached to the polymers by reaction of 4-(1-pyrenyl)butyl tosylate with predried cellulose ether in alkaline DMF.

Figure 2 illustrates the procedure for labeling Tylose. Because of the marked tendency of most cellulose ethers to form gels in DMF, it was necessary to invert the normal order of adding reagents in the Williamson ether synthesis. Instead of adding the tosylate to the preformed alkoxide, the tosylate was mixed first with a solution of dry

Table I. Substitution Patterns and Molecular-Weight Ranges of Polymers

Polymer	*DS*	*MS*	*MW (DP)*	*Ref.*
Tylose MH-300	1.5–1.6[a]	0.1–0.16[b]	130,000 (700)	—[c]
Methocel-50-F	1.6–1.8[a]	0.1–0.2[d]	65,000–80,000 (100)	14
Klucel-L	2.9	5.9[d]	100,000	24

NOTE: DS, degree of substitution; MS, molar substitution; MW, weight-average molecular weight; DP, degree of polymerization (average number of glucose units per polymer chain).
[a] Methyl substituent was used.
[b] Hydroxyethyl substituent was used.
[c] Information was provided by the manufacturer.
[d] Hydroxypropyl substituent was used.

cellulose ether in DMF. After several hours, when the mixture became homogeneous, sodium hydride was added to generate alkoxide groups on the polymer. Under these conditions, the syntheses could be carried out on a 1- to 10-g scale. Moreover, by changing the initial weight ratio of polymer to pyrene derivative, it was possible to vary the degree of incorporation of the chromophores. For example, two syntheses carried out with Methocel-50-F yielded samples with levels of pyrene incorporation of 4.1×10^{-5} mol (Py) g^{-1} of polymer (Me-Py/100) and 2.1×10^{-5} mol (Py) g^{-1} of polymer (Me-Py/250). In the acronyms of the polymers, the digit refers to the average number of anhydroglucose units per pyrene group (*see* ***Experimental Methods***).

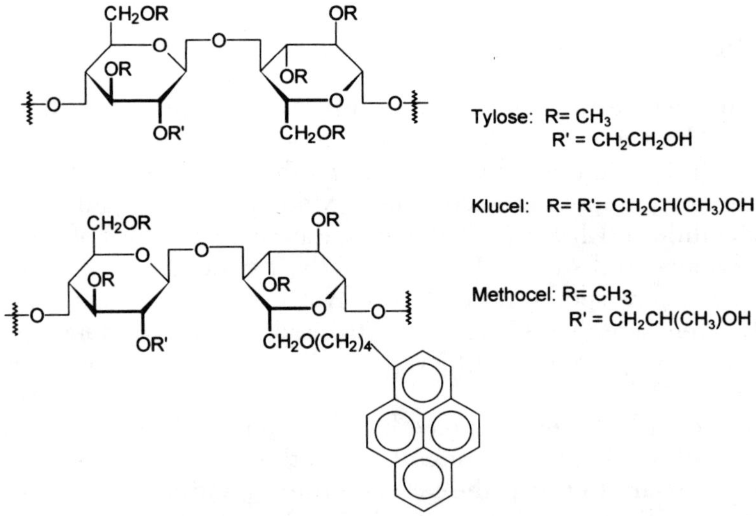

Figure 1. Structures of the cellulose ethers studied.

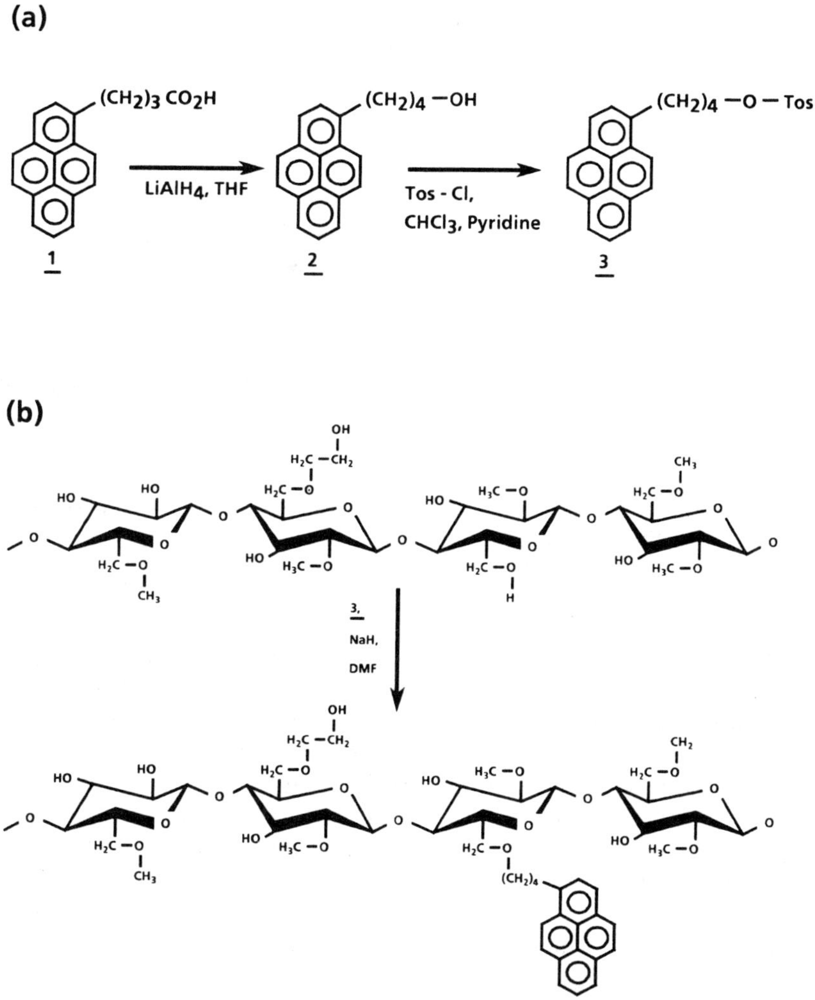

Figure 2. Synthetic scheme for the preparation of 4-(1-pyrenyl)butyl tosylate and pyrene-labeled Tylose (Ty-Py/190). THF is tetrahydrofuran.

For each sample the amount of pyrene incorporation was determined from UV absorption data of polymer solutions in CH_2Cl_2–MeOH; solutions of 4-(1-pyrenyl)butanol in the same solvent were used as standards (*see* ***Experimental Methods***). From these data and from the reported degrees of polymerization of the respective cellulose ethers, it can be estimated that there are on average approximately four pyrenes per macromolecule in Ty-Py/190 and one pyrene or fewer in Me-Py/100 and Me-Py/250.

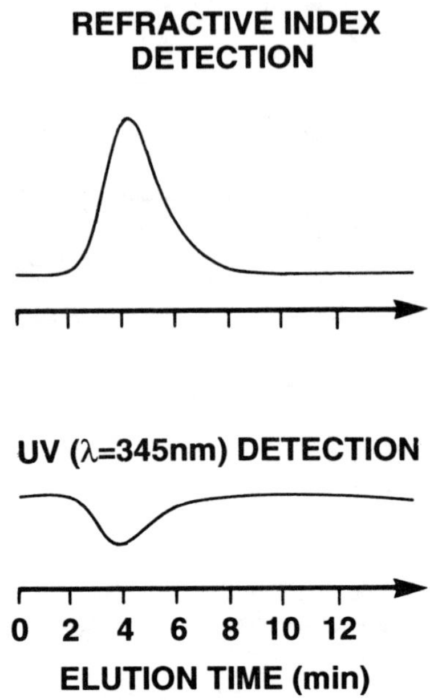

Figure 3. GPC elution profiles of pyrene-labeled Tylose (Ty-Py/190) monitored by refractive index detection and UV absorption detection (λ = 345 nm).

The purity of the pyrene-labeled polymers is an important aspect of the planned fluorescence experiments. The absence of low-molecular-weight chromophores has to be ascertained. This information was found by using UV–visible and refractive index detectors in tandem in the GPC analysis of the polymers. Thus it was established for each labeled sample that the pyrene groups were covalently attached to the polymers and that the chemical transformations did not alter the molecular weights and molecular-weight distributions of the polymers. The technique is also useful for monitoring the purification procedure and for ascertaining that the reprecipitations are effective in removing pyrene-containing low-molecular-weight impurities. Hydroxypropylcellulose is soluble in several organic solvents as well as in water; thus in this case, it is possible to carry out the GPC analysis either in tetrahydrofuran, with calibration of the data against polystyrene reference samples, or in water. Tylose and Methocel-50-F do not exhibit appreciable solubility in organic media. In such cases, one is limited to chromatography in aqueous medium. The use of water as an eluent gave irreproducible results. However, acceptable and reproducible GPC traces were obtained with aqueous sodium nitrate as the eluent, as shown in Figure 3 for the labeled Tylose (Ty-Py/190).

Spectroscopic Properties of Labeled Polymers. ***Pyrene-Labeled Tylose.*** The fluorescence spectra of Ty-Py/190 in water and in CH_2Cl_2–MeOH (3/1 vol/vol) are presented in Figure 3. Each

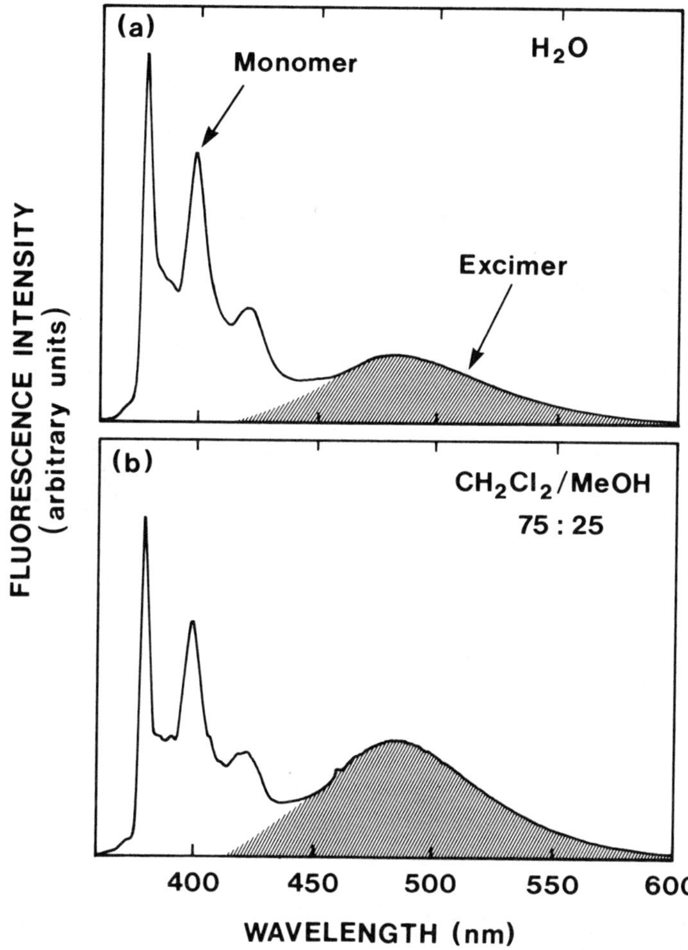

Figure 4. Fluorescence spectra of pyrene-labeled Tylose (Ty-Py/190) in water (a) and in CH_2Cl_2–MeOH (3/1 vol/vol) (b). λ_{exc} = 343 nm; polymer concentration, 0.1 g L^{-1}.

spectrum shows two emissions: One is due to locally excited pyrenes (intensity I_M, monomer emission) and has the (0,0) band located at 376 nm, and one is a broad emission centered at 480 nm that results from pyrene excimers (intensity I_E). For solutions of the same concentration, the relative contribution of the excimer to the total emission is stronger in the organic medium than in water. For samples in CH_2Cl_2–MeOH, identical excitation spectra were obtained for emissions monitored at 396 nm (monomer) and 480 nm (excimer), and the maxima corresponded to those in the UV absorption spectrum (Figure 4a). Time-dependent fluorescence measurements in the nanosecond

Table II. Spectroscopic Properties of Pyrene-Labeled Cellulose Ethers

Polymer	*Solvent*	A_{344}/A_{337}[a]	I_E/I_M	τ_M *(ns)*[b]	τ_E *(ns)*[b]
Ty-Py/190	water	1.32	0.21	96	12.6 (0.34) 63.1 (0.66) $\langle\tau\rangle = 58.4$
Ty-Py/190	CH_2Cl_2–MeOH (3/1 vol/vol)	2.10	0.40	6.4 (0.71) 97.3 (0.29) $\langle\tau\rangle = 84.7$	7 (−0.98) 58 (1.0)
Me-Py/100	water	1.32	0.10	26.1 (0.10) 102.0 (0.90) $\langle\tau\rangle = 94.2$	
Me-Py/100	CH_2Cl_2–MeOH (3/1 vol/vol)	2.18	0.01		
Me-Py/250	water	1.60	0.01		
HPC-Py/216[c]	water	2.16	0.10	28 (0.17) 116 (0.83) $\langle\tau\rangle = 112$	3.5 (0.64) 84 (0.36) $\langle\tau\rangle = 78$
HPC-Py/216[c]	MeOH	2.50	0.04	19 (0.17) 114 (0.83) $\langle\tau\rangle = 108$	

[a] Peak-to-valley ratios of the absorption spectra. *See* reference 19 for further information.
[b] Values in parentheses are prefactor values. τ_M represents fluorescence lifetime of the pyrene monomer emission. τ_E represents fluorescence lifetime of the pyrene excimer emission.
[c] Values are from reference 13.

time domain were performed for samples in the organic medium. The excimer time-dependent profile showed a growing-in component and a decaying component (7 and 58 ns, respectively). The monomer emission showed a nonexponential decay that could be fit to a sum of two exponential terms with decay times of 6.4 and 97.3 ns (Table II). Taken together, the data indicate that the pyrene excimer in organic solutions of Ty-Py/190 is formed by a dynamic mechanism via encounter of an excited pyrene and a ground-state pyrene, as described by Birks (*21*).

Time-dependent measurements were carried out next with solutions of Ty-Py/190 in water. In this situation it was not possible to detect a rising component in the excimer profile within the nanosecond time domain accessible to the instrumentation. The excimer decay could be fit satisfactorily to a sum of two exponential terms with decay times of 12.6 and 63.1 ns. The monomer profile obeyed an exponential decay with a lifetime of 96 ns. Further steady-state measurements revealed that the excitation spectra monitored for the monomer and excimer emissions were different. The monomer spectrum was blue-shifted by about 3 nm relative to the excimer spectrum

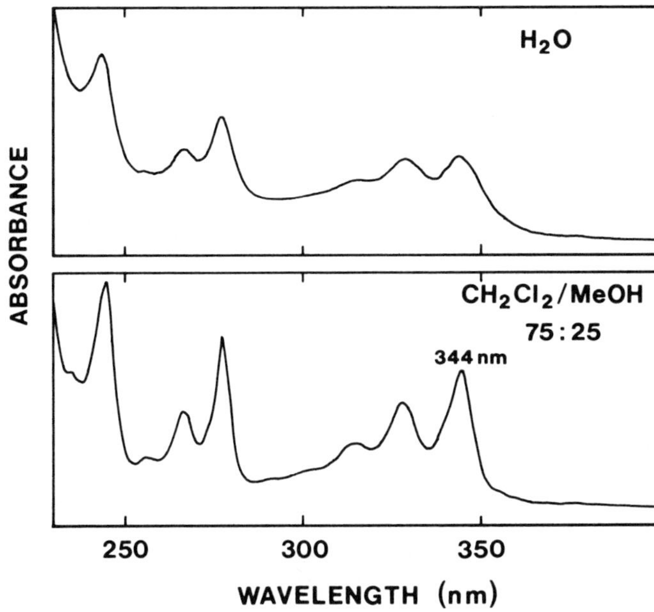

Figure 5. UV absorption spectra of pyrene-labeled Tylose (Ty-Py/190) in water (a) and in CH_2Cl_2–MeOH (3/1 vol/vol) (b); polymer concentration, 0.3 g L^{-1}.

and relative to the UV absorption spectrum of Ty-Py/190 in water (Figure 5).

Taken together, the experimental observations lead to the conclusion that the pyrene excimer emission originates from aggregates of pyrenes that exist prior to excitation, as is observed also in aqueous solutions of pyrene-labeled hydroxypropylcelluloses (*13*). In most conditions the formation of pyrene dimers or higher aggregates is precluded in solution. The situation in aqueous polymeric solutions is unique. It is postulated that the pyrene aggregates are stabilized by hydrophobic interactions. The nonpolar assemblies are surrounded by a cage of highly organized water molecules tightly bound through hydrogen bonding. This formation of dimers or higher aggregates has a positive entropy and a positive enthalpy. The entropic term is dominant, rendering the free energy of dimer formation favorable at room temperature. The spectroscopic information available does not allow one to distinguish between aggregates formed among pyrene groups attached to the same polymer or to several chains. It is expected that the latter situation predominates in view of the low level of pyrene incorporation along the macromolecules and of the restricted flexibility of the chains.

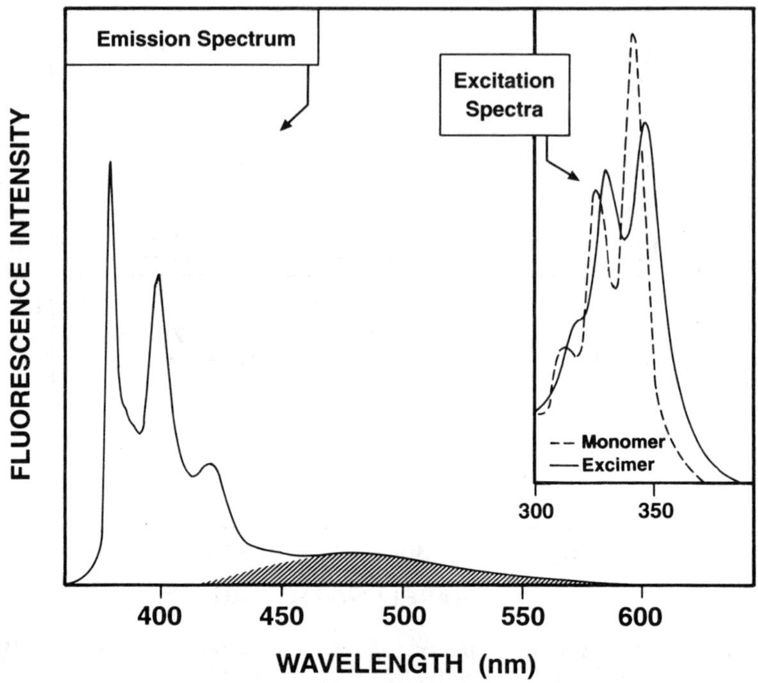

Figure 6. Fluorescence spectra of pyrene-labeled Methocel (Me-Py/100) in water; emission spectrum, λ_{em} = 343 nm; excitation spectra, λ_{exc} = 480 nm (excimer) and λ_{exc} = 396 nm (monomer); polymer concentration, 0.1 g L^{-1}.

Pyrene-Labeled Methocel Derivatives. The fluorescence spectrum of Me-Py/100 in water exhibits a strong emission attributed to isolated excited pyrenes as well as a contribution from pyrene excimers (Figure 6). As with Ty-Py/190, the excitation spectra monitored for the excimer and the monomer emissions are different: The excitation spectrum monitored at 480 nm (excimer) is red-shifted by ca. 4 nm and exhibits significant line broadening compared to the excitation spectrum monitored for the monomer emission (Figure 6). Time-resolved measurements (Table II) and absorption spectra offer further support to the steady-state fluorescence data to establish that the excimer emission originates from preassociated pyrenes. Thus, as with Ty-Py/190, the chromophores form hydrophobic clusters in aqueous solutions of Me-Py/100. The pyrene excimer contribution to the total emission is significantly weaker than that of Ty-Py/190, even though the degree of pyrene incorporation on the Methocel-50-F sample is higher (4.2×10^{-5} mol g^{-1} in Me-Py/100 versus 3.0×10^{-5} mol g^{-1} in Ty-Py/190). The two cellulose ethers exhibit different macroscopic properties in water prior to labeling. It may be inferred from the preliminary spectroscopic data that corresponding differences are maintained in the labeled materials.

Further work with the labeled Tylose and Methocel samples will focus on this issue. First, it is important to correlate the spectroscopic characteristics of the polymers to their macroscopic solution properties, such as viscosity and gelation. The effects of external stimuli, such as heat or the addition of surfactants, on pyrene photophysics will be monitored. Thus it should be possible to uncover detailed aspects of the molecular origin of macroscopic phenomena, such as the temperature-induced phase changes that take place in aqueous solutions of cellulose ethers. Key results for thermally induced changes in pyrene emission from aqueous solutions of pyrene-labeled hydroxypropylcellulose are reviewed next. A complete study of this system is reported elsewhere (*22*).

Application of Fluorescence Spectroscopy To Study Temperature-Induced Phase Changes in Aqueous Solutions of Cellulose Ethers. Hydroxypropylcellulose solutions in water exhibit a cloud point at 42 °C for polymer concentrations lower than 10 g L^{-1} (*3*). The cloud point of the labeled polymer HPC-Py/216 (Figure 1, Table I) occurs at a slightly lower temperature, 40 to 41 °C (*22*). This small but significant decrease is consistent with the usual effect of hydrophobic group substitution on the cloud point of water-soluble polymers (*23*). At 25 °C in water, HPC-Py/216 shows an emission due to pyrene monomers and excimers, with a ratio of excimer-to-monomer intensities (I_E/I_M) of 0.10. As the temperature of the solution is increased, the following events occur: First, the ratio of excimer-to-monomer intensity increases to a maximum at 35 °C. Above 35 °C the ratio decreases sharply to reach its limiting value of 0.01 at 50 °C. The midpoint of the transition is ca. 42 °C. When the sample temperature is decreased, a sharp increase in the ratio occurs, with the midpoint of the transition occurring at ca. 41 °C. At all temperatures lower than 50 °C, the I_E/I_M ratio measured upon cooling is slightly smaller than that measured upon heating. After a few hours at 25 °C, a cooled sample recovers the initial ratio value. Consecutive heating–cooling cycles performed with the same sample yield the same results.

The ratio of excimer to monomer intensity is a useful parameter for monitoring solution phase changes. This ratio is obtained from simple steady-state fluorescence measurements, and it can serve as a tool to correlate, at least qualitatively, heat-induced spectroscopic changes to changes in macroscopic solution properties, such as turbidity. The sharp decrease in I_E/I_M at the phase transition reflects an increase in the absolute pyrene monomer intensity at the expense of the pyrene excimer fluorescence intensity. The increase in pyrene monomer intensity must be due to an increase either in the number of emitting species or in the fluorescence lifetime of the pyrene. No significant changes were observed in the decay times of either the pyrene monomers or the pyrene excimers as the samples were heated through their cloud points. Therefore the changes in relative fluorescence intensities indicate changes in the relative numbers of isolated and aggregated pyrenes. During the phase transition, the aggregated pyrenes separate. They become accommodated within hydrophobic cavities in the polymer-rich phase.

Conclusion

Hydrophobic chromophores have been linked to cellulose ethers via chemical modifications akin to those employed in the preparation of hydrophobically modified cellulose ethers bearing alkyl or perfluoroalkyl groups. Spectroscopic measurements show that even at low solution concentrations, fluorescence-labeled cellulose ethers in water form interpolymeric aggregates via hydrophobic clusters of chromophores, a situation that parallels the behavior of commercial hydrophobically modified cellulose ethers. For pyrene-labeled cellulose ethers that exhibit a well-defined heat-induced phase transition in aqueous solutions, all spectral parameters point to the existence of organized assemblies of the pyrenes below the cloud point. Above the cloud point, the chromophores behave in patterns typical of those observed in isotropic solutions. Thus heat induces a complete reorganization of the interpolymeric association. Further experiments are required to answer questions related to temperature-triggered changes in the conformations of individual polymer chains associated with the macroscopic phase separation.

Acknowledgments

I thank the Xerox Research Center of Canada, Mississauga, Ontario, Canada, where part of the work was carried out. GPC measurements were performed by D. Boils-Boissier of the Xerox Research Center of Canada.

References

1. Just, E. K.; Majewicz, T. G. *Encyclopedia of Polymer Science and Engineering,* 2nd ed.; John Wiley: New York, 1985; Vol. 3, pp 226–269.
2. Kato, T.; Yokoyama, M.; Takahashi, A. *Colloid Polym. Sci.* **1978,** *265,* 15.
3. Fortin, S.; Charlet, G. *Macromolecules* **1989,** *22,* 2286.

4. Klug, E. D. *J. Polym. Sci. C* **1971,** *36,* 491.
5. Ibbett, R. N.; Philp, K.; Price, D. M. *Polymer* **1992,** *33,* 4087 and references therein.
6. Tanaka, R.; Meadows, J.; Phillips, G. O.; Williams, P. A. In *Cellulose: Structural and Functional Aspects;* Kennedy, J. F.; Phillips, G. O.; Williams, P. A., Eds.; Ellis Horwood: Chichester, United Kingdom, 1990; pp 323–327.
7. Robitaille, L.; Turcotte, N.; Fortin, S.; Charlet, G. *Macromolecules* **1991,** *24,* 2413.
8. Wirick, M. G.; Waldman, M. H. *J. Appl. Polym. Sci.* **1970,** *14,* 579.
9. Huikari, A.; Karlsson, A. *Acta Pharm. Fenn.* **1989,** *98,* 231.
10. Sarkar, N. *Polymer* **1984,** *25,* 481.
11. Varelas, C. G.; Steiner, C. A. *J. Polym. Sci., Polym. Phys. Ed.* **1992,** *30,* 1233.
12. Tanaka, R.; Meadows, J.; Phillips, G. O.; Williams, P. A. *Carbohydr. Polym.* **1990,** *12,* 443.
13. Winnik, F. M.; Winnik, M. A.; Tazuke, S.; Ober, C. K. *Macromolecules* **1987,** *20,* 38.
14. Winnik, F. M. *Macromolecules* **1987,** *20,* 2745.
15. *Polymers in Aqueous Media: Performance Through Association;* Glass, J. E., Ed.; Advances in Chemistry 223; American Chemical Society: Washington, DC, 1989.
16. Landoll, L. M. *J. Polym. Sci., Polym. Chem. Ed.* **1982,** *20,* 443.
17. Hwang, F. S.; Hogen-Esch, T. E. *Macromolecules* **1993,** *26,* 3156.
18. Dualeh, A. J.; Steiner, C. A. *Macromolecules* **1991,** *24,* 112.
19. Winnik, F. M. *Chem. Rev.* **1993,** *93,* 587 and references therein.
20. Schnieders, C; Müllen, K.; Huber, W. *Tetrahedron* **1984,** *40,* 1701.
21. Birks, J. B. *Photophysics of Aromatic Molecules;* John Wiley: New York, 1970; Chapter 7.
22. Winnik, F. M.; Tamai, N.; Yonezawa, J.; Nishimura, Y.; Yamazaki, I. *J. Phys. Chem.* **1992,** *96,* 1967.
23. Taylor, L. D.; Cerankowski, J. *J. Polym. Sci., Polym. Chem. Ed.* **1975,** *13,* 2551.
24. Sarkar, N. *J. Appl. Polym. Sci.* **1979,** *24,* 1073.

RECEIVED for review November 18, 1993. ACCEPTED revised manuscript September 27, 1994.

23

Influence of Alkali-Soluble Associative Emulsion Polymer Architecture on Rheology

Richard D. Jenkins,[1] L. Mark DeLong,[2] and David R. Bassett[1]

[1]Union Carbide Corporation, UCAR Emulsion Systems, Research and Development, 410 Gregson Drive, Cary, NC 27511
[2]Union Carbide Technical Center, 3200 Kanawha Turnpike, South Charleston, WV 25303

We examined the influence of the following synthesis parameters on the aqueous-solution rheology of alkali-soluble associative emulsion polymers in steady simple shear, oscillatory shear, and extensional shear: macromonomer structure (hydrophobe size and structure, double-bond structure, and moles of ethoxylation between hydrophobe and double bond), carboxylic acid monomer concentration, polymer glass transition temperature, and water solubility of the monomers in the polymer backbone. Polymers made with large-molar-volume hydrophobes of complex structure show enhanced thickening power, low shear viscosity, shear-thinning viscosity, and viscoelasticity (i.e., properties are less dependent on shear strain) compared to polymers made with smaller hydrophobes. Optimal values exist for concentrations of carboxylic acid and macromonomer as well as for water solubility and glass transition temperature of the polymer. Consistent with rheological data, static light-scattering data show that polymers containing complex hydrophobes aggregate, even in a mixed solvent chosen to enhance molecular dispersion, and have a much larger apparent molecular weight in solution than do polymers that do not contain associative macromonomer.

0065–2393/96/0248–0425$12.75/0

ALKALI-SOLUBLE ASSOCIATIVE POLYMERS thicken a variety of waterborne systems, such as architectural paints and paper coatings. These polymers are made by the emulsion polymerization of a carboxylic monomer, an associative "macromonomer" (i.e., a surfactant that has been capped with a polymerizable double bond), and a "nonassociative" flexibilizing monomer (*1*). The polymers are in the form of a latex after polymerization. At low pH, the carboxylic groups are uncharged and the polymer is not water soluble; at higher pH (usually greater than 6), the carboxylic acid groups ionize and the polymer goes into aqueous solution. These polymers thicken solutions by an associative mechanism and by an expansion of the high-molecular-weight polymer backbone at high pH that is due to electrostatic repulsion between the neutralized carboxylic acids. The technology can be generalized from the use of carboxyl (or other anionic groups) to cationic or amphoteric groups to control the pH of solubility.

The following are some of the synthesis parameters that control the steady shear viscosity, viscoelastic, and extensional properties of alkali-soluble associative polymer solutions:

1. the structure and concentration of the associative macromonomer in the polymer, including the size and structure of the hydrophobe; the moles of ethoxylation between hydrophobe and the double bond; the chemical nature of the bond between the ethoxylated portion and the reactive double bond (e.g., ester, ether, or urethane linkage); and the structure of the double bond itself (e.g., acrylic, methacrylic, crotonic, styrenic, etc.);
2. the structure and concentration of acid moiety in the polymer (e.g., acrylic, methacrylic, crotonic, itaconic, etc.);
3. the water solubility and glass transition temperature of the polymer backbone as controlled by chain-extending monomers (e.g., alkyl (meth)acrylates and styrene);
4. the structures and concentrations of monomers that cross-link with the polymer during polymerization (e.g., trimethylol propane triacrylate) and those that leave cross-linkable functionality in the associative polymer without cross-linking during polymerization (e.g., 2-hydroxyethylacrylate); and
5. the molecular weight of the polymer backbone.

This chapter presents some relationships between the rheological properties of alkali-soluble associative emulsion polymers in aqueous alkaline solution and their compositional variables as listed in 1, 2, 3 of the preceding paragraph.

Experimental Details

Hydrophobes with labile hydrogens were ethoxylated to 20, 40, and 80 mol in a pressure autoclave. The number-average molecular weights were determined by end-group analysis (hydroxyl number), and molecular weight distributions of the surfactants were determined by gel permeation chromatography. Conversion of surfactant into macromonomer followed standard synthetic preparative techniques: reaction with unsaturated anhydrides to make esters and reaction with unsaturated isocyanates to make urethanes. Specifically, urethane-containing macromonomers were made by reaction of the surfactant with either methacryloyl isocyanate, isocyanato ethyl methacrylate, or α, α-dimethyl *meta*-isopropenyl benzyl isocyanate (TMI); the ester-containing macromonomers were made by reaction of the surfactant with acrylic, methacrylic, maleic, or crotonic anhydride.

Alkali-soluble associative emulsion copolymers were prepared by the conventional semicontinuous emulsion polymerization of various weight fractions of methacrylic acid, associative macromonomer, alkyl (meth)-acrylates, or styrene. To convert the resulting latexes into solutions, the latexes were diluted to the desired concentration and neutralized to a pH of 9 with 2-amino 2-methyl 1-propanol (AMP-95, Angus Chemical Co.).

Steady shear and linear viscoelastic properties of alkaline thickener solutions were measured on a Bohlin VOR rheometer (Bohlin Instruments, Cranbury, NJ) using two fixtures: cone and plate (30-mm diameter, 2.5 ° cone angle) for medium and high shear rates and Mooney–Couette cup and bob for low shear rates at a temperature controlled at 25 ± 0.5 °C by a water circulator bath. Brookfield viscosity standards were run periodically to calibrate the instrument. In the steady-shear viscosity measurement, a delay time of 10 s was used between successive shear rate increments to allow the imposed flow and the fluid's response to it to reach steady state, and the steady-state data were averaged over a 10-s integration time. The strain amplitude employed in the oscillatory mode was chosen to be large enough to produce a measurable torque yet low enough to be in the linear viscoelastic region, where the measured properties were independent of strain. For most samples this strain amplitude was about 90 mrad (corresponding to about 20% strain). Data with less than a 0.5% torque range reading for steady-shear and oscillatory measurements were considered unreliable and were discarded from each data set.

Steady-state extensional viscosity measurements were made at 20 °C with a Rheometrics RFX Fluids Analyzer (Rheometrics, Piscataway, NJ) using opposing jets of 1-mm-i.d. set 1 mm apart. The RFX's autosensing routine determined when steady state had been achieved and how long the measurement of extensional viscosity should continue.

The light-scattering apparatus used a custom-built goniometer that has been described elsewhere (2). The incident radiation in all experiments was provided by a 5-W Lexel 3500 Argon ion laser operating at a wavelength of 488 nm. Static light-scattering data were acquired in UV geometry (unpolarized detection and vertically polarized incident light) using incoherent detection optics. All measurements were made at 25 ± 0.05 °C.

For the light-scattering study, the polymers were solubilized in a (by volume) 35% water/65% diethylene glycol monobutyl ether (Butyl Carbitol, Union Carbide Corp.) solution adjusted with ammonium hydroxide to pH 10. The polymer solutions studied ranged from 1 to 3 mg/ml. All polymer solutions were filtered through a 0.45-μm-pore-size Whatman poly(tetrafluoroethylene) filter. Large polymeric structures, larger than the filter pore size (as determined by examination of the hydrodynamic radius by quasielastic light scattering [QELS]), were visible in the microscope facility of the light-scattering apparatus at all concentrations of polymer containing complex hydrophobe (as defined below). These structures evidently undergo shear dissociation upon filtration and reform. This observation is consistent with the results of Seery et al., who, on an aqueous system of hydrophobically associating fluorocarbon-containing polymers, used QELS-measured structures with apparent radii much greater than 1 μm even though the solutions under study were filtered through a 1-μm-pore-size filter (*3*). Specific refractive index increments, necessary for determination of weight-average molecular weight, were determined at constant solvent composition, 25 ± 0.1 °C, and a wavelength of 488 nm for the incident radiation for each polymer; a C. N. Wood differential refractometer modified to accept laser line filters was used for these measurements.

Light-scattering molecular weights were obtained from so-called Berry plots (*4*) (a plot of the square root of the Zimm plot's ordinate plotted versus the Zimm plot's abscissa) instead of the customary Zimm plots (*5*) to eliminate the curvature in the constant angle lines and permit more reliable extrapolations to zero concentration to obtain the molecular weight. This curvature resulted because the third-order virial terms were nonnegligible; thus the polymers or their aggregates were highly branched and/or consisted of strongly interacting or charged particles. Molecular weight data were gathered over polymer dissolution times of up to 2 weeks, and no change in the molecular weight was measured after dissolution times of 3 days for the associative polymers. The polymer without associative macromonomer showed an increase in both measured molecular weight and radius of gyration up to 2 weeks after solvent was added to polymer.

Results and Discussion

Influence of Hydrophobe Structure. Hoy and Hoy (*6*) estimated the chemical potential of an associative polymer's hydrophobe $\Delta\mu$ from Scott–Hildebrand theory:

$$\Delta\mu = 2RT - \frac{V_s + V_p}{2}(\delta_s - \delta_p)^2 x^2 \tag{1}$$

where R is the gas constant; T is the absolute temperature; V_s and V_p are the molar volumes of the solvent and hydrophobe, respectively; δ_s and δ_p are the solubility parameters of the solvent and hydrophobe, respectively; and x is the volume fraction of the hydrophobe in solution. According to equation 1, the driving force for forming and sustaining intermolecular associations increases as the chemical potential of the hydrophobic groups becomes more negative; hence increasing the molar volume of the hydrophobe V_p, increasing the insolubility of the hydrophobe in aqueous media (i.e., decreasing δ_p), or increasing the number of hydrophobes on a given associative polymer backbone should promote intermolecular association in aqueous solution. Although equation 1 considers only the enthalpy of association, it often correctly predicts the qualitative relationship between hydrophobe structure and thickening power in aqueous alkaline solutions.

Chemically joining two or more traditional surfactant hydrophobes to form one larger composite hydrophobe (what we call a complex hydrophobe) is one way to increase the molar volume of the hydrophobe to increase the strength of association and thereby improve the thickening efficiency of an associative polymer (7). The complex hydrophobes act as "preassociated" versions of conventional hydrophobes, and this preassociation produces associative polymers with enhanced thickening power. Figure 1 compares the alkaline solution viscosities of polymers employing either linear hexadecyl, linear

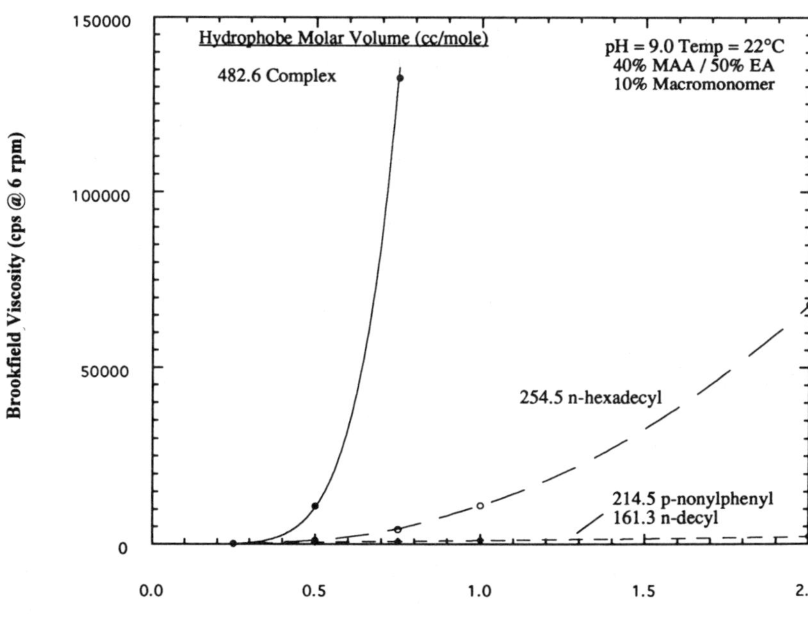

Figure 1. Influence of hydrophobe structure and molar volume on the alkaline-solution viscosity of alkali-soluble associative emulsion polymers. Abbreviations: MAA, methacrylic acid; EA, ethyl acrylate.

Table I. Calculated Molar Volumes and Solubility Parameters for Selected Hydrophobic Groups

Hydrophobe	*Molar Volume* V_p *(cm³/mol)*	*Solubility Parameter* δ_p *(cal/cm³)$^{1/2}$*
n-Decyl	161.3	8.25
p-Nonylphenyl	214.5	8.70
n-Hexadecyl	254.5	8.33
Complex[a]	482.6	8.18

[a] A proprietary bulky hydrophobic structure.

decyl, or *p*-nonylphenyl hydrophobes or a complex hydrophobic group, all else about the associative polymers being equal. As calculated from chemical structure by methods described in the literature (*8–10*), all of these hydrophobes have similar solubility parameters in the range of 8.2 to 8.7 (compared to 23.4 for water) but vary dramatically in their molar volumes, which range from 161.3 for *n*-decyl hydrophobe to 214.5 for the *p*-nonylphenyl hydrophobe to 482.6 for the complex hydrophobe (Table I). The particular hydrophobe illustrated in Figure 1 is composed of two *p*-nonylphenols joined chemically and has a molar volume more than twice that of two *p*-nonylphenyl hydrophobes. As expected from equation 1, the large-molar-volume complex hydrophobe imparts dramatically higher alkaline solution viscosity that increases more sharply with increasing polymer concentration than does that of any of the other hydrophobes. Combining two *p*-nonylphenyl hydrophobes in this manner produces an associative polymer that is an order of magnitude more effective than an associative polymer that uses several times as much *p*-nonylphenol-based macromonomer.

Comparative Rheological Responses. The structure of the hydrophobe also has a profound influence on the shear dependence of the rheological properties of the polymer solutions. Figures 2 through 4 compare the steady-shear and linear viscoelastic properties of alkaline aqueous solutions of two polymers that are identical except that one bears a *p*-nonylphenyl hydrophobe and the other bears the complex hydrophobe.

Steady-Simple-Shear Viscosity. Although the steady-simple-shear viscosity profiles for the polymers have low shear viscosities of widely different magnitudes in the limit of zero shear rate, they tend to asymptotically converge to a common viscosity at higher shear rates, although the data do not go high enough in shear rate to determine whether the solutions will obtain exactly the same high shear viscosity

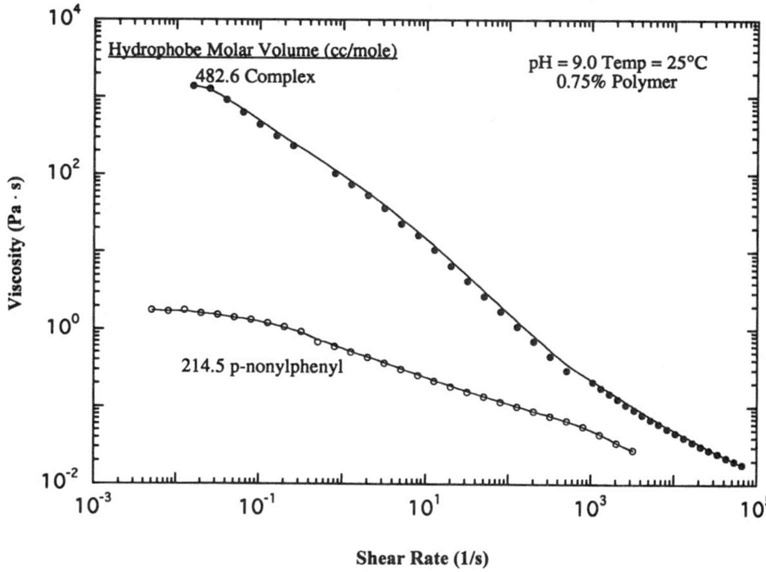

Figure 2. Influence of hydrophobe structure and molar volume on the steady-simple-shear viscosity profiles of alkaline aqueous solutions of associative emulsion polymers.

(Figure 2). The polymer containing the complex hydrophobe exhibits a large degree of shear thinning (even though no cross-linking monomer has been added to extend molecular weight during polymerization). Technologically, then, the low shear viscosity can be adjusted independently of the high shear viscosity by adjusting thickener composition.

The power law fluid model of Ostwald and de Waele (*11*) describes viscosity profiles that are linear on a doubly logarithmic plot with two parameters: a power law index n and a constancy index m (with units of Pa·s^n):

$$\eta(\dot{\gamma}) = m\dot{\gamma}^{n-1}$$

Shear-thinning materials have $n < 1$, and shear-thickening materials have $n > 1$; the larger m is, the larger the fluid's consistency is. The power law fluid model describes the shear-thinning regions of the viscosity profiles of alkaline solutions of complex-hydrophobe-bearing polymer quite well for shear rates between 1 and 10^5 s^{-1}. The power law index n and the constancy index m are in the ranges of 0–0.5

Table II. Influence of Polymer Concentration and Moles of Ethoxylation in the Associative Macromonomer on Power Law Model Parameters for Steady-Shear Viscosity Profiles of Alkaline Aqueous Solutions of Polymers Containing Complex Hydrophobe

	20 mol[a]		*40 mol*[a]		*80 mol*[a]	
Macromonomer %	m	n	m	n	m	n
0.75% polymer						
5	15	0.27	21	0.30	78	0.07
10			71	0.21	117	0.03
20	32	0.14	68	0.21	190	−0.01
30	51	0.14	23	0.29		
0.5% polymer						
5	1.2	0.52	4.6	0.40	20	0.20
10			5.7	0.32	20	0.20
20	4.8	0.28	6.8	0.33	43	0.07
30	9.8	0.28	1.2	0.48		

[a] Amount of ethoxylation at parameter values in $\eta(\dot{\gamma}) = m\,\dot{\gamma}^{n-1}$; $10^5 > \dot{\gamma}(s^{-1}) > 1$.

and 1–200, respectively (Table II). By comparison, a 0.5% solution of hydroxyethyl cellulose has $m = 0.8$ and $n = 0.51$, a 1% solution of poly(oxyethylene) has $m = 1$ and $n = 0.53$, and a 0.7% solution of carboxylmethyl cellulose has $m = 9.7$ and $n = 0.4$ (*12*), demonstrating that polymers bearing the complex hydrophobe exhibit comparatively more shear-thinning character in aqueous solution.

The physical chemistry that produces shear thinning in associative polymer solutions is still unknown, but at least two different mechanisms have been identified: the rupture of the network junctions under shear and the nonaffine deformation of the network (*13*).

Complex Shear Moduli. Consistent with the viscosity data, the storage and loss moduli measured at 1 Hz are much larger for the polymer bearing the complex hydrophobe than for the polymer bearing the *p*-nonylphenyl hydrophobe. Alkaline solutions of both polymers are linearly viscoelastic until the strain exceeds approximately 0.1 (Figure 3). The storage and loss moduli for the *p*-nonylphenyl-bearing polymer both decrease at strains larger than 0.1, whereas the storage modulus is essentially constant and the loss modulus increases at large strains for the complex-hydrophobe-bearing polymer. A decrease in storage modulus implies a decrease in the number density of network entanglements so that a decrease in both moduli for the *p*-nonylphenyl-bearing polymer suggests that large strains disrupt the association network. Because the loss modulus increases at large strain while the storage modulus remains essentially constant for the solu-

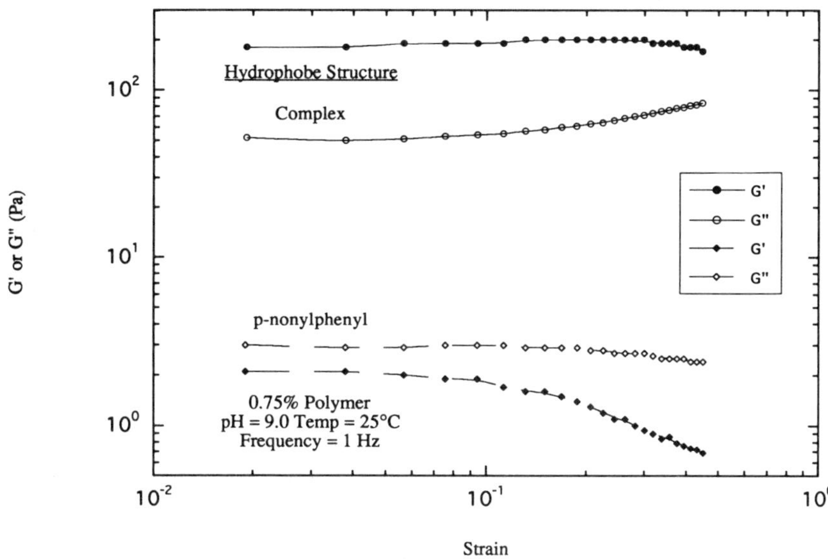

Figure 3. Influence of hydrophobe structure and molar volume on the dependence of complex-shear moduli on strain of alkaline aqueous solutions of associative emulsion polymers.

tion of the polymer bearing the complex hydrophobe, the network either rearranges so that it occupies a larger volume in solution or the intermolecular interactions become more viscous in character under strain. The strain hardening in the loss modulus may result from the large extension of the finitely extendible network chains.

The frequency response of the solution bearing the *p*-nonylphenyl hydrophobe shows that the solution approaches terminal behavior at low frequencies, where the storage modulus approaches a slope of 2 and the loss modulus approaches a slope of 1 on a doubly logarithmic plot (Figure 4). At higher frequencies, the storage modulus and loss modulus are similar in magnitude, although data at higher frequencies are needed to measure a pseudoequilibrium modulus. In contrast, alkaline solutions of polymers containing the complex hydrophobe are highly elastic; their storage moduli are much greater than their loss moduli for the entire range of frequencies measurable with the VOR (Figure 4). Although the loss modulus approaches the storage moduli in magnitude at low frequencies for the solution of the polymer bearing the complex hydrophobe, measurements at much lower frequencies are needed to obtain the characteristic relaxation time constant and to observe terminal behavior to determine whether the solution is truly an elastic gel or simply a highly viscoelastic fluid. A true gel

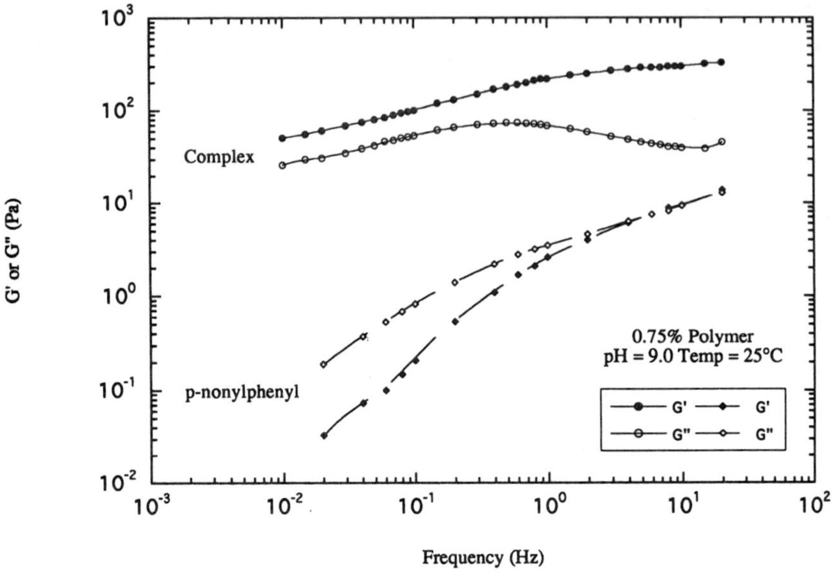

Figure 4. Influence of hydrophobe structure and molar volume on the dependence of complex-shear moduli on frequency of alkaline aqueous solutions of associative emulsion polymers.

would retain elastic character at low frequencies and would therefore exhibit a frequency-independent storage modulus at low frequencies as well as a yield point in the steady-shear viscosity profile.

That the viscous response for the solution of the polymer bearing the complex hydrophobe is shifted to frequencies lower than we can measure indicates that the characteristic relaxation time constant is large. Lacking the entire range of frequency dependence of the dynamic moduli, the order of magnitude of the characteristic relaxation time constant of the solution can be estimated by dividing the low shear viscosity, as approximated by the viscosity measured at a shear rate of 10^{-2}, by the pseudoequilibrium modulus, as determined from the value of the storage modulus at the inflection point in the storage modulus: $\lambda = \eta_0/G_N{}^\circ$. By this method, the relaxation time constant is on the order of 1–4 s for the alkaline solution of the polymer containing the complex hydrophobe, but only on the order of 0.1 s for the solution of the polymer containing the *p*-nonylphenyl hydrophobe.

The viscoelastic properties of these alkali-soluble associative polymers contrast with those of solutions of nonionic model associative polymers based on poly(oxyethylene) (*14*), whose shear moduli were well represented by a Maxwell model. In the Maxwell model, the spectrum of relaxation times was replaced with a single relaxation time constant (*15, 16*). This difference may arise from the relatively higher molecular weight of the polymers, from repulsion between charged groups in the backbone, or from their higher glass transition temperature compared to those of poly(oxyethylene)-based thickeners.

Extensional Viscosity. The extensional viscosity of an aqueous coating can be important in two-dimensional flow fields, such as those encountered in blade or roll coating, or in atomization through a spray nozzle. Low extensional viscosities have been correlated with the improved performance of associative polymers as a class of thickeners in aqueous coatings in geometries that involve converging flows, such as between a roller nip and a substrate, flow under a blade during paper coating, and atomization in a spray nozzle (*17*). Figure 5 presents the dependence of the extensional viscosity of a typical solution of the complex-hydrophobe-bearing polymer on the rate of extension. Tension thinning in extensional viscosity suggests that the polymer coils in solution are quite flexible.

Figure 5 also compares the extensional viscosity to the steady-shear viscosity and the complex viscosity. Because the extensional viscosity approaches 3 times the steady-shear viscosity for a inelastic non-Newtonian fluid, the extensional data in Figure 5 were normal-

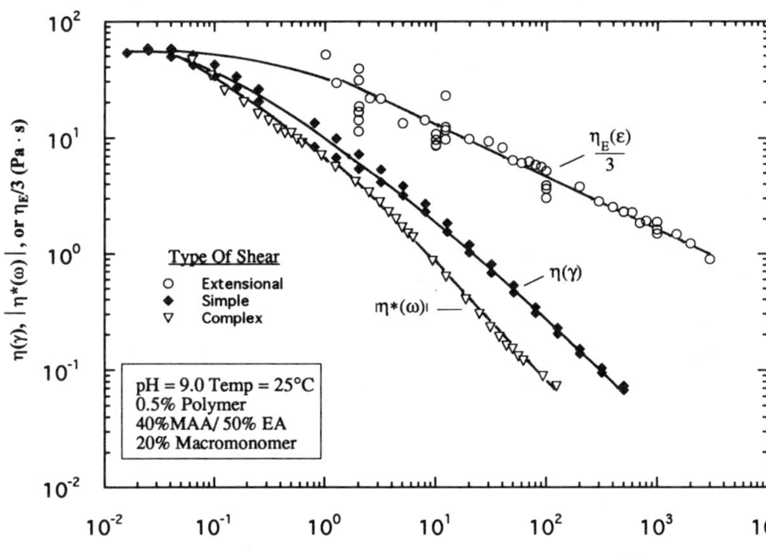

Figure 5. Comparison of steady-simple-shear, complex, and extensional viscosities of alkaline aqueous solutions of associative emulsion polymers. Abbreviations are as in Figure 1.

ized by 3. The extensional viscosity is larger than the steady-shear viscosity at high rates of extension because of viscoelasticity:

$$\frac{\partial \eta_E}{\partial \dot{\epsilon}}\Big|_{\dot{\epsilon}\to 0} = \frac{3}{2}\,[\Psi_1 + 2\Psi_2]\big|_{\dot{\gamma}\to 0} > 0$$

where Ψ_1 and Ψ_2 are the first and second normal stress difference coefficients, because $G'(\omega)/\omega^2 = \Psi_1(\gamma)/2$ in the limit of zero frequency and shear rate (*18*). The empirical Cox–Mertz rule (*19*) extrapolates the relationship between the complex viscosity and shear viscosity to higher frequencies and shear rates:

$$\eta(\dot{\gamma}) = |\eta^*(\omega)| = \eta'(\omega)\,\sqrt{1 + (\eta''/\eta')^2}\Big|_{\omega=\dot{\gamma}}$$

In the limit of zero shear rate, or zero frequency, the complex viscosity tends toward the zero-shear limiting viscosity:

$$\eta(\dot{\gamma})\big|_{\dot{\gamma}\to 0} = \eta'(\omega)\big|_{\omega\to 0}$$

As is typical for most polymer solutions, the complex viscosity decreases faster at high rates of shear than does the steady-shear viscosity. The difference between the steady-shear viscosity and the real component of the complex viscosity η' at large rates of shear can be approximated by a horizontal shift on the plot, which according to nonaffine network models measures the degree of nonaffine deformation (*20*). This nonaffine deformation also explains the strain hardening seen in the loss modulus for the solution of the complex-hydrophobe-bearing polymer.

Influence of Macromonomer Concentration and Double-Bond Structure. The relationship between alkaline-solution viscosity and macromonomer concentration depends on the structure of the macromonomer, because a change in the structure of the double bond in the macromonomer should change the monomer sequence distribution (i.e., from reactivity ratios), the degree of incorporation of the macromonomer, and the molecular weight of the resulting polymer. Table III describes the structures of the series of urethane-based and ester-based macromonomers used to study the characteristics of the double bond and the linking group between the double bond and the ethoxylated surfactant portion of the macromonomer.

The alkaline-solution viscosity can increase or decrease as associative macromonomer concentration changes (Figures 6 and 7). With either the adducts of methacryloyl isocyanate or the acrylic ester, the

Table III. Structures of Macromonomers

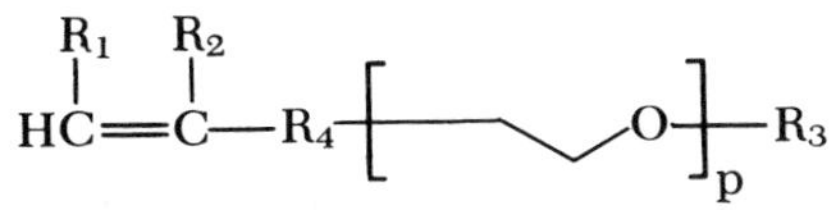

Unsaturated Compound	*Linkage*	*R_1*	*R_2*	*R_4*
Dimethyl *meta*-isopropenyl benzyl isocyanate	urethane	H	CH_3	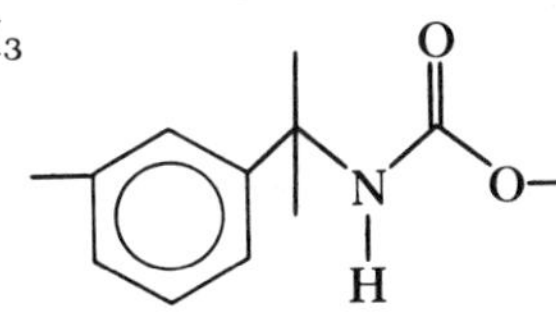
Methacryloyl isocyanate	urethane	H	CH_3	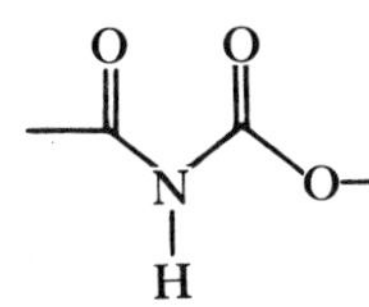
Isocyanato ethyl methacylate	ester/ urethane	H	CH_3	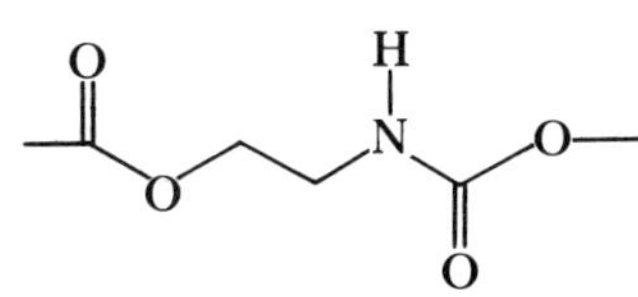
Methacrylic anhydride	ester	H	CH_3	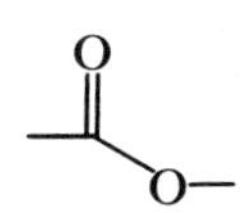
Crotonic anhydride	ester	CH_3	H	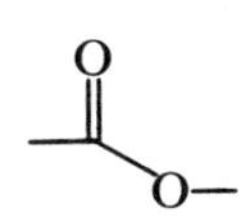
Acrylic anhydride	ester	H	H	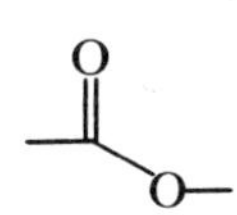
Maleic anhydride	acid half-ester	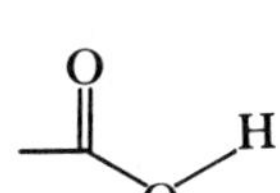	H	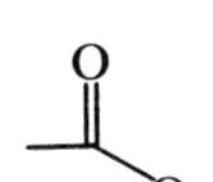

NOTE: R_3 is the hydrophobe.

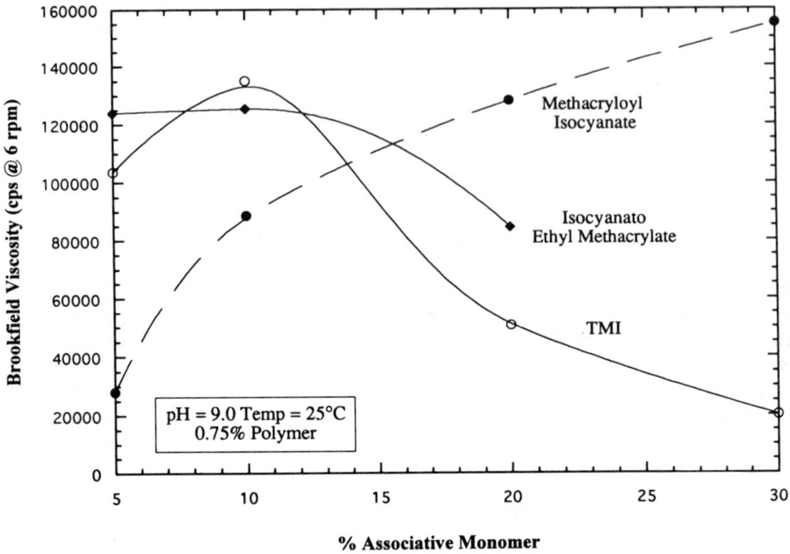

Figure 6. Influence of double-bond structure in urethane-based macromonomers on alkaline aqueous solution viscosity.

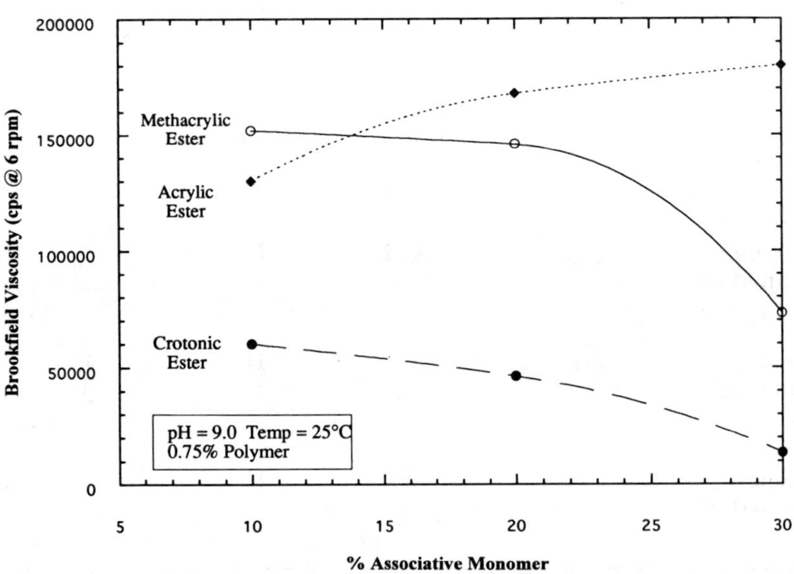

Figure 7. Influence of double-bond structure in ester-based macromonomers on alkaline aqueous solution viscosity.

solution viscosity increases as macromonomer concentration increases. With either the adduct of isocyanato ethyl methacrylate or the methacrylic and crotonic esters, the solution viscosity decreases at larger macromonomer concentrations. With the adduct of TMI, the solution viscosity increases to a maximum and subsequently decreases as macromonomer concentration increases. The alkaline-solution viscosities of polymers made with macromonomers based on acrylic esters are much larger than those based on crotonic esters at all concentrations. By analogy with conventional acrylic monomers, macromonomers based on acrylic and methacrylic esters should have copolymerized and incorporated with acrylic monomers better than macromonomers based on crotonic esters would.

As presented elsewhere, the structure of the double bond also affects rheological properties by imparting various degrees of leveling at low shear rates in viscosity profiles and by varying the elasticity of alkaline solutions and coatings (*21*). In agreement with the Brookfield viscosity data presented in Figure 6, the consistency index m for the power law model increases as concentration increases and goes through a maximum with respect to macromonomer concentration for adducts of TMI. The power law index n becomes smaller as concentration increases and passes through a minimum with respect to macromonomer concentration, so the degree of shear thinning is largest at the same macromonomer concentration that maximizes alkaline-solution viscosity (Table II). The viscoelastic properties (as measured by the value of the storage modulus at 10 Hz) of alkaline aqueous solutions of polymers bearing the complex hydrophobe also pass through a maximum with respect to macromonomer concentration at approximately 20%, confirming that viscosity enhancement results from associative network formation. These results are with macromonomers based on TMI; the concentration of macromonomer that maximizes elasticity depends on the structure of the macromonomer. The strain hardening in the loss modulus, the strain thinning in the storage modulus (a decrease in the number of intermolecular entanglements), and the characteristic relaxation time constant all increase as the macromonomer concentration increases.

The maximum in solution rheology described in the previous paragraph corresponded to a maximum in aggregate size, as evidenced by the large apparent molecular weights (Table IV) and non-negligible third virial coefficients measured by light scattering. The polymer containing 10% by weight of the TMI-based macromonomer bearing the complex hydrophobe had a much larger aggregate molecular weight than those polymers containing 0, 20, or 30% of the macromonomer, a finding that is consistent with solution viscosity results. The large estimated error in the apparent molecular weight of the polymer con-

Table IV. Light-Scattering Molecular Weights of Polymer Containing 40% Methacrylic Acid and Complex-Hydrophobe-Bearing Associative Macromonomer

Macromonomer (wt%)	*Apparent Molecular Weight[a] (Da)*
0	$2.04 \times 10^4 \pm 1\%$
10	$1.18 \times 10^7 \pm 40\%$
20	$2.15 \times 10^6 \pm 3\%$
30	$1.10 \times 10^5 \pm 5\%$

[a] Errors represent error to fit for lines extrapolated to zero concentration and scattering angle. Systematic error is estimated at 10%.

taining 10% macromonomer resulted from large, slow intensity fluctuations in the large polymeric structures recorded by QELS that were substantially stronger than that seen in the other complex-hydrophobe-containing polymers. This molecular weight is underestimated because intensity data were recorded in sets of 10; intensity values of more than twice the ratio of the apparent Poisson standard deviation to the Gaussian standard deviation were discarded, and the remaining data were then reaveraged. By discriminating against large intensity fluctuations, the results are biased toward lower molecular weights.

Because we were unable to measure the molecular weight of the unassociated polymer, we can only speculate as to how changes in synthesis parameters influence the fundamental structure of the associative polymer (e.g., molecular weight, monomer sequence distribution, etc.), and therefore influence the aggregation process and subsequent rheology. Possible explanations for the maxima in rheological properties with respect to macromonomer concentration include the following: incomplete incorporation of the associative macromonomer at large concentration, differences in reactivity ratios between the macromonomer and other monomers that lead to block polymerization, the competing effects of enriching the polymer with associative macromonomer while simultaneously decreasing molecular weight of the polymer due to chain transfer from macromonomer, differences in the glass transition temperature and water solubility of the polymer backbone, and changes in coil dimensions due to enriching the polymer with macromonomer while simultaneously decreasing the polymer's solubility and contracting the dimensions of the polymer coil in solution. NMR spectroscopy and further light-scattering studies to explain this result better are in progress.

Influence of Ethoxylation in the Macromonomer. As moles of ethoxylation increase, the viscosity increases through a maximum at approximately 80 mol of ethoxylation and subsequently decreases (Figure 8). The consistency parameter m and the power law index n rank in order of decreasing shear thinning: 80 > 20 > 40 mol of ethoxylation in the macromonomer (Table II). Although the qualitative nature of the strain and frequency dependencies do not depend strongly on the moles of ethoxylation in the macromonomer, the magnitudes of the storage modulus and the relaxation constant increase as the moles of ethoxylation increase. Comparisons based on constant weight fraction of macromonomer in the polymer are perhaps unfair, because the number of macromonomer units in a polymer decreases for a given weight fraction as the molecular weight of the macromonomer increases. Therefore on a molar basis, polymers made with 40 and 80 mol of ethoxylated macromonomers have a half and a quarter, respectively, of the active ingredient as those made with 20 mol of ethoxylates. Yet the thickening efficiency increases as moles of ethoxylation increases.

The maximum in alkaline-solution viscosity with respect to the degree of ethoxylation at approximately 80 ethylene oxide monomer units between the thickener hydrophobe and backbone depends on

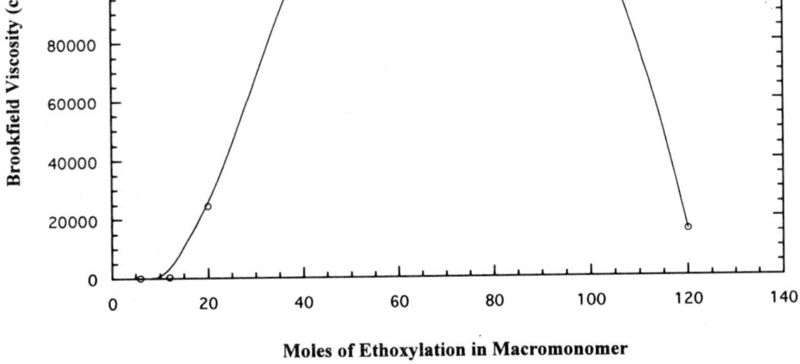

Figure 8. Influence of ethoxylation between the hydrophobe and the double bond in the associative macromonomer on alkaline aqueous solution viscosity.

the natures of the hydrophobic moiety and the polymer backbone and may result from a combination of the three following effects: First, neutralized carboxyl groups in the thickener backbone disrupt the hydrogen bonding and structure among water molecules near the thickener backbone to reduce the entropic driving force for association. Increasing the moles of ethoxylation between the thickener backbone and the hydrophobe extends the hydrophobe beyond the range of electrostatic interaction among segments of the hydrated thickener backbone so that the hydrophobes experience a more structured solvent environment. Viscosity should increase with increasing moles of ethoxylation because of this effect. Second, placing the hydrophobe on a flexible poly(oxyethylene) chain increases the mobility of the hydrophobe and allows the hydrophobe to diffuse to form associative clusters without dragging the thickener backbone along with it. Hence the ease with which the hydrophobes associate, and therefore the viscosity, increases as the moles of ethoxylation increase. Third, increasing the degree of ethoxylation dilutes the concentration of hydrophobes in solution and potentially shields the hydrophobe in a large coil of poly(oxyethylene) or permits more intramolecular association at the expense of intermolecular association because of the flexibility of the poly(oxyethylene) chain; these effects should reduce viscosity.

Influence of Water Solubility and Glass Transition Temperature. Figure 9 compares the influence of pH on the viscosities of polymers composed of 40% methacrylic acid, 30% ethyl acrylate, 10% associative monomer, and 20% "other" monomer by weight, where the "other" monomer is selected from the monomers listed in Table V. Decreasing the water solubility of the polymer backbone while holding all else (including glass transition temperature) constant decreases thickener efficiency, because the more hydrophobic polymer

Table V. Glass Transition Temperatures and Water Solubilities of Monomers Used in Alkali-Soluble Associative Polymers

Monomer	*Glass Transition Temperature (°C)*	*Monomer Solubility in Water (wt%)*
Butyl acrylate	−54	0.20–0.34
Ethyl acrylate	−22	1.50–1.84
Methyl acrylate	8	5.00–5.69
Methyl methacrylate	105	1.50
Styrene	100	0.03

SOURCE: Data are from references 1 and 26–28.

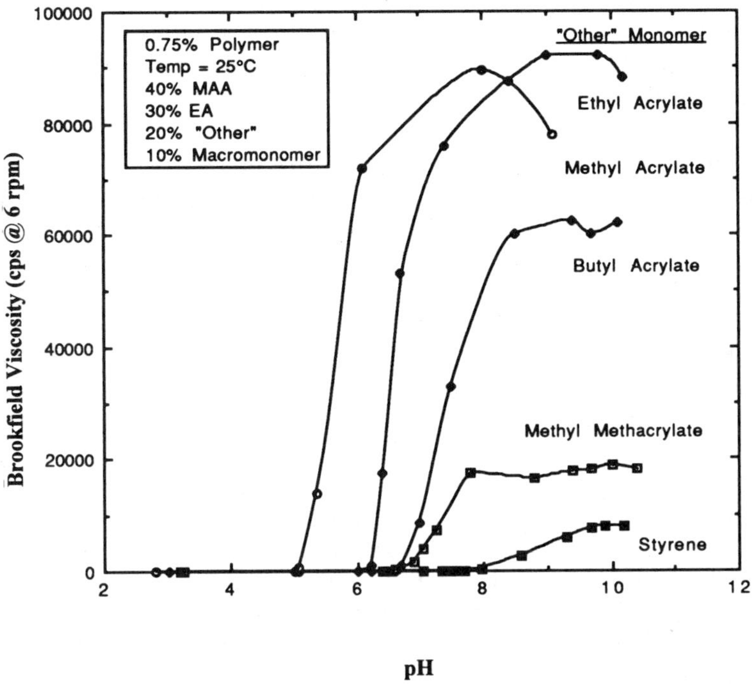

Figure 9. Influence of monomer water solubility and glass transition temperature on pH titration.

has a smaller coil diameter in solution and, because it is less soluble, delays solubilization of the latex polymer until a higher pH. For example, a polymer containing 10% styrene has the same glass transition temperature as one containing 10% methyl methacrylate but is more hydrophobic and therefore has a lower solution viscosity. As the glass transition temperature of the polymer increases, thickening efficiency decreases because the backbone is more coiled and more resistant to expansion because of increased rigidity. For example, ethyl acrylate has the same solubility in water as methyl methacrylate, but the glass transition temperature of a polymer containing methyl methacrylate is higher. Thus the viscosity of the ethyl acrylate-containing polymer is higher than that of the methyl methacrylate-containing polymer.

These trends follow the expansion behavior of carboxylic latexes studied previously (*1, 22–25*). From the perspective of maximizing alkaline-solution viscosity, ethyl acrylate exhibits an optimal water solubility and glass transition temperature: It is softer and less water-soluble than methyl acrylate but harder and more water-soluble than butyl acrylate. Although glass transition temperature and water solubility influence thickening efficiency and the magnitude of the viscosity and viscoelastic properties of alkaline solutions, they do not substantially alter the power law index that describes the degree of shear thinning in the viscosity profile.

Influence of Acid Monomer Concentration. The solution viscosity of alkali-soluble polymers increases as the concentration of methacrylic acid in the polymer increases until the methacrylic acid content exceeds approximately 40%; larger concentrations of methacrylic acid in the polymer do not significantly increase thickener efficiency (Figure 10). The qualitative nature of the steady-shear and viscoelastic alkaline-solution properties do not depend strongly on the amount of acid in the polymer. Thus probably only a minimal concentration of ionized carboxylic acid groups is needed to render the polymer water soluble (*22–24*). Increasing the water solubility of the polymer backbone enhances alkaline-solution viscosity, and increasing chain stiffness (i.e., glass transition temperature) decreases the hydrodynamic volume of polymer coil and therefore the solution viscosity. If we discount the methyl–methyl interaction of closely spaced methacrylic acid segments along the copolymer backbone, the stiffness contribution of acid groups makes very high concentrations of methacrylic

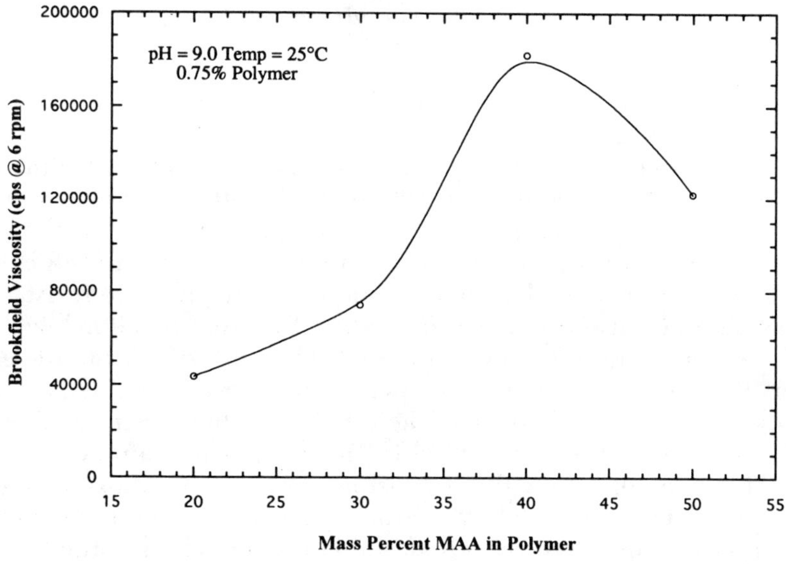

Figure 10. Influence of methacrylic acid (MAA) concentration on alkaline aqueous solution viscosity.

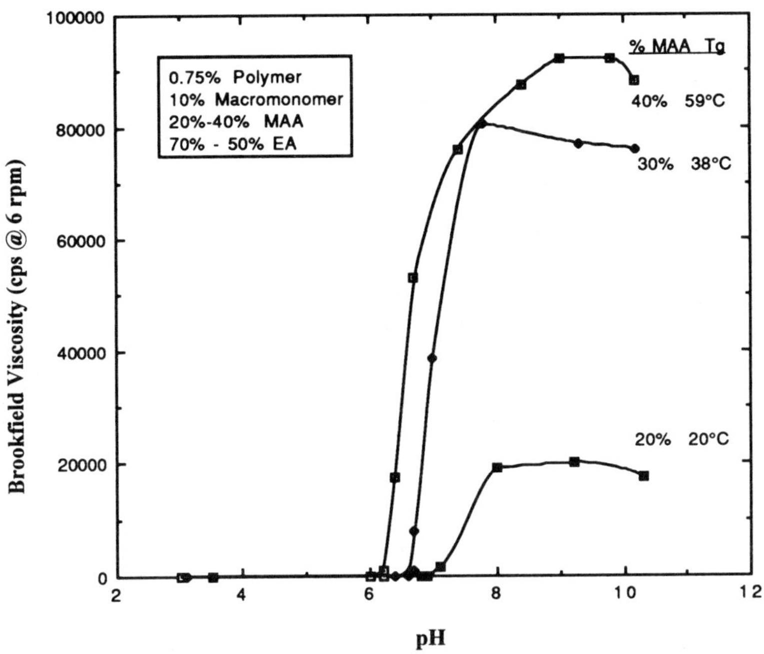

Figure 11. Influence of methacrylic acid concentration on the pH of solubility. Abbreviations are as in Figure 1. Tg, glass transition temperature.

acid counterproductive as far as thickening (i.e., chain extension) is concerned. Consequently, the competing effects of chain stiffness and hydrophilicity yield an optimum in thickening efficiency with respect to the concentration of acid groups in the polymer backbone.

The pH at which the polymers become solubilized depends on the concentration of acid in the polymer's backbone; polymers that have a high concentration of acid become soluble at a lower pH than polymers that have a low concentration of acid, a finding that is consistent with the data of Muroi et al. (23) (Figure 11). However, the mass of base required to neutralize the polymers is roughly the same (at about 0.15 g of AMP-95) for the three polymers presented in Figure 11. The pH required to solubilize the polymer decreases as the methacrylic acid content increases, because adding methacrylic acid groups to the polymer increases its water solubility, even though the calculated glass transition temperature of the polymers noted in the figure increases from 20 to 59 °C as the methacrylic acid content increases from 20 to 40%.

Summary and Conclusions

The large number of synthesis variables inherent with alkali-soluble associative emulsion polymers (both compositional and polymerization process variables) promotes the design of associative polymers that impart alkaline-aqueous-solution rheologies from nearly Newtonian (useful in architectural coatings) to extremely shear thinning (useful in spray-applied coatings, textured paints, paper coatings, and suspending fluids). The structure of the hydrophobe has the largest influence on the qualitative nature of the solution rheology. For example, large-molar-volume hydrophobes promote thickening power and elasticity and enhance the low shear viscosity better than do much smaller hydrophobes. From the perspective of thickening power, optimal levels for the concentration of carboxylic acid and macromonomer, the moles of ethoxylation in the macromonomer, and the water solubility and glass transition temperature of the polymer backbone exist. The optimal amount of macromonomer depends on the structure of the double bond in the macromonomer. Increasing the water solubility or decreasing the glass transition temperature of the polymer promotes thickening power and decreases the pH of solubility for the latex polymer. Because glass transition temperature and water solubility decrease as the number of carbons in the ester portion of the monomer increases, an optimal number of carbons in the ester portion in a series of acrylates or methacrylates results.

Acknowledgments

We thank Brij Sinha and Sherry Hoffman for their help with rheological measurements and David Paxson and Clarence Williams for their help in macromonomer and polymer synthesis.

References

1. Shay, G. D. In *Polymers in Aqueous Media: Performance Through Association;* Glass, J. E., Ed.; Advances in Chemistry 223, American Chemical Society: Washington, DC, 1989; p 457.
2. Russo, O. S.; Saunders, M. J.; DeLong, L. M.; Langley, K. H.; Detenbeck, R. W.; Kuehl, S. *Anal. Chem. Acta* **1986,** *189,* 69.
3. Seery, T. A. P.; Yassini, M.; Hogen-Esch, T. E.; Amis, E. J. *Macromolecules* **1992,** *25,* 4784.
4. Berry, G. C. *J. Chem. Phys.* **1966,** *44,* 4550.
5. Zimm, B. H. *J. Chem. Phys.* **1948,** *16,* 1093.
6. Hoy, K. L.; Hoy, R. C. U.S. Patent 4 426 485, 1984.
7. Jenkins, R. D.; Bassett, D. R.; Shay, G. D. U.S. Patent 5 292 843, 1994.
8. Hoy, K. L. *J. Paint Technol.* **1970,** *42*(541), 76.
9. Burrell, H. *Off. Dig.* **1955,** *27*(1369), 726.

10. Gardon, J. L. In *Encyclopedia of Polymer Science and Technology;* Mark, H. F.; Gaylord, N. G.; Bikales, N. M., Eds.; Wiley and Sons: New York, 1966; Vol. 3, p 833.
11. Bird, R. B.; Armstrong, R. C.; Hassager, O. *Dynamics of Polymeric Liquids;* Wiley and Sons: New York, 1987; Vol. 1, p 172.
12. Tanner, R. I. *Engineering Rheology;* Oxford Science: Oxford, England, 1988; p 14.
13. Jenkins, R. D.; Sinha, B. R.; Bassett, D. R. *Polym. Mater. Sci. Eng.* **1991,** *65,* 72.
14. Jenkins, R. D.; Silebi, C. A.; El-Aasser, M. S. In *Polymers as Rheology Modifiers;* Schultz, D. N.; Glass, J. E., Eds.; Advances in Chemistry 462; American Chemical Society: Washington, DC, 1991; p 222.
15. Jenkins, R. D. Ph.D. Dissertation, Lehigh University, 1990.
16. Annable, T.; Buscall, R.; Ettelaie, R.; Whittlestone, D. *J. Rheol.* **1993,** *37*(4), 695.
17. Fernando, R. H.; Lundberg, D. J.; Glass, J. E. In *Polymers in Aqueous Media: Performance Through Association;* Glass, J. E., Ed; Advances in Chemistry 223; American Chemical Society: Washington, DC, 1989; p 457.
18. Jones, D. M.; Walters, K.; Williams, P. R. *Rheol. Acta* **1987,** *26*(20–30), 170.
19. Cox, W. P.; Mertz, E. H. *J. Polym. Sci.* **1958,** *28,* 619.
20. Bird, R. B.; Curtiss, C. F.; Armstrong, R. C.; Hassager, O. *Dynamics of Polymeric Liquids;* Wiley and Sons: New York, 1987; Vol. 2, p 376.
21. Jenkins, R. D.; Shay, G. D.; Bassett, D. R. *Proceedings of the Third Asia-Pacific Conference,* 1993; Chapter 16.
22. Verbrugge, C. J. *J. Appl. Polym. Sci.* **1970,** *14,* 897 and 911.
23. Muroi, S.; Hosoi, K.; Ishikawa, T. *J. Appl. Polym. Sci.* **1967,** *11,* 1963.
24. Bassett, D. R.; Hoy, K. L. In *Polymer Colloids II;* Fitch, R. M., Ed.; Plenum: New York, 1980; p 1.
25. Hoy, K. L. *J. Coat. Technol.* **1979,** *51,* 27.
26. TMI (Meta) Product Literature, American Cyanamid, November 1990.
27. Yeliseyeva, V. I.; Zuikov, A. V. In *Emulsion Polymerization;* Piirma, I.; Gardon, J. L., Eds.; ACS Symposium Series 24; American Chemical Society: Washington, DC, 1976; p 71.
28. Peyser, P. In *Polymer Handbook,* 3rd ed.; Brandrup, J.; Immergut, E. H., Eds.; Wiley and Sons: New York, 1989; pp VI/209–VI/277.

RECEIVED for review February 8, 1994. ACCEPTED revised manuscript October 11, 1994.

24

Associative Thickeners in the Land of Commercial Reality: Coating Formulations

Ming-Ren Tarng,[1] Zeying Ma,[2] Karu Alahapperuma,[3] and J. Edward Glass

Department of Polymers and Coatings, North Dakota State University, Fargo, ND 58105

In this chapter we venture from the pristine world of aqueous solutions and those that contain only surfactant to the world of more complex fluids containing disperse phases, latex and the hiding pigment rutile TiO_2. The tendency of the latex to phase-separate (syneresis) in aqueous dispersions containing both commercial and model latices, nonassociating water-soluble polymers, and model and commercial associative thickeners is examined. With the introduction of TiO_2, an oligomeric polyacid (i.e., a dispersing aid), is added. The improved film gloss that results from the interaction of preadsorbed dispersants that possess hydrophobes with the hydrophobes of associative thickeners is discussed. In the final section, the influence of model associative thickeners (discussed for aqueous solutions in Chapter 17) on coating formulations is compared with that of selected commercial associative thickeners.

COATING FORMULATIONS are more than thickened solutions containing a uniform-molecular-weight surfactant and a monodisperse-parti-

[1]Current address: Dunn-Edwards Corporation, 4885 East 52nd Place, Los Angeles, CA 90040
[2]Current address: Hewlett Packard Company, 16399 West Bernardo Drive, San Diego, CA 92127–1899
[3]Current address: 3M Center, Building 544-1E-02, 1185 Wolters Boulevard, Vadnais Heights, MN 55110

0065–2393/96/0248–0449$16.50/0

cle-size, high-glass-transition-temperature (T_g) surfactant-stabilized latex. Heterodisperse materials that facilitate non-Newtonian flow at lower shear rates are employed. The primary monomer is methyl methacrylate (MMA), which has a proven exterior durability. In less demanding applications, lower-cost styrene or vinyl acetate is the primary monomer. To facilitate film formation with any of these primary monomers, an acrylate, generally butyl acrylate, is used to lower the T_g of the latex particle. The T_g of the copolymer is targeted high enough to ensure that the final film is not tacky so that erosion and dirt pickup in the final film will be minimal. To ensure adequate latex particle coalescence, a coalescing aid is added to the formulation. The coalescing aid acts as a temporary plasticizer. A low hydrogen-bonding solubility parameter, to ensure partitioning from the aqueous to the latex phase, and a low molar volume, to facilitate evaporation from the applied film, are characteristics (*1*) of an optimum coalescing aid. Commercial latices are seldom stabilized with only an anionic surfactant. The monomer feed generally includes methacrylic acid (MAA) (1–4 wt% of total monomer [2]); the acid provides oligomeric acids at the latex–water interface that give the latex additional electrosteric stability. This technique is discussed in greater historical detail in Chapter 23. Larger-particle latices (400–700 nm) are generally prepared in the presence of hydroxyethyl cellulose (HEC), which grafts to the monomer and ultimately resides at the surface of the final latex (*3*). The surface oligomers and the synthesis surfactants give the latex stability.

To impart hiding to the film, titanium dioxide (TiO_2), which has a high refractive index relative to the film and the air surface, is added to the formulation. Dispersion and stabilization of this high-energy surface require vigorous mechanical mixing and the addition of a surfactant (generally nonionic) and an oligomeric acid, referred to as a dispersant. The effectiveness of the dispersant is related to its ability to adsorb onto the surface of the pigment; this adsorption is dependent on the un-ionized acid concentrations of the dispersant in alkaline solutions (*4*). These comments provide a general description of only one segment of the coating industry, that involving architectural or trade sale paints. In this market, associative thickeners have achieved significant commercial acceptance. Several factors have been proposed as the sources of their success (*5, 6*):

1. Their ability to thicken coatings in the formulation range of 80–120 Kreb units (KU) (Stormer viscosity at ca. 50–100 s^{-1}) yet provide lower viscosities at lower shear rates and higher viscosities at higher shear rates (*7*). Discussion of how this is achieved with various hydrophobe-modified ethoxylated urethane (HEUR) associative

thickeners (*5, 6, 8*) and with hydrophobe-modified HEC (HMHEC) (*9*) is discussed in previous reviews (*5–9*).

2. Their ability to increase film gloss and improve the water sensitivity of applied films (*5, 7*). This ability is discussed later in this chapter.

One of the downsides to the use of HEUR associative thickeners has been phase separation, or syneresis. Phase separation and the flocculation due to interbridging of disperse phases has not been discussed in the context of coating formulations. This chapter begins with a discussion of syneresis and then considers how interaction of associative thickeners with the primary pigment rutile TiO_2 influences film gloss. Next, the influence of matrix interactions in a total coating system on the thickening efficiencies and rheology of formulations is discussed in terms of the thickener's structure and latex characteristics. In particular, the behavior of model associative thickeners in aqueous surfactant solutions (*see* Chapter 17) is compared with their influence in fully formulated coatings and with the influence of commercial associative thickeners.

Phase Separation

Combinations of model and commercial latices and model and commercial thickeners are examined (*10*) in this section in an attempt to understand the phenomenon of syneresis in architectural coatings.

Non-Hydrophobe-Modified Thickeners with Commercial Latex Dispersions. Control thickeners (HEC and poly(oxyethylene) [POE]) of different molecular weights were studied first to provide information on phase-separation phenomena without the complication of hydrophobes attached to the thickener. The latices used in this part of our study were three all-acrylic commercial binders: Rhoplex AC-234 (123 nm), AC-490 (an intermediate median-size latex; 390 nm with an intentional blend (5 wt%) of small particles for flow control), and AC-64 (a large binder; 560 nm). The smallest and largest latices have an excess of surfactant (ca. 0.5 wt%) that may vary with production batch. The degree of phase separation in the 700,000-molecular-weight HEC dispersions (Table I) correlates with the general volume-restricted flocculation observations of Sperry (*11*); the larger-particle latex exhibits phase separation. Considering that these latices contain different amounts and types of nonsurfactant stabilizers, correlation with the general concept of volume-restricted flocculation was not a given.

Table I. Phase Separation of 32% Nonvolatiles by Volume Commercial Acrylic Latex Dispersions Thickened with HEC

Latex Particle Size	*Amount on Aqueous Phase (wt%)*	*Volume Fraction of Separated Phase*	*Separated Phase at 2 Months*	
			Visual Viscosity	*Appearance*
123 nm	0.7	—	—	—
	1.0	—	—	—
	1.3	—	—	—
390 nm	0.7	—	—	—
	1.0	—	—	—
	1.3	—	—	—
560 nm	0.7	0.38	nonviscous	mostly clear
	1.0	0.21	intermediately viscous	less clear
	1.3	0.21	viscous	cloudy

NOTE: Dash indicates no visible phase separation.

The largest-particle latex exhibits the highest volume fraction of separated phase with the lowest concentration of HEC (Table I). As the HEC concentration is increased to 1.0 and then to 1.3 wt%, based on the aqueous phase, the volume of the separated phase decreases. This phenomenon, also observed by Sperry, is explained by the increasing rigidity of flocculent structures with increasing thickener concentration. Some of the separated phase is trapped in the lower phase layer. As the HEC concentration is increased, the viscosity and the cloudiness of the upper layer increase.

The HEUR thickeners (*see* Chapter 10 for structure and synthesis) are >95% POE by composition. The highest-M_v (viscosity-average molecular weight) POE used in this study has approximately the same molecular weight as two commercial urethane associative thickeners, HEUR-200 and HEUR-270; the 35,000-M_v POE is close to the molecular weight of HEUR-708. The 1.2×10^4 M_v POE represents the upper limit of POE that would reasonably be used in the production of a commercial HEUR. For Table II, the dispersions containing these linear unmodified POE thickeners were made at one concentration only (approximately that required to achieve a 90-KU coating viscosity, as would be the practice of a coating formulator). The amounts of POE are significantly greater than the concentration of HEC used, because the POE is significantly lower in molecular weight and more segmentally flexible. The extent of syneresis with the largest-particle latex is lower than that with HEC, but phase separation is observed with the two larger latices with all three unmodified POEs. With very large amounts of the low-molecular-weight POE ($M_v = 1.2 \times 10^4$), syneresis is observed with all three latices.

Table II. Phase Separation of 32% NVV Commercial Acrylic Latex Dispersions Thickened with POE

M_v (10^4) POE	Amount POE on Aqueous Phase (wt%)	Volume Fraction of Separated Phase	Separated Phase at 2 Months: Visual Viscosity	Separated Phase at 2 Months: Appearance
123-nm Latex				
10.0	6.5	—	—	separation in
3.5	9.0	—	—	pockets, cloudy
1.2	11.0	0.10	nonviscous	cloudy
390-nm Latex				
10.0	6.5	0.04	nonviscous	cloudy
3.5	9.0	0.08	nonviscous	cloudy
1.2	11.0	0.10	nonviscous	cloudy
560-nm Latex				
10.0	6.5	0.05	nonviscous	cloudy
3.5	9.0	0.08	nonviscous	cloudy
1.2	11.0	0.10	nonviscous	cloudy

NOTE: Dash indicates no visible phase separation.

Model HEUR Associative Thickeners and Latex Dispersions. Two model HEURs were examined in our studies (*12*). The first is NPIP$(EtO)_{595}$IPNP (in which EtO is ethylene oxide; Figure 1a), which is an analog of the HEUR-708 commercial associative thickener except that this model has no internal hydrophobes and has a narrower molecular weight distribution. The model NPIP$(EtO)_{595}$IPNP thickener is unique in that it exhibits no visible phase-separation behavior with any of the three latices. Dispersions containing this thickener exhibit structuring and extreme shear thinning (*13*). The gellike appearance (applesauce consistency), the presence of a yield stress in the thickened solution, and the high elasticity of this thickener in the presence of other coating formulation components (*14*) indicate strong intermolecular associations. Commercial associative thickeners associate in solution free of other components, but they do not exhibit the high solution yield stress observed with the NPIP$(EtO)_{595}$IPNP thickener; this absence is a consequence, in part, of the broad molecular weight distribution present in the step–growth synthesis of commercial HEURs (*15*). It may also be due to the formation of allophanates (contiguous hydrophobe placement) in our linear model.

The three-arm average HEUR synthesized by the addition of a nonylphenol with an average 100 oxyethylene units to the isocyanurate of isophorone (Figure 1b) is marginally soluble in aqueous solutions and exhibits a strong tendency to phase-separate even in the presence of surfactant. Phase separation is exhibited in each of the commercial latices containing this isocyanurate HEUR.

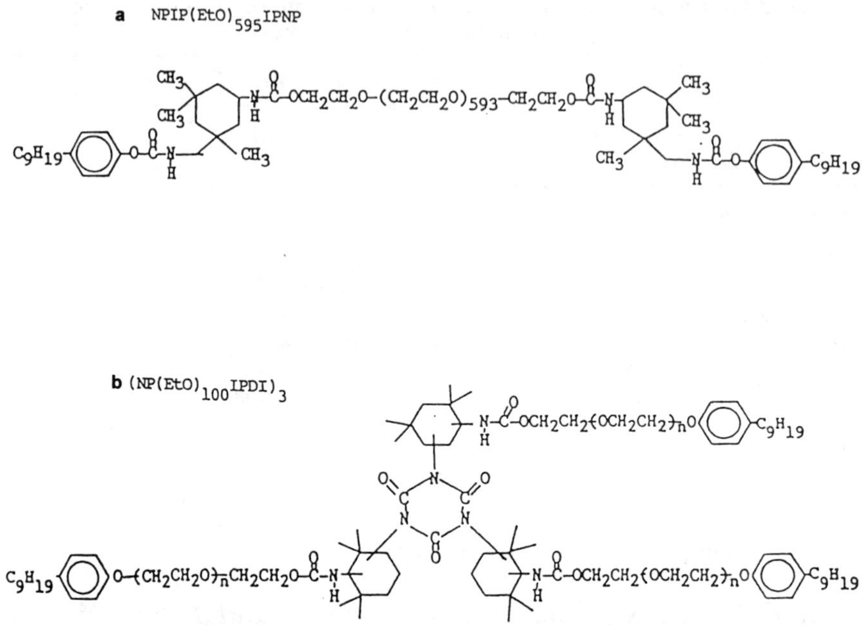

Figure 1. Model HEURs used in phase-separation and film gloss studies.

Commercial HEUR Associative Thickeners with Model Latex Dispersions. A generic structure for commercial HEUR associative thickeners is given in the step–growth polymerization section of Chapter 10. The degree of polymerization of commercial HEURs is determined by the reaction time and by the ratio of diisocyanate to polyetherdiol. For statistical and manufacturing reasons, none of the commercial HEUR thickeners would be made from either the 100,000- or the 35,000-M_v POEs discussed in the preceding section. In high-molecular-weight polymers, the end-group concentration is very low, and the viscosity of such solutions would make transfer and isolation of the modified polymer difficult in production. Diisocyanate coupling of lower-M_v POE (i.e., 6000 M_v) is the manufacturing technique used to increase the molecular weights of commercial HEURs.

The phase behavior of commercial HEURs is considered below with three distinct model poly(methyl methacrylate) latices. The first is a 220-nm latex that is not stabilized by oligomeric surface MAAs. The second, a 100-nm latex, requires a low concentration of surface acid segments to achieve stability. In the third set, both 100- and 220-nm latices are prepared with 4 wt% MAA monomer concentrations. These latices are characterized in Table IIIA.

Table III. Characteristics of Latices Used in Coating Formulations

MMA/MAA Ratio in Latex	*Particle Size (nm)*	*Surface Acid (meq $m^{-2} \cdot 10^2$)*	*Surfactant in Aqueous Phase (wt%)*
A. Latices Used in Syneresis Studies			
98/2	103	0.77	0.068
96/4	97	2.22	0.278
100/0	220	0.28	N.D.[a]
94/4	220	6.55	0.145
B. Latices Used in Adsorption Studies			
100/0	248	0.78	N.D.
98/2	205	1.42	N.D.
96/4	215	5.58	N.D.

[a] N.D. means not detectable.

220-nm Latices. When thickened with HEUR-708 and HEUR-270, the acid–monomer-free 220-nm latex dispersions were not uniform (*10*). The general appearance of the dispersions suggests phase separation with coagulated latex. HEUR-200, the most surfactant-sensitive associative thickener, gave a gellike consistency with this latex. Without surface acid stabilization and with an insufficient amount of synthesis surfactant to provide electrostatic stability, flocculation of the latex will occur. The hydrophobes of the HEUR thickener also could be expected to adsorb on the surface of such latices, perhaps interbridging particles. Commercial HEURs show distinct differences in their adsorption on latices with variable surface acids (Table IIIB, for latex characterization, Figures 2 and 3; these differences are discussed elsewhere [*16*]). They also show differences in their viscosity sensitivity to surfactant concentration (*17*). As the surface acid on the latex increased at a given synthesis surfactant concentration, the free-surfactant concentration increased and syneresis did not occur with the commercial HEURs studied.

100-nm Low-Surface-Acid Latex. A stable acid–monomer-free 100-nm latex could not by synthesized under conditions parallel to those for 220-nm latex synthesis. A small amount of surface acid (0.00077 meq/m^2) was necessary to achieve stability. The latex surface was saturated with 0.068% free surfactant, yet HEUR-708 dispersions with this latex exhibited nonuniform compositions (similar to that of cottage cheese), and the low viscosities at low shear rate (2 s^{-1}) were not reproducible because of a tendency toward phase separation. With HEUR-270, even at low concentrations (0.3%), strong gels were observed. Syneresis was not observed, probably because of the high vis-

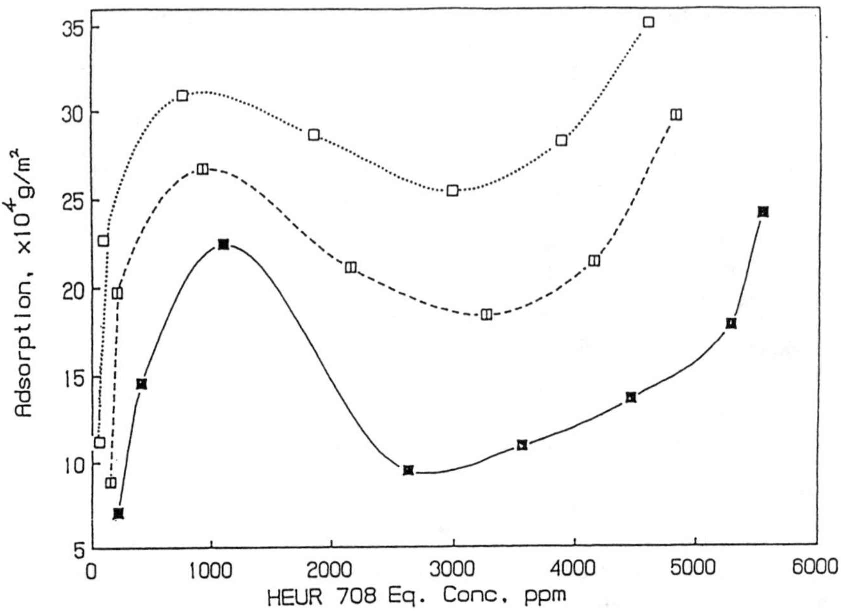

Figure 2. Adsorption isotherms of HEUR-708 from pH 6 dispersions on MMA/MAA copolymer latices. Adsorption of HEUR was determined by the β-cyclodextrin complex method. Key: □, 100/0 mol%; ◫, 98/2 mol%; ■, 96/4 mol%. (Reproduced with permission from reference 16. Copyright 1992.)

cosities of this dispersion. With HEUR-200, a gellike consistency was also observed at 0.32 volume fraction.

In contrast, these latices with no (220 nm) or low (100 nm) acid contents thickened with HEC, the nonassociative thickener, were uniform and less viscous and did not exhibit syneresis. Thus the phenomenon with HEUR thickeners is not due to depletion flocculation but may be due to surfactant demand and destabilization of the dispersions through the surfactant's interaction with the hydrophobic HEUR thickeners or through HEUR thickener adsorption and interbridging of latex particles.

High-Surface-Acid Latices. Uniform compositions with higher viscosities at a low shear rate were observed in the 0.32 volume fraction of the 220-nm MMA/MAA (96/4) latex at low HEUR concentrations. The higher-surface-acid MMA/MAA (96/4) 100-nm latex, however, gelled with HEUR-708 above 0.62 wt% thickener. Strong gels also were observed in HEUR-270 dispersions with the smaller-particle latices.

Addition of a nonionic surfactant to all nonuniform or gellike dis-

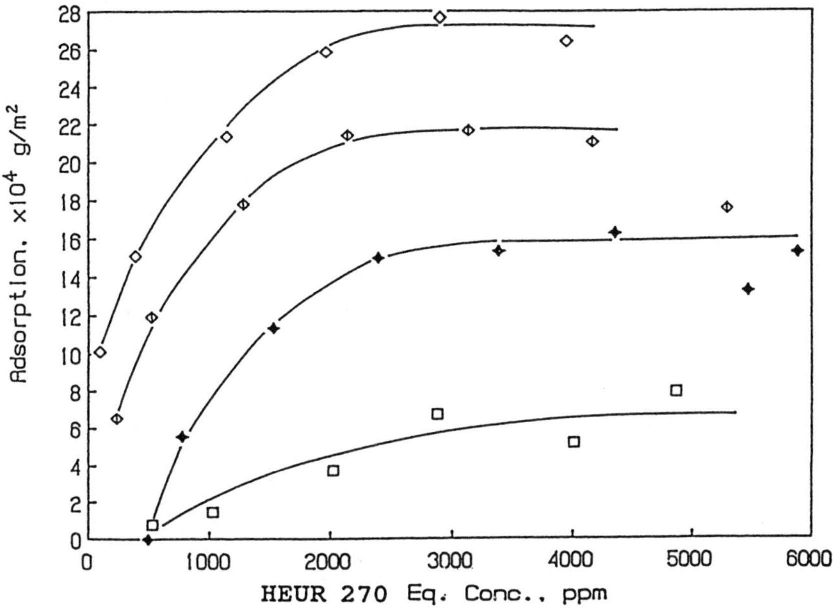

Figure 3. Adsorption isotherms of HEUR-270 on MMA/MAA and vinyl acetate/butyl acrylate copolymer latices; pH 6 dispersion. Adsorption of HEUR was determined by the UV absorbance of the aromatic ring at 276 nm. Key: ◇, 100/0 mol% MMA/MAA; ◇, 98/2 mol% MMA/MAA; ◆, 96/4 mol% MMA/MAA; □, poly(vinyl acetate/butyl acrylate).

persions provided uniform, free-flowing dispersions except at high concentrations of HEUR-270. An excess of surfactant statistically reduces the participation of hydrophobes from more than one HEUR thickener in a given micelle and thus minimizes the micellar network and viscosity increase at low deformation rates. A second factor would be the ability of the excess synthesis surfactant to inhibit the adsorption of the associative thickener on the latex. This ability would be influenced by the amount of each component present and by the relative sizes of the surfactant and the HEUR hydrophobes (discussed in Chapter 17).

Commercial HEUR Thickeners with Commercial Latex Dispersions. Use of commercial thickeners with commercial latices highlights the complexity of the phase-separation phenomenon (Table IV). The 390-nm median-size-particle latex with 5% small-particle (about 100-nm) latex is the only binder that phase-separated with the addition of commercial associative thickener HEUR-270. In contrast, HEUR-200 effects phase separation of the smaller- and larger-particle

Table IV. Phase Separation of 32% NVV Commercial Latex Dispersions Thickened with Commercial HEURs

Thickener and Latex Particle Size	*Amount on Aqueous Phase (wt%)*	*Volume Fraction of Separated Phase*	*Separated Phase at 2 Months*	
			Visual Viscosity	*Appearance*
HEUR-270				
123 nm	0.7	—	—	—
	1.7	—	—	—
390 nm	0.7	0.22	nonviscous	clear, sediment surface uneven
	1.7	—	—	
560 nm	0.7	—	—	—
	1.7	—	—	—
HEUR-200				
123 nm	0.7	0.15	nonviscous	cloudy, possibly small amount of latex present
	1.2	0.27	nonviscous	
	1.7	0.02	nonviscous	
	2.7	—	pockets of separated phase	
	3.2	—	pockets of separated phase	
390 nm	0.7	—	—	—
	1.2	—	—	—
	1.7	—	—	—
	2.7	—	—	—
	3.2	—	—	—
560 nm	0.7	0.48	nonviscous	cloudy, possibly small amount of latex present
	1.2	0.23	nonviscous	
	1.7	0.10	nonviscous	
	2.7	0.05	nonviscous	
HEUR-708				
123 nm	0.7	0.20	nonviscous	clear
	1.2	0.20	nonviscous	clear
	1.7	—	—	—
390 nm	0.7	—	—	—
	1.2	—	—	—
	1.7	—	—	—
560 nm	0.7	—	—	—
	1.2	—	—	—
	1.7	—	—	—

NOTE: Dash indicates no visible phase separation.

latices but not the intermediate-size-particle, bimodal latex. HEUR-708 effects syneresis with only the 123-nm latex. The studies described in the first two sections (i.e., model thickeners (HEC, POE, and model HEURs) with commercial latices and commercial HEURs with model latices) are not broad enough to provide an adequate understanding of the data provided in Table IV. Our studies on more complex HEUR structures to expand our database are continuing. A study (*18*) similar to those cited above but involving all of the components used in an architectural paint concluded with the simple statement that "At best this is a compromise, but in as complex a system as a latex paint coating, most parameters are." One of the points observed in both studies is the increased phase separation when both HEC and HEURs are used as thickeners; this increase suggests that polymer incompatability may also contribute to the syneresis.

We now turn to the other disperse phase, the primary pigment TiO_2. Less interaction of the surfactant or HEUR with this disperse phase provides a different matrix of interactions to address.

HEUR Interactions with TiO_2 and Film Gloss

Adsorption on TiO_2: Concepts. Although surfactant and polymer interactions with model latices have been studied over the past decades, studies of surface interactions of these components with the hiding pigment, titanium dioxide (TiO_2), have been sparse. The surface of TiO_2 is positively charged in low-pH solutions, neutral at the isoelectric point, and negatively charged in high-pH solutions. Prior art (*19, 20*) indicates that nonionic components with minimal proton-donating functions should not be expected to adsorb on TiO_2 in an alkaline medium. This lack of adsorption is indeed observed; neither the nonionic (ethoxylated) surfactant added to the formulation with TiO_2 nor a commercial associative thickener (HEUR-708) adsorbs on the surface of TiO_2 (Table V) under alkaline conditions (pH 9–9.5). This pH is the general pH of an exterior architectural coating.

Adsorption of anionic polyelectrolytes on a negatively charged pigment surface would also seem to be severely restricted by charge–charge repulsions. This restriction has been observed; the amount of a simple acid (i.e., glycolic acid, which has one acid and one alcohol group; the structures of this and other monomeric acids

Table V. Surfactant/Thickener Adsorption on TiO_2–32% NVV Al_2O_3/TiO_2 Unstabilized pH 9.3 Dispersions

Potential Adsorbate	*Thickener/ Surfactant Added to Dispersion*[a]	*Amount Adsorbed on Al_2O_3/TiO_2*[a]
Ethoxylated surfactant[b]	1.3	0
HEUR Thickener[c]	0.42	0

[a] Thickener level in grams per 100 g of pigment; experimental uncertainty, ±15%.
[b] *b*-$C_{13}H_{27}(EtO)_9OH$ surfactant.
[c] HEUR-708 thickener was separated from cosolvent by precipitation of the thickener into a nonsolvent.

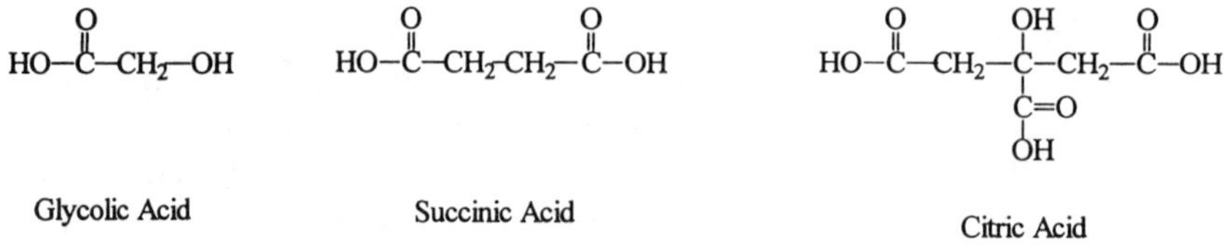

Figure 4. Structures of monomeric acids adsorbed on titanium dioxide.

studied are given in Figure 4) adsorbed from acidic or neutral media is significantly greater than that adsorbed from alkaline media (*17, 18*). No adsorption was noted at the higher pH with a comparable compound containing two contiguous carboxyl groups but no hydroxyl unit (succinic acid). At the higher pH, adsorption was observed with citric acid, which contains three carboxylate groups and one hydroxyl unit. The amount adsorbed at a pH of 9.4 approached that of glycolic acid adsorbed at pH 4. These differences were attributed (1) to the abilities of the acids with alcohol groups to hydrogen-bond to the substrate and therefore (2) to the adsorbent's pK_a. Compounds containing only acid functionalities adsorb on geothite (α-FeO·OH) and gibbsite (γ-$Al_2O_3 \cdot 3H_2O$) surfaces from alkaline media, and the importance of the "conjugate acid of the anion" was proposed for adsorption above the point of zero charge of the clay (*21, 22*).

Two factors favor the adsorption of higher-molecular-weight polyacids. The first is the close proximity of pendant acid functions, which inhibits ionization of all of the carboxylate groups because of electrostatic repulsions. With repeating acid units, un-ionized carboxylated groups would be present to facilitate hydrogen-bond or acid–base interactions with the pigment surface. The second factor favoring adsorption is that even with low enthalpies of interaction, the detachment of most interacting acid segments from the TiO_2 surface does not occur simultaneously for statistical reasons. Oligomeric acrylic acid (PAA), a negatively charged polyelectrolyte at high pH, adsorbs (*23–25*) on TiO_2 above the pigment's isoelectric point. The adsorption of our model MAA oligomer (PMAA, Table VI) on TiO_2 is similar to that reported for PAA.

The notable observation in our studies is the high adsorption from alkaline media of α-olefin/maleic acid [AO(C_x)/MA] co-oligomer dispersants (Table VI) relative to the adsorption of PAA and PMAA noted above. This high level of adsorption is due in part to the difference in the pK_as of 1,2 contiguous carboxyl units and 1,3 alternating acid groups. In part, the high adsorption level is also related to the presence of the repeating hydrophobe unit. The neighboring hydrophobe units

Table VI. Model Dispersant Adsorption on TiO_2 in 32% NVV TiO_2 Dispersions

Sample[a]	*pH*	*Adsorption on* TiO_2[b] *g/100 g*	*Adsorption on* TiO_2[b] *μeq/m²*	*Surfactant Adsorption on Pretreated* TiO_2
PMMA	9.6	0.086	0.60	0.1
AO(C_{10})/MA	10.1	0.293	1.60	0.7
AO(C_{14})/MA	10.2	0.309	1.40	1.0
AO(C_{18})/MA	10.2	0.314	1.22	1.2

[a] Isoelectric point, 8.7; 0.75 g of dispersant/100 g of pigment; PMMA, poly(methyl methacrylate).
[b] Experimental uncertainty, ±15%.

lower the dielectric constant in the area around the carboxyls contiguous to the hydrophobic groups (*26*). The pK_a of the first acid is 3.1; the pK_a of the second acid group is 10.8. Similar results were observed in smaller AO/MA copolymers (*27*). At pH 9.3, approximately 50% of the carboxyl groups from either set of AO/MA co-oligomers are unionized. As an independent entity, the AO/MA co-oligomers are an interesting family. For example, such copolymers with covalently bound anthryl groups (*28*) exhibit strong domain associations with sodium dodecyl sulfate. In other fluorescence studies (*29*), a PMAA (11,000 molecular weight) was not able to solubilize hydrophobic pyrene at a pH above 4; a 1-octadecene/MA copolymer (10,000 molecular weight) in the same study solubilized hydrophobic pyrene molecules across a pH range up to 10.

HEUR thickeners contain hydrophobes, and hydrophobes exist in alternating units in the AO/MA dispersant. An interaction between the hydrophobic segments of the thickener and the dispersant (illustrated in Figure 5) could provide steric stabilization complementing the electrostatic forces and a greater barrier to flocculation of the pigment, and thereby improvement in film gloss. Results of the experiments discussed next support the presence of such an interaction.

Adsorption of Nonionic Surfactants and HEURs on Model Dispersant-Pretreated TiO_2. When the pigment is pretreated with model dispersants, the nonionic surfactant "adsorbs" in proportion to the amount of dispersant adsorbed on the pigment surface (PMAA versus AO/MA) and in proportion to the hydrophobe size of the AO(C_x)/MA dispersants (Table VI). Adsorption studies were conducted with model HEURs (*4*), but in keeping with the spirit of this "applied reality" chapter, we emphasize the commercial HEUR aspects of this "association–adsorption" behavior on model PMAA and

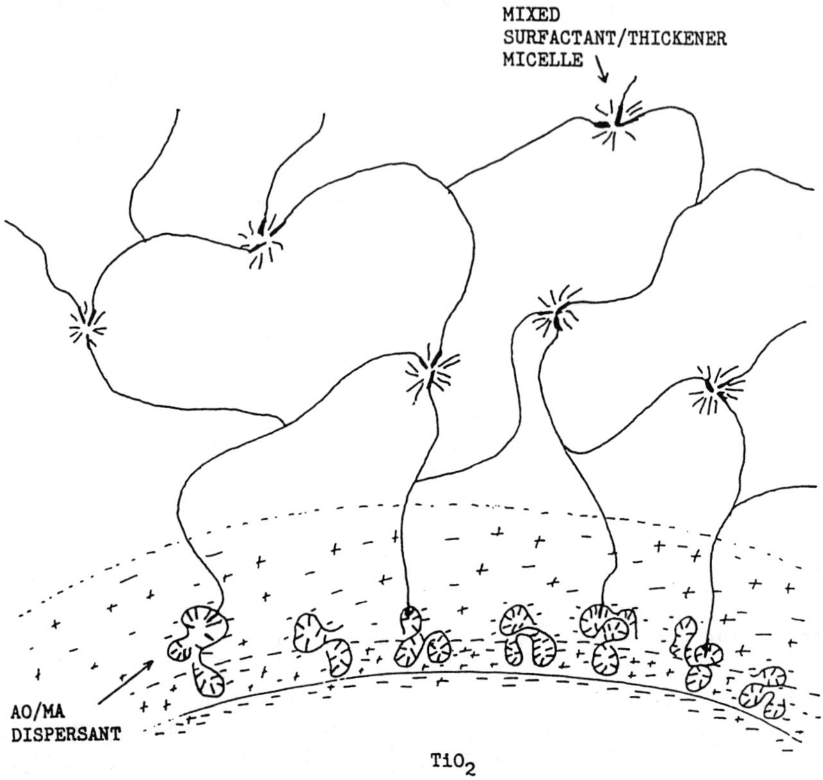

Figure 5. Schematic representation of preadsorbed (on TiO_2) hydrophobe containing polyacid dispersant with the hydrophobes of HEUR-thickened dispersions. (Reproduced with permission from ref. 14. Copyright 1990.)

AO/MA dispersant-stabilized titanium dioxide (Table VII). No significant adsorption of commercial HEUR thickeners on PMAA-stabilized pigment is observed, but in parallel with the studies just discussed, the adsorption of HEURs on AO(C_x)/MA-stabilized TiO_2 is significant. HEUR-200 exhibits the highest adsorption on the model AO/MA-treated TiO_2, a fact that is reflected in several exceptionally high-gloss formulations. The two currently most competitive thickeners, HEUR-708 and HEUR-270, exhibit similar adsorption. In these studies, a POE, $HO(EtO)_{595}H$, did not adsorb, and this result is not surprising.

The primary mechanism for stabilization of disperse phases in a medium of low dielectric constant is steric or entropic; in ion-containing water, electrostatic repulsive forces are generally too weak for particle stabilization (*30–32*). Those skilled in the art synthesize latex particles under conditions that produce nonsurfactant surface moieties

Table VII. HEUR Thickener Distribution in 32% NVV Al_2O_3/TiO_2 pH 9.3 Dispersions

Oligomer Used in Pretreatment[a]	*Thickener*	*HEUR*[b] *Added*	*HEUR*[b] *Concentration*
	HEUR-270	0.42	
PMAA			0.00
$AO(C_{10})/MA$			0.40
$AO(C_{14})/MA$			0.39
	HEUR-200	0.90	
PMAA			0.00
$AO(C_{10})/MA$			0.56
$AO(C_{14})/MA$			0.54
	HEUR-708	0.43	
PMAA			0.00
$AO(C_{10})/MA$			0.39
$AO(C_{14})/MA$			0.41
	control	0.00	
PMAA			0.06
$AO(C_{10})/MA$			0.10
$AO(C_{14})/MA$			0.05

[a] 0.75 g of dispersant/100 g of pigment.
[b] In grams per 100 g of pigment.

to stabilize latices by steric mechanisms in aqueous media. This procedure was discussed in the introduction to this chapter. The nonsurfactants (i.e., oligomeric surface acids) expand the electrical double layer and complement this mode of stabilization with steric stabilization. The surface of titanium dioxide, with or without its inorganic surface treatment (alumina and silica oxides), is higher in interfacial energy than any of the compositionally different latices (i.e., vinyl acetate, acrylics, or styrene) used in water-borne coatings, yet this high-energy inorganic surface is the most poorly stabilized in the crowded environment of a water-borne coating formulation. It follows that if any disperse phase was prone to flocculate by a depletion layer or by suppression of the electrical double-layer mechanism, that phase would be the high-energy pigment.

Dispersions stabilized by the hydrophobic diisobutylene/MA dispersant with the HEUR thickeners do not exhibit high complex viscosities at low oscillation frequencies that indicate flocculated pigment. An increasing complex viscosity is observed with the non-hydrophobe-HEC-thickened dispersions (Figures 6 and 7). This increase suggests a hydrophobic association between the diisobutylene/MA dispersant on the pigment and the HEUR thickener in solution that may reduce flocculation via increased electrosteric stabilization.

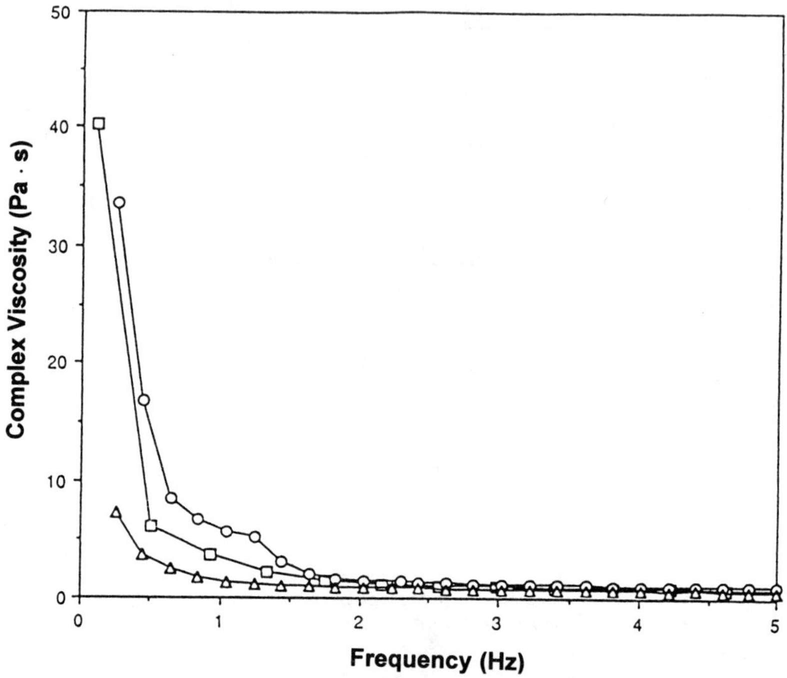

Figure 6. Complex viscosity (η') dependence on frequency of oscillation of 32% nonvolatiles by volume Al_2O_3/TiO_2 pigment dispersion stabilized with an oligomeric MAA dispersant and thickened with the following water-soluble polymers: □, HEC; ○, HEUR-708; △, HEUR-200. (Reproduced with permission from ref. 14. Copyright 1990.)

Gloss Studies. Many of the components added to a full coating formulation (i.e., the coalescing aid, freeze–thaw stabilizer, and antifoaming agent) were omitted to simplify the coating formulation. With all three commercial HEUR thickeners studied, 20° film gloss values were higher in model AO/MA-stabilized pigment formulations than in coatings containing model PMAA-stabilized pigments. The same trend is noted with model HEUR thickener/model dispersant-stabilized TiO_2 dispersions. Selected values for the commercial HEUR formulations are listed in Table VIII.

The total results indicate that the charge per se is not the primary factor effecting greater pigment stability. A similar conclusion has been observed with an alkyd/melamine water-borne system. No correlation between the amount of surface charge on the particle and its stabilization was evident. The important factor was the adsorption of the alkyd at the titanium dioxide interface (*33*).

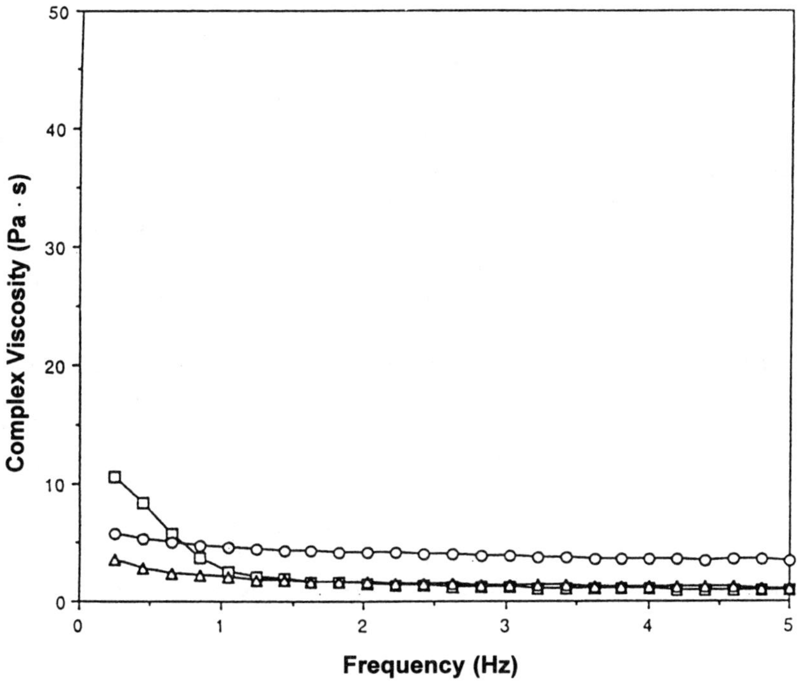

Figure 7. Complex viscosity (η') dependence on frequency of oscillation of 32% nonvolatiles by volume Al_2O_3/TiO_2 pigment dispersion stabilized with an AO/MA oligomer and thickened with the following water-soluble polymers: □, HEC; ○, HEUR-708; △, HEUR-200. (Reproduced with permission from ref. 14. Copyright 1990.)

Table VIII. Influence of Dispersant on 20° Gloss of 32% NVV Dispersions

		20° Gloss			
Thickener	*g of Thickener/ 100 g of Water*	*PMAA*	*AO(C_{10})/ MA*	*AO(C_{14})/ MA*	*AO(C_{18})/ MA*
HEUR-270	0.5	27	37	39	36
	0.8	30	40	41	41
HEUR-200	1.0	14	28	34	33
	1.7	25	32	34	36
HEUR-708	0.5	23	47	42	43
	0.8	28	41	43	42

NOTE: Components: 0.79 v/v fraction of 123-nm MMA/ethyl acrylate latex; 0.21 v/v fraction of Al_2O_3/TiO_2; 1.33 g of Tergitol 15-5-9 surfactant per 100 g of pigment; 0.75 g of dispersant per 100 g of pigment.

Commercial and Model HEURS in Coating Formulations

In this section associative thickeners in an interior architectural coating are examined. The influence of the thickener on the rheology of a small-particle-latex (the 123-nm latex studied in the section called **Phase Separation**) coating is considered. In the second part of this section, the latex in the coating formulation is varied to highlight the influence of the latex on the coating's rheology relative to that of the thickener. Both variations are compared to the model thickener responses in aqueous and surfactant solutions cited in Chapter 17.

Thickener Influences. In the work discussed in this subsection, all components in the formulation (*4*) except the water-soluble thickener are held constant. Enough thickener is added to achieve a 90-KU viscosity as measured by a Stormer indexer (i.e., a "viscometer" in which the velocity profile is ill-defined) (*34*). The latex in this formulation is 123 nm with surface acid stabilizers, and the dispersion contains an excess (residual from the latex synthesis) of nonionic surfactant. Hydrophobe-containing dispersants, the same as or similar to

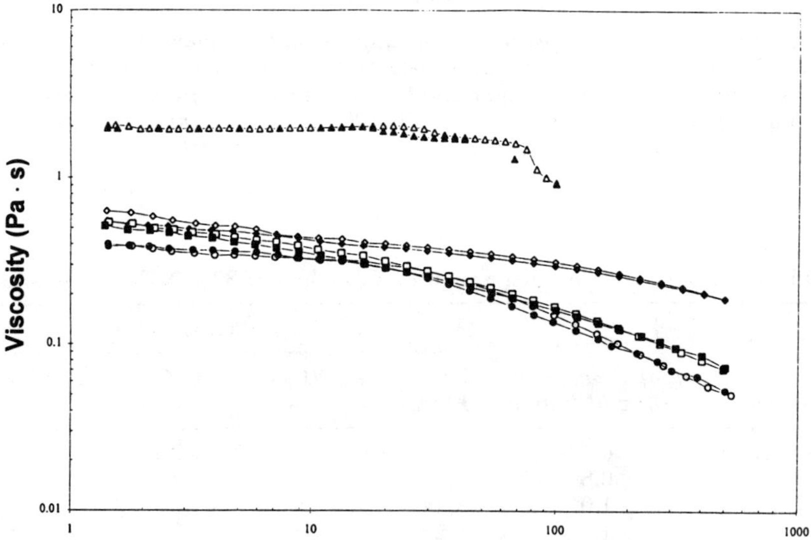

Figure 8. Viscosity dependence on shear rate of model thickeners in aqueous solution. Key: □, 1.0 wt% HEC; ○, 1.0 wt% HMHEC; ◇, 1.7 wt% HASE-35; △, 2 wt% $(C_{18}H_{37}NHCOO)_2$-$(EtO)_{531}$. Solid symbols represent viscosity during decreasing shear rate ramp.

the one discussed in the previous section, are used in the pigment grinds to stabilize TiO_2 before the TiO_2 is added to the other components (e.g., latex, coalescing aid) in the formulation. The thickeners studied include a standard nonassociating HEC, HMHEC, the three most prominent commercial HEUR thickeners (708, 270, and 200), and three commercial hydrophobe-modified alkali-swellable emulsion (HASE; discussed in Chapters 10 and 17) thickeners that have been commercially available for a decade. Model HASE and HEUR thickeners are also examined in this small-particle latex coating formulation. The structures of the associative thickener types are given in Chapters 10 and 17.

To begin this section, flow profiles of some of the model thickeners (*see* Chapter 17) and some of the commercial associative thickeners (*13*) in aqueous solution are given in Figures 8 and 9, respectively. The volume fraction of the disperse phase in the coating formulations is 0.32. Of this amount, ca. 80% is the latex and 20% is TiO_2. This amount of TiO_2 is a minimum amount for optimum hiding. In Figure 10, the flow profiles of most of the commercial thickeners containing 0.32 vol of TiO_2 *only* are illustrated. The relative order of the viscosities at low shear rates (and Newtonian flow) do not change when most of the pigment is replaced with the 123-nm latex. This result indicates

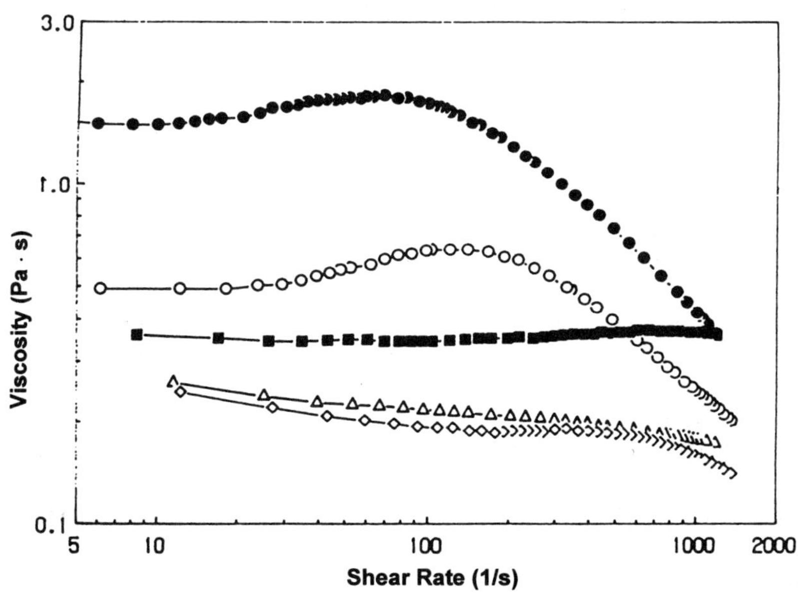

Figure 9. Viscosity dependence on shear rate of commercial thickeners in aqueous solution. Key: △, 9 wt% POE-N80; ■, 3 wt% HEUR-708; ◇, 1.5 wt% HEUR-270; ○, 2.0 wt% HEUR-200; ●, 2.5 wt% HEUR-200.

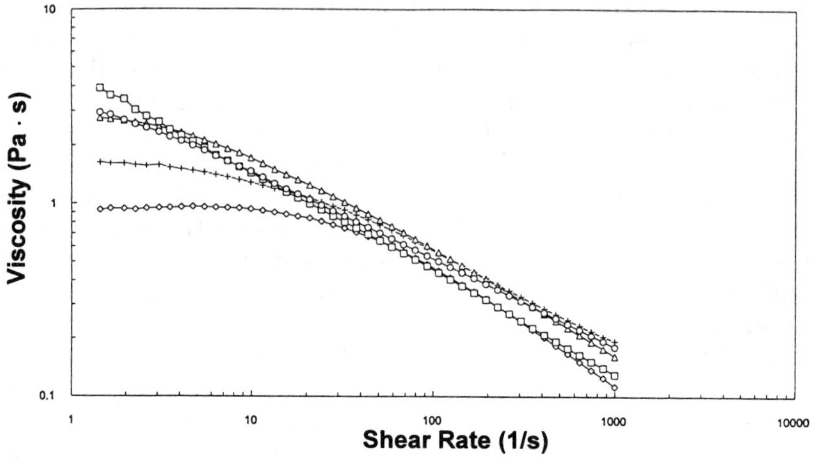

Figure 10. Viscosity dependence on shear rate of 32% nonvolatiles by volume, 90-KU Stormer viscosity Al_2O_3-treated TiO_2 pigment dispersion. Diisobutylene/MA copolymer dispersant. Thickeners: □, HEC, 0.39 wt%; ○, HMHEC, 0.36 wt%; ◇, HEUR-708, 0.28 wt%; △, HEUR-270, 0.33 wt%; +, HEUR-200, 0.96 wt%.

the dominant role the thickener has on the rheology of the dispersion (compare Figure 10 with Figures 11 and 12). The coatings containing the nonassociating HEC and the HMHEC thickener exhibit both shear thinning and thixotropy (i.e., the viscosity at a given shear rate is greater during the increasing shear rate ramp than during the decreasing shear rate ramp (*35, 36*); Figure 11). This rheological response is also observed with two of the three commercial HEUR thickener formulations (Figure 12). The thixotropy is related to the disruption of aggregates with increasing shear rate and reaggregation at a slower pace during the decreasing deformation. It may also be related, in part, to disentanglement of the HEC thickener. The amounts of thickener used to obtain the 90-KU formulation are given in Table IX. The low molecular weight of HEUR-708 and its relatively poor viscosity efficiency in water (*5*) would not project HEUR-708 as the most viscosity efficient among HEC, HEUR-270, and HEUR-200.

The activation energies for flow of the commercial HEURs in aqueous solutions (Table X) suggest that the hydrophobe size and/or concentration is larger in HEUR-270 and HEUR-200 than in HEUR-708. This difference would account for the larger difference in low shear rate viscosities. With the HEC and HMHEC of higher molecular weight, the effective volume fraction (*6*) and depletion flocculation may also influence the rheology observed. When the model HASE thickeners (discussed in Chapters 10 and 17) are used to thicken this

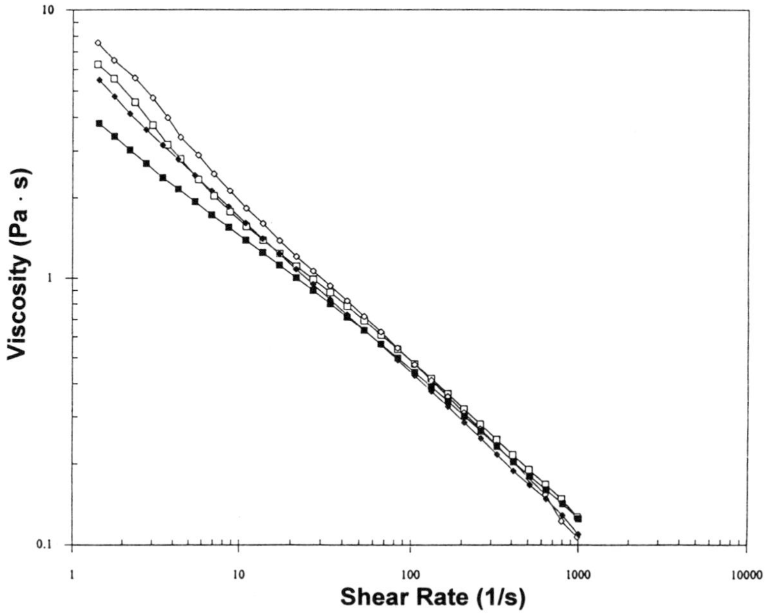

Figure 11. Viscosity dependence on shear rate of coating containing Al_2O_3/TiO_2, 123-nm MMA/ethyl acrylate copolymer latex, diisobutylene/MA dispersant and different commercial cellulosic thickeners. Key: □, HEC, 0.64 wt%; ◇, HMHEC, 0.69 wt%; open symbols, increasing shear rate; closed symbols, decreasing shear rate. The ramp cycle was completed in 6 min for all determinations in this chapter.

Table IX. Thickener Efficiency and Film Gloss of Paints When Al_2O_3/TiO_2 Pretreated with Diisobutylene/MA Copolymer Dispersant and 123-nm MMA/Ethyl Acrylate Copolymer Latex and Different Thickeners Are Used

Thickener	*Amount (wt%)*	*Stormer Viscosity (KU)*	*20° Gloss*
HEC	0.64	90	10
HMHEC	0.69	88	24
HEUR-708	0.33	91	29
HEUR-270	0.38	92	43
HEUR-200	1.18	87	29
HASE-RM5	3.89	89	57
HASE-935	0.47	86	39
HASE-615	0.34	98	36
HASE-15	0.41	87	48
HASE-25	0.50	87	38
HASE-35	0.41	88	22

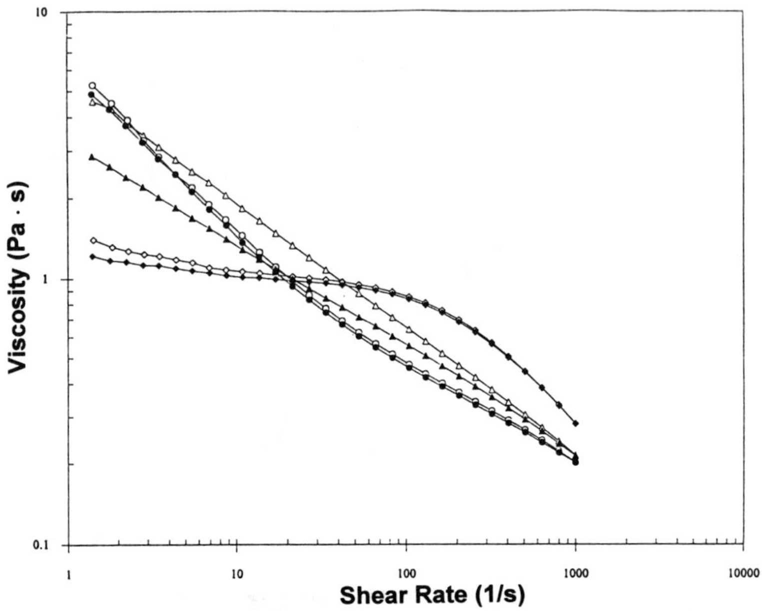

Figure 12. Viscosity dependence on shear rate of coating containing Al_2O_3/TiO_2, 123-nm MMA/ethyl acrylate copolymer latex, diisobutylene/MA copolymer dispersant, and different commercial HEUR thickeners. Key: ◇, HEUR-708, 0.33 wt%; △, HEUR-270, 0.38 wt%; ○, HEUR-200, 1.18 wt%; open symbols, increasing shear rate; closed symbols, decreasing shear rate.

small-particle latex formulation (Figure 13), shear-thinning dependence on the number of hydrophobes in the thickener is clearly indicated. Considering that the viscosity–shear rate profiles in this section are observed in coating formulations that contain smaller amounts of thickener that are more efficient in obtaining the 90-KU Stormer viscosity, this observation is indeed dramatic. The amounts used are listed in Table IX. According to these observations, the transition in

Table X. Activation Energy of Flow from Arrhenius Plots

Thickener	*Concentration Range (wt%)*	E_m *(kJ/mol)*
HEC	2–4	43.4
HMHEC	2–4	49.3
HEUR-708	4–6	62.4
HEUR-270	2–4	103.8
HEUR-200	4–6	81.7

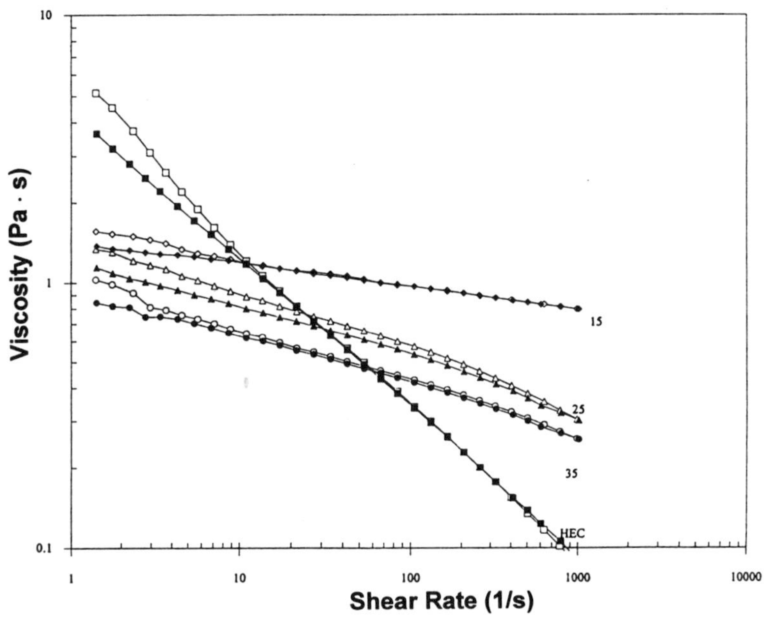

Figure 13. Viscosity dependence on shear rate of coating containing SiO_2/TiO_2, 123-nm MMA/ethyl acrylate copolymer latex, 1-decene/MA copolymer model dispersant, and different model HASE thickeners. Key: □, HEC, 0.51 wt%; ◇, HASE-15, 1.44 wt%; △, HASE-25, 1.00 wt%; ○, HASE-35, 0.84 wt%; open symbols, increasing shear rate; closed symbols, decreasing shear rate.

commercial HASE thickeners (i.e., to achieve high viscosities at low shear rates ($2\ s^{-1}$) observed with HASE-935 and HASE-615) indicates that more and/or larger hydrophobes must be involved (Figure 14), relative to the model HASE-35 thickener.

The viscosity of the formulation at relatively high shear rates ($>10{,}000\ s^{-1}$) is also an important parameter, for it influences the film thickness during application. This film thickness influences the hiding power and, in turn, the flowout of surface imperfections after application (*34*). In the small-particle latex formulation, a higher viscosity at $10^4\ s^{-1}$ is observed in the HEUR-708-thickened formulation. This viscosity is surprising, and according to the data in Figure 10 for TiO_2 only, the viscosity at high shear rates indicates a synergistic interaction between HEUR-708, the latex, and the total surfactant in the formulation. As noted in a previous review of this area (*5*), smaller hydrophobes require larger amounts of associative thickener to achieve a given KU viscosity, and viscosities at $10^4\ s^{-1}$ are proportional to the amount of thickener added. This relationship was first observed in

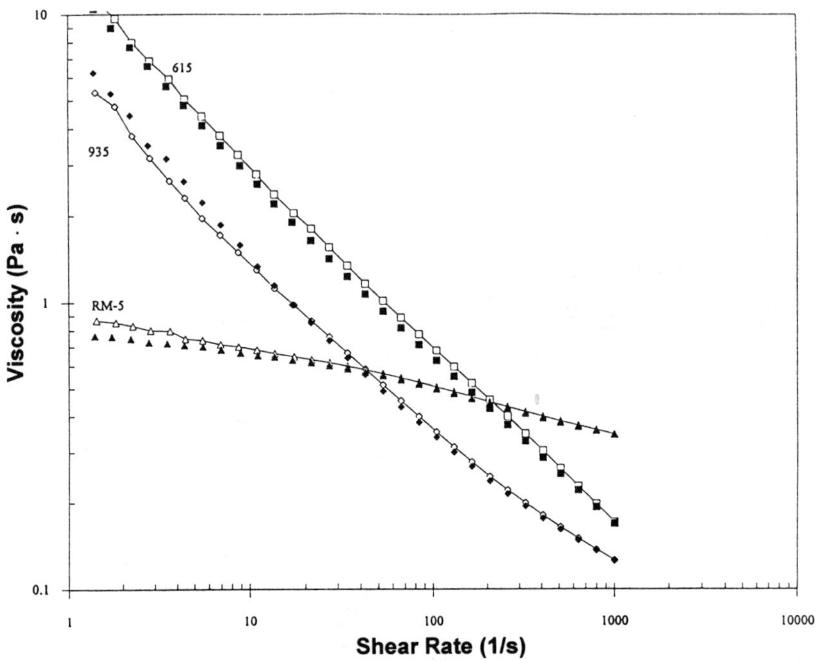

Figure 14. Viscosity dependence on shear rate of coating containing Al_2O_3/TiO_2, 123-nm MMA/ethyl acrylate copolymer latex, diisobutylene/MA copolymer dispersant, and different commercial HASE thickeners. Key: □, HASE-615, 0.34 wt%; ◇, HASE-935, 0.47 wt%; △, HASE-RM-5, 3.89 wt%; open symbols, increasing shear rate; closed symbols, decreasing shear rate.

conventional thickeners such as HEC, for which larger amounts could be added by using lower molecular weights (*37, 38*). Blending of different hydrophobe sizes (*8, 9*) as opposed to blending of molecular weights in HEC thickeners (*38*) is an effective way of balancing thickening efficiency with the total viscosity profile needed for optimum formulation performance. In Chapter 17, two series of *model* HEURs were examined. The first was prepared by the direct addition of $C_{18}H_{37}NCO$ to POEs of four molecular weights (ca. 6,000, 12,000, 20,000, and 35,000). The other series was synthesized by step-growth polymerization of the ca. 6,000-molecular-weight POE with OCN-$(CH_2)_6$-NCO and with OCN-C_6H_{10}-CH_2-C_6H_{10}-NCO. The viscosity dependence on shear rate of the coating formulations thickened with the $C_{18}H_{37}$-modified POE series is illustrated in Figure 15; corresponding data from the formulations thickened with the step-growth HEUR and model HASE thickeners (discussed in Chapters 10 and 17) and one of the $C_{18}H_{37}$-HEURs are given in Figure 16. Viscosity

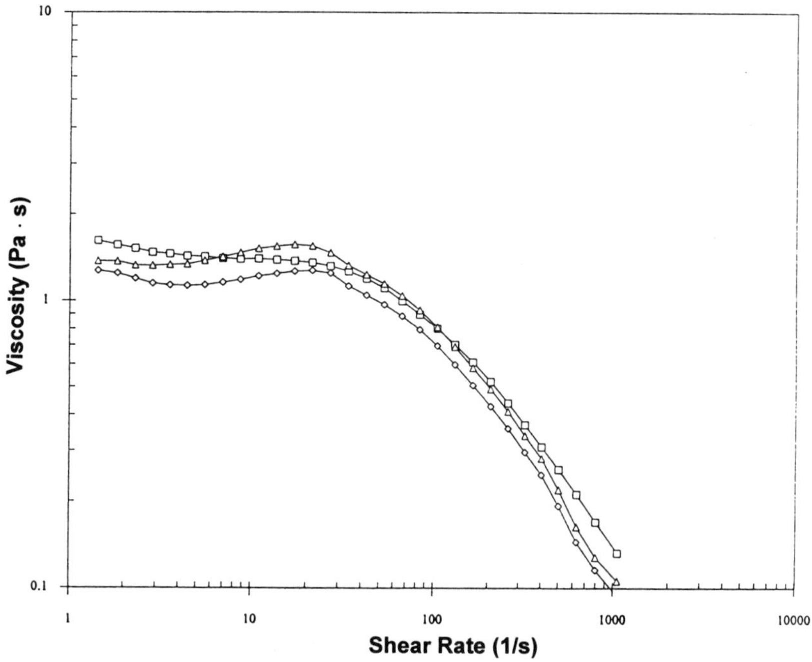

Figure 15. Viscosity dependence on shear rate of coating containing Al_2O_3/TiO_2, 123-nm MMA/ethyl acrylate copolymer latex, diisobutylene/MA copolymer dispersant and different model HEUR thickeners. Key: □, $(C_{18}H_{37}NHCOO)_2$-$(EtO)_{531}$, 0.28 wt%; ◇, $(C_{18}H_{37}NHCOO)_2$-$(EtO)_{331}$, 0.32 wt%; △, $(C_{18}H_{37}NHCOO)_2$-$(EtO)_{182}$, 0.28 wt%.

responses during the decreasing shear rate ramp are omitted in Figures 15 and 16 to avoid data overlap and congestion.

One might expect the $C_{18}H_{37}$-modified POE with the shorter oxyethylene chains to be less efficient because of their greater intrahydrophobic association and lower viscosity efficiency in water (discussed in Chapter 17). No lack of efficiency was observed in the small-particle coating formulation (Table IX). One could envision shear thickening in the lower-shear-rate behavior of the coating formulations illustrated in Figure 15. We examined a number of coating formulations (39), and the slight differences in this coating formulation at low shear rates are within experimental error. A coating formulation contains an excess of surfactants as well as cosolvents and coalescing aids that could influence the amounts of both intra- and interhydrophobic associations in a fully formulated coating. The difference in thickening efficiency and solution rheology noted in aqueous solutions is not clearly evident in the coatings.

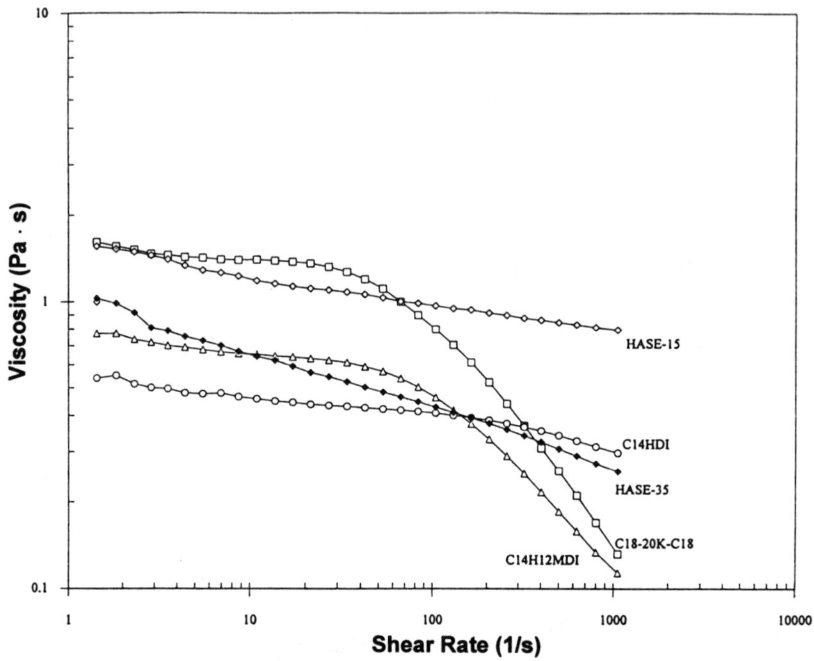

Figure 16. Viscosity dependence on shear rate of coating containing Al_2O_3/TiO_2, 123-nm MMA/ethyl acrylate copolymer latex, diisobutylene/MA copolymer dispersant and different model thickeners. Key: ○, S-G $C_{14}HDI$, 1.58 wt%; △, S-G $C_{14}H_{12}MDI$, 1.02 wt%; □, $(C_{18}H_{37}NHCOO)_2$-$(EtO)_{531}$, 0.28 wt%; ◆, HASE-35, 0.84 wt%; ◇, HASE-15, 1.44 wt%. HDI is hexamethylene diisocyanate. $H_{12}MDI$ is dicyclohexylmethane diisocyanate.

The data in Figure 16 compare model thickeners of different compositions and family types. Along with the rheology profiles of the commercial thickener formulations (Figures 11, 12, and 14), these data highlight similarities in response promoted by chemically different thickeners. For example, smaller terminal hydrophobes in the model HEUR families facilitate profiles similar to those of the relatively small and low-hydrophobe-content model HASE thickeners. With increasing hydrophobe content and size, the coating formulations become shear thinning. The rheology profiles for the model HASE-35 thickener and the commercial HEUR-708 thickener show similarities. The commercial HASE-935 thickener promotes a profile similar to that of the HEC-thickened formulation. The shear-thinning profile of a helical carbohydrate polymer (e.g., *Xanthomonas campestris* polysaccharide (*40*)) in aqueous solution is more pronounced than that of HEC (*41*), and this type of influence is observed in the HASE-615 solutions and formulations. Other factors also affect the selection of thickeners for a given latex coating; one such factor is the quality of the film. In exterior architectural coating formulations, whose films are exposed to more demanding environments (*42*), HASE thickeners and vinyl acetate latices generally are not acceptable. Lower-cost HASE and vinyl acetate latices are acceptable for the lower performance requirements of many interior architectural coatings.

Oscillatory rheometry (*35, 43*) has been definitive in defining elastic response differences among thickened aqueous dispersions (*6, 44*). The nonassociating HEC affects a formulation with an elastic character reflected by the crossover of the G′/G″ responses at a very low frequency (Figure 17). This elasticity (i.e., dominance of the storage modulus) is not surprising in view of the effective volume fraction and crowded environment in this type of formulation (*6*). In this study this crossover occurs with only one of the commercial associative thickeners, HASE-615 (Figure 18), with which it occurs at a much lower oscillatory frequency and thickener concentration than was noted with the HEC formulation. Similar responses occur with a change in latex, as discussed in the next subsection. The slow relaxation processes reflected in these formulations was not observed in the model associative thickener formulations.

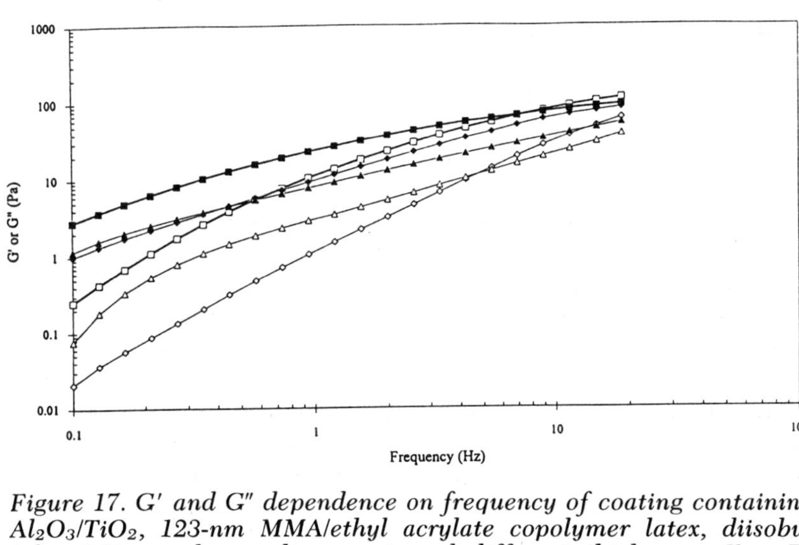

Figure 17. G′ and G″ dependence on frequency of coating containing Al_2O_3/TiO_2, 123-nm MMA/ethyl acrylate copolymer latex, diisobutylene/MA copolymer dispersant, and different thickeners. Key: □, HEC, 0.66 wt%; ◇, $(C_{18}H_{37}NHCOO)_2$-$(EtO)_{531}$, 0.28 wt%; △, HEUR-270, 0.37 wt%; open symbols, G′ (storage modulus); closed symbols, G″ (loss modulus).

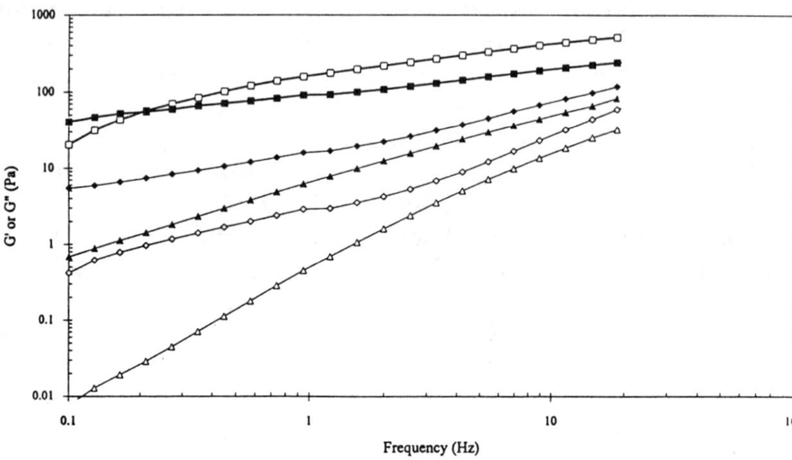

Figure 18. G′ and G″ dependence on frequency of coating containing Al_2O_3/TiO_2, 123-nm MMA/ethyl acrylate copolymer latex, diisobutylene/MA copolymer dispersant, and different commercial thickeners. Key: □, HASE-615, 0.34 wt%; ◇, HASE-935, 0.47 wt%; △, HEUR-708, 0.33 wt%; open symbols, G′ (storage modulus); closed symbols, G″ (loss modulus).

Influence of the Latex. The influence of the latex when the polymer colloids discussed in the phase-separation section were used was also examined. Larger amounts of thickener are usually required as the size of the latex particle increases (6), and this correlation was observed. The amounts of thickener required to achieve a 90-KU viscosity for the two larger latices are given in Table XI.

The rheology profiles of the HEC-thickened formulations are illustrated in Figure 19. The intermediate-bimodal-particle latex profile is distinguishable from the small- and large-particle-latex profiles; the intermediate-particle-size latex was tailored to given optimum film properties with the conventional HEC thickener. To achieve optimum latex performance, different components are often added. This may include a water-soluble polymer added to the commercial latex batch. No effort to overinterpret the results of any latex variations is given in this chapter. This section illustrates the influence that the latex can have on the rheology of a coating formulation containing the various thickeners discussed in previous sections.

The trend of greater efficiency with smaller latices is maintained in the HEUR-270-thickened formulations, but the rheology profile of the large-particle latex is distinctly different and very thixotropic (Fig-

Table XI. Thickener Efficiency and Film Gloss of Paints When Al_2O_3/TiO_2 Pretreated with Diisobutylene/MA Copolymer Dispersant and 390- and 560-nm MMA/Ethyl Acrylate Copolymer Latices and Different Thickeners Are Used

	Amount (wt%)		*Stormer Viscosity (KU)*		*20° Gloss*	
Thickener	*390 nm*	*560 nm*	*390 nm*	*560 nm*	*390 nm*	*560 nm*
HEC	0.78	1.02	89	90	26	8
HEUR-708	0.31	0.41	93	91	26	15
HEUR-270	0.42	0.50	88	91	19	22
HASE-RM5	2.51	2.60	94	90	56	45
HASE-935	0.34	0.45	108	84	45	26
HASE-615	0.27	0.34	98	92	31	7
HASE-15	0.54	0.56	90	88	29	25
HASE-25	0.34	0.35	90	86	14	16
HASE-35	0.30	0.31	88	86	11	10

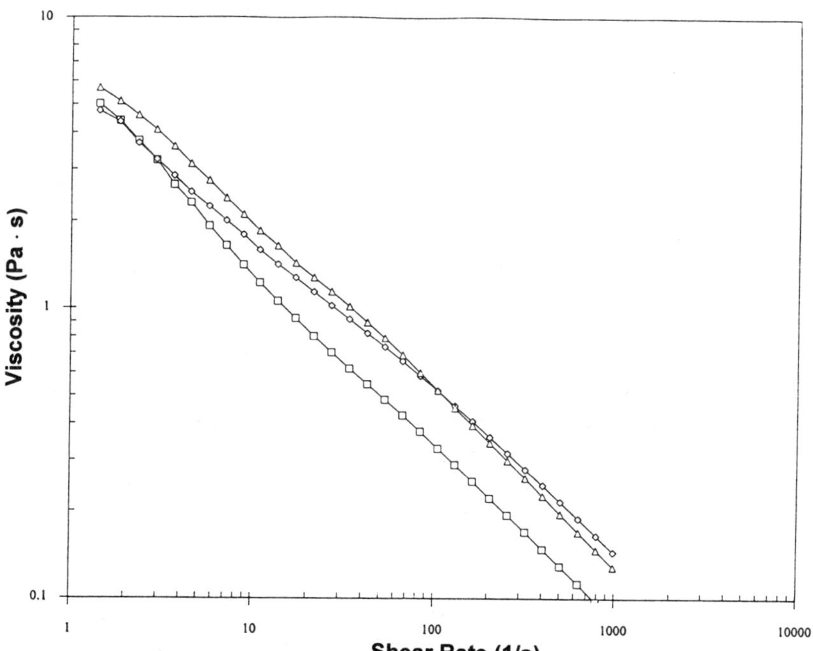

Figure 19. Viscosity dependence on shear rate of coating containing Al_2O_3/TiO_2, diisobutylene/MA copolymer dispersant, HEC thickener (amounts listed in Tables IX and XI), and different acrylic latices. Key: □, small particle (123 nm); ◇, medium particle (390 nm); △, large particle (560 nm).

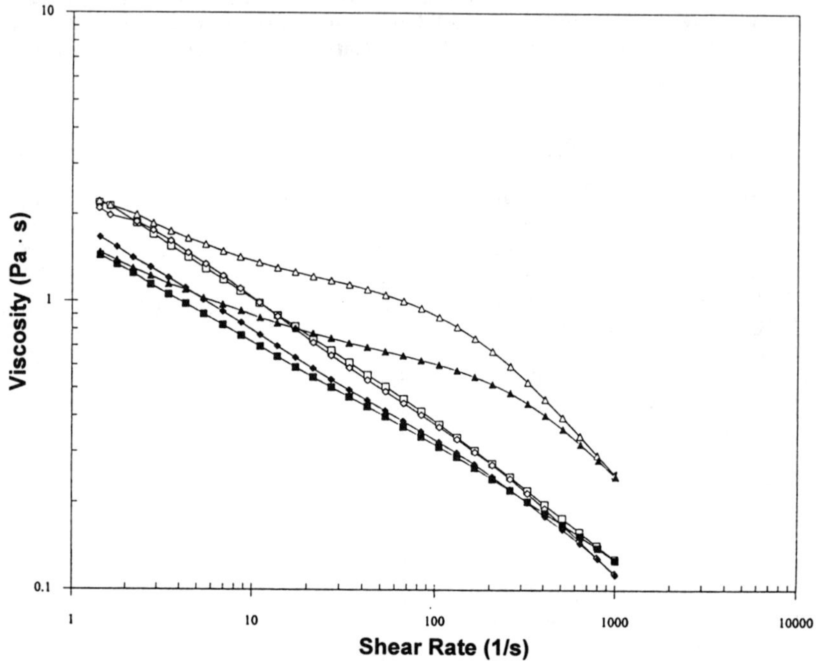

Figure 20. Viscosity dependence on shear rate of coating containing Al_2O_3/TiO_2, diisobutylene/MA copolymer dispersant, HEUR-270 thickener (amounts listed in Tables IX and XI), and different acrylic latices. Key: □, small particle (123 nm); ◇, medium particle (390 nm); △, large particle (560 nm); open symbols, increasing shear rate; closed symbols, decreasing shear rate.

ure 20). The three latices exhibit distinctly different responses in HEUR-708-thickened formulations (Figure 21), but the difference in responses is not as notable in thixotropy with the larger-particle latex. The viscosity efficiency with smaller-particle latices is not apparent (Tables IX and XI). This lack of thickening efficiency with the small-particle latex is distinctive in the HASE-thickened formulations.

With the model HASE-35 thickener, no marked differences in rheology profile are observed (Figure 22), but associations in the bimodal latex are indicated. With the larger number and size of the hydrophobes in HASE-935, this commercial thickener dominates the coating rheology, but the contribution of the latex (e.g., higher viscosities at higher shear rates) is evident with the smaller-particle latex (Figure 23). Interaction of the thickener with and aggregation of the disperse phase responsible for the higher viscosities in the lower-shear-rate range are more quantifiable in the storage (G′, reflecting

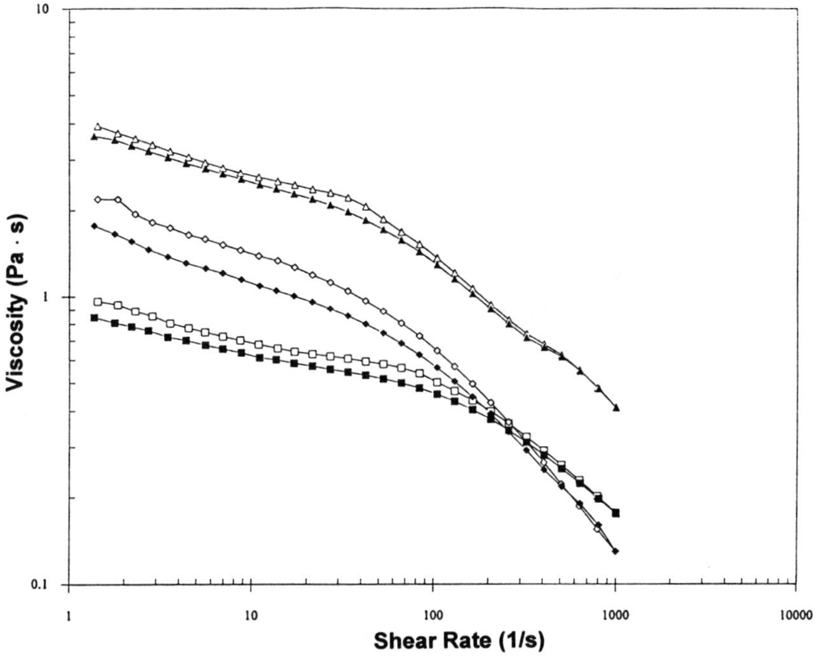

Figure 21. Viscosity dependence on shear rate of coating containing Al_2O_3/TiO_2, diisobutylene/MA copolymer dispersant, HEUR-708 thickener (amounts listed in Tables IX and XI), and different acrylic latices. Key: □, small particle (123 nm); ◇, medium particle (390 nm); △, large particle (560 nm); open symbols, increasing shear rate; closed symbols, decreasing shear rate.

the elasticity of the formulation) and loss (G″, reflecting the viscosity component) moduli data that are obtained in oscillatory rheometry. An early crossover of G′ and G″ reflects a longer relaxation time in the bimodal latex with the commercial HASE-35 (Figure 24) and HASE-935 (Figure 25), whereas the crossover in G″/G′ at low frequencies occurs with the small-particle latex thickened only with HASE-615 (Figure 18). Several mechanisms are apparently responsible for aggregation structures in coating formulations and in the formulation containing the bimodal latex with the less dominant HASE-35 and HASE-935 thickeners.

Influence of Other Components

Figure 19 reveals no significant influence of the latex on the rheology of HEC-thickened coatings. That lack of influence is one of the

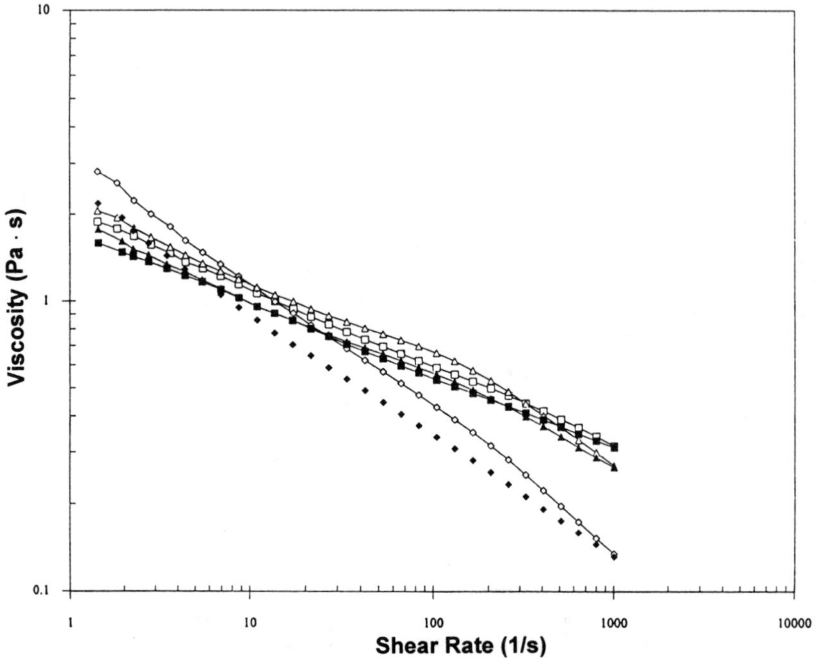

Figure 22. Viscosity dependence on shear rate of coating containing Al_2O_3/TiO_2, diisobutylene/MA copolymer dispersant, HASE-35 thickener (amounts listed in Tables IX and XI), and different acrylic latices. Key: □, small particle (123 nm); ◇, medium particle (390 nm); △, large particle (560 nm); open symbols, increasing shear rate; closed symbols, decreasing shear rate.

strengths of cellulose ether-thickened coatings and one of the downsides of most associative thickener formulations. The length of this chapter precludes discussion of other variable influences, but a few will be mentioned. For example, the size and type of the surfactant hydrophobe (*45*) and the extent of filler (calcium carbonate, talc, silica, etc.) (*46, 47*) have notable influences on coating rheology. In addition to improving the freeze–thaw and open times, cosolvents (*46*) have an influence on low- and high-shear-rate viscosities that is now achieved by mixing associative thickeners of different hydrophobe sizes. Cosolvents change the nature of the continuous media (*48*), and ineffective coalescing aids may form mixed micelles (*49*); both minimize the driving force for the interaction of thickener hydrophobes in aqueous solution. Effective coalescing aids may transport surfactant from the surface of the latex to its interior; such transport has been reported (*50*) for low-T_g vinyl acetate/butyl acrylate copolymers and

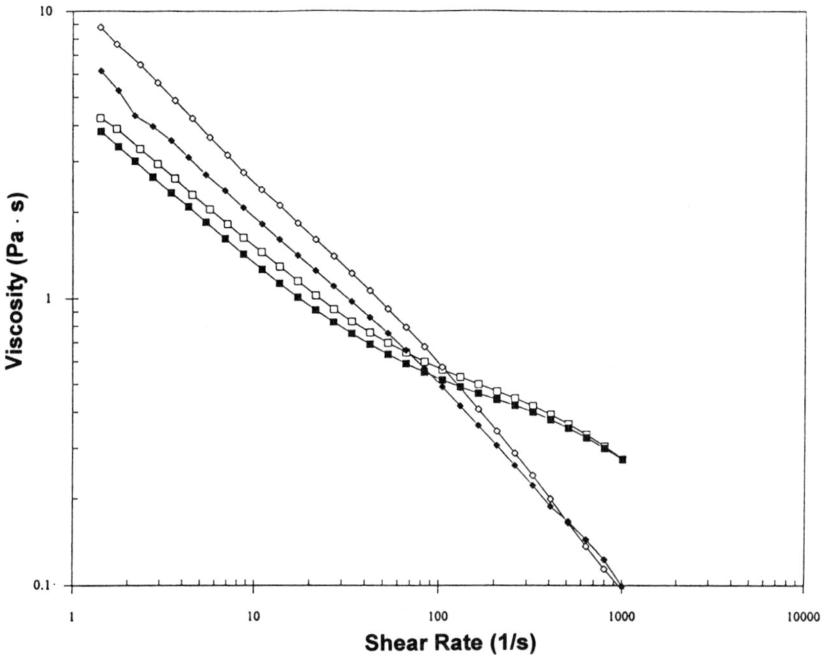

Figure 23. Viscosity dependence on shear rate of coating containing Al_2O_3/TiO_2 pigment slurry, HASE-935 thickener (amounts listed in Tables IX and XI), and different acrylic latices. Key: □, small particle (123 nm); ◇, medium particle (390 nm); open symbols, increasing shear rate; closed symbols, decreasing shear rate.

would dramatically influence the formulation's viscosity. Associative thickener/latex coatings are truly complex fluids.

Conclusions

In model thickener and model latex dispersions, the process of phase separation (or syneresis) can be clearly delineated by the chemical and physical stability of the materials. With nonassociating thickeners, syneresis is observed with increasing latex particle size, as predicted by the depletion flocculation concept. Model HEUR thickeners with marginal solubility in aqueous solution, due to extensive intrahydrophobic associations, promote syneresis in properly stabilized, small-particle latex dispersions. In model high-T_g latices thickened by commercial associative thickener, syneresis can be eliminated by increasing the oligomeric surface acid concentration and the surfactant concentration. In commercial latex dispersions thickened with

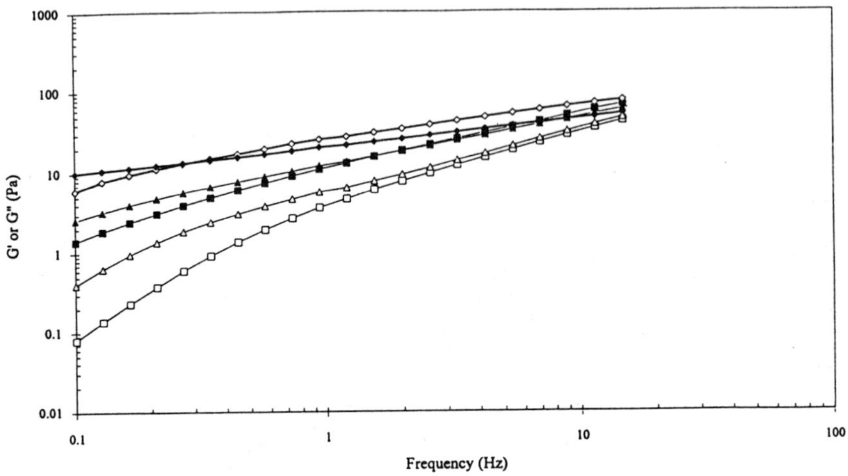

Figure 24. G′ and G″ dependence on frequency of coating containing Al_2O_3/TiO_2 pigment slurry, HASE-35 thickener, and different acrylic latices. Key: □, small particle (123 nm); ◇, medium particle (390 nm); △, large particle (560 nm); open symbols, G′ (storage modulus); closed symbols, G″ (loss modulus).

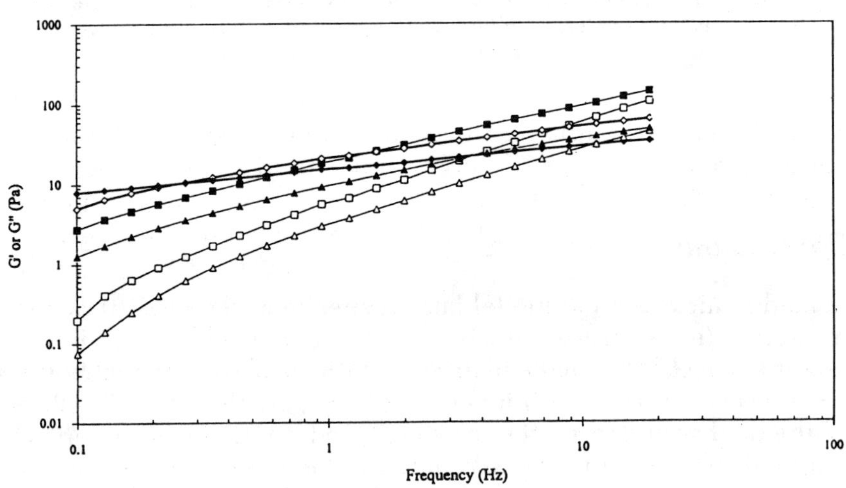

Figure 25. G′ and G″ dependence on frequency of coating containing Al_2O_3/TiO_2 pigment slurry, HASE-935 thickener, and different acrylic latices. Key: □, small particle (123 nm); ◇, medium particle (390 nm); △, large particle (560 nm); open symbols, G′ (storage modulus); closed symbols, G″ (loss modulus).

commercial associative thickener formulations, the many unknowns make it impossible to delineate phase-separation behavior. When such behavior is observed, using a secondary nonassociative thickener such as HEC only enhances the extent of separation, because polymer incompatibility contributes to the phase-separation problem.

An examination of a broad range of organic acid dispersants showed that the hiding pigment TiO_2 can be stabilized and final film gloss increased if the dispersant contains hydrophobic units to interact with the hydrophobes of the thickener; however, if the hydrophobe number and/or size is increased in HASE thickeners, the final film gloss decreases. This decrease may be due in part to adsorption of HASE thickeners independent of the dispersing agent and interbridging of the pigment particles due to HASE hydrophobic interaction.

As the matrix of components is increased in the move to a coating formulation, the interactions of the thickener with the pigment, that is, the minor disperse phase (0.07 volume fraction) in a formulation, are not reflected in the rheology of the final coating formulation. The viscosities at low shear rates are determined by the thickener when its molecular weight or interhydrophobic associations are high. When these parameters are low, the rheology of a formulation over the low- to moderate-shear-rate range can be influenced significantly by the latex.

In comparisons of thickeners, similar rheologies can be effected by thickeners of different chemical compositions by controlling the size and number of the hydrophobe and the molecular weight of the thickener. However, large differences in thickener efficiencies and film gloss occur within the HEUR and HASE thickener families. The thickeners are distinctly different within and between classes, and they influence coating formulations in distinctly different ways. With increasing hydrophobe strength, the thickening efficiency increases in both the HEUR and the HASE families. Increasing the hydrophobe strength in HEUR thickeners also leads to higher film gloss values; however, film gloss readings decrease with increasing hydrophobe concentration in both commercial and model HASE formulations, despite the very low concentration of HASE thickeners required to achieve 90 KU. HASE thickener contributions to viscosity are due in part to electrostatic repulsions. These differences are reflected in different responses in coating formulations. For example, the influence of hydration in smaller-particle latices provides a greater effective volume fraction of the disperse phase and lowers the amount of HEC or HEUR thickeners needed to achieve a given Stormer viscosity, but the particle size of the latex does not affect the efficiency of HASE thickeners.

Acknowledgment

Financial support of these studies by DuPont is gratefully acknowledged.

References

1. Alahapperuma, K.; Glass, J. E. *J. Coat. Technol.* **1991,** *63*(799), 69–78.
2. Karunasena, A.; Glass, J. E. *Prog. Org. Coat.* **1989,** *17*(3), 301–320.
3. Craig, D. H. In *Water-Soluble Polymers: Beauty with Performance;* Glass, J. E., Ed.; Advances in Chemistry 213; American Chemical Society: Washington, DC, 1986; Chapter 18.
4. Lundberg, D. J.; Glass, J. E. *J. Coat. Technol.* **1992,** *64*(807), 53–61.
5. Karunasena, A.; Brown, R. G.; Glass, J. E. In *Polymers in Aqueous Media: Performance Through Association;* Glass, J. E., Ed.; Advances in Chemistry 223; American Chemical Society: Washington, DC, 1989; Chapter 26.
6. Lundberg, D. J.; Ma, Z.; Alahapperuma, K.; Glass, J. E. In *Polymers as Rheology Modifiers;* Glass, J. E.; Schulz, D. N., Eds.; ACS Symposium Series 462; American Chemical Society: Washington, DC, 1991; Chapter 14.
7. Schwab, F. G. In *Water-Soluble Polymers; Beauty with Performance*; Glass, J. E., Ed.; Advances in Chemistry 213; American Chemical Society; Washington, DC, 1986; Chapter 19.
8. Howard, P. R.; Leasure, E. L.; Rosier, S. T.; Schaller, E. J. In *Polymers as Rheology Modifiers;* Glass, J. E.; Schulz, D. N., Eds.; ACS Symposium Series 462; American Chemical Society: Washington, DC, 1991; Chapter 12.
9. Anwari, F. M.; Schwab, F. G. In *Polymers in Aqueous Media: Performance Through Association;* Glass, J. E., Ed.; Advances in Chemistry 223; American Chemical Society: Washington, DC, 1989; Chapter 27.
10. Ma, Z.; Lundberg, D. J.; Roberts, S.; Glass, J. E. *J. Appl. Polym. Sci.* **1993,** *49,* 1509–1527.
11. Sperry, P. *J. Colloid Sci.* **1981,** *82,* 62. *Ibid.* **1982,** *87,* 375. *Ibid.* **1984,** *99,* 97.
12. Lundberg, D. J.; Glass, J. E.; Eley, R. R. *Polym. Mater. Sci. Eng.* **1989,** *61,* 533–538.
13. Lundberg, D. J.; Glass, J. E.; Eley, R. R. *J. Rheol.* **1991,** *35*(6), 1255–1274.
14. Lundberg, D. J. Ph.D. Thesis, North Dakota State University, 1990.
15. May, R.; Kaczmarski, J. P.; Glass, J. E. *Polym. Mater. Sci. Eng.* **1993,** *69,* 301.
16. Ma, Z. Ph.D. Thesis, North Dakota State University, 1992.
17. Bergh, J. S.; Lundberg, D. J.; Glass, J. E. *Prog. Org. Coat.* **1989,** *17,* 155–173.
18. Anwari, F.; Carlozzo, B. J.; Chokshi, K.; DiLarenzo, M.; Heble, M.; Knauss, C. J.; McCarthy, J.; Patterson, R.; Rozick, P.; Slifko, P. M.; Stipkovich, W.; Weaver, J. C.; Wolfe, M. *J. Coat. Technol.* **1993,** *65*(825), 123.
19. Morrison, W. H., Jr. *J. Colloid Interface Sci.* **1984,** *100,* 121–127.
20. Morrison, W. H., Jr. *J. Coat. Technol.* **1985,** *57*(721), 55–65.
21. Hingston, F. J.; Atkinson, R. J.; Posner, A. M.; Quirk, J. P. *Nature* (*London*) **1967,** *215,* 1459.
22. Hingston, F. J.; Posner, A. M.; Quirk, J. P. *J. Soil Sci.* **1972,** *23,* 177.

23. Losoi, T. *J. Coat. Technol.* **1989,** *61*(776), 57–63.
24. Foissy, A.; Attar, E. A.; Lamarche, J. M. *J. Colloid Interface Sci.* **1983,** *96*(1), 275–287.
25. Shih, L. B.; Sheu, E. Y.; Chen, S. H. *Macromolecules* **1989,** *36,* 1225.
26. Barone, G.; Di Virgilio, N.; Elia, V.; Rizzo, E. *J. Polym. Sci., Polym. Symp.* **1974,** *44,* 1–10.
27. Bianchi, E.; Ciferri, A.; Parodi, R.; Rampone, R.; Tealdi, A. *J. Phys. Chem.* **1970,** *74,* 1050.
28. McGlade, M. J.; Olufs, J. L. *Macromolecules* **1988,** *21,* 2346.
29. Chu, D.-Y.; Thomas, J. K. In *Polymers in Aqueous Media: Performance Through Association;* Glass, J. E., Ed.; Advances in Chemistry 223; American Chemical Society: Washington, DC, 1989; Chapter 18.
30. Verwey, E. J. W.; Overbeek, J. T. G. *Theory of the Stability of Lyophobic Colloids;* Elsevier: Amsterdam, 1948.
31. Overbeek, J. T. G. In *Colloid Science I;* Kruyt, H. R., Ed.; Elsevier: New York, 1952.
32. Hunter, R. J. *Zeta Potential in Colloid Science;* Academic: New York, 1981; pp 239–247.
33. Cremer, M. *Proc. XIXth FATIPEC Congr.* **1988,** *1,* 177–195.
34. Glass, J. E. In *Water-Soluble Polymers: Beauty with Performance*; Glass, J. E., Ed.; Advances in Chemistry 213; American Chemical Society: Washington, DC, 1986; Chapter 21.
35. Prud'homme, R. K. In *Polymers as Rheology Modifiers;* Glass, J. E.; Schulz, D. N., Eds.; ACS Symposium Series 462; American Chemical Society: Washington, DC, 1991; Chapter 2.
36. Van Wazer, J. R.; Lyons, J. W.; Kim, K. Y.; Colwell, R. E. *Viscosity and Flow Measurements;* Interscience: New York, 1966.
37. Glass, J. E. *J. Oil Colour Chemists Assoc.* **1975,** *58,* 169.
38. Arney, W. A., Glass, J. E. *Ibid.* **1976,** *59,* 372.
39. Tarng, M.-R. Ph.D. Thesis, North Dakota State University, 1995.
40. Glass, J. E. In *Water-Soluble Polymers: Beauty with Performance;* Glass, J. E., Ed.; Advances in Chemistry 213; American Chemical Society: Washington, DC, 1986; Chapter 1.
41. Fernando, R. H. Ph.D. Thesis, North Dakota State University, 1986.
42. LeSota, S.; Lewandowski, E. W.; Schaller, E. J. In *Polymers in Aqueous Media: Performance Through Association;* Glass, J. E., Ed.; Advances in Chemistry 223; American Chemical Society: Washington, DC, 1989; Chapter 28.
43. Kulicke, W. M.; Nottelmann, H. In *Polymers in Aqueous Media: Performance Through Association;* Glass, J. E., Ed.; Advances in Chemistry 223; American Chemical Society: Washington, DC, 1989; Chapter 2.
44. Kaczmarski, J. P.; Fernando, R. H.; Glass, J. E. *J. Coat. Technol.* **1993,** 65(818), 39.
45. Murakami, T.; Fernando, R. H.; Glass, J. E. *J. Oil Colour Chemists Assoc.* **1988,** *71*(10), 315–323.
46. Glass, J. E.; Fernando, R. H.; Egland-Jongewaard, S. K.; Brown, R. G. *J. Oil Colour Chemists Assoc.* **1984,** *67*(10), 256–261.
47. Fernando, R. H.; Glass, J. E. *J. Oil Colour Chemists Assoc.* **1984,** 67(11), 279–282.
48. Thibeault, J. C.; Sperry, P. R.; Schaller, E. J. In *Water-Soluble Polymers: Beauty with Performance;* Glass, J. E., Ed.; Advances in Chemistry 213; American Chemical Society: Washington, DC, 1986; Chapter 20.

49. Tanford, C. *The Hydrophobic Effect: Formation of Micelles and Biological Membranes;* John Wiley: New York, 1980.
50. Vijayendran, B. R.; Bone, T.; Gajria, C. In *Emulsion Polymers and Emulsion Polymerization;* Bassett, D. R.; Hamielec, A. E., Eds.; ACS Symposium Series 165; American Chemical Society: Washington, DC, 1981; p 225.

RECEIVED for review November 2, 1994. ACCEPTED revised manuscript May 31, 1995.

Conventional Water-Soluble Polymers

25

Synthesis of Amine Functional Homopolymers with *N*-Ethenylformamide

R. J. Badesso, A. F. Nordquist, R. K. Pinschmidt, Jr.,* and D. J. Sagl

Air Products and Chemicals, Inc., 7201 Hamilton Boulevard, Allentown, PA 18195

*Poly(vinylamine) is of interest because of its pH-dependent cationic nature and its reactive primary amine functionality. The preparation of poly(vinylamines) of various molecular weights via the synthesis and hydrolysis of poly(*N*-ethenylformamide) (PNEF) is readily accomplished. Solution properties of poly(vinylamine) and PNEF, including solution viscosity and compatibility with common electrolytes, were measured. Structural modifications of poly(vinylamine) were made via reaction of the primary amine groups pendant from the backbone. The effectiveness of poly(vinylamine) and its acid salts in several applications was evaluated.*

The practical, low-cost synthesis of polymers and copolymers containing stable primary amine functionality has long been a dream of polymer chemists, in part because of the high reactivity of the primary amine group for cross-linking and derivatization and in part because of the cationic nature of these polymers in appropriate pH regimes. Cationic polymers have demonstrated utility in a number of important applications such as wastewater treatment, petroleum production, and papermaking. To date, only the polymerization of aziridine to poly(ethyleneimine), a polymer containing a mixture of primary, secondary,

* Corresponding author.

0065–2393/96/0248–0489$12.00/0

and tertiary amines, has achieved a measure of commercial success (*1*).

Poly(vinylamine) (PVAm), a linear polymer with all primary amine groups, also has utility in a number of application areas. The preparation of PVAm has been attempted, sometimes successfully, by a number of researchers. Unfortunately, the simplest precursor to PVAm, vinylamine monomer, is unavailable, because it tautomerizes to acetaldehyde imine. PVAm can therefore be synthesized only via indirect routes. Attempts have been made to synthesize PVAm from polyacrylic acid via the Schmidt reaction (*2*) or from polyacrylamide via the Hofmann reaction (*3–7*). These methods suffer from the incomplete conversion of the carboxylic acid and amide moieties to amines, from chain scission, and from a number of side reactions that give incomplete amine formation. Attempts have also been made to synthesize PVAm via the hydrolysis of higher poly(*N*-vinylimides) (*8–10*), poly(*N*-vinylcarbamates) (*11*), and poly(*N*-vinylacetamide) (*12*). The commercial preparation of PVAm via these routes has been limited

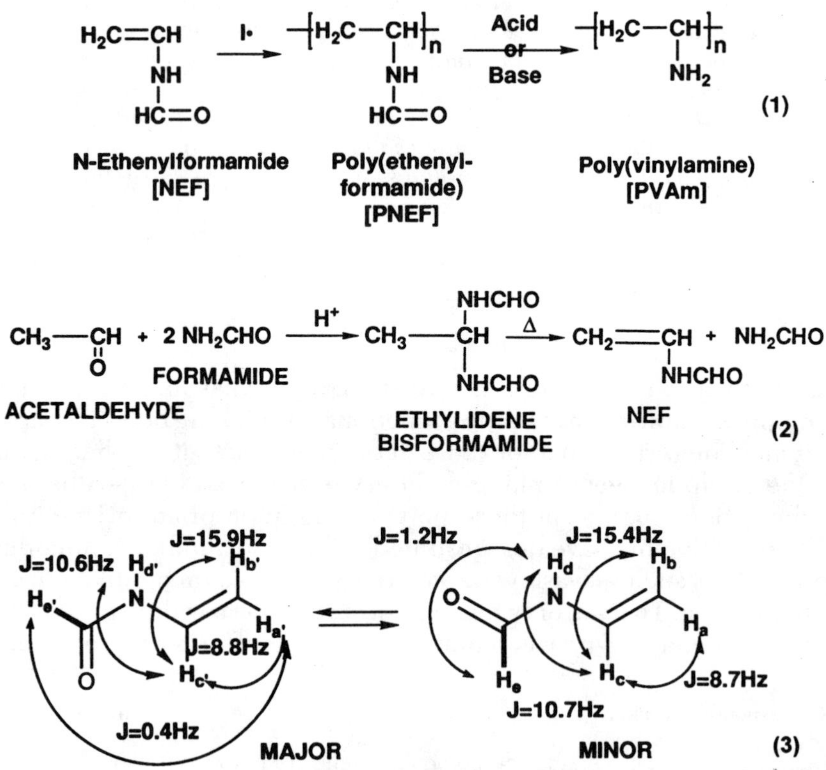

Chart I.

by the long reaction times and aggressive conditions required for hydrolysis of the precursor polymer and by the commercial unavailability of the monomers.

A readily hydrolyzed intermediate polymer would facilitate the synthesis of PVAm. The hydrolysis of formamides is relatively easy compared to that of imides or higher amide homologs such as acetamides (*13*). Poly(*N*-ethenylformamide) (PNEF) is readily prepared in a range of molecular weights from *N*-ethenylformamide (NEF) monomer (*14*) and can be easily and quantitatively hydrolyzed under basic or acidic conditions to PVAm (equation 1 in Chart I). PNEF is thus a nearly ideal precursor. This chapter describes the facile preparation of PVAm via the hydrolysis of PNEF prepared from NEF monomer. Various solution properties of these polymers are detailed. Modifications of PVAm are described, and data on the use of various amine functional polymers in selected applications are given.

Experimental Details

Materials. NEF was prepared via the vapor-phase pyrolysis of ethylidenebisformamide and purified via fractional distillation (equation 2 in Chart I) (*14*). Other reagents were obtained from commercial sources and used without purification.

Analytical Methods. NEF may be characterized by gas chromatography (25-m fused-silica 3% OV-1701 column, 3 min at 80 °C + 15 °C/min ramp to 200 °C). NEF response becomes severely nonlinear at small sample sizes.

NEF monomer levels can also be determined by adding a known weight of a test mixture to water and determining the UV absorbance of the resulting solution (NEF λ_{max} of 224 nm [ϵ = 13,900]). Comparison of this absorbance to the absorbance of a known standard allows calculation of the residual NEF monomer level. NEF monomer levels of 0.005 to 0.10% in polymers may also be assayed by UV spectroscopy by extracting 10 g of powdered polymer with 100 ml of acetonitrile, diluting by a factor of 10, and measuring the absorbance.

Molecular weights of water-soluble NEF homopolymers and hydrolyzed products were determined via aqueous size exclusion chromatography–multiple-angle laser light scattering (*15*). PNEF was conveniently analyzed on Toyo Soda TSK-PW columns at 35 °C with aqueous 0.05 M $NaNO_3$ eluent. PVAm was analyzed on SynChrom CATSEC columns with aqueous $NaNO_3$–0.1% CF_3CO_2H eluent at pH 2.5.

The intrinsic viscosity of PNEF was measured at 25 °C in deionized water. The intrinsic viscosity of PVAm was measured at 25 °C in 1.0 M aqueous sodium chloride. Aqueous solution viscosity measurements of PNEF and PVAm were made using a Haake VT-500 rheometer with an MV-DIN spindle. All T_g determinations were made on a Dupont 910 differential scanning calorimeter at a heating rate of 20 °C/min. The electrolyte compatibility studies (*see* Table III) were conducted by adding 10 parts of a 10 wt% solution of the electrolyte to 1 part of a 10 wt% solution of PNEF or PVAm·HCl and observing for signs of precipitation.

PNEF hydrolysis levels were determined by ^{13}C NMR.

^{1}H NMR of PNEF: (neat versus D_2O capillary) ∂ 4.80 (d, 1H, J = 8.7 Hz), 4.92 (dd, 1H, J = 1.2 Hz, J = 8.8 Hz), 5.13 (d, 1H, J = 15.4 Hz), 5.23 (d, 1H, J = 15.9 Hz), 7.08 (ddd, 1H, J = 8.7 Hz, J = 10.7 Hz, J = 15.5 Hz), 7.24 (dddd, 1H, J = 0.44 Hz, J = 8.8 Hz, J = 10.6 Hz, J = 15.9 Hz), 8.58 (s, 1H), 8.78 (d, 1H, J = 11.3 Hz), 9.99 ppm (br s, 1H).

^{1}H NMR of PNEF: ∂ (D_2O): 1.50 (br s, 2H, methylene), 3.22, 3.3 (2 low, br s, ~0.15H), 3.75 (br s, 0.85H, methine), 4.60 (NH + HOD) and 7.46, 7.6, 7.68 (m, 0.15H), 7.8 ppm (6 line m, 0.85H, formyl H).

^{13}C NMR of PNEF: ∂ (D_2O) 37.5–41 (m, 1C, methylene), 41–42.3 (m, ~0.85C), 46.5–48.5 (m, ~0.15C), 163.6 (s, ~0.85C, formyl), and 166–167.5 ppm (m, ~0.15C, formyl). NMR spectra of *p*-ethenylacetamide and PVAm have been reported (*16–19*).

Polymerization of NEF. Several methods of NEF polymerization were described briefly in a previous paper (*20*). Additional examples with more detail are presented below.

Preparation of PNEF ($\overline{M}_w$ = 60,000 and 430,000) by Precipitation Polymerization. A mixture of NEF (620 g), 2-propanol (2800 mL), toluene (700 mL), and Vazo 52 [2,2′-azobis(2,4-dimethylpentanenitrile)] (10.82 g) was sparged with nitrogen for 0.25 h. The reaction mixture was then heated to 65 °C and maintained at that temperature for 2 h. The resulting white powder was filtered, washed with additional 2-propanol, and dried in a vacuum oven (60 °C/133 Pa) to yield PNEF (605 g) with $\overline{M}_w$ = 59,700 and $\overline{M}_n$ = 27,400. Under similar conditions with *tert*-butanol as solvent and a smaller amount of Vazo 52 (4.19 g), heating to 45 °C for 10 h gave PNEF (557 g) with $\overline{M}_w$ = 432,000 and $\overline{M}_n$ = 133,000.

Preparation of PNEF ($\overline{M}_w$ = 2–4 × 10^6) via Inverse Emulsion Polymerization. In a cylindrical reactor at 40 °C with a dual-blade stirrer operating at 2000 rpm, sorbitan monostearate (Span 60, 45 g) was dissolved in cyclohexane (765 g) under nitrogen. A nitrogen-sparged mixture of NEF (375 g) and water (315 g) was added gradually to establish the inverse emulsion. The temperature was maintained at 40 °C, and Vazo 52 (0.78 g) was added. After 5 h, the temperature was increased slowly to 55 °C and held there for 5 h. A free-flowing PEF granular powder was recovered by simply drying the emulsion. Surfactant was removed by washing the solid polymer with ethyl acetate. The PNEF product had $\overline{M}_w$ = 2.9 × 10^6.

Preparation of PNEF via Aqueous Polymerization. A mixture of deionized water (1000 g), 2,2′-azobis(2-amidinopropane·hydrochloride) (4.34 g), and NEF (250 g) was sparged with nitrogen for 0.25 h. The reaction mixture was then heated to 55–60 °C over 0.5 h and held at this temperature for 4 h. The temperature was raised to 65 °C for 1 h to complete the polymerization. The PNEF product had $\overline{M}_w$ = 282,000 and $\overline{M}_n$

= 45,900. The same reaction with 750 g of water and 1.74 g of initiator gave a PNEF product with $\overline{M}_w$ = 420,000 and $\overline{M}_n$ = 178,000.

Higher-Molecular-Weight PNEF ($\overline{M}_w$ = 616,000) via Aqueous Polymerization. Deionized water (6600 g) and NEF (900 g) were added to a 10-L reactor and sparged with nitrogen for 1.0 h. 2,2′-Azobis(2-amidinopropane·hydrochloride) (11.25 g) was dissolved in water (150 g), and 15 g of this initiator solution was added to the reaction mixture. The reaction mixture was heated to 45 °C over 1.0 h. During this time, the remaining initiator solution was added. The reaction was maintained at 45 °C for 4 h. The temperature was then raised to 55 °C and held there for 2 h. Finally, the temperature was raised to 65 °C and held there overnight. The PNEF product had $\overline{M}_w$ = 616,000 and $\overline{M}_n$ = 241,000.

Acid Hydrolysis of PNEF. A stoichiometric amount of hydrochloric acid was added to a 15 wt% aqueous solution of PNEF. Heating for 6 h at 60 °C resulted in hydrolysis of 65% of the formamide groups.

Base Hydrolysis of PNEF. A solution of sodium hydroxide (66 g) in water (100 g) was added to a solution of PNEF (235 g) in water (1200 g) at 70 °C. The temperature was allowed to rise to 80 °C during the addition and was maintained there for 6 h. More than 98% of the formamide groups were hydrolyzed at that point. The product was isolated as PVAm·HCl by adding hydrochloric acid (520 mL of a 38 wt% solution) to the cold product solution. The precipitated PVAm·HCl was washed with methanol and dried (60 °C/133.3 Pa) to yield 211 g of product.

Nonaqueous Base Hydrolysis of PNEF. A suspension of high-molecular-weight PNEF powder (20–25 wt%) in cyclohexane–1-butanol (1.5:1 v/v) was hydrolyzed by adding finely ground dry potassium or sodium hydroxide to a well-mixed suspension at 70 °C. The mixture was stirred at that temperature for 4–6 h. The resulting product was isolated by centrifugation; it contained both the hydrolyzed polymer and formate salt. Alternatively, the mixture of high-molecular-weight PVAm and sodium formate can be swelled with water at 60 °C and then isolated as a gel by adding acetone. Further washing of this gel with 2:1 acetone–water mixtures followed by drying and grinding of the gel gave PVAm containing less than 1 wt% sodium formate (based on the polymer weight).

Alternative Nonaqueous Base Hydrolysis of PNEF. A stoichiometric amount of finely powdered sodium hydroxide was added to a slurry of low-molecular-weight PNEF powder (30 wt%) in 1-butanol at 80 °C. The resulting mixture was held at 80 °C for 7–8 h, during which time a large increase in the viscosity of the reaction mixture was observed. As hydrolysis proceeded, the PVAm product dissolved in the 1-butanol. The precipitated sodium formate was removed by centrifugation, and the resulting PVAm solution was concentrated on a rotary evaporator. The product contained less than 2 wt% of sodium formate (based on the weight of the PVAm), and 85% of the formamide groups were hydrolyzed.

Salt-Free PVAm. PVAm–sodium formate solution was added to cellulose acetate membranes with a molecular weight cutoff of 6000–8000.

Dialysis was conducted at room temperature for 2–3 days. A volume of deionized water equal to 4 to 5 times the volume of the polymer solution was used in the dialysis; the deionized water was changed 2 times per day. The level of sodium formate was reduced to less than 1 wt% of the polymer weight.

Derivatization of PVAm with Aldehydes. Derivatization with butyraldehyde is typical of reaction with aldehydes. Butyraldehyde (4.53 g) in methanol (10 mL) was added at 20 °C to PVAm·HCl (10 g) in water (90 mL). The reaction mixture was heated to 65 °C over 15 min and held at 65 °C for 2 h. After being cooled to room temperature, the reaction mix was slowly added to 400 mL of isopropanol with stirring; these steps precipitated the polymer. The tacky precipitate was soaked in isopropanol to dry, and solvent was removed in a vacuum oven (65 °C/33 kPa). The yield was 9.40 g.

Reactions with PVAm Free Base. The reaction of γ-butyrolactone with PVAm free base is typical of reaction with electrophiles. A methanolic solution of PVAm containing dissolved sodium formate was adjusted to pH 8.0 with formic acid, and a small amount of precipitated sodium formate was removed by centrifugation, producing 70 mL of a mixture containing 57.5 mmol of 83% hydrolyzed PNEF. γ-Butyrolactone (2.44 g) was added, the reaction was heated at 80 °C for 24 h, and the polymer was precipitated at room temperature by adding acetonitrile. Washing with isopropanol and drying at 60 °C yielded 4.42 g (70.5% yield) of a 24 mol% substituted polymer (sodium formate contaminated).

PVAm Propoxylation in *n*-Propanol. For nonaqueous hydrolysis of PNEF, PNEF (213.9 g, $\overline{M}_w$ = 56,000) in *n*-propanol (495 g) was heated to 60 °C over 20 min. Powdered sodium hydroxide (131 g) was added over a 2-h period, and a viscous paste formed. The temperature was raised to 75 °C and held there for 6 h. After cooling, the reaction mix was diluted with 1100 mL of *n*-propanol and neutralized with formic acid (29.1 g). Sodium formate was removed by centrifugation, and 82% of the PVAm in propanol solution was recovered. For propoxylation, PNEF hydrolyzate (containing 1.65 mol of PVAm in 1078 g of *n*-propanol) was heated to 100 °C in a pressure vessel. Propylene oxide (79.5 g) was pumped in over 0.5 h. After 2 h, the reaction mix was cooled and solvent was vacuum stripped at 75 °C, yielding 136.4 g of 52 mol% propoxylated product. (Product contained 1.5% ash, 6.8% *n*-propanol, and 3.5% propylene oxide.)

PVAm Propoxylation in Water. PNEF (80 g) was hydrolyzed with sodium hydroxide (92 g of 50%) in 500 mL of deionized water at 65 °C for 8 h at 80 °C. The reaction mixture was cooled to room temperature, and the pH was adjusted to 8.05 with concentrated HCl (58 mL). The aqueous PNEF hydrolyzate (1029 g, containing 1.09 mol of PVAm) was heated to 100 °C in a pressure reactor. Propylene oxide (50.3 g) was pumped in over 0.5 h. After 10 h, the reaction mix was cooled. The polymer was precipitated as an oil by adding the reaction product to 2400 mL of isopropanol and then adding 185 mL of concentrated HCl. The hydrochloride product was dried in isopropanol and solvent stripped at 60 °C; it yielded 94.5 g of 50 mol% propoxylated product (it also contained 10.1% ash and 20% isopropanol).

Results and Discussion

NEF Monomer. NEF monomer is a reasonably stable, low-volatility liquid with excellent polymerizability and copolymerizability. It can be purified by vacuum distillation. The molecular weight of the monomer is only 71 g/mol, and its nitrogen content is 20% on a weight basis (37% after hydrolysis and removal of formate byproduct). Thus the ionic character of protonated PVAm is extremely high relative to that of many other amine functional polymers, such as poly(diallyldimethylammonium chloride) or cationic polyacrylamides. Among the extant amine functional polymers, PVAm and isomeric poly(ethyleneimine) possess the highest known cationic charge density at low pH (*20*).

The known toxicological properties of NEF monomer appear to be quite favorable. NEF monomer is Ames negative and has an acute oral 50% lethal dose (rat) of 1444 mg/kg of body weight.

NEF is best stored below room temperature to maintain color and titer for extended periods. With appropriate inhibitors, it is stable for at least 3 months at 25 °C. It should not be stored above 40 °C. As expected for the monomer precursor of an easily hydrolyzed amide polymer, NEF is sensitive to acids and bases. Exposure of the neat monomer to strong acids and bases should therefore be avoided. Some of the physical properties of NEF monomer are tabulated in Table I.

Analytical. NEF monomer is a slowly interconverting 3:1 mixture of two carbonyl–nitrogen bond rotamers on the NMR time scale, giving rise to a surprisingly complex proton NMR (equation 3 in Chart I). The spectrum can be coalesced at elevated temperatures. The homopolymer likewise shows additional signals, and integration of the ∂3.71 peak is consistently low (14–18%) without the 3.1–3.4-ppm peaks. The ^{13}C spectra show similar effects, also not seen in *p*-ethenylacetamide (*16–19*) but reported for other ethenylamides and attributed to hindered rotation about the amide C–N bond (*21*).

NEF Polymerization. NEF can easily be converted to PNEF via precipitation, solution, or inverse emulsion polymerization. Most standard initiators, including azo, perester, and peroxide thermal initiators and redox systems, have been employed. Persulfate initiators, however, are highly unreliable for the polymerization of NEF and are not recommended. Chain transfer to solvent has a major impact on molecular weight. Variation of the solvent and solvent combinations

Table I. Physical Properties of NEF Monomer

Property	*Value*
Purity (%)	98–99+
Appearance	clear liquid
Color (APHA)	<300
Odor	mild, characteristic
Boiling point at 13.3 mbar	84 °C[a]
Melting point	~ −16 °C
Density (g/ml) at 25 °C	1.014
Refractive index at 25 °C	1.492
Vapor pressure (mbar) at 25 °C	~0.25
Flash point (°C) (ASTM D-3828)	102
pH, 5% aqueous solution	6.6
Viscosity (cps) at 25 °C	4
Surface tension (dyn/cm) at 21 °C	36.2
ΔH polymerization (kJ/mol)	~80
ΔH polymerization (kcal/mol)	~19
Miscibility	all common solvents (limited with alkanes)

[a] Theoretical: 210 °C at 1 bar.

and the initiator and monomer concentrations provides convenient control of molecular weight.

Precipitation polymerization of NEF can be carried out in a variety of solvents for NEF monomer that are nonsolvents for the PNEF produced. NEF monomer is soluble in water and in all common organic solvents except saturated hydrocarbons. PNEF is insoluble in most common solvents except water, formamide, ethylene glycol, and hot Me_2SO. The use of lower alcohols (fewer than four carbons) in the batch precipitation polymerization of NEF, however, tends to produce sticky masses of PNEF rather than free-flowing particles. This stickiness is probably due to plasticization of the PNEF particles at polymerization temperatures caused by a combination of these more polar alcohols and unreacted NEF monomer. Free-flowing particles can be obtained by using less polar solvents, such as toluene or *tert*-butanol, or by adding a small amount of a less polar solvent to a C-1 to C-4 alcohol.

Use of *tert*-butanol as a solvent allows the synthesis of reasonably high-molecular-weight PNEF ($\overline{M}_W$ up to ca. 450,000). Changing the solvent to 2-propanol–toluene mixtures provides PNEF with a lower molecular weight (typical $\overline{M}_W$ = 10,000–60,000).

NEF, like acrylamide, can be polymerized via inverse emulsion polymerization. PNEF of high molecular weight ($\overline{M}_W$ = 1,000,000–4,000,000) is most conveniently prepared via this method. An inverse emulsion using sorbitan monostearate in cyclohexane and an oil-soluble initiator has given good results. Solid PNEF can be isolated by drying the emulsion, and surfactant can be removed by washing with ethyl acetate.

The simplest method for the synthesis of PNEF, and frequently the one of choice, is solution polymerization in water. Water is the best solvent for both NEF and PNEF, and chain transfer to water is low. In addition, PNEF produced via aqueous-solution polymerization can be hydrolyzed directly, without isolation of the polymer, provided that the residual NEF level is sufficiently low. NEF is very stable in water at pH 6–8, providing a reasonable window in which to conduct polymerizations. Monomer stability decreases somewhat outside this range, particularly with declining pH ($t_{1/2} > 500$ min at 60 °C and pH 4). PNEF with $\overline{M}_W$ in the range of 30,000 to nearly 1,000,000 can be produced by aqueous-solution polymerization. The limiting factors are the high solution viscosities generated at higher molecular weights and the heat removal capacity of the reactor. The practical limit by this method is $\overline{M}_W \sim 600{,}000$.

PNEF Hydrolysis to PVAm. A particular advantage of NEF is the ease of hydrolysis of its polymers. PNEF easily undergoes 60–70% hydrolysis when a stoichiometric amount of acid under mild conditions (60 °C for 6 h) is used. Conversion above this level is slower because of charge repulsion effects but can be achieved by using additional acid under harsher conditions. Full hydrolysis of PNEF to PVAm is more easily accomplished by using stoichiometric base hydrolysis (80 °C for 8 h). Kinetic measurements of the hydrolysis reaction have revealed that ca. 95% of the formamide groups are hydrolyzed within 3 h when stoichiometric base at 60–80 °C is used. After 6–8 h, no formamide groups are detectable by NMR.

One of the major problems associated with the production of PVAm via the hydrolysis of PNEF is the generation of unwanted coproducts. Acid hydrolysis generates the acid salt of PVAm and an equivalent of formic acid. Base hydrolysis produces PVAm and an equivalent of a formate salt. The weight of coproduct generated when NaOH is used is slightly higher than the weight of polymer itself. These coproducts are of minimal concern when the polymer is used in applications at low add-on levels and when the salts do not affect the performance properties of the polymer. An example would be the use of PVAm in water treatment. These coproducts become a concern when polymer add-on levels are higher or when the coproducts adversely impact the performance of PVAm or its acid salt in an application. Several practical methods are currently being developed to deal with this issue.

At present, the coproducts of PNEF hydrolysis may be removed on a laboratory scale by dialysis of aqueous solutions of the polymer. Alternatively, low-molecular-weight PVAm can be prepared with low levels of salt coproduct by using sodium hydroxide to hydrolyze a suspension of PNEF in 1-butanol. The PVAm produced by hydrolysis dissolves in this solvent, and the sodium formate coproduct precipitates from it. Alcoholic solutions of PVAm containing less than 2 wt% sodium formate (based on the weight of the polymer) can be prepared. This method is limited to the production of lower-molecular-weight PVAm ($\overline{M}_W < 100{,}000$) because of viscosity buildup during the hydrolysis. Also, although aqueous solutions of PVAm have long shelf lives (more than 1 year), alcohol solutions of PVAm have a tendency to gel.

Higher-molecular-weight PVAm ($\overline{M}_W > 1{,}000{,}000$) may be isolated free of coproducts by taking advantage of the kinetically slow dissolution of high-molecular-weight PVAm in water.

Polymer Properties. Some of the typical properties of PNEF and PVAm·HCl are presented in Table II. PNEF forms tough, clear films when it is cast from water. Its T_g is approximately 150 °C and is strongly dependent on residual water content, which plasticizes PNEF and lowers the T_g. PNEF is soluble in water, ethylene glycol, formamide, and hot Me_2SO. PNEF is insoluble in other common solvents such as acetone, alcohols, dimethylformamide, and methylene chloride. The rheological properties are typical for a water-soluble polymer.

The solubility characteristics of PVAm·HCl are similar to those of PNEF. PVAm·HCl is soluble in water, formamide, ethylene glycol, and some alcohol–water mixtures. It is insoluble in the common solvents discussed above. A T_g for PVAm·HCl is not observed below its

Table II. Typical Polymer Properties

Property[a]	*Poly(ethenyl formamide)*			*PVAm·HCl*		
$\overline{M}_w \times 10^{-3}$ (GPC)	50–100	300–400	1000–3000	50–100	300–400	1000–3000
Intrinsic viscosity (dL/g), [η]	0.3–0.4	1.0–1.5	3.0–4.0	0.2–0.3	0.9–1.1	2.1–2.8
Viscosity (cps), 5% aqueous solution at 20 °C, Brookfield	6–10	70–80	2500–3500	5–9	60–75	2000–3000
T_g by DSC (°C)	150	150	150	decomposition observed from 220 to 250 °C		
pH, 5% aqueous solution at 20 °C	5.5–6.5	5.5–6.5	5.5–6.5	2–3	2–3	2–3

[a] Abbreviations: GPC, gel permeation chromatography; DSC, differential scanning calorimetry.

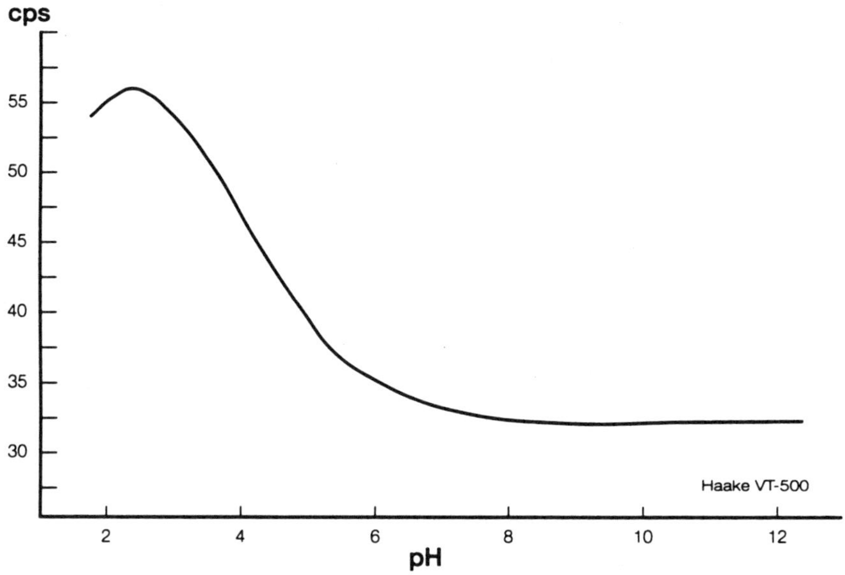

Figure 1. PVAm·HCl viscosity versus pH in 5% solution with $\overline{M}_w$ = 325,000 at 20 °C.

decomposition temperature (220–250 °C). PVAm·HCl exhibits good stability in high-salt environments, in strongly acidic environments, and at elevated temperatures. The solution viscosity of PVAm·HCl is high. This viscosity is attributed to the high cationic charge density on the polymer backbone, which should cause the protonated polymer to adopt a relatively uncoiled conformation. Support for this possibility comes from data in Figure 1. As the pH increases, the cationic character of PVAm·HCl decreases, resulting in decreased viscosity. This decrease in viscosity is not observed for the nonionic water-soluble polymers investigated.

The solubility of PNEF and PVAm·HCl in the presence of several electrolytes is summarized in Table III. As expected, PNEF, an uncharged polymer, tolerates all of the electrolytes tested. PVAm·HCl does not tolerate electrolytes that contain divalent sulfate ion as the anionic portion of the electrolyte.

Reactive Chemistries and Cross-linking. The reactions of PVAm are those expected for a primary amine functional polymer. Figure 2 summarizes results with several cross-linking agents in aqueous solutions at various pH levels. As expected, dialdehydes, diisocyanates, and epichlorohydrin induce PVAm cross-linking in appropriate pH ranges. In addition, metals that form complexes with PVAm can

Table III. Electrolyte Compatibility of PNEF and PVAm·HCl

Electrolyte	*PNEF*	*PVAm·HCl*
Aluminum sulfate	+	−
Calcium chloride	+	+
Potassium sulfate	+	−
Copper sulfate	+	−
Sodium nitrate	+	−
Boric acid[a]	+	+
Ferric chloride	+	+
Zinc sulfate	+	−
Sodium bicarbonate	+	+
Sodium borate[a]	+	+

NOTE: Symbols: +, polymer remains in solution; −, polymer precipitates.
[a]At 50 °C because of borate insolubility at 20 °C.

be used to cross-link the polymer. Di- and trivalent anions also induce cross-linking. The ease with which PVAm can be cross-linked opens up a number of possible applications for this polymer.

PVAm can undergo a number of reactions that modify the polymer structure without cross-linking. These modifications may be made to introduce a hydrophobic or an amphoteric character to the polymer or to increase the water solubility of some hydrophobic functional groups. Figure 3 provides a graphical summary of reactions that use PVAm. In essence, any reaction that can be carried out using a primary amine should be feasible with PVAm.

Modification with Aldehydes. PVAm reacts with aldehydes in water under acidic or neutral conditions, producing cationic poly-

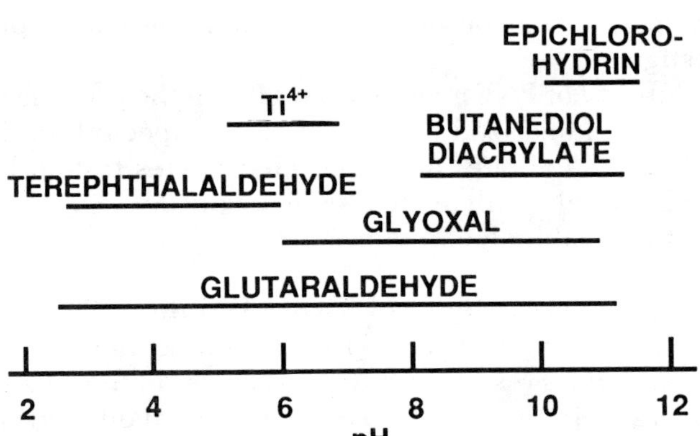

Figure 2. Cross-linking of PVAm versus pH. Cross-linkers are shown.

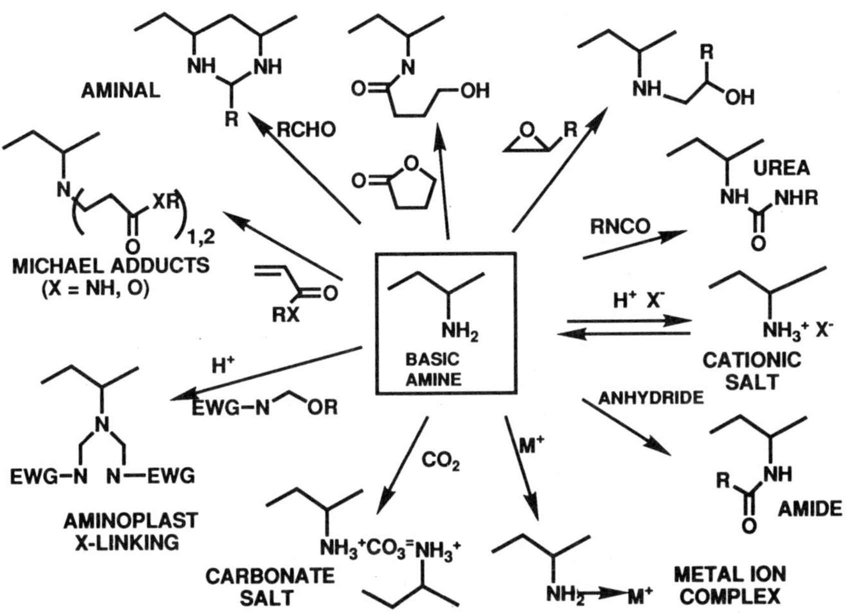

Figure 3. Amine functional polymer reactions. EWG is electron withdrawing group.

mers that have a hydrophobe bound to the polymer via an aminal linkage (Figure 3). No additional catalyst is required. The product can be isolated by precipitation of the hydrochloride salt.

The reaction may be performed with isolated PVAm or with PVAm synthesis mixtures containing hydrolysis byproducts. When synthesis mixtures are used, NEF is polymerized, the PNEF is base hydrolyzed to greater than 95%, the PVAm is precipitated with HCl, and the wet PVAm·HCl is dissolved in water for reaction with aldehyde. The PNEF must be highly hydrolyzed.

Butyraldehyde-modified polymers have been made with up to 30 mol% aldehyde per nitrogen. Selectivity to aminal formation is higher at lower temperatures (20 °C), and higher temperatures (80 °C) promote aldehyde self-condensation and dehydration. The polymers are water-soluble but have lower intrinsic viscosity than the parent PVAm, presumably because of coiling or chain–chain complexation (*see* Figure 4).

Reaction of PVAm ($\overline{M}_w$ = 600,000) with low levels (0.8 mol%) of long-chain aldehydes (C_8–C_{12}) produces associative thickeners with low solubility in water (*see* Figure 5). Low-molecular-weight ($\overline{M}_w$ = 60,000) PVAm at the same level of substitution is not an associative

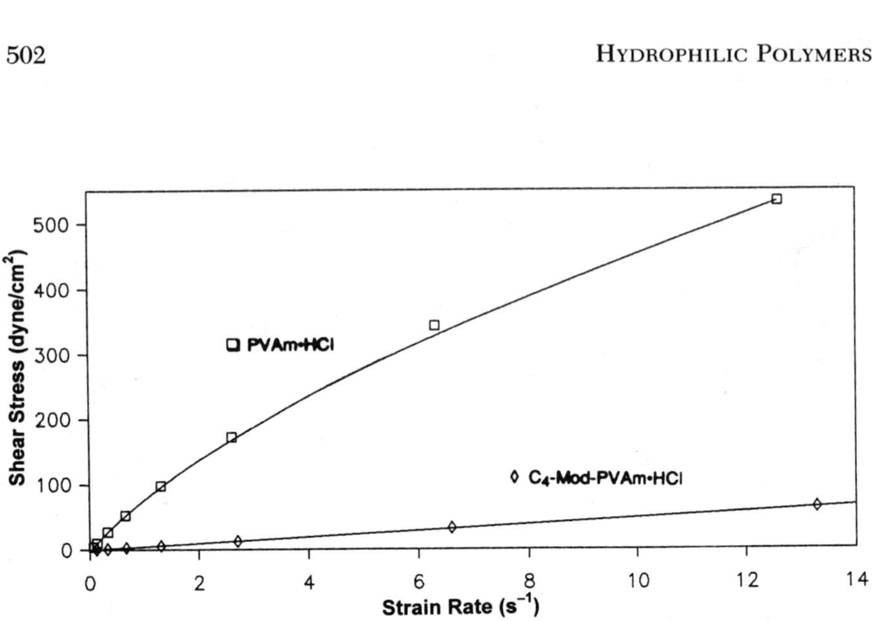

Figure 4. Rheology of butyraldehyde-modified PVAm·HCl versus unmodified PVAm·HCl. Key: □, PVAm·HCl ($\overline{M}_w$ = 9 × 10^5), 3.3% solution, ~6000 cps; ◇, C_4-modified PVAm·HCl ($\overline{M}_w$ = 1.3 × 10^6), ca. one C_4H_8 group per 3 mer units, 3.3% solution, ~484 cps. Brookfield viscosities, cylindrical no. 1.

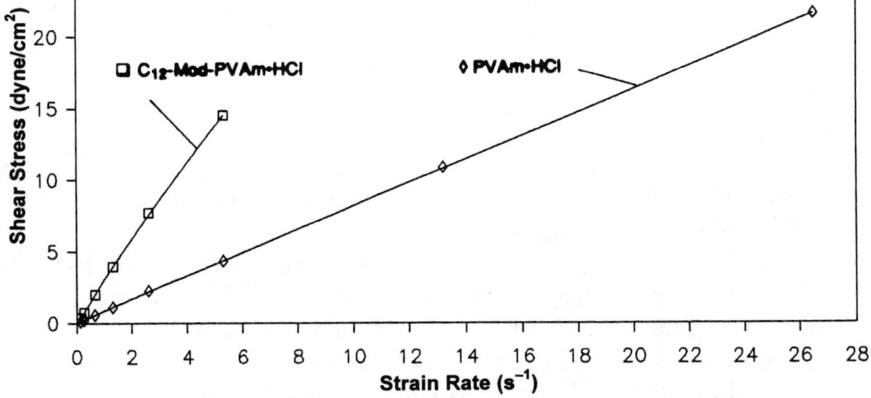

Figure 5. Rheology of dodecylaldehyde-modified PVAm·HCl versus unmodified PVAm·HCl. Key: ◇, PVAm·HCl ($\overline{M}_w$ = 6.42 × 10^5, 1.25% solution, 275 cps; □, C_{12}-modified PVAm·HCl, ca. one n-$C_{12}H_{24}$ group per 100 mer units, 1.25% solution, 82 cps. Brookfield viscosities, cylindrical no. 1.

thickener. Modification with benzaldehyde (3.5 mol%) has a negligible effect on rheology.

Reactions with PVAm Free Base. Water-soluble derivatives are formed by Michael addition of acrylamide to the amine and by reaction of propane sultone or lactones to form the sulfonamides or amides. Acrylates reacting by both Michael addition and amide formation cross-link PVAm.

Reaction with propylene oxide (50 mol%) or butyl glycidyl ether (16 mol%) yields amino alcohol products that are water- and alcohol-soluble. Reaction with 1 mol% dodecylsuccinic anhydride yields a water-swellable but insoluble product. Reaction with alkylketene dimer produces an amide with low solubility in water.

Polymer Applications. The applications of PVAm are quite numerous. PVAm may find use in any application that requires a cationic water-soluble polymer. PVAm·HCl has two advantages over older commercially available cationic polymers: high cationic charge density and high reactivity of all primary amine pendant groups. PNEF and its hydrolysis products are of interest in water treatment, papermaking, textiles, personal care, adhesives and coatings, rheology modifiers, oil field chemicals, and metal complexation.

In wastewater treatment, PVAm·HCl and butyraldehyde-modified PVAm·HCl are useful agents for flocculation, sedimentation, and dewatering of solids (*22*). These polymers are most effective at $\overline{M}_w >$ 300,000, and butyraldehyde modification generally improves the performance. Similarly, hydrophobic modification with lower (C-2, C-6, and C-8) aldehydes, butyl glycidyl ether, alkylketene dimer, γ-butyrolactone, or acrylamide produces polymers that are useful in separation of solids.

In papermaking, PVAm·HCl, particularly with $\overline{M}_w > 100{,}000$, and PVAm·HCl modified with lower aldehydes are useful as fines and filler retention agents, especially for use with recycled fiber (*23*). Acrylamide and γ-butyrolactone-modified polymers are also effective.

For enhanced oil recovery, high-molecular-weight PVAm·HCl ($\overline{M}_w > 1{,}000{,}000$) has useful properties. It maintains high viscosity under simulated drilling conditions, as is required for acidized fracturing fluids and seawater drilling muds (*24*). PVAm derivatives are also useful in metal complexation (*25*).

Acknowledgments

W. E. Carroll, C. A. Costello, A. S. Drayton-Elder, T. W. Lai, T. P. McAndrew, B. A. Muratore, T. L. Pickering, L. M. Robeson, S. R.

Salvanorich, B. R. Vijayendran, and K. J. White-Gaebe, among many other people, made significant contributions to this work.

References

1. Tomalia, D. A.; Killat, G. R. In *Encyclopedia of Polymer Science and Engineering;* Mark, H. F.; Bikales, N. M.; Overberger, C. G.; Menges, G., Eds.; Wiley Interscience: New York, 1985; Vol. 1, pp 680–739.
2. Rath, H.; Hilscher, E. German Patent 1 153 528, 1963.
3. Jones, G. D.; Zomlefen, J.; Hawkins, K. *J. Org. Chem.* **1944,** *9,* 500.
4. Arcus, C. L. *J. Polym. Sci.* **1952,** *8,* 365.
5. Mullier, M.; Smets, G. *J. Polym. Sci.* **1957,** *23,* 915.
6. Tanaka, H. *J. Polym. Sci., Polym. Chem. Ed.* **1979,** *17,* 1239.
7. Achari, A. E.; Coqueret, X.; Lablache-Combier, A.; Loucheux, C. *Makromol. Chem.* **1993,** *194,* 1879.
8. Reynolds, D. D.; Kenyon, W. O. *J. Am. Chem. Soc.* **1947,** *69,* 911.
9. Katchalsky, A.; Mazur, J.; Spitnik, P. *J. Polym. Sci.* **1957,** *23,* 513.
10. Bayer, E.; Geckler, K.; Weingartner, K. *Makromol. Chem.* **1980,** *181,* 585.
11. Hart, R. *Makromol. Chem.* **1959,** *32,* 51.
12. Dawson, D. J.; Glass, R. D.; Wingard, R. E., Jr. *J. Am. Chem. Soc.* **1976,** *19,* 5996.
13. Greene, T. W. In *Protective Groups in Organic Synthesis;* Wiley: New York, 1981; pp 249–253.
14. Dawson, D. J.; Otteson, K. M. U.S. Patent 4 578 515, 1986.
15. Nagy, D. J. *Proc. 1994 Int. GPC Symp.,* Orlando, Florida, 1994.
16. Murano, M.; Harwood, H. J. *Macromolecules* **1970,** *3,* 605.
17. Rinaldi, P. L.; Yu, C.; Levy, G. C. *Macromolecules* **1981,** *14,* 551–554.
18. St. Pierre, T.; Lewis, E. A.; Levy, G. C. In *Polymeric Amines and Ammonium Salts* (Proc. Int. Symp. Polymeric Amines and Ammonium Salts, Ghent, Belgium, Sept. 24–26, 1976); Goethals, E. J., Ed.; Pergamon: Oxford, England, 1980; pp 245–248 and references therein.
19. Chang, C.; Muccio, D. D.; St. Pierre, T.; Chen, C. C.; Overberger, C. G. *Macromolecules* **1986,** *19,* 913–916.
20. Badesso, R. J.; Lai, T. W.; Pinschmidt, R. K., Jr.; Sagl, D. J.; Vijayendran, B. R. *Polym. Prepr. (Am. Chem. Soc., Polym. Tech. Conf.)* **1991,** *32,* 110.
21. Kirsh, Yu. E., *Vysokomol. Soedin.* **1993,** *35,* 98–114.
22. Smigo, J. G.; McAndrew, T. P.; Pinschmidt, R. K.; Nordquist, A. F. U.S. Patent 5 185 083, 1993.
23. Smigo, J. G.; Pinschmidt, R. K.; Nordquist, A. F.; Pickering, T. L. U.S. Patent 5 232 5532, 1993.
24. Lai, T.-W.; Vijayendran, B. R. U.S. Patent 4 843 118, 1989.
25. Geckeler, K.; Weingärtner, K.; Bayer, E. In *Polymeric Amines and Ammonium Salts* (Proc. Int. Symp. Polymeric Amines and Ammonium Salts, Ghent, Belgium, Sept. 24–26, 1976); Goethals, E. J., Ed.; Pergamon: Oxford, England, 1980; pp 277–285.

RECEIVED for review November 11, 1993. ACCEPTED revised manuscript January 10, 1995.

Indexes

Author Index

Affiliation Index

Subject Index

D

E

F

G

H

Production: Paula M. Bérard
Acquisition: Rhonda Bitterli
Cover design: Amy O'Donnell

Typeset by Maryland Composition Company, Inc., Glen Burnie, MD
Printed by United Book Press, Baltimore, MD